Undergraduate Algebra
A First Course

Undergraduate Algebra
A First Course

C. W. NORMAN

Department of Mathematics
Royal Holloway and Bedford New College, University of London

CLARENDON PRESS · OXFORD

1986

Oxford University Press, Walton Street, Oxford OX2 6DP

Oxford New York Toronto
Delhi Bombay Calcutta Madras Karachi
Petaling Jaya Singapore Hong Kong Tokyo
Nairobi Dar es Salaam Cape Town
Melbourne Auckland

and associated companies in
Beirut Berlin Ibadan Nicosia

Oxford is a trade mark of Oxford University Press

Published in the United States
by Oxford University Press, New York

British Library Cataloguing in Publication Data

Norman, C. W.
Undergraduate algebra: a first course.
1. Algebra
I. Title
512   QA155
ISBN 0-19-853249-0
ISBN 0-19-853248-2 Pbk

Library of Congress Cataloging in Publication Data
Norman, C. W.
Undergraduate algebra.
Bibliography: p.
Includes index.
1. Algebra. I. Title.
QA154.2.N65 1986   512   85-31057
ISBN 0-19-853249-0
ISBN 0-19-853248-2 (pbk.)

Typeset and printed by The Universities Press (Belfast) Ltd

To Lucy, Tessa, and Timmy

Preface

Arithmetic is part of everyone's education: at an early age we learn how to add, multiply, and perform rote calculations such as 'long' division; not only are these exercises a help to our understanding—we may even enjoy doing them! The present book aims to make its readers feel equally at home with the basic techniques of contemporary algebra, especially with matrix manipulation—skills entirely analogous to those of elementary arithmetic and just as useful.

Here is a first course in algebra which, though written primarily for mathematics students at college or university, will I hope be useful to aspiring engineers and scientists generally; much of the material is particularly relevant to computer science. The subject lends itself to a virtually self-contained treatment, and the amount of knowledge presupposed is indeed small; however, the reader is expected to have some familiarity with calculus, co-ordinate geometry, and trigonometry.

A glance at the list of contents will convey the scope of the book. After a preliminary chapter on sets, Part One introduces the concept of a ring (which is no more than arithmetic in an abstract setting) and leads the reader gently but firmly through the basic theory, including complex numbers, integers, and polynomials. He or she is now well prepared to meet vector spaces (which generalize everyday 3-dimensional space), matrices, and groups in Part Two. Throughout, systematic techniques are given pride of place: the Euclidean algorithm for finding greatest common divisors is at the heart of Chapters 3 and 4; many of the problems which linear algebra sets out to solve are dealt with in a practical way by the row-reduction algorithm in Chapter 8. Determinants, matrix diagonalization, and quadratic and hermitian forms—topics which have wide application—are thoroughly discussed, and a final chapter with a geometric flavour treats Euclidean and unitary spaces. In short, this first course is comprehensive and suitable for students with a clear commitment to algebra; for those who want a direct route to the rudiments of linear algebra, Part Two may be tackled as soon as the concepts of ring, field, and polynomial have been grasped.

Starred sections are optional. However, each section ends with exercises arranged roughly in order of difficulty; these are ignored at the reader's peril!

As it is all too easy to be discouraged by the abstract nature of algebra, I have tried to keep in mind the standpoint of a student meeting the subject for the first time; as it is all too easy for a university teacher to forget what this entails, I gratefully acknowledge the reminders supplied by my classes at Westfield College, University of London. I must thank an ex-student, Geoffrey G. Silver, for encouraging me to 'go into print'. Many improvements to the text were made by Professor B. C. Mortimer, Dr M. Walker, Colleen Farrow, John Bentin, and others who read preliminary drafts; I am grateful to them all and also to Mrs G. A. Place for her first-class typing of the manuscript.

Royal Holloway and Bedford New College   C. W. N.
August 1985

Contents

Notation

1 Preliminary concepts

Our first chapter is an introduction to **sets**, **mappings**, and **equivalence relations**; these general concepts form a framework within which modern algebra may be constructed.

We begin with an informal account of set theory, 'set' being the technical name used for any collection of objects; thus one might speak of the set of leopards in London Zoo or the set of planets in the solar system—although we shall be more concerned with sets having an arithmetic flavour: the set of whole numbers or the set of prime numbers, for instance. As our interest lies more in the manipulation of sets and their uses rather than in sets for themselves, a formal treatment of set theory would be out of place; in fact we need sets primarily for the description of certain algebraic concepts, which begins in earnest with the next chapter.

It is important to have the means of comparing one set with another, and as we shall see later, this is especially true when algebraic systems are involved. The most significant comparisons between sets are furnished by **mappings**; we leave the details until later, but, roughly speaking, a mapping is a way of moving from one set to another.

It is also important to have the means of sorting out the objects (or **elements** to use the technical term) which make up a given set; when there is a definite criterion for the sorting process, we speak of an **equivalence relation** on the set. Equivalence relations arise naturally and frequently in mathematics and many of the constructions we undertake are best described using them.

Sets

A **set** is any collection of objects, each object in the collection being called an **element** of the set.

For example the set of days of the week has elements:

Monday, Tuesday, Wednesday, Thursday,
Friday, Saturday, Sunday.

The above intuitive description of a set will be adequate for our purposes, though the reader should realize that it is, at best, merely conveying the right impression; for in spite of being suggestive, the terms 'collection' and 'object' are themselves undefined. In effect we regard the concept of a set as being **primitive**, that is, it is taken for granted and not defined—but this will not stop us defining everything else in terms of sets!

It is often helpful to regard a collection of objects as being itself a single entity and denote it by a single symbol. Generally, capital letters $A, B, C, \ldots, X, Y, Z$ are used to denote sets, while $a, b, c, \ldots, x, y, z$ stand for elements. We write

$$x \in X$$

to indicate that x *is* an element of the set X, in which case we say also that x belongs to X, that x is contained in X, that x is a member of X, or simply that x is in X. The notation

$$x \notin X$$

means that x is *not* an element of the set X.

It is usual to reserve certain symbols for certain frequently-occurring sets; foremost among these is the set $\mathbb{N}$ of **natural numbers**:

$$1, 2, 3, 4, \ldots$$

and so, for instance, $7 \in \mathbb{N}$ but $0 \notin \mathbb{N}$.

The symbol $\mathbb{Z}$ (from the German word 'Zahl' meaning 'number') denotes the set of all **integers**:

$$\ldots, -2, -1, 0, 1, 2, \ldots$$

that is, the set of whole numbers, positive, negative and zero.

Similarly $\mathbb{Q}$ (from 'quotient') will always stand for the set of all **rational numbers** (numbers expressible as m/n, where $m \in \mathbb{Z}, n \in \mathbb{N}$).

Fig. 1.1

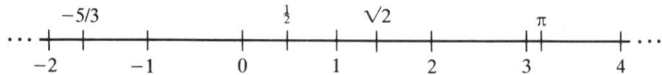

The symbol $\mathbb{R}$ is reserved for the set of all **real numbers**; this set may be pictured geometrically as a line (Fig. 1.1), each point of this line representing a real number. In particular we have

$$3 \in \mathbb{N}, \qquad \tfrac{1}{3} \notin \mathbb{Z}, \qquad \tfrac{1}{3} \in \mathbb{Q}, \qquad \sqrt{2} \notin \mathbb{Q}, \qquad \sqrt{2} \in \mathbb{R}, \qquad \pi \in \mathbb{R}.$$

Definition 1.1

Let X and Y denote sets. If each element of X is also an element of Y, then X is called a **subset** of Y and we write $X \subseteq Y$.

As each natural number is also an integer, we have $\mathbb{N} \subseteq \mathbb{Z}$. Similarly $\mathbb{Z} \subseteq \mathbb{Q}$, as $m = m/1$ shows that each integer m is also a rational number. Also $\mathbb{Q} \subseteq \mathbb{R}$, as every rational number is real.

If X is not a subset of Y, which means there is at least one element of X which does not belong to Y, then we write $X \not\subseteq Y$. For instance $\mathbb{Z} \not\subseteq \mathbb{N}$ as $0 \in \mathbb{Z}$ but $0 \notin \mathbb{N}$.

The next definition is implicit in our intuitive description of a set.

Definition 1.2 Let X and Y denote sets. If $X \subseteq Y$ and $Y \subseteq X$, then X and Y are called **equal** and we write $X = Y$.

Therefore $X = Y$ means each element of X belongs also to Y and each element of Y belongs also to X; in other words, two sets are equal if they consist of the same elements.

Some sets may be specified by listing their elements, it being customary to enclose this list between braces (curly brackets); so

$Y = \{1, 2, 3, 4, 6, 12\}$

means that Y is the set consisting of the six natural numbers in the above list. The *order* in which the elements are listed is of no importance; so, for example

$\{1, 2, 3\} = \{2, 3, 1\}$.

Nor is it of any significance if an element occurs more than once in the listing; so, for instance

$\{2, 2\} = \{2\}$.

A more common method of specifying a set is by means of a **characteristic property** of its element, that is, a set consists of those elements and only those elements having a certain property. We make no attempt to define the word 'property', but use it in the following intuitive way: a **property** is something which elements either have or do not have—sitting on the fence is not allowed! We write

$X = \{x : x \text{ has property } P\}$

for the set of all elements x having the given property P. In a similar way we use the notation

$Y = \{x \in X : x \text{ has property } Q\}$

for the subset Y of X consisting of those elements having property Q; so

$Y = \{1, 2, 3, 4, 6, 12\} = \{n \in \mathbb{N} : n \text{ is a divisor of } 12\}$

as the six elements of Y are precisely those natural numbers which are divisors of 12. Of course it may be possible to specify the same set in many different ways; for example

$$\{-1, 0, 1\} = \{m \in \mathbb{Z} : -1 \leqslant m \leqslant 1\}$$
$$= \{m \in \mathbb{Z} : m \text{ is not a prime or a product of primes}\}$$
$$= \{x \in \mathbb{R} : x^3 = x\}.$$

Definition 1.3

Let X_1 and X_2 be sets. The set of those elements which belong to *both* X_1 and X_2 is called the **intersection** of X_1 and X_2 and is denoted by $X_1 \cap X_2$.

Therefore $X_1 \cap X_2 = \{x : x \in X_1 \text{ and } x \in X_2\}$. For instance if $X_1 = \{1, 2, 3, 4, 6, 12\}$ and $X_2 = \{1, 2, 3, 6, 9, 18\}$, then

$$X_1 \cap X_2 = \{1, 2, 3, 6\}.$$

Definition 1.4

Let X_1 and X_2 be sets. The set of those elements which belong to *at least one* of X_1 and X_2 is called the **union** of X_1 and X_2 and is denoted by $X_1 \cup X_2$.

Therefore $X_1 \cup X_2 = \{x : \text{either } x \in X_1 \text{ or } x \in X_2\}$ where 'either . . . or . . .' includes the case of belonging to both sets, that is,

$$(X_1 \cap X_2) \subseteq (X_1 \cup X_2).$$

If $X_1 = \{1, 2, 4, 8\}$ and $X_2 = \{1, 2, 3, 6\}$, then

$$X_1 \cup X_2 = \{1, 2, 3, 4, 6, 8\}.$$

A practical and visual aid to the study of sets is provided by **Venn diagrams**, in which sets are pictured as plane regions bounded by simple closed curves (Fig. 1.2).

Fig. 1.2

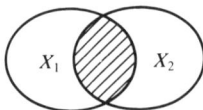

The shaded region represents $X_1 \cap X_2$

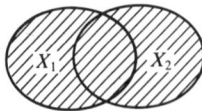

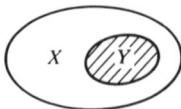

The shaded region represents $X_1 \cap X_2$

The Venn diagram of a set X having subset Y.

Definition 1.5   Let X be a subset of the set U. The set

$$X' = \{x \in U : x \notin X\}$$

is called the **complement** of X in U.

So X' consists of those elements of U which do *not* belong to X. For example, the set of odd integers is the complement in $\mathbb{Z}$ of the set of even integers.

Suppose $U = \{1, 2, 3, 4\}$, $X_1 = \{1, 2\}$, $X_2 = \{1, 3\}$. Then $X_1 \cup X_2 = \{1, 2, 3\}$ and so $(X_1 \cup X_2)' = \{4\}$. But as $X_1' = \{3, 4\}$ and $X_2' = \{2, 4\}$, we see $X_1' \cap X_2' = \{4\}$ also, and so

$$(X_1 \cup X_2)' = X_1' \cap X_2'$$

in this case. The Venn diagram (Fig. 1.3) *suggests* that the above set equality is valid for all subsets X_1 and X_2 of an arbitrary set U, for the double-shaded region represents both $(X_1 \cup X_2)'$ and $X_1' \cap X_2'$; however, a Venn diagram has its limitations and does not in itself constitute a proof—after all, the sets U, X_1, X_2 may *not* be plane regions, and even if they were, one might happen to draw a misleading diagram. Rather, as below, we must devise a proof made up of logical deductions from the definitions of the terms involved.

Fig. 1.3

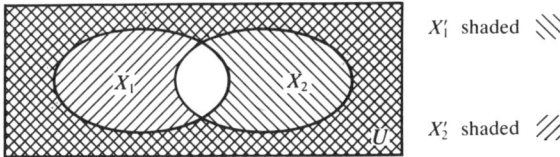

X_1' shaded

X_2' shaded

Theorem 1.6   (De Morgan's laws). Let X_1 and X_2 be subsets of the set U. Then

(a) $(X_1 \cup X_2)' = X_1' \cap X_2'$,   (b) $(X_1 \cap X_2)' = X_1' \cup X_2'$.

Proof   (a) To establish the first of the above set equalities we must show, by (1.2), that the left-hand set is a subset of the right-hand set and vice-versa. So let $x \in (X_1 \cup X_2)'$, which means $x \in U$ and $x \notin X_1 \cup X_2$ by (1.5). By (1.4), $x \notin X_1 \cup X_2$ means x does not belong to either X_1 or X_2, that is, $x \notin X_1$ *and* $x \notin X_2$. Therefore $x \in X_1'$ and $x \in X_2'$ by (1.5), which means $x \in X_1' \cap X_2'$ by (1.3). Using (1.1) we conclude

$$(X_1 \cup X_2)' \subseteq X_1' \cap X_2'.$$

To prove that the right-hand set is a subset of the left-hand set, let $x \in X_1' \cap X_2'$, which means $x \in X_1'$ and $x \in X_2'$ by (1.3). Using (1.5) we see $x \in U$, $x \notin X_1$, and $x \notin X_2$. So x does *not* belong to either X_1

or X_2, which means $x \notin X_1 \cup X_2$ by (1.4). Using (1.5) we conclude $x \in (X_1 \cup X_2)'$ and so by (1.1) we have

$$X_1' \cap X_2' \subseteq (X_1 \cup X_2)'.$$

The two set inclusions can now be combined by (1.2) to give the set equality of (1.6)(a).

(b) The second of De Morgan's laws may be proved in a similar way; alternatively, one may apply the first law to the subsets X_1' and X_2' of U, obtaining

$$(X_1' \cup X_2')' = (X_1')' \cap (X_2')'.$$

Now X_1 is the complement of X_1' in U, that is, $X_1 = (X_1')'$. Similarly $X_2 = (X_2')'$ and so the above set equality becomes

$$(X_1' \cup X_2')' = X_1 \cap X_2.$$

Taking the complement in U of both sides of this equation gives the set equality of (1.6)(b), for $X_1' \cup X_2'$ is the complement of $(X_1' \cup X_2')'$ in U. $\square$

The *end* of each proof will be marked, as above, by the symbol $\square$

The **empty** set, that is, the set having no elements, is denoted by $\varnothing$. In many ways the role of $\varnothing$ in set theory is analogous to that of zero in arithmetic. As $\varnothing$ has no elements, it is certainly true that each element of $\varnothing$ belongs to every set; therefore $\varnothing$ is a subset of every set by (1.1).

Definition 1.7

The sets X_1 and X_2 are called **disjoint** if $X_1 \cap X_2 = \varnothing$.

Therefore two sets are disjoint if they have no elements in common. For example $X_1 = \{1, 3, 5, 7\}$ and $X_2 = \{0, 2, 4, 6, 8\}$ are disjoint. Let X be a subset of the set U and let X' be the complement of X in U; then $X \cap X' = \varnothing$ and $X \cup X' = U$. In fact the subsets of a non-empty set U occur in complementary pairs X, X'.

We next explain what is meant by the intersection and the union of several sets.

Definition 1.8

Let $X_1, X_2, \ldots, X_n$ be sets. Their **intersection** $\bigcap_{i=1}^{n} X_i$ is the set of elements which belong to *all* of $X_1, X_2, \ldots, X_n$.

For instance taking $n = 3$, $X_1 \cap X_2 \cap X_3 = \bigcap_{i=1}^{3} X_i$ is the set of elements common to X_1, X_2, and X_3 (Fig. 1.4).

Fig. 1.4

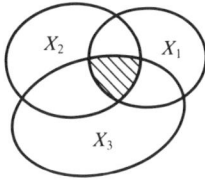

The shaded region represents $X_1 \cap X_2 \cap X_3$

Definition 1.9

Let $X_1, X_2, \ldots, X_n$ be sets. Their **union** $\bigcup_{i=1}^{n} X_i$ is the set of elements which belong to *at least one* of $X_1, X_2, \ldots, X_n$.

Again taking $n = 3$, we see that $X_1 \cup X_2 \cup X_3 = \bigcup_{i=1}^{3} X_i$ is the set of elements which belong either to X_1 or to X_2 or to X_3 (Fig. 1.5).

More generally, the **intersection** of a non-empty collection of sets consists of the elements belonging to *every* set in the collection, and the **union** of such a collection of sets consists of the elements belonging to *at least one* of the sets in the collection.

Fig. 1.5

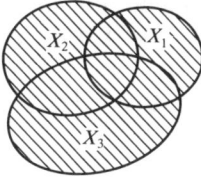

The shaded region represents $X_1 \cup X_2 \cup X_3$

Example 1.10

In a survey of drinking habits, 100 people are interviewed. It is found that 47 drink tea, 33 drink wine, 42 drink coffee, 9 drink tea and coffee, 5 drink wine and coffee, and 8 drink tea and wine.

Let X_t denote the set of people interviewed who drink tea, X_w those who drink wine, and X_c those who drink coffee. Suppose $X_t \cap X_w \cap X_c$ contains exactly m elements, that is, m of the people interviewed drink all three of tea, wine, and coffee. Working backwards through the data, we see that $8 - m$ people drink tea and wine but not coffee, $5 - m$ people drink wine and coffee but not tea, and $9 - m$ people drink tea and coffee but not wine (Fig. 1.6). The

Fig. 1.6

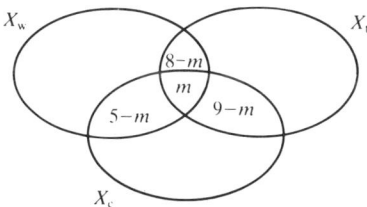

Fig. 1.7

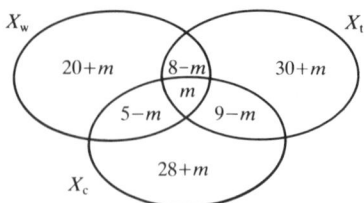

remaining numbers can be found in terms of m, as we are given the sizes of the sets X_t, X_w, X_c (Fig. 1.7). For instance, $28 + m$ people drink coffee but neither tea nor wine. Adding all the numbers together, we see that $X_t \cup X_w \cup X_c$ contains exactly $100 + m$ elements; but as only 100 people were interviewed, we deduce $m = 0$. Therefore everyone interviewed drinks at least one of tea, wine, and coffee, but no one drinks all three.

Definition 1.11

Let the set X contain exactly n elements for some non-negative integer n. Then X is called a **finite** set and we write $|X| = n$.

Therefore a finite set X contains precisely $|X|$ elements. If $X = \{0, 2, 4, 6, 8\}$, then $|X| = 5$.

Definition 1.12

Let X and Y be sets. The symbol (x, y) is called the **ordered pair** with first entry x and second entry y. The set of *all* ordered pairs (x, y) for $x \in X$ and $y \in Y$ is called the **cartesian product** of X and Y and denoted by $X \times Y$.

An ordered pair is itself a single element having two entries: as in co-ordinate geometry, two ordered pairs (two points) are equal if and only if the first entries agree and the second entries agree, that is,

$$(x, y) = (x', y') \quad \text{means} \quad x = x' \text{ and } y = y'.$$

In particular $(x, y) \neq (y, x)$ for $x \neq y$, showing that the ordering of the entries is important. The cartesian product $X \times X$ is often denoted by X^2. Suppose $X = \{0, 1\}$ and $Y = \{1, 2, 3\}$; then

$$X \times Y = \{(0, 1), (0, 2), (0, 3), (1, 1), (1, 2), (1, 3)\},$$
$$Y \times X = \{(1, 0), (1, 1), (2, 0), (2, 1), (3, 0), (3, 1)\},$$
$$X^2 = X \times X = \{(0, 0), (0, 1), (1, 0), (1, 1)\}.$$

More generally, suppose X and Y are finite sets. In forming the elements (x, y) of the cartesian product $X \times Y$, there are $|X|$

Fig. 1.8

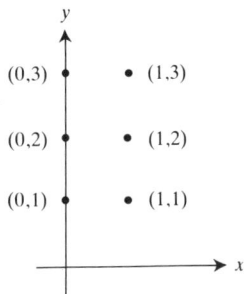

choices for x and $|Y|$ choices for y; therefore $X \times Y$ contains exactly $|X|\,|Y|$ elements, that is,

$$|X \times Y| = |X|\,|Y|.$$

We saw earlier that the set $\mathbb{R}$ of real numbers can be thought of as a line. In the same way

$$\mathbb{R}^2 = \mathbb{R} \times \mathbb{R} = \{(x, y) : x, y \in \mathbb{R}\}$$

may be pictured as the **cartesian plane**, the ordered pair (x, y) being regarded in the usual way as the point with coordinates x and y; in fact the cartesian product $X \times Y$ of the arbitrary sets X, Y is modelled on this familiar example. If X and Y are subsets of $\mathbb{R}$, then $X \times Y$ is a subset of $\mathbb{R}^2$; for instance, if $X = \{0, 1\}$ and $Y = \{1, 2, 3\}$, then $X \times Y$ is the set of six points shown in Fig. 1.8.

Definition 1.13   Let X be a set and n a natural number. An n-**tuple** of elements belonging to X is a symbol of the form $(x_1, x_2, \ldots, x_n)$ where each x_i is in X. The set of all such n-tuples is called the n-**fold cartesian product** of X and denoted by X^n.

The elements $x_1, x_2, \ldots, x_n$ appearing in an n-tuple need not be all different, but their order is important, because as with ordered pairs, equality of n-tuples means equality of corresponding entries, that is,

$$(x_1, x_2, \ldots, x_n) = (y_1, y_2, \ldots, y_n)$$

means

$$x_1 = y_1, x_2 = y_2, \ldots, x_n = y_n.$$

In fact $X^n = X \times X \times \ldots \times X$ is the cartesian product of n copies of X.

Let $X = \{0, 1\}$; in this case X^3 has eight elements, namely

$(0, 0, 0)$, $(1, 0, 0)$, $(0, 1, 0)$, $(0, 0, 1)$
$(0, 1, 1)$, $(1, 0, 1)$, $(1, 1, 0)$, $(1, 1, 1)$,

and X^4 has sixteen elements. More generally, if X is a finite set, then so is X^n; in fact $|X^n| = |X|^n$ as there are $|X|$ choices for each n-tuple entry.

Just as $\mathbb{R}$ and $\mathbb{R}^2$ may be pictured as a line and a plane, so $\mathbb{R}^3$ can be thought of as cartesian 3-dimensional space and the triple (x, y, z) as the point with coordinates x, y, z.

Exercises 1.1

1. (a) Let $X = \{1, 2, 3, 4\}$, $Y = \{1, 4, 5, 6\}$, and $Z = \{1, 2, 6, 7\}$. Express the following sets in terms of X, Y, and Z using $\cap$ and $\cup$:

$\{1\}$, $\{1, 6\}$, $\{1, 2, 6\}$, $\{1, 2, 3, 4, 6\}$, $\{1, 2, 4, 6, 7\}$, $\{1, 2, 4, 6\}$.

(b) A set U has subsets X, Y, and Z. Draw a Venn diagram and shade the regions representing $X' \cup Y \cup Z$, $X \cap (Y \cup Z)'$, and $X' \cap Y' \cap Z$.

(c) Let $U = \{1, 2, 3\}$. List the eight subsets of U.

(d) Let P denote the set of all professors, and let R denote the set of all researchers. Interpret in colloquial English:

$P \not\subseteq R$, $\quad R \not\subseteq P$, $\quad P \cap R \neq \emptyset$.

(e) Let X, Y, Z, and T be subsets of the set U. Draw a Venn diagram of these subsets (the diagram should have sixteen regions).

Let $U = \{n \in \mathbb{N} : 1 \leqslant n \leqslant 16\}$. Construct an example of subsets X, Y, Z, and T of U such that all sixteen regions in their Venn diagram are non-empty.

2. (a) In a certain college, 70% of the staff are blue-eyed, 80% are blonde, and 90% are female. What percentage, at least, of the staff are blue-eyed, blonde, and female?

(b) Would you believe a survey of 1000 people which reported that 517 liked sweets, 597 liked ice-cream, 458 liked cake, 243 liked sweets and ice-cream, 197 liked sweets and cake, and 224 liked ice-cream and cake, while 93 liked all three?

3. Let U be a finite set.

(a) Let u be an element of U. By considering complements, show that exactly half of the subsets of U contain u.

(b) Show that U has exactly 2^n subsets, where $n = |U|$. (Hint: subsets X of U are constructed by choosing either $u \in X$ or $u \notin X$ for each $u \in U$.)

(c) Determine the number of subsets X of U with $|X|$ even.

4. (a) Let $X = \{\text{John, Susan}\}$ and $Y = \{\text{Doe, Jones, Smith}\}$. List the elements of $X \times Y$.

(b) Let X, Y, Z be finite sets. With the help of a Venn diagram, justify the equation $|X \cup Y| = |X| + |Y| - |X \cap Y|$. Express $|X \cup Y \cup Z|$ in terms of $|X|$, $|Y|$, $|Z|$, $|X \cap Y|$, $|X \cap Z|$, $|Y \cap Z|$, $|X \cap Y \cap Z|$.

(c) Let X and Y be finite sets. Find a formula for the number of subsets of type $\{x, y\}$ where $x \in X$ and $y \in Y$, in terms of $|X|$, $|Y|$, and $|X \cap Y|$. List these subsets in the case $X = \{1, 2, 3, 4\}$ and $Y = \{2, 3, 4, 5\}$.

5. (a) Let X_1 and X_2 be subsets of the set U. Write out a proof of De Morgan's second law $(X_1 \cap X_2)' = X_1' \cup X_2'$ by appealing directly to (1.1)–(1.5).

(b) Let X, Y, Z be sets. Prove the distributive law:

$$X \cap (Y \cup Z) = (X \cap Y) \cup (X \cap Z).$$

Is the dual distributive law $X \cup (Y \cap Z) = (X \cup Y) \cap (X \cup Z)$ also true?

6. Let X_1, X_2, Y_1, Y_2 be sets. Prove
   (a) $(X_1 \times Y_1) \cap (X_2 \times Y_2) = (X_1 \cap X_2) \times (Y_1 \cap Y_2)$
   (b) $(X_1 \times Y_1) \cup (X_2 \times Y_2) \subseteq (X_1 \cup X_2) \times (Y_1 \cup Y_2)$.

Under what conditions on X_1, X_2, Y_1, Y_2 does equality hold in (b)?

7. Let U be a finite set. Write $n = |U|$ and let $\binom{n}{k}$ denote the number of subsets X of U with $|X| = k$. Evaluate $\binom{4}{1}$, $\binom{4}{2}$, and $\binom{4}{3}$.

By counting pairs (x, X) with $x \in X \subseteq U$ and $|X| = k > 0$, prove that

$$\binom{n}{k} k = n \binom{n-1}{k-1}.$$

Using induction on n, deduce the formula

$$\binom{n}{k} = \frac{n!}{(n-k)!\, k!}.$$

(The method of induction is explained in Chapter 3.)

8. Let $X_1, X_2, \ldots, X_n$ be sets and let S be a non-empty subset of $\{1, 2, \ldots, n\}$. The sets I_S and J_S are defined as follows:

$$I_S = \bigcap_{i \in S} X_i, \qquad J_S = \bigcup_{i \in S} X_i.$$

Let k be a natural number with $1 \leqslant k \leqslant n$. Let $I_k = \bigcup_{|S|=k} I_S$ and $J_k = \bigcap_{|S|=k} J_S$, where in both cases S ranges over all subsets of $\{1, 2, \ldots, n\}$ having exactly k elements.
   (a) Taking $n = 3$, verify by means of a Venn diagram that $I_2 = J_2$.
   (b) Prove that, in general, $I_k = J_{n-k+1}$.

Mappings

Two sets may be related to each other in various ways; in this section we deal with the most important type of relation, namely mappings of one set to another.

To get the idea, consider the set X of all persons with an account at a certain bank. Person x belonging to X has a bank account number which we take to be a positive integer and denote by $(x)\alpha$; for instance, if Sally Jones belongs to X and 199 is her bank account number, then (Sally Jones)$\alpha = 199$. In fact each element x of X gives rise to a specific element $(x)\alpha$ in $\mathbb{N}$; this is an example of a **mapping** of X to $\mathbb{N}$. We call α the account number mapping and use the notation

$$\alpha : X \to \mathbb{N}$$

for, as the arrow suggests, α is a way of moving from the set X to the set $\mathbb{N}$.

Definition 1.14

Let X and Y be sets. Any subset S of the cartesian product $X \times Y$ is called a **relation** between X and Y.

A relation between X and Y is therefore a collection S of certain ordered pairs (x, y) where $x \in X$ and $y \in Y$; the concept of a mapping involves relations of a special type.

Definition 1.15

A **mapping** (**function** or **transformation**) consists of an ordered pair of sets, first X and secondly Y say, together with a relation S between them, such that for each x in X there is a *unique* y in Y with $(x, y) \in S$.

Without further ado, we introduce a notation for mappings which is both practical and suggestive. Suppose, given a mapping as in (1.15); we now call this mapping α (generally, Greek letters $\alpha, \beta, \gamma, \ldots$ will be used for mappings). We say α **maps** X to Y and write $\alpha : X \to Y$; as the element y depends only on x and α, we write $y = (x)\alpha$. So in other words, α consists of two sets X and Y together with a rule which assigns, to each element x of the first set X, a unique element $(x)\alpha$ of the second set Y.

For instance, suppose $X = \{1, 2, 3, 4, 6, 12\}$, $Y = \{1, 2, 4, 8\}$, and $S = \{(1, 1), (2, 2), (3, 1), (4, 4), (6, 2), (12, 4)\}$. Notice that each element of X occurs in the first place of a unique (exactly one) pair in S, and so the condition of (1.15) is satisfied. Using the notation

Fig. 1.9

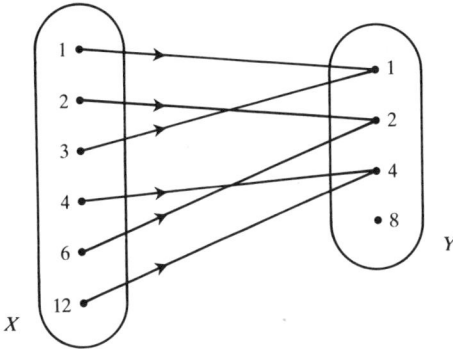

$\alpha : X \to Y$, we have

(1)$\alpha = 1$,   (2)$\alpha = 2$,   (3)$\alpha = 1$,

(4)$\alpha = 4$,   (6)$\alpha = 2$,   (12)$\alpha = 4$,

corresponding to the six ordered pairs in S. This mapping is pictured in Fig. 1.9; an arrow is drawn from each $x \in X$ to $(x)\alpha \in Y$, and so there is exactly one arrow beginning at each element of X, these arrows ending at elements of Y.

Definition 1.16   Let $\alpha : X \to Y$ be a mapping. For $x \in X$, the element $(x)\alpha$ is called the **image** of x by α; we say α **maps** x to $(x)\alpha$. The set X is called the **domain** of α and Y is called the **codomain** of α. The set $S = \{(x, (x)\alpha) : x \in X\}$ is called the **graph** of α.

For example consider the mapping $\alpha : \mathbb{R} \to \mathbb{R}$ given by $(x)\alpha = x^3$ for each $x \in \mathbb{R}$; in other words, α is the operation of 'cubing' applied to real numbers. As $(2)\alpha = 2^3 = 8$, we see that the image of 2 by α is 8; as $(\frac{1}{5})\alpha = (\frac{1}{5})^3 = \frac{1}{125}$, α maps $\frac{1}{5}$ to $\frac{1}{125}$. In this case domain α = codomain $\alpha = \mathbb{R}$, and the graph $S = \{(x, x^3) : x \in \mathbb{R}\}$ of α is the graph of $y = x^3$ in the familiar sense (Fig. 1.10).

Fig. 1.10

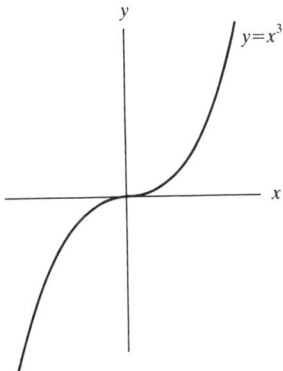

Definition 1.17
Let $\alpha_1 : X_1 \to Y_1$ and $\alpha_2 : X_2 \to Y_2$ be mappings. Then α_1 and α_2 are called **equal**, and the notation $\alpha_1 = \alpha_2$ is used, if $X_1 = X_2$, $Y_1 = Y_2$, and $(x)\alpha_1 = (x)\alpha_2$ for all $x \in X_1$.

So two mappings are equal if and only if they have the same domain, the same codomain, and they agree on each element of their common domain. We illustrate (1.17) by an example of *unequal* mappings: take $X = \{1, 2\}$, $Y = \{3, 4\}$ and consider the mappings α and β of X to Y defined by

$$(1)\alpha = 3, \quad (2)\alpha = 3; \quad (1)\beta = 3, \quad (2)\beta = 4.$$

Although α and β have the same domain X, the same codomain Y, and $(1)\alpha = (1)\beta$, because $(2)\alpha \neq (2)\beta$ we see $\alpha \neq \beta$ by (1.17). In fact there are four mappings of X to Y, namely α and β as above together with γ and δ defined by

$$(1)\gamma = 4, \quad (2)\gamma = 4; \quad (1)\delta = 4, \quad (2)\delta = 3.$$

More generally, let X and Y denote finite sets. Each mapping $\alpha : X \to Y$ is constructed by choosing, for each of the $|X|$ elements x of X, any one of the $|Y|$ elements of Y to be $(x)\alpha$; so there are $|Y|^{|X|}$ mappings of X to Y.

Definition 1.18
Let $\alpha : X \to Y$ and $\beta : Y \to Z$ be mappings. The **composition** of α and β is the mapping $\alpha\beta : X \to Z$ defined by

$$(x)(\alpha\beta) = ((x)\alpha)\beta \quad \text{for all} \quad x \in X.$$

Therefore $\alpha\beta$ is the mapping which results from applying *first* α and *secondly* β. Notice that the composition of α and β is defined only if codomain $\alpha = $ domain β in which case α and β are called **compatible**.

Let $\alpha : \mathbb{R} \to \mathbb{Z}$ be the mapping defined by $(x)\alpha = \lfloor x \rfloor$ for all $x \in \mathbb{R}$, where $\lfloor x \rfloor$ denotes the greatest integer not larger than x; $\lfloor x \rfloor$ is called the **integer part** of x. For example $\lfloor 5.162 \rfloor = 5$, $\lfloor \pi \rfloor = 3$, $\lfloor \frac{1}{3} \rfloor = 0$, $\lfloor -3.25 \rfloor = -4$. Let $\beta : \mathbb{Z} \to \mathbb{N}$ be defined by $(m)\beta = m^2 + 1$ for all $m \in \mathbb{Z}$. Then α and β are compatible as codomain $\alpha = \mathbb{Z} = $ domain β; their composition $\alpha\beta : \mathbb{R} \to \mathbb{N}$ can therefore be formed and is given by

$$(x)(\alpha\beta) = ((x)\alpha)\beta = (\lfloor x \rfloor)\beta = \lfloor x \rfloor^2 + 1 \quad \text{for all } x \in \mathbb{R}.$$

So, for instance, the image of 5.162 by $\alpha\beta$ is 26, as $(5.162)\alpha = 5$ and $(5)\beta = 26$.

The mapping $\alpha : \mathbb{R} \to \mathbb{R}$, defined by $(x)\alpha = x^3$ for all $x \in \mathbb{R}$, may be composed with itself to produce $\alpha^2 : \mathbb{R} \to \mathbb{R}$; by (1.18)

$$(x)\alpha^2 = ((x)\alpha)\alpha = (x^3)\alpha = (x^3)^3 = x^9 \quad \text{for all real numbers } x.$$

Similarly $\alpha^3 : \mathbb{R} \to \mathbb{R}$ is the mapping given by $(x)\alpha^3 = x^{27}$ for all $x \in \mathbb{R}$.

Proposition 1.19 (The associative law for mappings.) Let α and β be compatible mappings, and let β and γ be compatible mappings. Then

$$(\alpha\beta)\gamma = \alpha(\beta\gamma).$$

Proof As codomain $\alpha\beta = $ codomain $\beta = $ domain γ, we see that $\alpha\beta$ and γ are compatible mappings; their composition $(\alpha\beta)\gamma$ is a mapping of domain α to codomain γ. Similarly $\alpha(\beta\gamma)$ is also a mapping of domain α to codomain γ. Using (1.18) we obtain

$$(x)((\alpha\beta)\gamma) = ((x)(\alpha\beta))\gamma = (((x)\alpha)\beta)\gamma = ((x)\alpha)(\beta\gamma)$$
$$= (x)(\alpha(\beta\gamma)) \quad \text{for all } x \in \text{domain } \alpha.$$

By (1.17), the definition of equality of mappings, we conclude $(\alpha\beta)\gamma = \alpha(\beta\gamma)$. □

Under the hypothesis of (1.19), we may therefore refer unambiguously to $\alpha\beta\gamma$, meaning the combined effect of α, β, and γ, taken in that order.

Consider now mappings $\alpha : X \to Y$ and $\beta : Y \to X$. In this case $\alpha\beta : X \to X$ and $\beta\alpha : Y \to Y$ can both be formed; but even if $X = Y$, it is usually true that $\alpha\beta \neq \beta\alpha$. This fact is expressed by saying that composition of mappings is, in general, **non-commutative**.

For example, let $\alpha, \beta : \mathbb{R} \to \mathbb{R}$ be the mappings defined by: $(x)\alpha = x + 1$ and $(x)\beta = x^2$ for all $x \in \mathbb{R}$. Then $(x)(\alpha\beta) = (x + 1)^2$ and $(x)(\beta\alpha) = x^2 + 1$. Now $(x + 1)^2 = x^2 + 1$ holds only for $x = 0$, that is, $(x)(\alpha\beta) \neq (x)(\beta\alpha)$ for $x \neq 0$. Therefore $\alpha\beta \neq \beta\alpha$ by (1.17).

Definition 1.20 The mapping $\alpha : X \to Y$ is called **injective (one–one)** if for each $y \in Y$ there is at *most* one $x \in X$ with $(x)\alpha = y$.

The mapping $\alpha : \mathbb{Z} \to \mathbb{N}$, defined by $(m)\alpha = m^2 + 1$ for all $m \in \mathbb{Z}$, is *not* injective as $(1)\alpha = (-1)\alpha$; for a mapping is injective only if every pair of distinct (different) elements of its domain have distinct images by the mapping. The mapping $\beta : \mathbb{N} \to \mathbb{N}$, defined by $(n)\beta = n^2 + 1$ for all $n \in \mathbb{N}$ is injective: for suppose $(n_1)\beta = (n_2)\beta$ where $n_1, n_2 \in \mathbb{N}$. Therefore $n_1^2 + 1 = n_2^2 + 1$ and so $n_1 = \pm n_2$, which gives $n_1 = n_2$ as n_1 and n_2 are both positive. Comparing α and β we see

Fig. 1.11

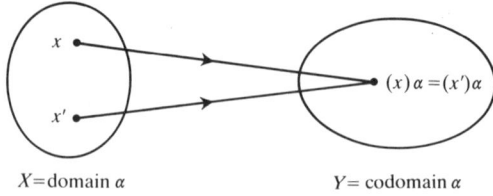

X = domain α Y = codomain α

that the domain of a mapping must be clearly specified when discussing injectivity.

The account number mapping mentioned in the introduction is an injection (an injective mapping) provided there are no joint accounts.

In the diagram of an injective mapping, no two or more arrows point to the same element; in other words the situation depicted in Fig. 1.11 with $x \neq x'$ *cannot* occur if $\alpha : X \rightarrow Y$ is injective. So α being injective is the same as the condition:

$$(x)\alpha = (x')\alpha \Rightarrow x = x' \quad \text{where} \quad x, x' \in X.$$

Here the symbol $\Rightarrow$ is read '*implies*' or '*only if*'; later we shall use $\Leftarrow$ which means '*is implied by*' or simply '*if*', and the double-headed arrow $\Leftrightarrow$ which is an abbreviation for '*if and only if*'.

Definition 1.21
The mapping $\alpha : X \rightarrow Y$ is called **surjective** (**onto**) if for each $y \in Y$ there is at *least* one $x \in X$ with $(x)\alpha = y$.

So $\alpha : X \rightarrow Y$ is surjective if every element of Y is of the form $(x)\alpha$ for some $x \in X$; in the diagram of a surjective mapping there is an arrow pointing to each element of the codomain.

Consider again $\alpha : \mathbb{R} \rightarrow \mathbb{Z}$, where $(x)\alpha = \lfloor x \rfloor$ for all $x \in \mathbb{R}$ and, as before, $\lfloor x \rfloor$ stands for the integer part of the real number x. As $(m)\alpha = m$ for every integer m, we see that α is surjective; in other words, given any integer, there is a real number having that integer as its integer part. Incidentally $(0)\alpha = (\frac{1}{2})\alpha$, showing that α is not injective. We may compare α with $\beta : \mathbb{R} \rightarrow \mathbb{R}$, defined by $(x)\beta = \lfloor x \rfloor$ for all $x \in \mathbb{R}$. As α and β have different codomains, we have $\alpha \neq \beta$ by (1.17); in fact β is not surjective because there is no real number x with $(x)\beta = \frac{1}{2}$. Therefore it is important to specify the codomain before discussing surjectivity.

Using the concepts of injectivity and surjectivity, mappings can be divided into four types; these types are listed in Fig. 1.12 together with the diagram of a simple example of each type.

Fig. 1.12

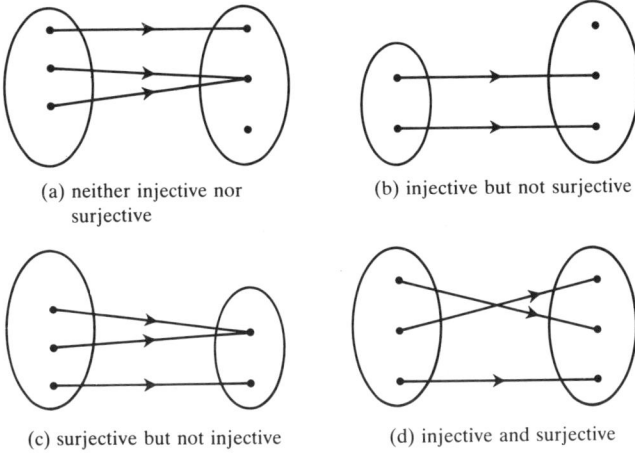

(a) neither injective nor
    surjective

(b) injective but not surjective

(c) surjective but not injective

(d) injective and surjective

**Definition
1.22**

A mapping which is both injective and surjective is called
bijective or a **bijection** (**one–one correspondence**).

Let $\alpha : X \to Y$ be a bijection; combining (1.20) and (1.21), for
each $y \in Y$ there is *exactly* one $x \in X$ with $(x)\alpha = y$.

Consider $\alpha : \mathbb{R} \to \mathbb{R}$ where $(x)\alpha = 2x + 1$ for all $x \in \mathbb{R}$. For each
real number y, the equation $(x)\alpha = y$, that is, $2x + 1 = y$, has a
unique solution $x = (y - 1)/2$; therefore α is bijective.

**Definition
1.23**

Let $\alpha : X \to Y$ be bijective. The mapping $\alpha^{-1} : Y \to X$, defined by
$(y)\alpha^{-1} = x$ where $(x)\alpha = y$, for all $y \in Y$, is called the **inverse** of
α.

The above bijection $\alpha : \mathbb{R} \to \mathbb{R}$, where $(x)\alpha = 2x + 1$ for all $x \in \mathbb{R}$,
has inverse $\alpha^{-1} : \mathbb{R} \to \mathbb{R}$, where $(y)\alpha^{-1} = (y - 1)/2$ for all $y \in \mathbb{R}$; in
particular α maps 6 to 13, and so α^{-1} maps 13 to 6.

Let $X = \{1, 2, 3\}$, $Y = \{4, 5, 6\}$ and let $\beta : X \to Y$ be defined by
$(1)\beta = 4$, $(2)\beta = 5$, $(3)\beta = 6$; then β is bijective and has inverse
$\beta^{-1} : Y \to X$ where $(4)\beta^{-1} = 1$, $(5)\beta^{-1} = 2$, $(6)\beta^{-1} = 3$.

The mapping $\gamma : \mathbb{R} \to \mathbb{R}$, defined by $(x)\gamma = x^3$ for all $x \in \mathbb{R}$, is
bijective and $\gamma^{-1} : \mathbb{R} \to \mathbb{R}$ is given by $(y)\gamma^{-1} = y^{\frac{1}{3}}$ for all $y \in \mathbb{R}$. It is
worth noticing that the graphs of γ and γ^{-1} are related by
interchanging the roles of x and y, that is, by reflection in the line
$y = x$ (Fig. 1.13).

Generally, whatever the effect of the bijection α, the *reverse*
effect is produced by α^{-1}. In the diagram of the bijection
$\alpha : X \to Y$, for each $y \in Y$ there is exactly one arrow pointing to y,

Fig. 1.13

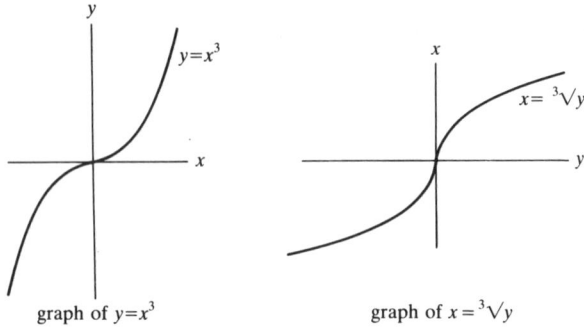

graph of $y=x^3$ graph of $x=\sqrt[3]{y}$

as in Fig. 1.12(d); the diagram of α^{-1} is obtained by reversing the direction of all these arrows (Fig. 1.14).

Fig. 1.14

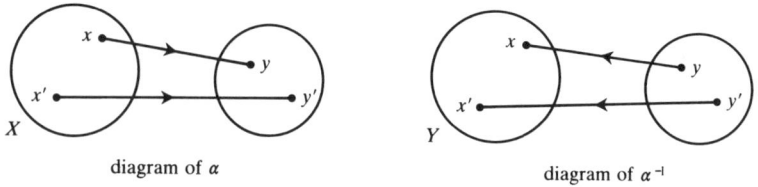

diagram of α diagram of α^{-1}

Definition 1.24   The mapping $\iota_X : X \to X$, defined by $(x)\iota_X = x$ for all $x \in X$, is called the **identity mapping** of the set X.

The equations

$$\iota_X\alpha = \alpha = \alpha\iota_Y$$

which are valid for all mappings $\alpha : X \to Y$ show how ineffective identity mappings are; indeed it is hard to imagine duller mappings! However, just as the empty set has a role to play, the same is true of identity mappings; for instance, the connection between a bijection $\alpha : X \to Y$ and its inverse $\alpha^{-1} : Y \to X$ is expressed by the equations:

$$\alpha\alpha^{-1} = \iota_X, \qquad \alpha^{-1}\alpha = \iota_Y$$

for the composition of a bijection with its inverse is the identity mapping. Notice that the inverse α^{-1} of a bijection α is also a bijection and $(\alpha^{-1})^{-1} = \alpha$ for the inverse of α^{-1} is the original bijection α.

We end this section with a word about notation. The image of x by α is denoted here by $(x)\alpha$ because this notation is suited to the discussion of composite mappings. However many texts, especially those on calculus, adopt instead the *functional* notation $f(x)$ for the image of x by the mapping (or function) f, in which case gf denotes the composite formed by applying the mapping f *before* the mapping g. Generally we shall adhere to the notation $(x)\alpha$, although from time to time the functional notation and other variants fitted to the context will be used.

Exercises 1.2

1. Let $X = \{1, 2\}$ and $Y = \{3, 4, 5\}$. List the nine mappings of X to Y and state which are injective. List the mappings of Y to X and state which are surjective.

2. Let α, $\beta : \mathbb{Q} \to \mathbb{Q}$ be defined by $(x)\alpha = 2x - 1$, $(x)\beta = x^2$, for all $x \in \mathbb{Q}$. Write down the image of x by each of:

$\alpha\beta, \quad \beta\alpha, \quad \alpha^2, \quad \beta^2, \quad \beta^3.$

Find the rational number x satisfying $(x)\alpha\beta = (x)\beta\alpha$.
Show that α is bijective and write down the image of x by:

$\alpha^{-1}, \quad \alpha^{-2}, \quad \alpha\beta\alpha^{-1}.$

3. I have bought a set X of presents, each of which is earmarked for one of the set Y of my colleagues. Suppose present x is earmarked for colleague $(x)\alpha$; then α is a mapping of X to Y. Express in colloquial English:

(i) α is injective, (ii) α is surjective, (iii) α is bijective.

If I have two presents for my boss but nothing for my junior colleague, what type of mapping is α?

4. Test the mappings $\alpha : \mathbb{Z} \to \mathbb{Z}$ for injectivity and surjectivity, where $(m)\alpha$ is defined for all $m \in \mathbb{Z}$ as follows:

(a) $(m)\alpha = m^2$, (b) $(m)\alpha = m^3$, (c) $(m)\alpha = 1 - m$,
(d) $(m)\alpha = \lfloor m/2 \rfloor$ (the integer part of $m/2$).

5. Let $\alpha : \mathbb{Z} \to \mathbb{Z}$ be defined by $(m)\alpha = m + 1$ for all $m \in \mathbb{Z}$. Show that there is a unique mapping $\beta : \mathbb{Z} \to \mathbb{Z}$ which commutes with α (that is, $\alpha\beta = \beta\alpha$) and satisfies $(1)\beta = 0$; what is the connection between α and β? Determine all the mappings $\beta : \mathbb{Z} \to \mathbb{Z}$ which commute with α.

6. Test the following mappings $\alpha : \mathbb{R} \to \mathbb{R}$ for injectivity and surjectivity and sketch their graphs:

(a) $(x)\alpha = x^2 - 1$ for all $x \in \mathbb{R}$,
(b) $(x)\alpha = 1/x$ for all non-zero real numbers x, and $(0)\alpha = 0$,
(c) $(x)\alpha = \lfloor x \rfloor$ (the integer part of x) for all $x \in \mathbb{R}$.

7. Let $\mathbb{R}_+$ denote the set of positive real numbers. Let $\alpha : \mathbb{R} \to \mathbb{R}_+$ and $\beta : \mathbb{R}_+ \to \mathbb{R}$ be defined by $(x)\alpha = e^x$ for all $x \in \mathbb{R}$, $(y)\beta = \log_e y$ for all $y \in \mathbb{R}_+$. Sketch the graphs of α and β. What is the connection between α and β?

8. Let $\alpha : X \to Y$, $\beta : Y \to Z$ be mappings.

(a) If α and β are injective, show that $\alpha\beta$ is injective; if $\alpha\beta$ is injective, show that α is injective. Give an example of α and β with $\alpha\beta$ injective but β not injective.

(b) If α and β are surjective, show that $\alpha\beta$ is surjective; if $\alpha\beta$ is surjective, show that β is surjective. Give an example of α and β with $\alpha\beta$ surjective but α not surjective.

(c) Suppose $X = Z$ and let $x \in X$ and $y \in Y$. If $\alpha\beta = \iota_X$ and $(x)\alpha = y$, show that $(y)\beta = x$; if $\beta\alpha = \iota_Y$ and $(y)\beta = x$, show that $(x)\alpha = y$. Deduce that if $\alpha\beta = \iota_X$ and $\beta\alpha = \iota_Y$, then α is bijective and $\beta = \alpha^{-1}$.

(d) Let α and β be bijective. Show that $\alpha\beta$ is bijective and

$$(\alpha\beta)^{-1} = \beta^{-1}\alpha^{-1}.$$

(Hint: consider the mappings $\alpha\beta\beta^{-1}\alpha^{-1}$ and $\beta^{-1}\alpha^{-1}\alpha\beta$.)

9. Let X and Y be finite sets. Write $m = |X|$ and $n = |Y|$.

(a) If $m \leqslant n$, determine the number of injections $\alpha : X \to Y$.

(b) If $n = 2$, determine the number of surjections $\alpha : X \to Y$. (Hint: any non-constant mapping is surjective.)

(c) Use the fact that, if $n > 0$, exactly half of the subsets Z of Y are such that $|Z|$ is even, to show that

$$\sum_{k=0}^{n} (-1)^k \binom{n}{k} = \begin{cases} 1 & \text{if } n = 0, \\ 0 & \text{if } n > 0. \end{cases}$$

(d) Let $\sigma(m, n)$ denote the number of surjections $\alpha : X \to Y$, taking $\sigma(0, 0) = 1$. Let x_0 and y_0 be given elements of X and Y respectively. Show that there are

$$\sigma(m - 1, n - 1) + \sigma(m - 1, n)$$

surjections of X to Y with $(x_0)\alpha = y_0$ $(m, n \geqslant 1)$. Deduce the equation

$$\sigma(m, n) = n(\sigma(m - 1, n - 1) + \sigma(m - 1, n)).$$

Use the above relation to prove, by induction on m, that

$$\sigma(m, n) = \sum_{k=0}^{n} (-1)^k \binom{n}{k} (n - k)^m.$$

(Hint: the case $m = 0$ is covered by (c) above.)

Equivalence relations

Suppose we wish to look at the elements of a given set X from a certain point of view; two elements are called equivalent if, from this viewpoint, they appear to be the same. Equivalent elements are then amalgamated to form the elements of a new set, which is often of more interest than X.

For example, consider a stamp-collector who has a set X of stamps and let us suppose that his only interest is in the country of each stamp. From his point of view all French stamps are identical, all Italian stamps are identical etc., but French stamps are different to Italian ones; in short, two stamps are equivalent if and only if they are of the same country. It is natural for the collection X to be sorted out according to the countries represented, which we assume correspond to the pages of an album: so all the French stamps (assuming that there is at least one) in X are amalgamated (that is, stuck on a certain page of the album) and similarly for the other countries. The various amalgamations (the album pages) then form the elements of a new set (the album itself).

Definition 1.25

Let X be a set. A subset S of $X \times X$ is called an **equivalence relation** on X if

(i) $(x, x) \in S$ for all $x \in X$,
(ii) $(x, x') \in S \Rightarrow (x', x) \in S$,
(iii) $(x, x') \in S$ and $(x', x'') \in S \Rightarrow (x, x'') \in S$.

In the above example where X is the stamp collection, then $(x, x') \in S$ means x and x' are stamps of the same country, while $(x, x') \notin S$ means x and x' are stamps of different countries; it is clear that the conditions of (1.25) are satisfied in this case.

As another illustration, let $X = \{1, 2, 3\}$ and $S = \{(1, 1), (1, 2), (2, 1), (2, 2), (3, 3)\}$. By inspection we see that S satisfies the conditions of (1.25) and so S is an equivalence relation on X. On the other hand $T = \{(1, 1), (1, 2), (1, 3), (2, 2), (2, 3), (3, 3)\}$ does not satisfy (1.25)(ii) as $(1, 2) \in T$ but $(2, 1) \notin T$; therefore T is *not* an equivalence relation on X.

We now introduce some helpful notation and terminology. Let S be an equivalence relation on X. In place of $(x, x') \in S$ we write $x \equiv x'$ and say that x is **equivalent** to x'. The conditions of (1.25) now become

(i) **the reflexive law**: $x \equiv x$ for all $x \in X$ (that is, every element is equivalent to itself),

(ii) **the symmetric law**: $x \equiv x' \Rightarrow x' \equiv x$ (that is, whenever x is equivalent to x', then x' is equivalent to x),

(iii) **the transitive law**: $x \equiv x'$ and $x' \equiv x'' \Rightarrow x \equiv x''$ (that is, whenever x is equivalent to x' and x' is equivalent to x'', then x is equivalent to x'').

In practice $x \equiv x'$ means that x and x' have some particular property in common.

Definition 1.26

Let S be an equivalence relation on the set X and let $x_0 \in X$. The subset $\bar{x}_0 = \{x \in X : x \equiv x_0\}$ of X is called the **equivalence class** of x_0.

Therefore $\bar{x}_0$ consists of those elements of X which are equivalent to the given element x_0. In the case of the stamp collection, if x_0 is a Canadian stamp in X, then $\bar{x}_0$ is the set of all Canadian stamps in X.

Example 1.27

For $m, m' \in \mathbb{Z}$, write $m \equiv m'$ if m and m' are of the same **parity** (that is, m and m' are either both even or both odd). This relation is an equivalence relation on $\mathbb{Z}$. As 0 is an even integer, $\bar{0}$ is the set of all even integers:

$$\bar{0} = \{\ldots, -4, -2, 0, 2, 4, \ldots\}.$$

As 2 is also an even integer, $\bar{2}$ is again the set of all even integers and so $\bar{0} = \bar{2}$; more generally $\bar{0} = \bar{m}$ for all even integers m. Similarly, as 1 is an odd integer, $\bar{1}$ is the set of all odd integers:

$$\bar{1} = \{\ldots, -3, -1, 1, 3, 5, \ldots\}.$$

If m is any odd integer, $\bar{m}$ is the set of all odd integers and so $\bar{m} = \bar{1}$. In this example there are just two equivalence classes since there are only two possibilities for the parity of an integer; notice however that a given equivalence class can be denoted in as many different ways as it has elements.

We now derive properties of an arbitrary equivalence relation.

Proposition 1.28

Suppose given an equivalence relation on the set X and let $x_0, x_1 \in X$. Then the following statements are logically equivalent:

(i) $x_0 \equiv x_1$,   (ii) $\bar{x}_0 = \bar{x}_1$,   (iii) $\bar{x}_0 \cap \bar{x}_1 \neq \emptyset$.

Proof

First we establish: (i) $\Rightarrow$ (ii). Therefore suppose $x_0 \equiv x_1$. In order to verify the set equality $\bar{x}_0 = \bar{x}_1$, let $x \in \bar{x}_0$. This means $x \equiv x_0$ by

(1.26), which together with $x_0 \equiv x_1$ gives $x \equiv x_1$ using the transitive law. So $x \in \bar{x}_1$ and so $\bar{x}_0 \subseteq \bar{x}_1$. Applying the symmetric law to $x_0 \equiv x_1$ gives $x_1 \equiv x_0$; using the preceding argument again, we obtain $\bar{x}_1 \subseteq \bar{x}_0$ and so $\bar{x}_0 = \bar{x}_1$.

Now we prove: (ii) $\Rightarrow$ (iii). So suppose $\bar{x}_0 = \bar{x}_1$. By the reflexive law $x_0 \equiv x_0$, which means $x_0 \in \bar{x}_0$. Therefore $x_0 \in \bar{x}_1$ also. So $x_0 \in \bar{x}_0 \cap \bar{x}_1$ showing $\bar{x}_0 \cap \bar{x}_1 \neq \varnothing$.

Finally we establish: (iii) $\Rightarrow$ (i). To do this suppose $\bar{x}_0 \cap \bar{x}_1 \neq \varnothing$. Therefore there is an element $x_2 \in \bar{x}_0 \cap \bar{x}_1$. This means $x_2 \in \bar{x}_0$ and $x_2 \in \bar{x}_1$, that is, $x_2 \equiv x_0$ and $x_2 \equiv x_1$. Using the symmetric and transitive laws, we obtain $x_0 \equiv x_1$.

The cycle of implications (i) $\Rightarrow$ (ii) $\Rightarrow$ (iii) $\Rightarrow$ (i) shows that the statements (i), (ii), and (iii) are logically equivalent, meaning that if one statement is true, then so are the other two. $\qquad\square$

We now introduce an alternative approach to equivalence relations.

Definition 1.29

Let X be a set. A **partition** of X is a family of non-empty subsets of X such that each element of X belongs to exactly one member of the family. (We use the terms 'member' and 'family' instead of 'element' and 'set' in order to give a clearer definition.)

In other words, a partition of X is a family of non-overlapping non-empty subsets of X, the union of the family being X. Figure 1.15 shows the Venn diagram of a set partitioned into seven subsets.

Let $\mathbb{R}_+$ denote the set of positive real numbers and $\mathbb{R}_-$ the set of negative real numbers; then $\mathbb{R}_+$, $\mathbb{R}_-$, and $\{0\}$ form a partition of $\mathbb{R}$.

The equivalence relation (1.27) of parity on $\mathbb{Z}$ gives rise to a partition of $\mathbb{Z}$, for it splits $\mathbb{Z}$ into two parts: $\bar{0}$ and $\bar{1}$; in other words

$$\mathbb{Z} = \{\text{all even integers}\} \cup \{\text{all odd integers}\}.$$

Fig. 1.15

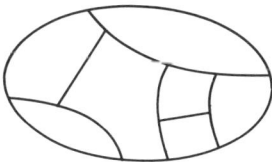

The following corollary shows how every equivalence relation gives rise to a partition.

Corollary 1.30

Suppose that an equivalence relation on the set X is given. Then the family of equivalence classes $\bar{x}$, for $x \in X$, forms a partition of X.

Proof
For $x \in X$ we have $x \equiv x$; therefore $x \in \bar{x}$, showing that x belongs to a member of the family of equivalence classes. But if $x \in \bar{x}_0$ and $x \in \bar{x}_1$ then $\bar{x}_0 \cap \bar{x}_1 \neq \varnothing$, and so $\bar{x}_0 = \bar{x}_1$ by (1.28). Therefore x belongs to exactly one equivalence class. By (1.29), the equivalence classes form a partition of X. $\qquad\qquad\square$

As another illustration, consider the set P of all people, past and present! For $x, x' \in P$ let us write $x \equiv x'$ if x and x' have the same birthday (we assume that nobody is born exactly at midnight). Then $\equiv$ is an equivalence relation on P and $\bar{x}_0$ is the set of all people having the same birthday as x_0; for instance knowing that W. Shakespeare was born on April 23rd, we deduce

$\overline{\text{W. Shakespeare}}$ = set of all people born on April 23rd.

This equivalence relation partitions P into 366 equivalence classes, one class for each day of the year (including February 29th); therefore

$P = \{$all people born on January 1st$\} \cup \ldots$

$\ldots \cup \{$all people born on December 31st$\}$.

Notice that a partition of X gives rise to an equivalence relation on X: for $x, x' \in X$ we write $x \equiv x'$ if x and x' belong to the same member of the partition; then $\equiv$ is an equivalence relation on X. Together with (1.30) this shows that each equivalence relation on X corresponds to a partition of X, for the one determines the other.

For instance the equivalence relation $S = \{(1, 1), (1, 2), (2, 1), (2, 2), (3, 3)\}$ on $X = \{1, 2, 3\}$ corresponds to the partition $\{\{1, 2\}, \{3\}\}$ of X, as these two subsets of X are the equivalence classes of S. There are five partitions of X, namely

$\{\{1\}, \{2\}, \{3\}\}, \quad \{\{1, 2\}, \{3\}\}, \quad \{\{1, 3\}, \{2\}\},$
$\{\{1\}, \{2, 3\}\}, \quad \{\{1, 2, 3\}\},$

and so there are five equivalence relations on $\{1, 2, 3\}$.

Suppose now that X is any set and that an equivalence relation on X is given. We denote the set of equivalence classes by $\bar{X}$; therefore

$\bar{X} = \{\bar{x} : x \in X\}$.

The set $\bar{X}$ is the new set referred to at the beginning of this section; each equivalence class $\bar{x}$ forms a single element of $\bar{X}$, that is, the elements of each equivalence class are amalgamated into a single element of $\bar{X}$.

For example, consider the equivalence relation $\equiv$ on $X = \{-2, -1, 0, 1, 2\}$ where $x \equiv x'$ means $x^2 = (x')^2$; so two elements of X are equivalent if and only if their squares are equal. The equivalence classes, in this case, are

$$\bar{0} = \{0\}, \qquad \bar{1} = \{1, -1\}, \qquad \bar{2} = \{2, -2\}$$

and so $\bar{X} = \{\bar{0}, \bar{1}, \bar{2}\} = \{\{0\}, \{1, -1\}, \{2, -2\}\}$.

Definition 1.31

Suppose given an equivalence relation on X. The surjection

$$\eta : X \to \bar{X} \quad \text{defined by} \quad (x)\eta = \bar{x} \text{ for all } x \in X$$

is called the **natural mapping** relative to the given equivalence relation.

Therefore η maps each element to its own equivalence class; of course one must have a specific equivalence relation in mind when referring to η.

If $X = \{-2, -1, 0, 1, 2\}$ and $\bar{X} = \{\{0\}, \{1, -1\}, \{2, -2\}\}$, as above, then $\eta : X \to \bar{X}$ is the mapping with $(0)\eta = \{0\}$, $(1)\eta = (-1)\eta = \{1, -1\}$, $(2)\eta = (-2)\eta = \{2, -2\}$.

Example 1.32

We consider again the equivalence relation (1.27) of parity on $\mathbb{Z}$. Notice that the integers m and m' have the same parity if and only if their difference $m - m'$ is even; in other words $m \equiv m'$ means 2 is a divisor of $m - m'$. In this case we denote the set of equivalence classes by $\mathbb{Z}_2$; therefore

$$\mathbb{Z}_2 = \{\bar{0}, \bar{1}\}$$

is the set consisting of the two elements $\bar{0}$ and $\bar{1}$.

Now the sum of any two even integers is itself an even integer; this fact is expressed by the equation

$$\bar{0} + \bar{0} = \bar{0}$$

since $\bar{0}$ stands for the set of even integers. As the sum of any even integer and any odd integer is an odd integer, we write

$$\bar{0} + \bar{1} = \bar{1} \quad \text{and} \quad \bar{1} + \bar{0} = \bar{1}$$

since $\bar{1}$ stands for the set of odd integers. The sum of any two odd integers being an even integer is expressed by the equation

$$\bar{1} + \bar{1} = \bar{0}.$$

All these facts are summarized by writing

$$\overline{m + m'} = \bar{m} + \bar{m'} \quad \text{for all } m, m' \in \mathbb{Z}.$$

In other words, the parity of $m + m'$ depends only on the parity of m and the parity of m'; this is expressed by saying that the equivalence relation of parity is **compatible with addition**. Similarly the product of two integers, one or both being even, is itself even; the product of two odd integers is odd. Therefore the parity of mm' depends only on the parity of m and the parity of m'; summarizing, we write

$$(\overline{m})(\overline{m'}) = \overline{mm'} \quad \text{for all } m, m' \in \mathbb{Z}$$

and say that the equivalence relation of parity is **compatible with multiplication**. So the two elements of $\mathbb{Z}_2$ add and multiply as in the tables:

+	$\bar{0}$	$\bar{1}$
$\bar{0}$	$\bar{0}$	$\bar{1}$
$\bar{1}$	$\bar{1}$	$\bar{0}$

×	$\bar{0}$	$\bar{1}$
$\bar{0}$	$\bar{0}$	$\bar{0}$
$\bar{1}$	$\bar{0}$	$\bar{1}$

In this case the natural mapping $\eta : \mathbb{Z} \to \mathbb{Z}_2$ is defined by:

$$(m)\eta = \begin{cases} \bar{0} & \text{if } m \text{ is even,} \\ \bar{1} & \text{if } m \text{ is odd.} \end{cases}$$

The above equations summarizing the rules of parity now become

$$\left.\begin{aligned} (m + m')\eta &= (m)\eta + (m')\eta \\ (mm')\eta &= ((m)\eta)((m')\eta) \end{aligned}\right\} \quad \text{for all } m, m' \in \mathbb{Z}.$$

In fact $\mathbb{Z}_2$ is the smallest example of an algebraic structure called a **field** (see (2.18) and (3.18)) and η is an example of a **ring homomorphism** (see (5.1)).

The above example is typical of many we shall meet in our journey through algebra; generally, given an operation on X, such as addition or multiplication, and an equivalence relation on X compatible with this operation, then $\bar{X}$ 'inherits' the operation from X. Often the structure of $\bar{X}$ is more interesting than that of X, but in any case η provides the link between them.

Exercises 1.3
1. Let x and x' belong to the set X of all living people. In each of the following cases, decide which of the reflexive, symmetric, and transitive laws hold; in the case of an equivalence relation, describe, in colloquial English, your own equivalence class and the partitioning of X.

(a) $x \equiv x'$ means x and x' live within 100 miles of each other.
(b) $x \equiv x'$ means x and x' live in the same country.

(c) $x \equiv x'$ means x and x' have the same sex.
(d) $x \equiv x'$ means x is not taller than x'.

(Make the assumption that everybody lives somewhere and has a definite sex and height.)

2. Let x, x' belong to the set $\mathbb{R}$ of all real numbers. In each case decide whether or not $\equiv$ is an equivalence relation on $\mathbb{R}$, and if so describe the equivalence classes.

(a) $x \equiv x'$ means $x - x'$ is an integer.
(b) $x \equiv x'$ means xx' is positive.
(c) $x \equiv x'$ means $x^2 = x'^2$.
(d) $x \equiv x'$ means $\lfloor x \rfloor = \lfloor x' \rfloor$, where $\lfloor x \rfloor$ denotes the integer part of x.

3. For each of the following equivalence relations on the cartesian plane $\mathbb{R}^2$, describe, by a diagram, the equivalence class of the point $(1\frac{1}{2}, 3\frac{1}{2})$ and the partitioning of $\mathbb{R}^2$.

(a) $(x, y) \equiv (x', y')$ means $x^2 + y^2 = x'^2 + y'^2$.
(b) $(x, y) \equiv (x', y')$ means $x = x'$.
(c) $(x, y) \equiv (x', y')$ means $x + y = x' + y'$.
(d) $(x, y) \equiv (x', y')$ means $\lfloor x \rfloor = \lfloor x' \rfloor$ and $\lfloor y \rfloor = \lfloor y' \rfloor$, where $\lfloor x \rfloor$ denotes the integer part of x.

4. List the fifteen partitions of $\{1, 2, 3, 4\}$. How many equivalence relations are there on $\{1, 2, 3, 4\}$? How many equivalence relations are there on $\{1, 2, 3, 4, 5\}$? Find the number of equivalence relations on $\{1, 2, 3, 4, 5, 6, 7, 8\}$ such that none of 1, 2, 3, 4 is equivalent to any of 5, 6, 7, 8.

5. (a) Let $X = \{1, 2, 3\}$. For each of the reflexive, symmetric, and transitive laws, find a subset S of $X \times X$ for which that law is false but the remaining two laws are true. (These laws are therefore **independent**, that is, no law can be deduced from the others.)

(b) Find the mistake in the following 'proof' that the symmetric and transitive laws imply the reflexive law. For $x \in X$ let $x' \in X$ be any element with $x \equiv x'$. By symmetry $x' \equiv x$ and so by transitivity $x \equiv x$ for all $x \in X$, showing that the reflexive law holds.

6. Let S be a subset of the cartesian plane $\mathbb{R}^2$. Describe geometrically (pictorially) the following conditions on S

(i) $(x, x) \in S$ for all $x \in \mathbb{R}$.
(ii) $(x, x') \in S$ whenever $(x', x) \in S$.
(iii) $(x, x'') \in S$ whenever $(x, x') \in S$ and $(x', x'') \in S$.

7. (a) Let $\alpha : X \rightarrow Y$ be a mapping. For $x, x' \in X$, write $x \equiv x'$ if $(x)\alpha = (x')\alpha$. Verify that $\equiv$ is an equivalence relation on X and that $\bar{\alpha} : \bar{X} \rightarrow Y$, defined by $(\bar{x})\bar{\alpha} = (x)\alpha$ for all $\bar{x} \in \bar{X}$, is injective.

(b) Describe $\bar{\mathbb{R}}$ and $\bar{\alpha}$ (as above) for $\alpha : \mathbb{R} \rightarrow \mathbb{R}$ defined by $(x)\alpha = x^2$ for all $x \in \mathbb{R}$.

(c) Show that every mapping can be expressed in the form $\beta\gamma$ where β is surjective and γ is injective. (Hint: use part (a) above.)

Part I
Rings and fields

2 Rings, fields, and complex numbers

From our standpoint algebra begins with the study of sets within which operations of addition and multiplication can be carried out, or **systems** as we shall call them; systems having operations which obey the familiar rules of manipulation will be of special concern to us.

The most general type of system we shall meet is called a **ring**; the sum and product of ring elements are again ring elements and although many familiar laws must be obeyed (e.g. $x + y = y + x$), some equally familiar laws are not imposed (for instance $xy \neq yx$ *may* occur in a ring). It is perhaps comforting to know that the laws of a ring are not arbitrarily decreed by a mathematical dictator! On the contrary, they arise out of practical experience being modelled on the properties of matrices (see Chapter 7), for the laws which govern matrix addition and matrix multiplication are precisely the ring laws.

The least general (but most important) type of system we shall deal with is called a **field**, for in a field *all* the familiar laws of arithmetic are required to hold; in particular every non-zero element of a field has an inverse within the field. The rational numbers $\mathbb{Q}$ form a field as do the real numbers $\mathbb{R}$; on the other hand, the system $\mathbb{Z}$ of integers is *not* a field (the integer 2 has no integer inverse). We shall show in the next chapter that the system $\mathbb{Z}_2$ introduced in (1.32) is the smallest field; this field is used in the theory of electric circuits, the element $\bar{0}$ corresponding to 'current off' and $\bar{1}$ corresponding to 'current on'.

In the second half of the chapter we concentrate on one particular field, the field $\mathbb{C}$ of **complex numbers**; a working knowledge of this field is indispensable, for it is no exaggeration to say that $\mathbb{C}$ is the most useful field of all!

Definition 2.1 Let R denote a set. A mapping $\alpha : R \times R \rightarrow R$ is called a **binary operation** on R.

A binary operation on R is therefore a rule which associates, with each ordered pair (x, y) of elements from R, a single element $(x, y)\alpha$ of R.

For example, let $\alpha : \mathbb{Z} \times \mathbb{Z} \to \mathbb{Z}$ be the binary operation of **integer addition**, that is,

$$(x, y)\alpha = x + y \quad \text{for all } x, y \in \mathbb{Z}.$$

Similarly let $\mu : \mathbb{Z} \times \mathbb{Z} \to \mathbb{Z}$ be the binary operation of **integer multiplication**, that is,

$$(x, y)\mu = xy \quad \text{for all } x, y \in \mathbb{Z}.$$

The binary operations α and μ are sensible in as much as they obey familiar laws of manipulation. The familiar **associative** law of addition

$$(x + y) + z = x + (y + z) \quad \text{for all } x, y, z \in \mathbb{Z}$$

can be expressed using α as

$$((x, y)\alpha, z)\alpha = (x, (y, z)\alpha)\alpha \quad \text{for all } x, y, z \in \mathbb{Z}.$$

Similarly the familiar **distributive** law

$$(x + y)z = xz + yz \quad \text{for all } x, y, z \in \mathbb{Z}$$

can be expressed in cumbersome form using α and μ as

$$((x, y)\alpha, z)\mu = ((x, z)\mu, (y, z)\mu)\alpha \quad \text{for all } x, y, z \in \mathbb{Z}.$$

However we are not trying to make life complicated for the sake of it! Rather our aim is to render abstract manipulations more natural by expressing them in familiar notation, and at the same time analyse the foundations of ordinary arithmetic.

Let α and μ be binary operations on the set R. To make life as easy as possible for ourselves, we proceed to interpret α and μ as 'addition' and 'multiplication' on R. Therefore we *introduce* addition by writing $x + y = (x, y)\alpha$ for all $x, y \in R$; as α maps $R \times R$ to R, we obtain

$$x + y \in R \quad \text{for all } x, y \in R$$

which is expressed by saying that R is **closed under addition**, that is, the *sum* of every pair of elements from R is itself an element of R. In the same way, multiplication is *introduced* on R by writing $xy = (x, y)\mu$ for all $x, y \in R$; as μ maps $R \times R$ to R, we obtain

$$xy \in R \quad \text{for all } x, y \in R$$

which is expressed by saying that R is **closed under multiplication**, that is, the *product* of every pair of elements from R is itself an element of R.

The set R, together with the binary operations α and μ, is referred to as the **system** (R, α, μ); when α and μ are interpreted as addition and multiplication we write $(R, +, \times)$ for this system. When it is clear from the context which binary operations on R we have in mind, we refer simply to the system R.

For instance, when referring to the system $\mathbb{Z}$ it is understood that we have in mind the usual operations of integer addition and integer multiplication; similarly by $\mathbb{Q}$ (or $\mathbb{R}$) we understand the system of rational (or real) numbers with the usual addition and multiplication.

Definition 2.2

Suppose given two binary operations on the set R. Interpreting these operations as addition and multiplication, the system $(R, +, \times)$ is called a **ring** if laws 1–7 below hold:

1. **Associative law of addition:** $(x + y) + z = x + (y + z)$ for all $x, y, z \in R$.

2. **Existence of 0-element:** there is an element $0 \in R$ satisfying

$0 + x = x$ for all $x \in R$.

3. **Existence of negatives:** for each $x \in R$ there is $-x \in R$ satisfying

$-x + x = 0$.

4. **Commutative law of addition:** $x + y = y + x$ for all $x, y \in R$.

5. **Distributive laws:** $\left.\begin{array}{l} x(y + z) = xy + xz \\ (x + y)z = xz + yz \end{array}\right\}$ for all $x, y, z \in R$.

6. **Associative law of multiplication:** $(xy)z = x(yz)$ for all $x, y, z \in R$.

7. **Existence of 1-element:** there is an element $1 \in R$ such that

$1x = x = x1$ for all $x \in R$.

The ring $(R, +, \times)$ is called **commutative** if the following law also holds:

8. **Commutative law of multiplication:** $xy = yx$ for all $x, y \in R$.

The reader will have made use, subconsciously, of the above laws many times already, for they are involved in every routine calculation; so here one must guard against their unwitting use—familiar laws should not be treated with contempt!

Matrices will provide us with many examples of non-commutative rings, that is, rings for which law 8 above is false; for the present we shall be concerned mainly with commutative rings. We take it for granted that the familiar systems $\mathbb{Z}, \mathbb{Q}, \mathbb{R}$ are commutative rings.

However one should not assume that the elements x, y, z of a ring are necessarily real numbers; in particular the elements 0 and 1 referred to in laws 1 and 7 may not be the integers 0 and 1. In other words, the above familiar laws do sometimes hold in unfamiliar situations; we now discuss a case in point of a ring having sets as elements.

Definition 2.3 Let U be a set. The set $P(U)$ of all subsets of U is called the **power set** of U.

If U is a finite set with $|U| = n$, then $|P(U)| = 2^n$, for a set having exactly n elements has exactly 2^n subsets.

Definition 2.4 Let X and Y be sets. The **symmetric difference** $X + Y$ of X and Y is the set of elements belonging to *exactly one* of X and Y.

Fig. 2.1

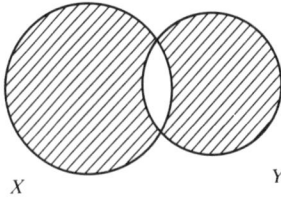

So for $X = \{1, 2, 3, 4\}$, $Y = \{2, 3, 4, 5, 6\}$, we have $X + Y = \{1, 5, 6\}$. The shaded region in the Venn diagram (Fig. 2.1) represents $X + Y$.

Starting with a set U, we form the system $(P(U), +, \times)$ as follows: the elements of the system are the subsets of U; the sum of the subsets X and Y of U is $X + Y$ as defined in (2.4); the product of X and Y is their intersection, and so $XY = X \cap Y$. Notice that $P(U)$ is closed under addition and multiplication, because if X and Y are subsets of U, then $X + Y$ and $X \cap Y$ are also subsets of U.

Example 2.5 Let $U = \{a, b\}$ where $a \neq b$. Then $P(U) = \{\varnothing, \{a\}, \{b\}, \{a, b\}\}$. The **addition** and **multiplication tables** of the system $(P(U), +, \times)$ are written out in Table 2.1.

Table 2.1

+	$\varnothing$	$\{a\}$	$\{b\}$	$\{a, b\}$
$\varnothing$	$\varnothing$	$\{a\}$	$\{b\}$	$\{a, b\}$
$\{a\}$	$\{a\}$	$\varnothing$	$\{a, b\}$	$\{b\}$
$\{b\}$	$\{b\}$	$\{a, b\}$	$\varnothing$	$\{a\}$
$\{a, b\}$	$\{a, b\}$	$\{b\}$	$\{a\}$	$\varnothing$

$\times$	$\varnothing$	$\{a\}$	$\{b\}$	$\{a, b\}$
$\varnothing$	$\varnothing$	$\varnothing$	$\varnothing$	$\varnothing$
$\{a\}$	$\varnothing$	$\{a\}$	$\varnothing$	$\{a\}$
$\{b\}$	$\varnothing$	$\varnothing$	$\{b\}$	$\{b\}$
$\{a, b\}$	$\varnothing$	$\{a\}$	$\{b\}$	$\{a, b\}$

Table 2.2

+	...	y	...
$\vdots$		$\vdots$	
x	...	$x+y$	...
$\vdots$		$\vdots$	

×	...	y	...
$\vdots$		$\vdots$	
x	...	xy	...
$\vdots$		$\vdots$	

More generally, to form the addition table of a finite ring (one with only a finite number of elements), list the elements down the left-hand side and (in the same order) across the top of the table; then insert $x + y$ where row x meets column y. The multiplication table is laid out in the same way, but with xy at the meet of row x and column y (Table 2.2).

Our next theorem explains why we have adopted the notation $X + Y$ for the symmetric difference and XY for the intersection of X and Y.

Theorem 2.6

Let U be a set. The system $(P(U), +, \times)$, where addition is symmetric difference and multiplication is intersection, forms a commutative ring.

Proof

We verify laws 1–8 of (2.2). Let X, Y, and Z be subsets of U.

1. It is straightforward to verify that $(X + Y) + Z$ is the set of elements belonging to an *odd* number of X, Y, and Z, that is, $(X + Y) + Z$ consists of those elements which belong to exactly one or all three of X, Y, and Z. As the set $X + (Y + Z)$ has the same description, we have $(X + Y) + Z = X + (Y + Z)$.

2. The equation $\emptyset + X = X$ for all subsets X of U, tells us that $\emptyset$ is the 0-element of the system.

3. As $X + X = \emptyset$, we see $-X = X$, that is, X is equal to its negative.

4. The set equality $X + Y = Y + X$ follows directly from the symmetry of the symmetric difference.

6. $(XY)Z = (X \cap Y) \cap Z = X \cap (Y \cap Z) = X(YZ)$ as each of these sets consists of the elements belonging to all three of X, Y, and Z.

7. $UX = U \cap X = X = X \cap U = XU$ for all subsets X of U, and so U is the 1-element of the system.

8. $XY = X \cap Y = Y \cap X = YX$ using the symmetry of intersection.

5. As law 8 holds, it is enough to verify the distributive law $X(Y + Z) = XY + XZ$; as both sides stand for the set· of elements

belonging to X and exactly one of Y and Z, we see that this set equality holds.

So laws 1–8 hold in the system $(P(U), +, \times)$ which is therefore a commutative ring. ☐

Definition 2.7   The system $(P(U), +, \times)$ is called the **Boolean** ring of subsets of U.

Although the elements of the system $(P(U), +, \times)$ are sets, (2.6) tells us that these elements may be manipulated using the familiar laws of ordinary arithmetic.

We now return to the general case of a ring and derive some simple consequences from laws 1–7 of (2.2).

Proposition 2.8   Let $(R, +, \times)$ be a ring.

(a) The 0-element of R is unique.
(b) Each element of R has a unique negative.
(c) Given $x, y \in R$ there is a unique $z \in R$ with $z + x = y$.
(d) $0x = 0 = x0$ for all $x \in R$.
(e) $\left. \begin{aligned} (-x)y &= -xy = x(-y) \\ (-x)(-y) &= xy \end{aligned} \right\}$ for all $x, y \in R$.
(f) The 1-element of R is unique.

Proof   (a) We show that 0 is the only element of R satisfying $0 + x = x$ for all $x \in R$. So suppose $0' \in R$ satisfies $0' + x = x$ for all $x \in R$. Setting $x = 0'$ in the first of these equations and $x = 0$ in the second equation gives

$$0 + 0' = 0', \qquad 0' + 0 = 0.$$

But $0 + 0' = 0' + 0$ by law 4 and so $0 = 0'$.

(b) Let $x \in R$. Suppose $y \in R$ satisfies $y + x = 0$. Using laws 1–4 we have

$$y = 0 + y = y + 0 = y + (-x + x) = y + (x + (-x))$$
$$= (y + x) + (-x) = 0 + (-x) = -x$$

showing that x has a unique negative $-x$. (Incidentally, by law 4 we obtain $x + (-x) = 0$, and so x is the negative of $-x$, that is, $-(-x) = x$.)

(c) Suppose first that $z + x = y$. By laws 1–4 we have

$$z = z + 0 = z + (x + (-x)) = (z + x) + (-x) = y + (-x);$$

it is usual to write $y - x$ instead of $y + (-x)$. As

$$(y - x) + x = y + (-x + x) = y + 0 = y$$

we see that $z = y - x$ does satisfy $z + x = y$, and so $z = y - x$ is the unique element of R satisfying $z + x = y$.

(d) By laws 2 and 5 we have $0x + 0x = (0 + 0)x = 0x$; comparing this equation with $0 + 0x = 0x$, by part (c) above we may deduce $0x = 0$. In a similar way, from $x0 + x0 = x0$, it follows that $x0 = 0$.

(e) By part (d) above we have

$$(-x)y + xy = (-x + x)y = 0y = 0$$

showing that $(-x)y$ is the negative of xy, that is, $(-x)y = -xy$. Similarly $x(-y) + xy = 0$ and so $x(-y) = -xy$ also.

Replacing x by $-x$ in the equation $(-x)y = x(-y)$ and using $-(-x) = x$ gives $xy = (-(-x))y = (-x)(-y)$.

(f) The proof is analogous to the proof of part (a) above, using multiplication in place of addition: we know $1x = x = x1$ for all $x \in R$. Let $1' \in R$ satisfy $1'x = x = x1'$ for all $x \in R$. Setting $x = 1'$ in the first equation and $x = 1$ in the last equation gives $1' = 1 \times 1' = 1$. So $1 = 1'$ showing that the 1-element of R is unique. □

Definition 2.9 A ring with only one element is called **trivial**.

Therefore a non-trivial ring has at least two elements; we shall see in (3.18) that the system $\mathbb{Z}_2$ of (1.32) is the smallest non-trivial ring.

Corollary 2.10 A ring is trivial if and only if its 0-element and its 1-element are equal.

Proof Let $(R, +, \times)$ be a ring. Now $0, 1 \in R$ by laws 1 and 7 of (2.2). If R is trivial, then $0 = 1$ because R contains only one element. Conversely suppose $0 = 1$; then $x = 1x = 0x = 0$ for all $x \in R$ using (2.8)(d). So $R = \{0\}$, showing that R is trivial. □

Definition 2.11 Let x be an element of the ring $(R, +, \times)$. If there is an element x^{-1} in R satisfying $x^{-1}x = 1 = xx^{-1}$, then x is called a **unit** (or an **invertible** element) of R and x^{-1} is called the **inverse** of x.

So a unit of a ring is an element having an inverse in the ring; for example, the units of the ring $\mathbb{Z}$ are 1 and -1, for these are the only integers with integer inverses. The 1-element of each ring R is a unit of R, because it is self-inverse (equal its own inverse). It is easy to show that a ring element cannot have more than one inverse: for let x be a unit of the ring R and suppose $xy = 1$ where $y \in R$. Then

$$y = 1y = (x^{-1}x)y = x^{-1}(xy) = x^{-1}1 = x^{-1}$$

and so it is reasonable to call x^{-1} *the* inverse of x.

Let x_1, x_2, x_3, x_4 be elements of a ring $(R, +, \times)$; using law 6 of (2.2), the associative law of multiplication, we have

$$((x_1x_2)x_3)x_4 = (x_1(x_2x_3))x_4 = x_1((x_2x_3)x_4)$$
$$= x_1(x_2(x_3x_4)) = (x_1x_2)(x_3x_4)$$

and so all ways of multiplying these elements together, in the given order, produce the same element of R, which may therefore be denoted unambiguously by $x_1x_2x_3x_4$. This is an instance of the **generalized associative law** (we omit the exact statement and proof) which tells us that brackets can be omitted in the product, in order, of any finite number of ring elements; similarly, brackets are not required in the sum of any finite number of ring elements. Further, by law 4 of (2.2), the commutative law of addition, we see

$$x_1 + x_2 + x_3 = x_2 + x_1 + x_3 = x_2 + x_3 + x_1$$
$$= x_3 + x_2 + x_1 = x_3 + x_1 + x_2 = x_1 + x_3 + x_2$$

showing that the sum of three ring elements is independent of the order in which these elements are taken. The **generalized commutative law** assures us that in forming the sum of any finite number of ring elements, the order in which they are added together does not matter; similarly, if R is a commutative ring, the product of a finite number of ring elements is independent of the order of the factors. At this point, the reader should not be too surprised to learn that the distributive law has a generalization: let $x_1, x_2, \ldots, x_m$ and $y_1, y_2, \ldots, y_n$ be elements of a ring R; as usual we write

$$\sum_{i=1}^{m} x_i = x_1 + x_2 + \ldots + x_m \quad \text{and} \quad \sum_{j=1}^{n} y_j = y_1 + y_2 + \ldots + y_n.$$

The **generalized distributive law** asserts

$$\left(\sum_{i=1}^{m} x_i\right)\left(\sum_{j=1}^{n} y_j\right) = \sum_{i,j} x_iy_j$$

where the right-hand side is the sum of the mn elements x_iy_j; in other words, products of sums of ring elements can be 'multiplied out' in the normal way.

The generalized associative law allows integer multiples and integer powers of ring elements to be formed; these elements are again ring elements.

Notation Let $(R, +, \times)$ be a ring and let $x \in R$, $n \in \mathbb{N}$. Write

$$nx = \overset{\xleftarrow{\hspace{1em}n\hspace{1em}}}{x + x + \ldots + x}, \qquad 0x = 0, \qquad (-n)x = n(-x),$$

$$x^n = \overset{\xleftarrow{\hspace{0.5em}n\hspace{0.5em}}}{xx \ldots x}, \qquad x^0 = 1.$$

If x has inverse x^{-1}, write $x^{-n} = (x^{-1})^n$.

So nx is the result of adding up n elements x, and $(-n)x$ is the sum of n elements each being equal $-x$. Similarly x^n is the product of n elements equal x; x^{-n} only has meaning if x has an inverse, in which case it is the product of n elements each equal to x^{-1}. Our next proposition is stated, without proof, for reference.

Proposition 2.12 (The laws of indices.) Let x, y be elements of a ring.

(a) $\left.\begin{aligned}(m+n)x &= mx + nx \\ (mn)x &= m(nx) \\ m(x+y) &= mx + my\end{aligned}\right\}$ for all $m, n \in \mathbb{Z}$.

(b) $\left.\begin{aligned}x^{m+n} &= x^m x^n \\ x^{mn} &= (x^m)^n \\ \text{and if} \quad xy &= yx \\ \text{then} \quad (xy)^m &= x^m y^m\end{aligned}\right\}$ for all $m, n \in \mathbb{N}$.

If x and y have inverses, then (b) is valid for all $m, n \in \mathbb{Z}$.

We may sum up (2.12) by saying that it is 'business as usual' as far as integer multiples and integer powers of ring elements are concerned. Nevertheless, strange things can happen in rings: for instance, let X belong to the Boolean ring $P(U)$ and let $m \in \mathbb{Z}$. Using (2.4), the definition of symmetric difference, we obtain

$$mX = \begin{cases} X & \text{if } m \text{ is odd,} \\ \varnothing & \text{if } m \text{ is even.} \end{cases}$$

In particular $2U = \varnothing$, and so twice the 1-element of $P(U)$ gives the 0-element of $P(U)$.

We now discuss **cancellation** in the context of rings.

Definition 2.13 Let x, y, z be elements of the ring $(R, +, \times)$ and suppose $x \neq 0$. **Left cancellation** is said to be valid in R if whenever $xy = xz$ then $y = z$. Similarly **right cancellation** being valid in R means that whenever $yx = zx$ then $y = z$.

When confronted with an equation such as $\sqrt{2}\,y = \sqrt{2}\,z$, where y, z are real numbers, it is second nature to cancel the non-zero factor $\sqrt{2}$ from both sides and conclude $y = z$. On the other hand, in a Boolean ring $P(U)$ it is possible to have an equation $XY = XZ$ with $X \neq \varnothing$ and $Y \neq Z$ showing that cancellation is *not* valid in $P(U)$; for instance, taking $U = \{a, b\}$ where $a \neq b$, $X = \{a\}$, $Y = \varnothing$, $Z = \{b\}$, we have $XY = X \cap Y = \varnothing = X \cap Z = XZ$ although $Y \neq Z$.

Notation   Let $(R, +, \times)$ be a ring. Write $R^* = \{x \in R : x \neq 0\}$.

So R^* stands for the set of *non-zero* elements of R. For instance, $\mathbb{Z}^*$ denotes the set of non-zero integers, $\mathbb{Q}^*$ the set of non-zero rational numbers; as $\varnothing$ is the 0-element of the Boolean ring $P(U)$, we see that $P(U)^*$ consists of all *non-empty* subsets of U.

Our next proposition shows, whether the ring R is commutative or not, that there is a close connection between left and right cancellation in R—in fact, if one is valid then so is the other. As we shall see, the set R^* plays an important and impartial role in the proof.

Proposition   Let R be a ring. Left cancellation is valid in R if and only if right
2.14   cancellation is valid in R; both are logically equivalent to R^* being closed under multiplication.

Proof   We show first that the validity of left cancellation in R is logically equivalent to R^* being closed under multiplication. Suppose therefore that left cancellation is valid in R and let $x, y \in R^*$. To show $xy \in R^*$, we argue by contradiction: suppose $xy \notin R^*$, which means $xy = 0$. By (2.8)(d) we have $x0 = 0$ and so $xy = x0$; cancelling the non-zero left factor x produces $y = 0$, which is a contradiction as $y \in R^*$. Therefore $xy \in R^*$, showing that R^* is closed under multiplication.

Now suppose that R^* is closed under multiplication and that $x \in R^*$ and $y, z \in R$ are such that $xy = xz$. By (2.8)(e), this equation can be rewritten $x(y - z) = 0$. To show $y = z$, we argue by contradiction again: suppose $y \neq z$, which means $y - z \neq 0$; as $x \in R^*$ and $y - z \in R^*$, we deduce $x(y - z) \in R^*$, as the product of elements in R^* belongs itself to R^*. Therefore $x(y - z) \neq 0$, which is contrary to $x(y - z) = 0$. So $y = z$, showing that left cancellation is valid in R.

The proof is completed by an appeal to the reader's common sense. We have just shown that left cancellation is valid in R if and only if R^* is closed under multiplication. But the closure of R^*

under multiplication is an unbiased condition—it is biased neither to the left nor to the right. Therefore it is equally true that right cancellation is valid in R if and only if R^* is closed under multiplication.   □

As the only element of the ring R which is not in R^* is the 0-element, the closure of R^* under multiplication can be expressed:

$xy = 0$ *only* if either $x = 0$ or $y = 0$ $(x, y \in R)$.

Definition 2.15   The ring R is said to have **zero-divisors** if there are elements $x, y \in R^*$ with $xy = 0$.

So a ring R has zero-divisors if it contains non-zero elements with product zero, that is, R^* is not closed under multiplication. By (2.14), cancellation is not valid in a ring having zero-divisors, and conversely, a ring in which cancellation is not valid does have zero-divisors. The Boolean ring P(U), where $U = \{a, b\}$ with $a \neq b$, has zero-divisors because $\{a\}\{b\} = \varnothing$ although $\{a\} \neq \varnothing$ and $\{b\} \neq \varnothing$; notice that (2.15) must be interpreted in context—in this case the ring elements are subsets of U, multiplication of ring elements is intersection of subsets, and $\varnothing$ is the 0-element of P(U). We shall see that matrix rings generally have zero-divisors and that the direct sum of rings (discussed in Chapter 5) is itself a ring which generally has zero-divisors; so it is unusual for a ring not to have zero-divisors and we now introduce a special type of ring with this property.

Definition 2.16   A non-trivial commutative ring R is called an **integral domain** if R^* is closed under multiplication.

We take it for granted that the systems $\mathbb{Z}$, $\mathbb{Q}$, and $\mathbb{R}$ are integral domains; the integral domain $\mathbb{Z}$ is the best example to keep in mind and we shall give it special attention in the next chapter. Notice that cancellation is valid in each integral domain, for integral domains have no zero-divisors.

Example 2.17   Consider the set $\mathbb{Z}[\sqrt{2}]$ of all real numbers of the form $m + n\sqrt{2}$ where m, n are integers. From the equations

$$(m + n\sqrt{2}) + (m' + n'\sqrt{2}) = (m + m') + (n + n')\sqrt{2}$$
$$(m + n\sqrt{2})(m' + n'\sqrt{2}) \quad = (mm' + 2nn') + (mn' + nm')\sqrt{2}$$

where m, m', n, n' are integers, we see that $\mathbb{Z}[\sqrt{2}]$ is closed under addition and multiplication. Therefore the binary operations of addition and multiplication on the set $\mathbb{R}$ of real numbers give, on restriction, binary operations on the subset $\mathbb{Z}[\sqrt{2}]$; we say $\mathbb{Z}[\sqrt{2}]$ **inherits** these operations from $\mathbb{R}$. Using these inherited operations, it makes sense to investigate the system $\mathbb{Z}[\sqrt{2}]$: the integers $0 = 0 + 0\sqrt{2}$ and $1 = 1 + 0\sqrt{2}$ belong to $\mathbb{Z}[\sqrt{2}]$ and they play the roles of 0-element and 1-element, and also the negative of $m + n\sqrt{2}$ is $(-m) + (-n)\sqrt{2}$ which belongs to $\mathbb{Z}[\sqrt{2}]$; therefore laws 2, 3, 7 of (2.2) hold in the system $\mathbb{Z}[\sqrt{2}]$. Now

$$\mathbb{Z} \subseteq \mathbb{Z}[\sqrt{2}] \subseteq \mathbb{R}$$

as $m = m + 0\sqrt{2}$ for all integers m; so the system $\mathbb{Z}[\sqrt{2}]$ is non-trivial. What is more, the remaining laws of (2.2) hold in $\mathbb{Z}[\sqrt{2}]$ simply because they hold in the larger system $\mathbb{R}$. For instance multiplication is commutative in $\mathbb{Z}[\sqrt{2}]$ as multiplication is commutative in $\mathbb{R}$. So $\mathbb{Z}[\sqrt{2}]$ is a non-trivial commutative ring; as the ring operations on $\mathbb{Z}[\sqrt{2}]$ are inherited from those on $\mathbb{R}$, we say $\mathbb{Z}[\sqrt{2}]$ is a **subring** of $\mathbb{R}$. Finally $\mathbb{Z}[\sqrt{2}]$ is an integral domain because $\mathbb{R}$ is an integral domain, for the product of each pair of non-zero real numbers being non-zero implies that the same is true of the real numbers in $\mathbb{Z}[\sqrt{2}]$.

We now introduce one of the most important concepts of algebra, namely that of a **field**. Fields occur in all branches of algebra and have played a crucial role in the solution of many classical mathematical problems; yet they are nothing to be frightened of, for fields are merely systems in which operations of addition, subtraction, multiplication, and division can be carried out, and *all* the familiar laws of manipulation hold!

Definition 2.18 A non-trivial commutative ring $(R, +, \times)$ is called a **field** if for each non-zero element x in R there is x^{-1} in R with $x^{-1}x = 1$.

So a field is a non-trivial commutative ring such that every non-zero element has an inverse within the system. Many fields will be constructed in the following two chapters and we devote our next section to a detailed discussion of the most important field of all—the field $\mathbb{C}$ of complex numbers. For the moment we may keep in mind the field $\mathbb{Q}$ of rational numbers, the field $\mathbb{R}$ of real numbers, and the field $\mathbb{Z}_2$ of (1.32).

We show next that fields cannot have zero-divisors.

Proposition 2.19 Every field is also an integral domain.

Proof　　Let R be a field; so R is a non-trivial commutative ring by (2.18). Suppose that R is *not* an integral domain; by (2.16) there are $x, y \in R^*$ such that $xy = 0$. By (2.18) the field elements x and y have inverses $x^{-1}, y^{-1} \in R^*$, and so, using (2.11) and (2.8)(d), we have

$$1 = (x^{-1}x)(yy^{-1}) = x^{-1}(xy)y^{-1} = x^{-1}0y^{-1} = 0.$$

But as R is non-trivial, by (2.10) we know $1 \neq 0$. This contradiction shows that our supposition about R is *false*, and so R is an integral domain. □

　　Therefore the types of system we have studied can be arranged in order of merit: fields are the best type, integral domains come next, then commutative rings, and lastly rings:

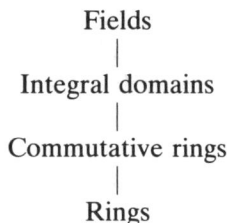

Fields
|
Integral domains
|
Commutative rings
|
Rings

Exercises 2.1　　1. (a) The binary operation $+$ on the set $\{x, y\}$ has addition table:

+	x	y
x	x	y
y	y	y

Show that $+$ is associative (law 1 of (2.2) holds). Which of laws 2, 3, and 4 of (2.2) also hold?

(b) The binary operation $+$ on the set $\{x, y, z\}$ is associative. Find the missing elements in its addition table:

+	x	y	z
x	y	z	x
y	*	*	*
z	*	*	*

Which of laws 2, 3, and 4 of (2.2) also hold?

(c) Find the missing elements in the addition table

+	x	y	z
x	*	y	*
y	*	*	z
z	x	*	*

of the associative binary operation $+$ on the set $\{x, y, z\}$. Show that laws 2 and 4 of (2.2) do not hold.

2. (a) Write down the addition table of a binary operation $+$ on the set $\{x, y\}$ which is associative but not commutative, that is, law 1 of (2.2) holds but law 4 of (2.2) does not hold.

(b) Write down the addition table of a binary operation $+$ on the set $\{x, y\}$ which is commutative but not associative.

(c) How many binary operations on the set $\{x, y\}$ are there? Interpreting these operations as addition, how many of them are such that

(i) law 4 of (2.2) holds?
(ii) laws 2 and 4 of (2.2) hold?
(iii) laws 2, 3, and 4 of (2.2) hold?
(iv) law 1 of (2.2) holds?

3. (a) Let X_1, X_2, X_3, X_4 be subsets of U and let $+$ denote symmetric difference (2.4). Draw the Venn diagrams of

$$X_1 + (X_2 + X_3), \qquad U + X_1, \qquad (X_1 + X_2) + (X_3 + X_4).$$

(b) Show that the Boolean ring $(P(U), +, \times)$ is an integral domain if and only if U has exactly one element. Is it possible for the Boolean ring $P(U)$ to be a field? Determine the units (2.11) of the ring $P(U)$.

(c) Let $X_1, X_2, \ldots, X_n$ be sets. Prove by induction on n that their symmetric difference $X_1 + X_2 + \ldots + X_n$ consists of those elements which belong to X_i for an odd number of integers i $(1 \leqslant i \leqslant n)$.

(d) Let U be a non-empty set. Decide which of laws 1–8 of (2.2) hold in the system $(P(U), \cup, \cap)$; here $X + Y = X \cup Y$ and $XY = X \cap Y$ for all subsets X and Y of U.

4. Let R denote the set of rational numbers expressible in the form m/n with n odd $(m, n \in \mathbb{Z})$. Show that R is closed under addition and multiplication. Using the fact that $\mathbb{Q}$ is a field, prove, as in (2.17), that R is an integral domain. Describe the units of R. Is R a field?

5. Which of the following subsets of $\mathbb{R}$ are closed under addition and multiplication?

(a) $\{m + n\sqrt{3} : m, n \in \mathbb{Z}\}$,
(b) $\{m + n\sqrt{3} : m, n \in \mathbb{N}\}$,
(c) $\{l + m(\sqrt[3]{2}) + n(\sqrt[3]{4}) : l, m, n \in \mathbb{Z}\}$,
(d) $\{l + m\sqrt{2} + n\sqrt{3} : l, m, n \in \mathbb{Z}\}$,
(e) $\{k + l\sqrt{2} + m\sqrt{3} + n\sqrt{6} : k, l, m, n \in \mathbb{Z}\}$.

Which of these subsets, together with the binary operations of $+$ and $\times$ inherited from $\mathbb{R}$ (see (2.17)), form integral domains?

6. (a) Let x and y be units of the ring R. Verify, using (2.11), that xy is a unit of R by showing that its inverse is $y^{-1}x^{-1}$.

(b) Find integers m and n such that $(3+2\sqrt{2})^{-1} = m + n\sqrt{2}$. Hence show that the integral domain $\mathbb{Z}[\sqrt{2}]$ of (2.17) has an infinite number of units.

7. (a) The binary operations $\oplus$ and $\otimes$ on $\mathbb{Z}$ are defined in terms of the usual addition and multiplication by the rules:

$$x \oplus y = x + y + 1, \qquad x \otimes y = xy + x + y,$$

for all $x, y \in \mathbb{Z}$. Show that the system $(\mathbb{Z}, \oplus, \otimes)$ is an integral domain. Is this system a field?

(b) Let $(\mathbb{Z}, +, \otimes)$ denote the system consisting of the set $\mathbb{Z}$ of integers together with the usual binary operation $+$ of integer addition and an 'unknown' binary operation $\otimes$ on $\mathbb{Z}$. If $(\mathbb{Z}, +, \otimes)$ is a ring, show that there are two possibilities for its 1-element, and deduce that there are two possibilities for $\otimes$.

8. (a) Let x, y be elements of the ring $(R, +, \times)$. Using the distributive law (law 5 of (2.2)), verify that

$$(x + y)^2 = x^2 + xy + yx + y^2$$

and expand $(x + y)^3$ as a sum of eight terms of the form $z_1 z_2 z_3$, where either $z_i = x$ or $z_i = y$ ($i = 1, 2, 3$).

If R^* is closed under multiplication and x and y satisfy

$$x^3 + 2x^2 y + 2xyx + 4xy^2 + 2yx^2 + 4yxy + 4y^2 x + 8y^3 = 0,$$

express x in terms of y.

(b) Let x and y be elements of the ring $(R, +, \times)$. Prove, by induction on the natural number n, that $(x + y)^n$ is the sum of the 2^n ring elements of the form $z_1 z_2 \ldots z_n$, where either $z_i = x$ or $z_i = y$ ($i = 1, 2, \ldots, n$).

If $xy = yx$, deduce the binomial theorem:

$$(x + y)^n = \sum_{i=0}^{n} \binom{n}{i} x^i y^{n-i}.$$

9. A ring $(R, +, \times)$ satisfies $x^2 = x$ for all $x \in R$. By expanding $(x + y)^2$ where $x, y \in R$, show that $2x = 0$ for all $x \in R$. Hence show that R is commutative.

10. Let R be an integral domain and let $a \in R^*$. Show that the mapping $\alpha : R \to R$, defined by $(x)\alpha = xa$ for all x in R, is injective. If R is finite, deduce that α is surjective and that R is a field.

The complex field

Here we introduce the reader to the famous field $\mathbb{C}$ of complex numbers; this field plays a fundamental role in analysis and many branches of mathematical physics, notably electro-magnetic theory and quantum mechanics. We shall be concerned with its equally important place in algebra, for $\mathbb{C}$ has many remarkable properties which make it preferable to the real field $\mathbb{R}$; for instance, the roots of every quadratic equation (with real or complex coefficients) are themselves complex numbers, and so, in particular, there is a complex number i satisfying

$$i^2 = -1.$$

This equation (which can form a mental barrier to the acceptance of complex numbers) is therefore in no way inconsistent with the laws of a field, and leads one naturally from the real field $\mathbb{R}$ to the complex field $\mathbb{C}$. What is more, the complex field represents, for all practical purposes, the 'end of the line', for as we shall see in Chapter 4, $\mathbb{C}$ is so perfect that the path which led us to it cannot possibly lead on to anything better!

Basic properties of $\mathbb{C}$

We begin with the construction of the system $\mathbb{C}$, making use of the real field $\mathbb{R}$.

Definition 2.20 The **complex field** $\mathbb{C}$ is the system consisting of the set $\mathbb{R}^2$, of all ordered pairs (x, y) of real numbers, together with the following binary operations of addition and multiplication on $\mathbb{R}^2$:

$$\left.\begin{array}{l} (x, y) + (x', y') = (x + x', y + y') \\ (x, y)(x', y') = (xx' - yy', xy' + yx') \end{array}\right\} \text{ for all } x, x', y, y' \in \mathbb{R}.$$

The above rule of addition of ordered pairs occurs throughout algebra, especially in connection with vectors (see Chapter 6); it amounts to performing addition *componentwise*. For instance

$$(2, 3) + (5, 8) = (7, 11)$$

and generally, the first entry in the sum is the sum of the individual first entries, and the second entry in the sum is the sum of the individual second entries. On the other hand, the rule of multiplication of ordered pairs given in (2.20) is the distinguishing mark of the system $\mathbb{C}$ and its significance will soon become clear. As examples of multiplication in $\mathbb{C}$ we have

$$(2, 3)(4, 5) = (8 - 15, 10 + 12) = (-7, 22),$$
$$(0, 1)(4, 5) = (0 - 5, 0 + 4) = (-5, 4).$$

Theorem 2.21   The system $\mathbb{C}$ is a field.

Proof   Consider the elements $z = (x, y)$, $z' = (x', y')$, $z'' = (x'', y'')$ of $\mathbb{C}$. We show first that multiplication in $\mathbb{C}$ is commutative:

$$zz' = (x, y)(x', y') = (xx' - yy', xy' + yx')$$
$$= (x'x - y'y, x'y + y'x) = (x', y')(x, y) = z'z$$

using (2.20) and the fact that $\mathbb{R}$ is a field. Similarly the first (left) distributive law holds in $\mathbb{C}$:

$$z(z' + z'') = (x, y)(x' + x'', y' + y'')$$
$$= (x(x' + x'') - y(y' + y''), x(y' + y'') + y(x' + x''))$$
$$= (xx' - yy', xy' + yx') + (xx'' - yy'', xy'' + yx'')$$
$$= zz' + zz''.$$

Therefore laws 5 and 8 of (2.2) hold in the system $\mathbb{C}$ and laws 1, 4, 6 may be verified in the same way. The ordered pair $(0, 0)$ is the 0-element of $\mathbb{C}$ as

$$(0, 0) + (x, y) = (0 + x, 0 + y) = (x, y) \quad \text{for all } (x, y) \in \mathbb{C}.$$

Similarly, we see that $(-x, -y)$ is the negative of (x, y). The 1-element of $\mathbb{C}$ is $(1, 0)$, because using (2.20):

$$(1, 0)(x, y) = (1x - 0y, 1y + 0x) = (x, y).$$

Therefore laws 2, 3, and 7 hold in the system $\mathbb{C}$, and so $\mathbb{C}$ is a commutative ring. As the integers 0 and 1 are distinct, we see $(0, 0) \neq (1, 0)$ showing that $\mathbb{C}$ is non-trivial.

Finally, let $z = (x, y)$ be a non-zero element of $\mathbb{C}$; this means $(x, y) \neq (0, 0)$, that is, x and y are not *both* zero, and so $x^2 + y^2 > 0$. To show that z has an inverse in the system $\mathbb{C}$, we require $z^{-1} = (x', y')$ such that zz^{-1} is the 1-element of $\mathbb{C}$, that is,

$$(x, y)(x', y') = (xx' - yy', xy' + yx') = (1, 0).$$

Comparing first the second entries in the above ordered pairs gives the simultaneous equations

$$xx' - yy' = 1, \qquad xy' + yx' = 0.$$

Eliminating y' (multiplying the first equation by x and the second by y) gives $(x^2 + y^2)x' = x$ and so $x' = x/(x^2 + y^2)$ as $x^2 + y^2 \neq 0$. Similarly eliminating x' leads to $y' = -y/(x^2 + y^2)$. This calculation tells us how to complete the proof: given the non-zero element $z = (x, y)$ of $\mathbb{C}$, we may form

$$z^{-1} = \left(\frac{x}{x^2 + y^2}, \frac{-y}{x^2 + y^2} \right)$$

as $x^2 + y^2 \neq 0$. Using the rule (2.20) of complex multiplication, we obtain

$$zz^{-1} = \left(\frac{x^2 + y^2}{x^2 + y^2}, \frac{-xy + yx}{x^2 + y^2} \right) = (1, 0)$$

showing that z^{-1} is indeed the inverse of z. By (2.18), the system $\mathbb{C}$ is a field. □

Elements (x, y) of the complex field $\mathbb{C}$ are called **complex numbers** and we now introduce the customary notation for such numbers. Setting $y = y' = 0$ in (2.20) produces

$$\left. \begin{array}{l} (x, 0) + (x', 0) = (x + x', 0) \\ (x, 0)(x', 0) = (xx', 0) \end{array} \right\} \quad \text{for all } x, x' \in \mathbb{R},$$

which tell us that complex numbers having second entry zero add and multiply in the same way as real numbers; what is more, we may write x in place of $(x, 0)$, which in effect *identifies* the real number x with the complex number $(x, 0)$, this identification being consistent with the operations of addition and multiplication on $\mathbb{R}$ and $\mathbb{C}$.

The complex number $(0, 1)$ has a special property: using the multiplication rule (2.20) we obtain

$$(0, 1)^2 = (0, 1)(0, 1) = (-1, 0) = -1.$$

Writing i in place of $(0, 1)$, as is the custom, the above equation becomes

$$i^2 = -1.$$

Further, the system $\mathbb{C}$ can be described concisely using *only* the real field $\mathbb{R}$ and the complex number i as above, for

$$(x, y) = (x, 0) + (0, y) = (x, 0) + (0, 1)(y, 0) = x + iy$$

by the rules of addition and multiplication in (2.20). Therefore

$$(x, y) = x + iy \quad \text{where } x, y \in \mathbb{R}$$

which is the practical notation for complex numbers; the complex number $x + iy$ is said to have **real part** x and **imaginary part** y. To say that $x + iy$ is real means $y = 0$, while complex numbers of the form iy are called **(pure) imaginary**. (Notice that the imaginary part of a complex number is actually real, and so mathematical terminology is not always as reasonable as one might expect!) In this notation, calculations become a matter of routine; for instance

$$(2 + i)(3 + i4) = 2(3 + i4) + i(3 + i4) = 6 + i8 + i3 + i^2 4$$
$$= 6 + i11 - 4 = 2 + i11.$$

More generally we can calculate the product of any pair of complex numbers:

$$(x + iy)(x' + iy') = x(x' + iy') + iy(x' + iy')$$
$$= xx' + ixy' + iyx' + i^2 yy' = (xx' - yy') + i(xy' + yx')$$

which is the multiplication rule (2.20) expressed in the new notation—so perhaps this rule is not so mysterious after all! Anyway it can be forgotten, so long as one remembers that complex numbers are uniquely expressible in the form $x + iy$, where x and y are real and i satisfies $i^2 = -1$.

Definition 2.22 The complex number $z^* = x - iy$ is called the **conjugate** of $z = x + iy$ where $x, y \in \mathbb{R}$.

Replacing i by $-i$ changes each complex number into its conjugate, and this process is called **conjugation**. Conjugation can be used to find inverses of complex numbers; for instance

$$(2 + i3)^{-1} = \frac{1}{2 + i3} = \frac{2 - i3}{(2 + i3)(2 - i3)} = \frac{2 - i3}{13}$$

on multiplying numerator and denominator of $1/(2 + i3)$ by $2 - i3$. In fact the long-winded process of finding the inverse of the non-zero complex number $x + iy$, as in the proof of (2.21), can be replaced by:

$$(x + iy)^{-1} = \frac{1}{x + iy} = \frac{x - iy}{(x + iy)(x - iy)} = \frac{x - iy}{x^2 + y^2}.$$

We show next that conjugation is well-behaved with respect to addition and multiplication.

Lemma 2.23 (Properties of complex conjugation). Let z and w be complex numbers. Then $(z + w)^* = z^* + w^*$, $(zw)^* = z^* w^*$, $(z^*)^* = z$.

Further $z^* = z$ if and only if z is real, $z^* = -z$ if and only if z is imaginary.

Proof

Let $z = x + iy$ and $w = u + iv$ where x, y, u, v are real. Then

$$(z + w)^* = ((x + u) + i(y + v))^* = (x + u) - i(y + v)$$
$$= (x - iy) + (u - iv) = z^* + w^*,$$
$$(zw)^* = ((x + iy)(u + iv))^* = ((xu - yv) + i(xv + yu))^*$$
$$= (xu - yv) - i(xv + yu) = (x - iy)(u - iv) = z^*w^*.$$

As the conjugate of $z^* = x - iy$ is $z = x + iy$, we obtain $(z^*)^* = z$. As $z - z^* = 2iy$, we see $z^* = z$ if and only if $y = 0$, that is, z is real. Similarly $z + z^* = 2x$ and so $z^* = -z$ if and only if $x = 0$, that is, z is imaginary. $\qquad\square$

We turn now to the geometric description of the complex field $\mathbb{C}$, which rests on the equation

$$(x, y) = x + iy$$

for we regard the point (x, y) in the cartesian plane $\mathbb{R}^2$ as representing the complex number $x + iy$. The real numbers are included as the points of the x-axis, which in this context is called the **real axis** (so $\mathbb{R}$ is still regarded as a horizontal line); the imaginary numbers iy are the points of the y-axis, now called the **imaginary axis**. The usual picture of $\mathbb{R}^2$, but with the points labelled in the form $x + iy$, is called the **Argand diagram** (Fig. 2.2).

Fig. 2.2

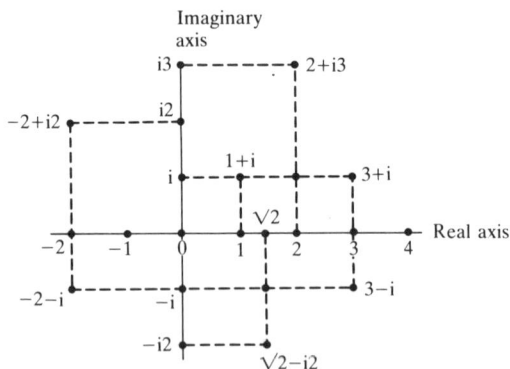

Definition 2.24

The **modulus (absolute value)** of the complex number $z = x + iy$, where $x, y \in \mathbb{R}$, is the non-negative real number $|z| = \sqrt{(x^2 + y^2)}$.

Fig. 2.3

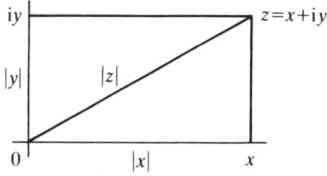

For example, $|3 + i4| = \sqrt{(3^2 + 4^2)} = \sqrt{25} = 5$,

$|1 + i| = \sqrt{(1^2 + 1^2)} = \sqrt{2}$, $|1 - i| = \sqrt{(1^2 + (-1)^2)} = \sqrt{2}$, $|i| = |-i| = 1$.

In the case of a real number x we have

$$|x| = \sqrt{x^2} = \begin{cases} x & \text{for } x \geq 0, \\ -x & \text{for } x < 0, \end{cases}$$

and so $|x|$ can be thought of geometrically as the **distance** of the point x from the origin. Using Pythagoras' theorem, the same interpretation holds in the Argand diagram: the point z is at distance $|z|$ from the point 0 (Fig. 2.3); for $|z| = \sqrt{(x^2 + y^2)}$ is the length of the hypotenuse of a right-angled triangle with other sides of lengths $|x|$ and $|y|$.

The set of complex numbers z with $|z| = 1$, that is, those complex numbers at distance 1 from 0, forms the **unit circle** in $\mathbb{C}$; for instance, $z = (1 + i\sqrt{3})/2$ lies on the unit circle as $|z| = \sqrt{((1/2)^2 + (\sqrt{3}/2)^2)} = \sqrt{(1/4 + 3/4)} = 1$. If $z = x + iy$ is any non-zero complex number, then $z' = z/|z|$ lies on the unit circle because

$$|z'| = |(x + iy)/\sqrt{(x^2 + y^2)}| = \sqrt{((x^2 + y^2)/(x^2 + y^2))} = 1.$$

We say $z' = z/|z|$ is the result of **normalizing** z (Fig. 2.4); for example, normalizing $z = 3 - i4$ produces $z' = 3/5 - i4/5$.

Notice that $zz^* = (x + iy)(x - iy) = x^2 + y^2 = |z|^2$ for all complex numbers z. Therefore, using $(zw)^* = z^*w^*$ from (2.23) we obtain

$$|z|^2 |w|^2 = (zz^*)(ww^*) = (zw)(zw)^* = |zw|^2$$

for all $z, w \in \mathbb{C}$. Taking the non-negative square root of the above

Fig. 2.4

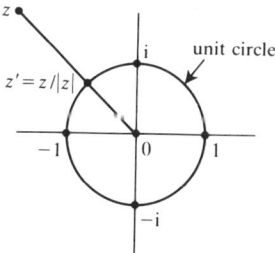

equation produces the important **multiplicative property of the modulus**:

$$|zw| = |z| |w| \quad \text{for all } z, w \in \mathbb{C}.$$

In words, the modulus of a product is the product of the moduli. For instance

$$|(2 + i3)(1 - i4)| = |2 + i3| \, |1 - i4| = (\sqrt{13})(\sqrt{17}) = \sqrt{221}$$

which is verified directly below:

$$|(2 + i3)(1 - i4)| = |14 - i5| = \sqrt{(14^2 + 5^2)} = \sqrt{221}.$$

It is sometimes preferable to use *polar* co-ordinates (r, θ) to specify points of the cartesian plane $\mathbb{R}^2$, and this is equally true of complex numbers; here the non-negative real number r measures distance from the origin and so is the same as the modulus. The arbitrary real number θ is used as follows: starting at the real number 1, we measure the distance θ *along the unit circle* in the *anti-clockwise* sense, and let us suppose this brings us to the normalized complex number z' (Fig. 2.5); if $\theta \geqslant 2\pi$ (the circumference of the unit circle), then the measured distance will make at least one anti-clockwise circuit, while negative values of θ correspond to distances from 1 along the unit circle in the clockwise sense. The angle at 0, traced out by the path of length θ on the unit circle (beginning at 1 and ending at z') is said to be θ **radians**. As z' is at distance 1 from 0, resolving horizontally and vertically we obtain

$$z' = \cos \theta + i \sin \theta$$

as the parametric form of a general complex number z' of modulus 1. (Incidentally, we see immediately that $\cos^2 \theta + \sin^2 \theta = |z'|^2 = 1$, and more trigonometric formulae will be deduced from this resolution.) Multiplying the above expression for z' by the non-

Fig. 2.5

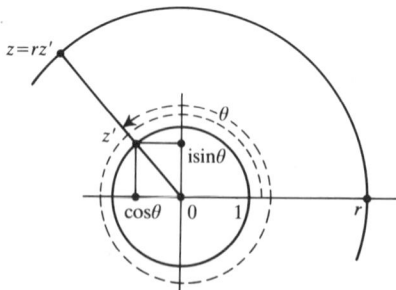

negative real number r, we arrive at the **polar form** of a general complex number $z = rz'$ of modulus r (Fig. 2.5), namely

$$z = r(\cos\theta + \mathrm{i}\sin\theta)$$

Definition 2.25   Let the complex number z be expressed as above, where r and θ are real numbers, r being non-negative. Then z is said to have **argument** θ.

The argument of a complex number tells us the **direction**, from the origin, in which that complex number lies. For instance $1 + \mathrm{i}$ has argument $\pi/4$, i has argument $\pi/2$, -5 has argument π, $-\mathrm{i}$ has argument $3\pi/2$, 5 has argument 0. However, it is also true that $1 + \mathrm{i}$ has argument $\pi/4 + 2\pi = 9\pi/4$, for adding 2π to the argument alters the direction by a complete revolution and so, in fact, leaves it unchanged; for the same reason $-\mathrm{i}$ has infinitely many arguments (see Fig. 2.6) since the angle between the positive real axis and the negative imaginary axis may be specified in an infinite number of ways: $-\pi/2 + 2n\pi = (4n - 1)\pi/2$ where n is any integer. More generally, suppose

$$r(\cos\theta + \mathrm{i}\sin\theta) = r'(\cos\theta' + \mathrm{i}\sin\theta')$$

where r, r', θ, θ' are real and r, r' are non-negative. Comparing moduli, we see $r = r'$; if $r \neq 0$ we may compare arguments of the non-zero complex numbers in the above equation obtaining

$$\theta - \theta' = 2n\pi \quad \text{for some integer } n$$

as θ and θ' both define the same direction and so their difference is an integer multiple of 2π. Let us call the real numbers θ and θ' **equivalent** and write $\theta \equiv \theta'$ if the above equation holds; so $\theta \equiv \theta'$ means that θ and θ' define the *same* direction. The conditions of (1.25) are straightforward to check and so $\equiv$ is an equivalence relation on $\mathbb{R}$; the equivalence class $\bar\theta$ is simply the direction defined by θ. For example

$$\overline{3\pi/2} = \{\ldots, -9\pi/2, -5\pi/2, -\pi/2, 3\pi/2, 7\pi/2, \ldots\}$$

which is the set of all arguments of $-\mathrm{i}$, and so it is reasonable to write $\arg(-\mathrm{i}) = \overline{3\pi/2}$.

Fig. 2.6

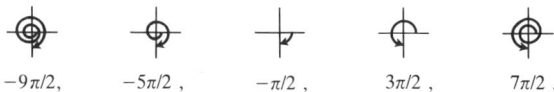

$-9\pi/2,$ $-5\pi/2,$ $-\pi/2,$ $3\pi/2,$ $7\pi/2,$

Notation   Let z be a non-zero complex number with argument θ. We write $\arg(z) = \bar{\theta}$.

Notice that each complex number $z \neq 0$ is *uniquely* specified by $|z|$ and $\arg(z)$, for $|z|$ is the distance of z from 0 and $\arg(z)$ is the .direction of z from 0.

We show now that directions may be added in a sensible way; this is because the above equivalence relation on $\mathbb{R}$ is compatible with addition of real numbers, that is, $\theta \equiv \theta'$ and $\phi \equiv \phi'$ together imply $\theta + \theta' \equiv \phi + \phi'$. For suppose $\theta - \theta' = 2n\pi$ and $\phi - \phi' = 2m\pi$ where $m, n \in \mathbb{Z}$; adding these equations gives

$$(\theta + \phi) - (\theta' + \phi') = 2(m + n)\pi$$

showing that $\theta + \phi \equiv \theta' + \phi'$. Therefore it is legitimate to define **addition** of **directions** by the rule:

$$\bar{\theta} + \bar{\phi} = \overline{\theta + \phi} \quad \text{where } \theta, \phi \in \mathbb{R}.$$

For instance $\overline{3\pi/2} + \bar{\pi} = \overline{5\pi/2} = \overline{\pi/2}$, as the above equation tells us that addition of directions is no more than addition of real numbers excepting that integer multiples of 2π may be *ignored*. The reader should realize that this process is essentially the same as '*telling the time*' in the familiar sense, for one is used to ignoring complete revolutions of a clock's hour hand, that is, integer multiples of 12 are ignored when telling the time in hours; because of this analogy we shall refer to addition of directions as **clock addition**.

Exercises 2.2

1. (a) Verify the associative law (law 6 of (2.2)) for complex multiplication of ordered pairs of real numbers as given in (2.20).

   (b) Using (2.20) and (2.21), express the following complex numbers as ordered pairs of real numbers:

   $$(0, 1)^2, \quad (0, 1)^4, \quad (1, -1)^{-1}, \quad (1, -1)^{-2}, \quad (\tfrac{3}{5}, \tfrac{4}{5})^{-1}.$$

2. (a) Find the real and imaginary parts of

   $$(1 + i)(2 + i) + 3 + i, \quad (1 + i2)^2, \quad (1 + i\sqrt{3})^2, \quad (1 + i)^4;$$

   write down the moduli of these complex numbers and normalize each of them.

   (b) Verify the properties (2.23) of conjugation and also verify $|zw| = |z|\,|w|$, if $z = 1 + i3$ and $w = 4 + i$.

   (c) Describe the set of all complex numbers z such that

   (i) $z = iz^*$,   (ii) $z - z^* = 2i$,   (iii) $\sqrt{2}\,z = (1 + i)z^*$,

   (iv) $(z + 1)/(z - 1)$ has real part equal to 1, where $z \neq 1$.

(d) Determine the inverses of the complex numbers:

$1+i, \quad (1-i\sqrt{3})/2, \quad 3+i4, \quad 3-i4.$

3. (a) Find arguments of the complex numbers:

$-1+i, \quad (-1+i)^2, \quad 1-i, \quad (1-i)^2, \quad 1+i\sqrt{3}.$

(b) Using clock addition, express each of the following directions in the form $\bar{\theta}$, where $0 \leqslant \theta < 2\pi$:

$\overline{3\pi/2 + 7\pi/4}, \quad \overline{(-7\pi/2)}, \quad \overline{15\pi/2 + 27\pi/4 + 39\pi/8}.$

(c) Let z be a non-zero complex number. In each of the following cases, state the connection between the two given quantities and sketch their relationship using the Argand diagram.

(i) z and z^*, (ii) $\arg(z)$ and $\arg(z^*)$,
(iii) $\arg(z)$ and $\arg(-z)$,
(iv) $\arg(z^*)$ and $\arg(z^{-1})$.

(d) The complex number $z \neq \pm 1$ is such that $(z+1)/(z-1)$ has argument $\pi/4$. Show that z has negative imaginary part and lies on the circle with centre $-i$ and radius $\sqrt{2}$. Determine $\arg((z+1)/(z-1))$ if z lies on this circle and has positive imaginary part.

Geometric properties of $\mathbb{C}$

The operations of addition and multiplication of complex numbers have geometric interpretations as we now explain; the resulting interplay between algebra and geometry will do us nothing but good—because the geometry provides insights into the more formal algebraic aspects of $\mathbb{C}$, while many facts of plane geometry are best explained in the context of the complex field.

Definition 2.26   Let w be a given complex number. The mapping $\tau_w : \mathbb{C} \to \mathbb{C}$, where $(z)\tau_w = z + w$ for all $z \in \mathbb{C}$, is called **translation** by w.

In terms of real and imaginary parts, writing $w = x_0 + iy_0$ and $z = x + iy$, we have

$$(x + iy)\tau_w = x + iy + x_0 + iy_0 = (x + x_0) + i(y + y_0)$$

showing that translation by w increases the real part of z by x_0 and the imaginary part of z by y_0; in other words, τ_w moves each complex number through a distance $\sqrt{(x_0^2 + y_0^2)} = |w|$ in the direction $\arg(w)$. For instance τ_w, where $w = 1 + i$, moves each z through $\sqrt{2}$ in a 'north-easterly' direction. As $z + w = w + z = (w)\tau_z$, we see

Fig. 2.7

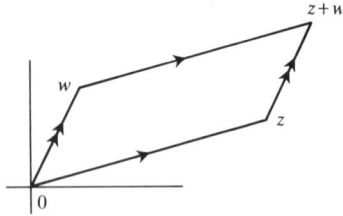

that $z + w$ is also the result of moving w through the distance $|z|$ in the direction $\arg(z)$; this fact leads to the **parallelogram construction** (Fig. 2.7) for the sum $z + w$: the parallelogram with vertex 0 and opposite vertices z and w has fourth vertex $z + w$. Since $(z - w) + w = z$, we may use the parallelogram construction (Fig. 2.8) to locate the difference $z - w$: the parallelogram with vertex w and opposite vertices 0 and z has fourth vertex $z - w$. As $z - w$ is at distance $|z - w|$ from 0, using the parallelogram in Fig. 2.8 we obtain:

$|z - w|$ is the distance between z and w.

Fig. 2.8

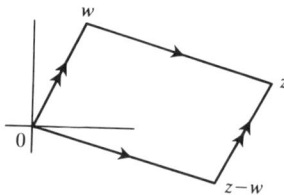

The above distance formula is the corner-stone of complex analysis, for in the discussion of limiting processes one is concerned with the 'nearness' of one complex number to another. As an illustration, the inequality $|z - (3 + i)| < \frac{1}{2}$ is satisfied by all complex numbers z which are within $\frac{1}{2}$ of $3 + i$; the set of such complex numbers (shaded in Fig. 2.9) is the interior of the disk with centre $3 + i$ and radius $\frac{1}{2}$.

We show next that distance between complex numbers satisfies the **triangle inequality**

$$|z_1 - z_3| \leq |z_1 - z_2| + |z_2 - z_3| \quad \text{for all } z_1, z_2, z_3 \in \mathbb{C}$$

Fig. 2.9

Fig. 2.10

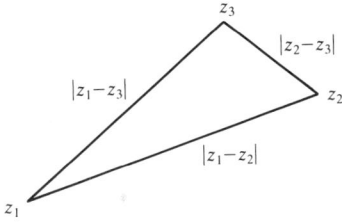

that is, the length of one side of a triangle cannot be greater than the sum of the lengths of its other two sides (Fig. 2.10).

Proposition Let z and w be complex numbers. Then
2.27
$$|z + w| \leqslant |z| + |w|.$$

Proof (Notice first that the triangle inequality is obtained on setting $z = z_1 - z_2$, $w = z_2 - z_3$.) Let $u = zw^* + z^*w$ and $v = zw^* - z^*w$; using the properties of conjugation (2.23), we see that u is real and v is imaginary, for $u^* = z^*w + zw^* = u$ and $v^* = z^*w - zw^* = -v$. As $u + v = 2zw^*$ and $u - v = 2z^*w$, we obtain

$$|u|^2 \leqslant |u|^2 + |v|^2 = u^2 - v^2 = (u + v)(u - v)$$
$$= 4 |z|^2 |w|^2$$

and so $|u| \leqslant 2 |z| |w|$ on taking positive square roots. Therefore

$$|z + w|^2 = (z + w)(z^* + w^*)$$
$$= zz^* + u + ww^* \leqslant zz^* + |u| + ww^* \leqslant zz^* + 2 |z| |w| + ww^*$$
$$= |z|^2 + 2 |z| |w| + |w|^2 = (|z| + |w|)^2.$$

So $|z + w|^2 \leqslant (|z| + |w|)^2$, and taking the positive square root of this inequality completes the proof. □

The manipulation of inequalities is discussed at the start of Chapter 3; meanwhile the reader should guard against using inequalities between *non-real* complex numbers, for these have no meaning.

Definition Let r be a positive real number. The mapping $\mu_r : \mathbb{C} \to \mathbb{C}$, where
2.28 $(z)\mu_r = zr$ for all $z \in \mathbb{C}$, is called **radial expansion** by r.

So μ_r multiplies each complex number z by the positive real number r; writing $z = r'(\cos \theta + i \sin \theta)$ in polar form, we obtain

Fig. 2.11

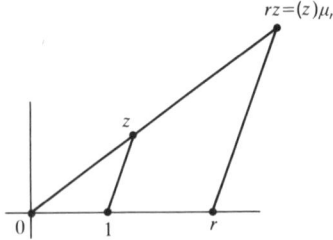

$rz = (z)\mu_r$

the polar form of the image of z by μ_r, namely $(z)\mu_r = rr'(\cos\theta + i\sin\theta)$, showing that $|(z)\mu_r| = r|z|$ and that z and $(z)\mu_r$ have the same arguments. The effect of μ_r is shown in Fig. 2.11, each complex number being 'blown up' by the factor r; so strictly, μ_r is only an expansion for $r > 1$, μ_r being actually a contraction for $0 < r < 1$, while μ_1 is the identity mapping of $\mathbb{C}$. For example, $\mu_{\frac{1}{2}}$ maps each z to the complex number $z/2$ which is halfway between 0 and z.

Definition 2.29

Let ϕ be a real number. The mapping $\rho_\phi : \mathbb{C} \to \mathbb{C}$, called **rotation** through ϕ, is defined as follows: $(0)\rho_\phi = 0$, $(z)\rho_\phi = w$ where $|w| = |z|$ and $\arg(w) = \arg(z) + \phi$ (clock addition) for all $z \neq 0$.

The effect of ρ_ϕ is that of rotating the cartesian plane $\mathbb{C}$ through ϕ radians about 0 (Fig. 2.12); resolving horizontally and vertically we obtain

$$(z)\rho_\phi = r(\cos(\theta + \phi) + i\sin(\theta + \phi))$$

where $z = r(\cos\theta + i\sin\theta)$. As a whole number n of complete revolutions about 0 amounts to fixing every z, we see that $\rho_{2n\pi}$ is the identity mapping of $\mathbb{C}$. More generally $\rho_\phi = \rho_{\phi'}$ if and only if $\phi \equiv \phi'$; for instance $\rho_{\pi/4} = \rho_{-7\pi/4}$, as the effect of an anti-clockwise rotation through $\pi/4$ is equal to that of a clockwise rotation through $7\pi/4$.

Fig. 2.12

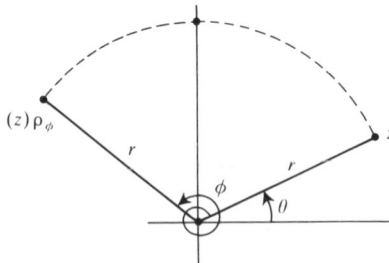

$(z)\rho_\phi$

Fig. 2.13

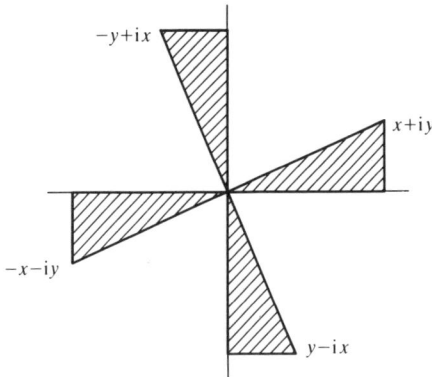

The rotation ρ_π has a particularly simple form, for turning $\mathbb{C}$ through half a complete revolution maps each complex number z into its negative $-z$, that is,

$$(z)\rho_\pi = -z \quad \text{for all } z \in \mathbb{C}$$

and so rotation through π is the same as multiplication by -1. It is perhaps surprising that $\rho_{\pi/2}$ can also be simply expressed: for if the real numbers x, y are non-negative and not both zero, then the four complex numbers $x + iy$, $-y + ix$, $-x - iy$, $y - ix$ form the vertices of a square (Fig. 2.13), each of the shaded triangles being right-angled with sides of lengths x, y, and $\sqrt{(x^2 + y^2)}$. But if each of these complex numbers is multiplied by i, the next (anti-clockwisely speaking) of them is obtained, that is $(x + iy)i = -y + ix$, $(-y + ix)i = -x - iy$, $(-x - iy)i = y - ix$, and $(y - ix)i = x + iy$; on the other hand, if z denotes any one of these complex numbers, then $(z)\rho_{\pi/2}$ is *also* the next of them! Putting the pieces together we obtain

$$(z)\rho_{\pi/2} = zi \quad \text{for all } z \in \mathbb{C}$$

that is:

Rotation through $\pi/2$ is the same as multiplication by i.

Of course i has argument $\pi/2$, and we shall see shortly that rotation through θ is the same as multiplication by $\cos\theta + i\sin\theta$.

We turn now to the geometric construction for the product wz where $w, z \in \mathbb{C}$; we may assume $w \neq 0$ and write $z = x + iy$ where $x, y \in \mathbb{R}$. Just as the numbers 1 and i define the system of cartesian co-ordinates we have used throughout (1 is on the positive real axis, i is on the positive imaginary axis, and both 1 and i are at distance 1

Fig. 2.14

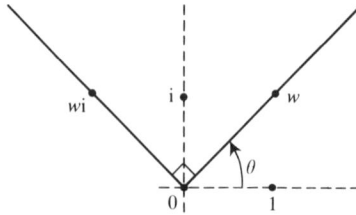

from 0), so the numbers w and wi define another system of cartesian co-ordinates (Fig. 2.14). The system defined by w and wi is related to the familiar system by rotation through the argument θ of w; as w and wi are at distance $|w|$ from 0, we see that a change of scale is also involved, namely radial expansion by $|w|$. The equation

$$wz = w(x + iy) = wx + wiy$$

tells us that wz has the same co-ordinates (namely x and y) in the system defined by w and wi as z has in the familiar system defined by 1 and i; therefore the rectangle with vertices 0, wx, wz, wiy is obtained from the rectangle 0, x, z, iy by applying the composite mapping $\rho_\theta \mu_{|w|}$ (Fig. 2.15), that is, by rotating through θ and expanding by $|w|$. Therefore multiplication by w amounts to rotation through $\arg(w)$ together with radial expansion by $|w|$, that is:

$$wz = (z)\rho_\theta \mu_{|w|} \quad \text{for all } z \text{ in } \mathbb{C} \text{ where } w \text{ has argument } \theta.$$

The above interpretation of complex multiplication has a number of important consequences. As ρ_θ increases arguments by $\bar{\theta} = \arg(w)$ in the sense of clock addition and $\mu_{|w|}$ leaves arguments unchanged, comparing arguments in the above equation gives the **argument**

Fig. 2.15

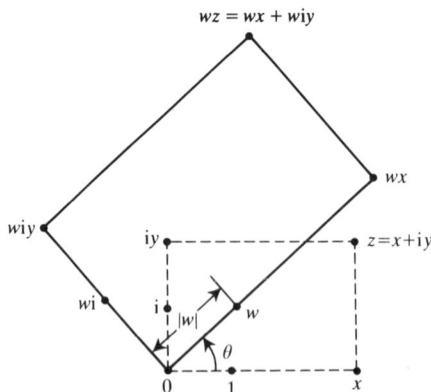

$$wz = wx + wiy$$

formula:

$$\arg(wz) = \arg(w) + \arg(z) \quad \text{for all } w, z \in \mathbb{C}^*$$

In other words, the argument of a product is the clock sum of the arguments of the individual factors; for instance, as $\arg(-1+i) = \overline{3\pi/4}$, we see $\arg((-1+i)^2) = \overline{3\pi/4 + 3\pi/4} = \overline{3\pi/2}$, $\arg((-1+i)^3) = \overline{3\pi/2 + 3\pi/4} = \overline{9\pi/4} = \overline{\pi/4}$, and $(-1+i)^4$ is a negative real number since $\arg((-1+i)^4) = \bar{\pi}$.

Now suppose $|w| = 1$; as $\mu_{|w|}$ is the identity mapping of $\mathbb{C}$, we obtain:

$$wz = (z)\rho_\theta \quad \text{for all } z \in \mathbb{C} \text{ where } w = \cos\theta + i\sin\theta.$$

That is, multiplication by the complex number w of modulus 1 and argument θ has the effect of rotation through θ. For instance, multiplication by $w = (1+i)/\sqrt{2}$ amounts to rotation through $\pi/4$, multiplication by $w^2 = i$ is rotation through $\pi/2$, multiplication by w^3 is rotation through $3\pi/4$, and so on until we arrive at multiplication by $w^8 = 1$ which is rotation through 2π, that is, the identity mapping of $\mathbb{C}$; in fact $w, w^2, \ldots, w^7, w^8 = 1$ are the eighth roots of 1, that is, they are the solutions of $z^8 = 1$, and they form the vertices of a regular octagon inscribed in the unit circle (Fig. 2.16).

Suppose now that w and z are both of modulus 1, and so $w = \cos\theta + i\sin\theta$, $z = \cos\phi + i\sin\phi$; as wz has modulus 1 and argument $\theta + \phi$, we have

$$\cos(\theta + \phi) + i\sin(\theta + \phi) = wz$$
$$= (\cos\theta + i\sin\theta)(\cos\phi + i\sin\phi).$$

Comparing real and imaginary parts in the above equation produces the **trigonometric formulae**:

$$\cos(\theta + \phi) = \cos\theta\cos\phi - \sin\theta\sin\phi,$$
$$\sin(\theta + \phi) = \sin\theta\cos\phi + \cos\theta\sin\phi.$$

Fig. 2.16

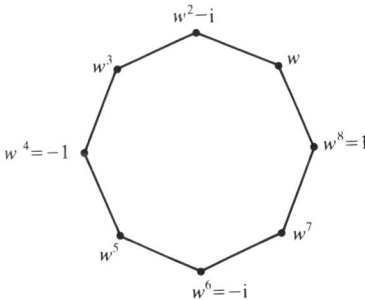

Lastly, we establish a remarkable property of $\mathbb{C}$: given a positive integer n and a non-zero complex number w, there are exactly n complex numbers z satisfying $z^n = w$, that is, within the system $\mathbb{C}$, w has exactly two square roots, exactly three cube roots, exactly four fourth roots, and so on. In order to prove this property of $\mathbb{C}$, we assume one further fact about the real field $\mathbb{R}$:

for each positive integer n and each positive real number x there is a unique positive real number y with $y^n = x$,

in other words, each positive real number x has a unique real and positive n th root $y = x^{1/n}$. (The existence of y follows from the intermediate-value theorem in real analysis.) We also require the following consequence of the argument formula:

Corollary 2.30

(De Moivre's theorem). Let n be a positive integer and θ a real number. Then $(\cos \theta + i \sin \theta)^n = \cos n\theta + i \sin n\theta$.

Proof

The result is clearly true for $n = 1$. Using induction, we take $n > 1$ and assume $(\cos \theta + i \sin \theta)^{n-1} = \cos(n-1)\theta + i \sin(n-1)\theta$; multiplying this equation by $\cos \theta + i \sin \theta$ gives

$$(\cos \theta + i \sin \theta)^n$$
$$= (\cos \theta + i \sin \theta)(\cos(n-1)\theta + i \sin(n-1)\theta)$$
$$= \cos n\theta + i \sin n\theta$$

as the product of two complex numbers, both of modulus 1, with arguments θ and $(n-1)\theta$, is the complex number of modulus 1 and argument $\theta + (n-1)\theta = n\theta$. The inductive step from $n-1$ to n has now been verified, and so, by the principle of induction (3.5), De Moivre's theorem is proved. $\qquad\square$

Let n be a given positive integer and w a given non-zero complex number. We use the polar form of $w = r'(\cos \theta' + i \sin \theta')$ where $0 \leqslant \theta' < 2\pi$, and similarly write $z = r(\cos \theta + i \sin \theta)$ where $0 \leqslant \theta < 2\pi$. The equation

$$z^n = w$$

becomes $r^n(\cos \theta + i \sin \theta)^n = r'(\cos \theta' + i \sin \theta')$ which, by De Moivre's theorem, gives

$$r^n(\cos n\theta + i \sin n\theta) = r'(\cos \theta' + i \sin \theta').$$

Comparing moduli, we see $r^n = r'$ and so z has modulus $r = (r')^{\frac{1}{n}}$. Comparing arguments, we have $n\theta - \theta' = 2m\pi$ for some integer

m, and so $\theta = (2m\pi + \theta')/n$. However adding the inequalities $0 \leqslant n\theta < 2n\pi$ and $-2\pi < -\theta' \leqslant 0$ gives $-2\pi < n\theta - \theta' < 2n\pi$; therefore $-2\pi < 2m\pi < 2n\pi$ and so $-1 < m < n$. The n solutions of $z^n = w$ are therefore

$$z_m = |w|^{1/n}\left(\cos\frac{2m\pi + \theta'}{n} + \mathrm{i}\sin\frac{2m\pi + \theta'}{n}\right)$$

where $m = 0, 1, \ldots, n-1$. These solutions $z_0, z_1, \ldots, z_{n-1}$ form the vertices of a regular n-gon inscribed in the circle of radius $|w|^{1/n}$ with centre 0, the first vertex z_0 (proceeding in the anticlockwise sense from the positive real axis) having argument θ'/n.

Example 2.31

Suppose we wish to find the sixth roots of 64i, that is, the complex numbers z satisfying $z^6 = 64\mathrm{i}$. Taking moduli: $|z|^6 = 64$, and so $|z| = 2$, showing that all the sixth roots of 64 have modulus 2. As 64i has argument $\pi/2$, on dividing this argument by 6 we obtain the argument $\pi/12$ of one of the sixth roots of 64i; but $2\pi + \pi/2 = 5\pi/2$ is also an argument of 64i, and so dividing by 6 we obtain the argument $5\pi/12$ of another of the sixth roots of 64i. In a similar way, starting with the arguments $4\pi + \pi/2$, $6\pi + \pi/2$, $8\pi + \pi/2$, $10\pi + \pi/2$ of 64i, we obtain the arguments $9\pi/12$, $13\pi/12$, $17\pi/12$, $21\pi/12$ of the remaining sixth roots of 64i. Writing z_m for the sixth root of 64i having argument $(2m\pi + \pi/2)/6 = (4m+1)\pi/12$ $(m = 0, 1, 2, 3, 4, 5)$, we see that $z^6 = 64\mathrm{i}$ has solutions $z_0, z_1, z_2, z_3, z_4, z_5$ forming the vertices of a regular hexagon.

We can now find the roots z of the quadratic equation

$$z^2 + az + b = 0 \quad \text{where } a, b \in \mathbb{C}.$$

Adding $a^2/4 - b$ to both sides of the above equation gives

$$z^2 + az + a^2/4 = a^2/4 - b$$

this process being called '**completing the square**' because the resulting equation is

$$(z + a/2)^2 = a^2/4 - b.$$

Therefore $z + a/2 = \pm\sqrt{(a^2/4 - b)}$ and so

$$z^2 + az + b = 0 \quad \text{has roots} \quad z = -a/2 \pm \sqrt{(a^2/4 - b)}.$$

So the familiar formula for the roots of a quadratic equation always works when complex numbers are allowed.

Applying the above formula to the quadratic equation $z^2 + 6z + 13 = 0$ gives $z = -3 \pm \sqrt{(-4)} = -3 \pm i2$. The equation $z^2 + (4 + i2)z + 3 + i2 = 0$ may be treated in the same way: as $(1 + i)^2 = i2$, this quadratic has roots $z = -(2 + i) \pm \sqrt{(i2)} = -(2 + i) \pm (1 + i)$, that is, -1 and $-3 - i2$.

Exercises 2.3

1. (a) Find the distance between

   (i) $2 + i5$ and $-1 + i9$,   (ii) $2 + i$ and $-2 - i$.

   (b) Find the complex numbers z such that $|z - 1 - i3| = 2$ and

   (i) z has argument $\pi/4$,   (ii) z has argument $\pi/2$.

   (c) Verify the parallelogram construction by sketching the positions of z, w, $z + w$ if

   (i) $z = 1 + i3$, $w = 4 + i$,   (ii) $z = -2 + i$, $w = 4 - i$.

   (d) Describe the set of all complex numbers z such that

   (i) $|z - 1 - i| = \frac{1}{2}$,   (ii) $|z + 1 + i| \leqslant \frac{1}{3}$ (iii) $|z + 1 + i| = |z| + \sqrt{2}$.

   (e) Show that the complex numbers z satisfying

   $$|z - 1 - i3| = |z - 4 - i2|$$

   lie on a certain line; what is the connection between this line and the line segment joining $1 + i3$ to $4 + i2$?

   (f) Show that the complex numbers $2 + i2$ and $(1 - \sqrt{3}) + (1 + \sqrt{3})i$, together with 0, are the vertices of an equilateral triangle.

   (g) Determine the complex numbers z and w such that 0, $1 + i3$, z, and w are the vertices of a square having

   (i) diagonally opposite vertices 0 and $1 + i3$,
   (ii) adjacent vertices 0 and $1 + i3$.

2. (a) Let $w = (-1 + i\sqrt{3})/2$. Show that 1, w, and w^2 are the cube roots of 1 (that is, they satisfy $z^3 = 1$) and sketch their positions. Verify that $1 + w + w^2 = 0$. Find all the complex numbers z satisfying $z^2 = z^*$.

   (b) Find the real and imaginary parts of the complex numbers z such that $z^4 = -1$ and sketch their positions in the Argand diagram.

   (c) Find the polar forms of the roots of $z^5 = 1$ and sketch their positions.

   (d) Sketch the positions of $1 + i$, $(1 + i)^2, \ldots, (1 + i)^8$ in the Argand diagram. The positive real numbers x, y are such that $z = x + iy$ satisfies $z^2 = 1 + i$. Sketch, in the same diagram, the positions of z, z^2, $z^3, \ldots, z^{16}$ and determine x and y.

   (e) Show that the complex numbers 0, z, w ($z \neq 0 \neq w$) are the vertices of an equilateral triangle if and only if $z^2 + w^2 = zw$.

3. (a) Find the real and imaginary parts of the roots of $z^2 - z + 1 = 0$. Verify that these roots are also cube roots of -1.

(b) Find the roots of

(i) $z^2 + 6z + 18 = 0$ (ii) $z^2 + (2 + i2)z + (1 + i2) = 0$.

(c) Find the real and imaginary parts of $(1 + i3)^2$ and hence find the roots of

(i) $z^2 + (6 - i4)z + (13 - i18) = 0$,   (ii) $z^2 + (2 + i2)z + (4 + i5) = 0$.

(d) Find the real and imaginary parts of $(1 + i2)^3$ and hence find the roots of $z^3 + 11 + 2i = 0$. By 'completing the cube' find the roots of the cubic equations

$$z^3 + 3z^2 + 3z + 12 + i2 = 0, \qquad z^3 + (3 + i3)z^2 + i6z + 8 = 0.$$

(e) Find the fourth roots of i and hence determine the roots of $(z + 1)^4 = iz^4$. If $|w| = 1$, show that the roots z of $(z + 1)^n = wz^n$ all have real part $-\frac{1}{2}$.

(f) Let n be a positive integer and write $w = \cos(2\pi/n) + i\sin(2\pi/n)$. Show that $w, w^2, \ldots, w^n$ are the n roots of $z^n = 1$. If z_0 is a given non-zero complex number, show that $wz_0, w^2z_0, \ldots, w^nz_0$ are the n roots of $z^n = z_0^n$.

4. (a) Write $z = \cos\theta + i\sin\theta$. Use De Moivre's theorem to prove

$$z^n + z^{-n} = 2\cos n\theta \quad \text{and} \quad z^n - z^{-n} = i2\sin n\theta.$$

Expanding $(z + z^{-1})^6$ by the binomial theorem and collecting together terms involving z^m and z^{-m} ($m = 0, 2, 4, 6$), show that

$$2^6\cos^6\theta = 2\cos 6\theta + 12\cos 4\theta + 30\cos 2\theta + 20$$

and use this formula to evaluate $\int_0^{\pi/2}\cos^6\theta\,d\theta$. Derive a similar formula for $\sin^6\theta$ by expanding $(z - z^{-1})^6$.

(b) Express $\cos^7\theta$ in terms of $\cos 7\theta$, $\cos 5\theta$, $\cos 3\theta$, $\cos\theta$.

(c) Express $\sin^7\theta$ in terms of $\sin 7\theta$, $\sin 5\theta$, $\sin 3\theta$, $\sin\theta$.

5. (a) Show that the mapping $\alpha : \mathbb{C}^* \to \mathbb{C}$, defined by $(z)\alpha = z - z^{-1}$ for all non-zero complex numbers z, is surjective but not injective. Determine the complex numbers w such that there is a unique $z \in \mathbb{C}^*$ with $(z)\alpha = w$.

(b) Verify the following equations involving translations τ_w, radial expansions μ_r, and rotations ρ_θ:

(i) $\tau_w\tau_{w'} = \tau_{w+w'}$ for all $w, w' \in \mathbb{C}$.

(ii) $\mu_r\mu_{r'} = \mu_{rr'}$ for all positive real numbers r and r'.

(iii) $\rho_\phi\rho_{\phi'} = \rho_{\phi+\phi'}$ for all $\phi, \phi' \in \mathbb{R}$.

(iv) $\tau_w^{-1} = \tau_{-w}$, $\mu_r^{-1} = \mu_{r^{-1}}$, $\rho_\phi^{-1} = \rho_{-\phi}$.

Show that $\rho_\phi\mu_r = \tau_w$ if and only if $w = 0$, $r = 1$, $\phi = 2m\pi$ for some integer m.

6. (a) Sketch the positions of z, w, zw, where $z = 1 + i2$, $w = 3 + i4$. Find the lengths of the sides of the triangle with vertices 0, w, zw and show that they are proportional to the lengths of the sides of the triangle with vertices 0, 1, z. (Triangles with this property are called **similar**.)

(b) Prove the **parallelogram rule**:

$$|z + w|^2 + |z - w|^2 = 2|z|^2 + 2|w|^2$$

where z, $w \in \mathbb{C}$, and use the parallelogram with vertices 0, z, w, $z + w$ to interpret this equation geometrically.

(c) Let z and w be non-zero complex numbers such that

$$|z + w| = |z| + |w|.$$

Show that $zw^* + z^*w$ is positive and $zw^* - z^*w = 0$ (see the proof of (2.27)). Deduce that $\arg(z) = \arg(w)$.

3 Integers

It should come as no surprise that the integral domain $\mathbb{Z}$ of integers occupies a unique position in algebra. We begin this chapter with a discussion of the properties of **order** (that is, the do's and don'ts concerning inequalities) in the context of an arbitrary integral domain; this leads on to the particular properties—closely connected to mathematical induction—which distinguish $\mathbb{Z}$ from all other integral domains.

However, our main concern is with more commonplace aspects of $\mathbb{Z}$, such as the division of one integer by another obtaining an integer quotient and integer remainder (the reader should be on familiar ground here) and Euclid's practical method of finding **greatest common divisors**. Once mastered, these basic techniques are used to investigate the **residue class rings** $\mathbb{Z}_n$, which arise from $\mathbb{Z}$ on 'throwing away' all multiples of the given integer n, that is, by working **modulo** n. We are used to working modulo 7 when reckoning the days of the week: will New Year's Day in the year 2000 fall on a Sunday? In answering this question, complete weeks, that is, multiples of seven, can be ignored. Similarly when 'telling the time' in hours we are used to working modulo 12 (or modulo 24); it is often sufficient to know that 'it's nine o'clock'—one might not be interested in the exact number of hours which have elapsed since the world began, even if it could be calculated! Of particular importance among the residue class rings are the **prime fields** $\mathbb{Z}_p$, obtained by working modulo the prime number p (we have already met $\mathbb{Z}_2$ in (1.32)); these fields are the simplest of the **finite** fields, which have many applications especially to problems of a combinatorial kind. We shall meet further examples of finite fields when we study polynomials in the next chapter.

Order properties

Here we study inequalities between the elements of an integral domain D. We remind the reader that D is, by (2.16), a non-trivial commutative ring having no zero-divisors.

Definition 3.1   The integral domain D is called **ordered** if there is given a subset D_+ of its element such that

(i) D_+ is closed under addition and multiplication
(ii) for each $x \in D$, *exactly* one of the following holds:

$$x \in D_+, \qquad x = 0, \qquad -x \in D_+.$$

As we shall see, an integral domain D may have several subsets D_+ satisfying the conditions of (3.1), it may have exactly one, or none at all. Throughout the following discussion we suppose D *is ordered*, which means that a particular subset D_+, as above, has been singled out.

Notation   Write $x > y$ if and only if $x - y \in D_+$.

We call D_+ the set of **positive** elements of D, for $x > 0$ if and only if $x = x - 0 \in D_+$; similarly $0 > x$ if and only if $-x = 0 - x \in D_+$ and hence $D_- = \{x \in D : -x \in D_+\}$ is called the set of **negative** elements of D. The conditions of (3.1) can now be restated:

(i) $x + y > 0$ and $xy > 0$ for all $x, y > 0$

(that is, the sum and product of positive elements are positive).

(ii) for each $x \in D$, exactly one of the following holds:

$$x > 0, \qquad x = 0, \qquad 0 > x$$

(that is, each element of D is either positive, zero, or negative).

Condition (3.1)(ii) is known as the **trichotomy law**, for it effectively partitions D into the three disjoint subsets D_+, $\{0\}$, D_-. As usual, if x is positive and y is negative, then xy is negative: for $x \in D_+$ and $-y \in D_+$ imply $-(xy) = x(-y) \in D_+$, as D_+ is closed under multiplication. Similarly $-x \in D_+$ and $-y \in D_+$ imply $xy = (-x)(-y) \in D_+$, and so the product of two negative elements is itself positive.

We take it for granted that $\mathbb{Z}$, $\mathbb{Q}$, and $\mathbb{R}$ are ordered, 'positive' having its usual meaning; indeed, our discussion of complex numbers in Chapter 2 relies on the fact that the real field $\mathbb{R}$ is ordered. Notice that each **subdomain** D of $\mathbb{R}$, such as the integral domain $\mathbb{Z}[\sqrt{2}]$ of (2.17), inherits an ordering from $\mathbb{R}$; for $D_+ = D \cap \mathbb{R}_+$ satisfies the conditions of (3.1). On the other hand we shall see that the complex field $\mathbb{C}$ *cannot* be ordered.

Our next proposition deals with the properties and manipulation of inequalities.

**Proposition
3.2**

Let x, x', y, y', z belong to the ordered integral domain D.

    (a) The inequalities $x > y$, $-y > -x$, $x + z > y + z$ are logically equivalent (that is, if one holds then all three hold).

    (b) Exactly one of $x > y$, $x = y$, $y > x$ holds.

    (c) $x > y$ and $y > z$ together imply $x > z$.

    (d) $x > y$ and $x' > y'$ together imply $x + x' > y + y'$.

    (e) $x > y$ and $z > 0$ together imply $xz > yz$.

    (f) $x > y$ and $0 > z$ together imply $yz > xz$.

    (g) $x^2 > 0$ for all $x \neq 0$.

    (h) $1 > 0$.

Proof

These properties are direct consequences of (3.1).

    (a) If the element $x - y = -y - (-x) = x + z - (y + z)$ belongs to D_+, then all three inequalities hold; otherwise, none of them hold.

    (b) Applying (3.1)(ii) to the element $x - y$ of D, exactly one of $x - y \in D_+$, $x - y = 0$, $y - x \in D_+$ holds; therefore exactly one of $x > y$, $x = y$, $y > x$ holds.

    (c) We use the closure of D_+ under addition. Since $x > y$ and $y > z$ mean $x - y \in D_+$ and $y - z \in D_+$, we have $x - z = (x - y) + (y - z) \in D_+$, that is, $x > z$. The proof of (3.2)(d) is similar.

    (e) We use the closure of D_+ under multiplication. Since $x > y$ and $z > 0$ mean $x - y \in D_+$ and $z \in D_+$, we have $xz - yz = (x - y)z \in D_+$, that is, $xz > yz$. The proof of (3.2)(f) is similar.

    (g) By (3.1)(ii), either $x \in D_+$ or $-x \in D_+$; as $x^2 = (x)(x) = (-x)(-x)$ and D_+ is closed under multiplication, we see $x^2 \in D_+$ in any case.

    (h) As $1 \neq 0$ and $1 = 1^2$, we have $1 \in D_+$ by (3.2)(g).   □

    It is legitimate to add inequalities by (3.2)(d); it is also legitimate to multiply inequalities between *positive* elements: for suppose $x > y > 0$ and $x' > y' > 0$. By (3.2)(e) we deduce $xx' > yx'$ and $yx' > yy'$; therefore $xx' > yy' > 0$ by (3.2)(c).

    Suppose it possible to order the complex field $\mathbb{C}$. Then $1 \in \mathbb{C}_+$ by (3.2)(h) and also $-1 = i^2 \in \mathbb{C}_+$ by (3.2)(g); this is contrary to (3.1)(ii) (1 and -1 cannot both be positive), and so $\mathbb{C}$ cannot be ordered.

    As is customary, we use $y < x$ as an alternative to $x > y$; similarly $x \geq y$ (or equivalently $y \leq x$) means either $x > y$ or $x = y$.

    Notice that the set $D_+ \cup \{0\} = \{x \in D : x \geq 0\}$ of **non-negative** elements of D is also closed under addition and multiplication.

Further D has, by (3.2)(g), the **positive-definite property**:

$$x_1^2 + x_2^2 + \ldots + x_n^2 \geqslant 0 \quad \text{for all } x_1, x_2, \ldots, x_n \in D$$

and

$$x_1^2 + x_2^2 + \ldots + x_n^2 = 0 \quad \text{if and only if} \quad x_1 = x_2 = \ldots = x_n = 0$$

for $x_1^2 + x_2^2 + \ldots + x_n^2$ is positive if at least one of $x_1, x_2, \ldots, x_n$ is non-zero.

Definition 3.3

Let D be an ordered integral domain. The set D_+ of positive elements is said to be **well-ordered** if every non-empty subset of D_+ contains a least element.

So if D_+ is well-ordered and $\varnothing \neq X \subseteq D_+$, there is an element $l \in X$ such that $l \leqslant x$ for all $x \in X$ (l is the least element in X). Our next proposition deals with the properties of an arbitrary integral domain satisfying (3.3); these properties should suggest to the reader that D is very similar to $\mathbb{Z}$. In fact it is only a short step (which we take in Chapter 5) to show that $\mathbb{Z}$ is, for all practical purposes, the *only* example of such an integral domain.

Proposition 3.4

Let D be an ordered integral domain with D_+ well-ordered.

(a) The 1-element of D is the least element in D_+.

(b) If $xy = 1$, where $x, y \in D$, then $x = y = \pm 1$.

(c) All elements in D are expressible as integer multiples of the 1-element of D.

Proof

(a) By (3.2)(h), we have $1 \in D_+$ and so D_+ is non-empty; we now apply (3.3) taking the non-empty subset mentioned there to be D_+ itself: D_+ contains a least element l; as $1 \in D_+$ we see $1 \geqslant l$. In order to prove $1 = l$, we assume to the contrary that $1 > l$; multiplying this inequality by $l > 0$ gives $l > l^2 > 0$, which is impossible (l^2 being positive and less than l, the least of all the positive elements). The assumption $1 > l$ is therefore false, and so $1 = l$, that is 1 is the least element of D_+.

(b) As 1 is positive and $xy = 1$, we see that x and y have the same sign; suppose first that x and y are both positive. Then $x \geqslant 1$ and $y \geqslant 1$ as 1 is the least of the positive elements; but $x > 1$ leads to $xy > y$, that is, $1 > y$ which is contrary to $y \geqslant 1$. So in fact $x = 1$ and hence $y = 1y = xy = 1$ also.

Suppose that x and y are both negative; then $(-x)(-y) = xy = 1$. As $-x$ and $-y$ are both positive, we may deduce $-x = -y = 1$ as in the above paragraph; therefore $x = y = -1$.

(c) To avoid confusion with the integer 1, let e denote the 1-element of D. We show first, by contradiction, that every element of D_+ is of the form

$$\overleftarrow{\quad\quad n\quad\quad}$$
$$ne = e + e + \ldots + e$$

for some positive integer n; so suppose this is not true and let X denote those elements of D_+ which are *not* of the form ne for any $n \in \mathbb{N}$. Therefore X is non-empty and contains a least element l by (3.3); now $l > e$ as e is the least element in D_+, and so $l - e \in D_+$. However $l - e < l$, and so $l - e \notin X$, which means $l - e = ne$ for some $n \in \mathbb{N}$; but then $l = (n + 1)e$, which is a contradiction as $l \in X$ and $n + 1 \in \mathbb{N}$. Therefore our supposition $X \neq \varnothing$ is false; so $X = \varnothing$, that is, every element of D_+ is of the form ne.

Finally let $x \in D$. By (3.1)(ii) either $x \in D_+$, or $x = 0$, or $-x \in D_+$. In the first case, $x = ne$ for some $n \in \mathbb{N}$, as above. In the second case, $x = 0 = 0e$. In the third case, $-x = ne$ for some $n \in \mathbb{N}$, as above; hence $x = (-n)e$. So in any case, x is an integer multiple of e. $\qquad \square$

We make the following assumption:

The well-ordering principle. The set $\mathbb{N}$ of natural numbers is well-ordered.

As $\mathbb{N} = \mathbb{Z}_+$, we may apply (3.4) to the integral domain $\mathbb{Z}$: (3.4)(a) tells us that there is no integer between 0 and 1 (for if there were, 1 would not be the least of the positive integers), and (3.4)(b) says that ± 1 are the only integer divisors (factors) of 1; we shall need these 'obvious' facts in our discussion of $\mathbb{Z}$.

The reader has now had some experience of proof *by induction* (see for instance De Moivre's theorem (2.30)); we now discuss the general form of such a proof and the principle it relies on. Suppose that, for each natural number n, there is given a statement $P(n)$; for example, let $P(n)$ be the assertion:

$$1^2 + 2^2 + 3^2 + \ldots + n^2 = \tfrac{1}{6}n(n + 1)(2n + 1).$$

One may have the feeling that $P(n)$ is true for all $n \in \mathbb{N}$ (this hunch could come from verifying $P(n)$ for $n = 1, 2, 3, 4$ say). The principle of induction is a method of *proving* that $P(n)$ is always true, that is, of clinching one's hunch!

Theorem 3.5 (The principle of induction). Suppose given a statement $P(n)$ for each $n \in \mathbb{N}$. Then $P(n)$ is true for all $n \in \mathbb{N}$ if
(i) $P(1)$ is true and

(ii) the truth of $P(n-1)$ implies the truth of $P(n)$, for all $n \in \mathbb{N}$ with $n > 1$.

Proof

We argue by contradiction again: suppose $P(n)$ is false for some $n \in \mathbb{N}$ even though conditions (3.5)(i) and (ii) are satisfied. Then

$$X = \{n \in \mathbb{N} : P(n) \text{ is false}\}$$

is a non-empty set of natural numbers, and so by the well-ordering principle there is a least integer l in X. By (3.5)(i) we see $1 \notin X$ and so $l > 1$. Therefore $l - 1 \in \mathbb{N}$, but $l - 1 \notin X$ as l is the least integer in X; this means $P(l-1)$ is true and so $P(l)$ is also true on applying (3.5)(ii) with $n = l$. But $P(l)$ is false as $l \in X$, which makes a contradiction as $P(l)$ cannot be both true and false. Our original supposition '$P(n)$ is false for some $n \in \mathbb{N}$' has led to a contradiction; therefore that supposition must have been false! So $P(n)$ is true for all $n \in \mathbb{N}$. $\qquad\square$

A more down-to-earth explanation of induction goes as follows: $P(1)$ is true by (3.5)(i) and so $P(2)$ is true by (3.5)(ii) with $n = 2$; hence $P(3)$ is true using (3.5)(ii) with $n = 3$; hence $P(4)$ is true using (3.5)(ii) with $n = 4$. Carrying on in this repetitive way, we see that the truth of $P(n)$ can be established for all natural numbers n; in other words (3.5)(i) allows us to 'kick off' while (3.5)(ii) ensures that at every subsequent stage 'the ball keeps rolling'.

We now set out a specimen proof of this type; so consider again the statement $P(n) : 1^2 + 2^2 + 3^2 + \ldots + n^2 = \frac{1}{6}n(n+1)(2n+1)$. Setting $n = 1$ we see that $P(1)$ says that $1 = 1$, establishing (3.5)(i). Next assume $P(n-1)$ is true for some $n \in \mathbb{N}$ with $n > 1$ (don't worry—we are not assuming the very thing we are trying to prove); so we assume

$$1^2 + 2^2 + 3^2 + \ldots + (n-1)^2 = \frac{1}{6}(n-1)n(2n-1).$$

Adding n^2 to both sides of the above equation and factorizing the resulting right-hand side produces

$$1^2 + 2^2 + 3^2 + \ldots + (n-1)^2 + n^2$$
$$= \frac{1}{6}(n-1)n(2n-1) + n^2$$
$$= \frac{1}{6}n[(n-1)(2n-1) + 6n] = \frac{1}{6}n[2n^2 + 3n + 1]$$
$$= \frac{1}{6}n(n+1)(2n+1)$$

showing that $P(n)$ is then also true. The truth of $P(n)$ has been deduced from that of $P(n-1)$; as this deduction is valid for all natural numbers $n > 1$, we have successfully completed the

inductive step, that is, we have established (3.5)(ii). The conclusion of (3.5) is therefore valid, namely $P(n): 1^2 + 2^2 + 3^2 + \ldots + n^2 = \frac{1}{6}n(n+1)(2n+1)$ *is* true for all $n \in \mathbb{N}$.

Exercises 3.1 Throughout these exercises x, y, z are elements of the ordered integral domain D.

1. If $xy \in D_+$, use (3.1) to show either $x, y \in D_+$ or $x, y \in D_-$. If xyz is negative, list the (four) possibilities for the signs of x, y, and z.
   Determine the range of values of the real number x such that

   (a) $(x-1)(x-2) > 0$, (b) $x^2 + 3x < 4$,
   (c) $x^3 + 3x < x + 3$, (d) $x^3 + 2 \leqslant 2x^2 + x$.

2. If $xz > yz$ and $z > 0$, prove by contradiction that $x > y$.

3. If x and y are positive elements with $x^2 \geqslant y^2$, prove by contradiction that $x \geqslant y$.

4. Show that $x^2 + y^2 \geqslant 2xy$ with equality if and only if $x = y$. Hence establish an inequality between $(x+y)(x^3+y^3)$ and $(x^2+y^2)^2$ which is valid

   (a) for all $x, y \in D_+$ (b) for all $x \in D_+, y \in D_-$.

5. Write out a proof of (3.2)(d).

6. Write out a proof of (3.2)(f).

7. Let x be positive; show that the mapping $\alpha : \mathbb{N} \to D_+$, defined by $(n)\alpha = nx$ for all natural numbers n, is injective. Deduce that D_+ has an infinite number of elements.

8. Use induction to establish the following formulae for all $n \in \mathbb{N}$:

   (a) $1 + 2 + 3 + \ldots + n = n(n+1)/2$.
   (b) $1^3 + 2^3 + 3^3 + \ldots + n^3 = [n(n+1)/2]^2$.
   (c) $\sum_{r=1}^{n} r(r+1) = \frac{1}{3}n(n+1)(n+2)$.

9. For each of the sums below, make a guess at the appropriate formula and either confirm your guess by induction or guess again:

   (a) $1 + 3 + 5 + \ldots + (2n-1)$
   (b) $1 - 3 + 7 - 9 + \ldots + (-1)^{n+1}(2n-1)$.
   (c) $1 - 2 + 3 - 4 + \ldots + (-1)^{n+1}n$.
   (d) $\sum_{r=1}^{n} r(r+1)(r+2)$. (e) $\sum_{r=1}^{n} 1/(r^2 + r)$.

10. Use (3.5) to establish the following modification of the principle of induction:

For each $n \in \mathbb{N}$ let $P(n)$ be a statement such that $P(1)$ is true, and for all $n \in \mathbb{N}$ with $n > 1$ the truth of $P(1)$, $P(2)$, ..., $P(n-1)$ implies the truth of $P(n)$. Then $P(n)$ is true for all $n \in \mathbb{N}$.

Division properties

We now look at certain computational aspects of whole numbers, some of which will be familiar to the reader, beginning with the well-known process of dividing the integer m by the natural number n. For example, let $m = 19665$ and $n = 79$; we can either use a calculator or carry out a 'long division' sum as follows:

$$
\begin{array}{r}
248 \\
79)\overline{19665} \\
158 \\
\overline{386} \\
316 \\
\overline{705} \\
632 \\
\overline{73}
\end{array}
$$

The conclusion is that 79 divides 248 times into 19665 with 73 left over; the process is summarized by the equation

$$19665 = 248 \times 79 + 73$$

the integer 248 being called the **quotient** (it is the integer part of 19665/79) and 73 the **remainder** (which is non-negative and less than 79). The reader's experience should suggest that the division of m by n can always be carried out to give a quotient q and remainder r; without going into the mechanics of the process, we now apply the well-ordering principle to show that this is indeed so.

Proposition 3.6 (The division law) Let m and n be integers with $n > 0$. Then there are unique integers q and r, with $0 \leqslant r < n$, such that

$$m = qn + r.$$

Proof Let X denote the set of non-negative integers of the form $m - zn$, where $z \in \mathbb{Z}$ (so X consists of the non-negative differences between m and integer multiples of n). As

$$m + |m| n \geqslant m + |m| \geqslant 0$$

we see that $m + |m|n$ belongs to X (take $z = -|m|$) and so $X \neq \emptyset$. If $0 \in X$, then 0 is the least integer in X; if $0 \notin X$, then X is a non-empty set of positive integers and so contains a least integer by the well-ordering principle. So in any case X contains a least integer, which we denote by r. As $r \in X$ we have $0 \leq r$ and $r = m - qn$ where $q \in \mathbb{Z}$. Now $r - n \notin X$ as $r - n < r$ and r is the least integer in X; so $r - n = m - (q + 1)n$ is negative, for otherwise this integer would belong to X (take $z = q + 1$). Therefore $r - n < 0$, that is, $r < n$, showing that there are integers q and r satisfying $0 \leq r < n$ and $m = qn + r$.

To prove the uniqueness of q and r, suppose $m = qn + r$ and $m = q'n + r'$, where $q, q', r, r' \in \mathbb{Z}$ and $0 \leq r, r' < n$. As q and r have the same status as q' and r', we may assume $r \leq r'$; subtracting the above equations gives

$$0 \leq (q - q')n = r' - r < n$$

which on multiplying by the positive rational $1/n$ produces

$$0 \leq (q - q') < 1.$$

As 1 is the least of the positive integers, there is only one possibility for the integer $q - q'$, namely $q - q' = 0$; so $q = q'$ and hence $r = r'$. $\qquad\qquad\square$

It is customary to represent integers in the scale of 10; for example

$$19665 = 10^4 + 9 \times 10^3 + 6 \times 10^2 + 6 \times 10 + 5$$

the digit form on the left being shorthand for the right-hand side. However the choice of 10 for this role is arbitrary (in spite of most people having ten fingers to help them count) as any integer $b > 1$ can be used instead; more precisely, every positive integer m can be expressed uniquely in the form

$$m = r_k b^k + \ldots + r_2 b^2 + r_1 b + r_0$$

where

$$0 \leq r_i < b \quad \text{and} \quad r_k \neq 0$$

the symbol $r_k \ldots r_2 r_1 r_0$ being called the **representation** of m in the **scale** of b. A method of finding this representation, by successive applications of the division law (3.6), is outlined below:

divide m by b : $m = q_0 b + r_0$ with $0 \leq r_0 < b$

divide q_0 by b : $q_0 = q_1 b + r_1$ with $0 \leq r_1 < b$

divide q_1 by b : $q_1 = q_2 b + r_2$ with $0 \leq r_2 < b$

(The quotients $q_0, q_1, q_2, \ldots$ form a decreasing sequence of non-negative integers, and so ultimately a zero quotient is obtained, $q_k = 0$ say, and the process terminates.)

divide q_{k-2} by $b : q_{k-2} = q_{k-1}b + r_{k-1}$ with $0 \leqslant r_{k-1} < b$

divide q_{k-1} by $b : q_{k-1} = 0b + r_k$ with $0 < r_k < b$.

Eliminating the quotients $q_0, q_1, \ldots, q_{k-1}$ from the above $k+1$ equations then produces $m = r_k b^k + \ldots + r_2 b^2 + r_1 b + r_0$ as above, showing that the remainders in the above process make up the representation of m in the scale of b. One may think of this process as a recursive 'machine' which moves from state z to next state $\lfloor z/b \rfloor$ (the integer part of z/b) with output $z - \lfloor z/b \rfloor b$; the machine is initially in state m and subsequently moves through states $q_0, q_1, \ldots, q_{k-1}$ until it comes to rest in state 0, the outputs being $r_0, r_1, \ldots, r_k$.

For instance, to express 643 in the scale of 6:

divide 643 by 6 : $643 = 107 \times 6 + 1$

divide 107 by 6 : $107 = 17 \times 6 + 5$

divide 17 by 6 : $17 = 2 \times 6 + 5$

divide   2 by 6 : $2 = 0 \times 6 + 2$.

Therefore 2551 is the representation of 643 in the scale of 6.

Binary numbers, that is, representations in the scale of 2, are particularly important because of their relevance to computing and their uses within mathematics. Applying the above technique with $m = 19665$ and $b = 2$, we find that 19665 has binary representation 100110011010001.

Definition 3.7
The integer n is said to be a **divisor (factor)** of the integer m if there is an integer q with $m = qn$, in which case we write $n \mid m$.

Notice that $1 \mid m$ for all integers m as $m = m \times 1$, that is, 1 is a divisor of every integer; also $n \mid 0$ for all integers n, as $0 = 0 \times n$,

Fig. 3.1

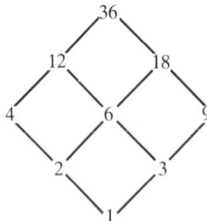

and so we arrive at the surprising (but not insignificant) fact that every integer is a divisor of zero, although $\mathbb{Z}$ is a ring without zero-divisors (2.15). The positive divisors of 36 are 1, 2, 3, 4, 6, 9, 12, 18, 36 and may be arranged in the form of a lattice (Fig. 3.1); notice that each integer is a divisor of all integers above it in the diagram.

Definition 3.8

Let m and n be integers. An integer d such that $d \mid m$ and $d \mid n$ is called a **common divisor** of m and n.

A positive integer d is called the **greatest common divisor** (**highest common factor**) of m and n if d is a common divisor of m and n, and also $d' \mid d$ for all common divisors d' of m and n.

Notice that the greatest common divisor (g.c.d.) is specified in terms of *divisor* properties, rather than *order* properties; this makes for a workable definition which then extends with only minor changes to polynomials, as we shall see in the next chapter. The common divisors of 24 and 36 are ± 1, ± 2, ± 3, ± 4, ± 6, ± 12, and 12 is the g.c.d. of 24 and 36; the common divisors of 0 and -20 are ± 1, ± 2, ± 4, ± 5, ± 10, and ± 20, and 20 is the g.c.d. of 0 and -20.

A given pair of integers cannot have more than one g.c.d.; for let d and d_1 be g.c.d.s of m and n. As d and d_1 are also common divisors of m and n, we see $d \mid d_1$ and $d_1 \mid d$ by (3.8); so $d_1 = zd$ and $d = z_1 d_1$ where z, $z_1 \in \mathbb{Z}$ by (3.7). Therefore $z_1 zd = d$ and so $z_1 z = 1$ on cancelling $d \neq 0$; from (3.4)(b) we deduce $z = \pm 1$, and so $d_1 = \pm d$. As d and d_1 are positive, we conclude $d = d_1$.

The reader is probably predisposed to believe that g.c.d.s exist for most pairs of integers; we show next that this is so, although their existence is not immediately clear from (3.8); as a by-product we obtain a useful expression of the g.c.d. in terms of the original integers m and n.

Proposition 3.9

Let m and n be integers which are not both zero. Then m and n have a unique greatest common divisor d. Further, d can be expressed in the form

$$d = sm + tn \quad \text{where } s, t \in \mathbb{Z}.$$

Proof

We 'kill two birds with one stone', showing that d exists and has the above form by means of the well-ordering principle. Let X denote the set of positive integers of the form $xm + yn$ where $x, y \in \mathbb{Z}$. As m and n are not both zero, $m^2 + n^2$ is positive and so belongs to X; therefore X is non-empty and so contains a least integer which we denote by d (as this integer is, in fact, the one we are looking for). Since d belongs to X, we see $d = sm + tn$ for some integers s and t.

The next step is to divide an arbitrary integer (not necessarily positive) of the form $xm + yn$ by the positive integer d: there are integers q and r with

$$xm + yn = qd + r \quad \text{where } 0 \leqslant r < d.$$

Substituting for d gives $r = (x - sq)m + (y - tq)n$, which, if positive, belongs to X as $x - sq$ and $y - tq$ are integers; but r is less than the least element d in X and so $r \notin X$. The conclusion is inescapable: r *cannot* be positive, and so $r = 0$ as r is non-negative. Therefore d is a divisor of every integer of the form $xm + yn$.

Setting $x = 1$ with $y = 0$ in the above paragraph gives $d \mid m$; setting $x = 0$ with $y = 1$ gives $d \mid n$. Therefore d is a common divisor of m and n. Finally, let d' be a common divisor of m and n; then $m = m'd'$ and $n = n'd'$ where $m', n' \in \mathbb{Z}$. Therefore $d = sm + tn = (sm' + tn')d'$, showing $d' \mid d$ as $sm' + tn' \in \mathbb{Z}$; so d satisfies the conditions of (3.8). As m and n cannot have two g.c.d.s, we see that d is the unique g.c.d. of m and n. $\square$

It is at first sight curious that the *greatest* common divisor should turn up as the *least* integer in the set X. However many things are more easily identified from above (an aeroplane pilot can pick out the mountain peak which may be out of view to an ascending mountaineer); g.c.d.s are defined 'from below' in (3.8), but are seen to exist 'from above' in (3.9).

Notation   Let m and n be integers, not both zero. Write (m, n) for the g.c.d. of m and n.

Therefore $(24, 36) = 12$, $(54, 36) = 18$, and $(0, -10) = 10$; more generally $(0, n) = |n|$ for all non-zero integers n, and $(m, n) = (n, m) = (-n, m)$ where m and n are integers, not both zero. We extend the definition of g.c.d. to all pairs of integers by writing $(0, 0) = 0$.

Although (3.9) tells us that g.c.d.s exist, it does not tell us how to find them; nevertheless the basic idea (that of 'descending from above' by means of the division law) is common to the proof of (3.9) and the following method of determining g.c.d.s.

An **algorithm** is a mathematical sausage machine—it consists of a certain process which, if repeated sufficiently many times, produces the solution to the problem in hand! The long division process is an algorithm, and so is the process which produces the representation of a given positive integer in a given scale; now we meet another, the **Euclidean algorithm**, which starts with the positive integers m and n, and culminates in their g.c.d. (m, n) after a finite number of applications of the division law (3.6). Before discussing the theory, we work an example.

Example 3.10

To find the g.c.d. $(2782, 2249)$ carry out the following sequence of divisions:

divide 2782 by 2249 : $2782 = 2249 + 533$

divide 2249 by   533 : $2249 = 4 \times 533 + 117$

divide   533 by 117 : $533 = 4 \times 117 + 65$

divide   117 by   65 : $117 = 65 + 52$

divide   65 by 52 : $65 = 52 + \mathbf{13}$

divide   52 by   13 : $52 = 4 \times 13.$

The last non-zero remainder (in bold type) in this process is the g.c.d. of the original pair of positive integers; in this case $13 = (2782, 2249)$. In fact all the non-zero remainders (as well as the original two positive integers) in the above process belong to the set X of positive integers of the form $2782x + 2249y$ where $x, y \in \mathbb{Z}$ (see the proof of (3.9)); to find s and t as in (3.9), work backwards through the algorithm, substituting for each of the remainders:

$13 = 65 - 52$

$\quad = 65 - (117 - 65) = 2 \times 65 - 117$

$\quad = 2 \times (533 - 4 \times 117) - 117 = 2 \times 533 - 9 \times 117$

$\quad = 2 \times 533 - 9 \times (2249 - 4 \times 533) = 38 \times 533 - 9 \times 2249$

$\quad = 38 \times (2782 - 2249) - 9 \times 2249 = 38 \times 2782 - 47 \times 2249.$

Therefore $2782s + 2249t = 13$ is satisfied by $s = 38$ and $t = -47$; calculations of this type are used in the solution of **congruences**, which we discuss later in this chapter.

The theory of the Euclidean algorithm rests on the following fact about g.c.d.s.

Lemma 3.11

Let m, n, q, r be integers such that $m = qn + r$. Then

$(m, n) = (n, r)$.

Proof   Let d be a common divisor of m and n; then $m = m'd$ and $n = n'd$ where $m', n' \in \mathbb{Z}$. Hence $r = m - qn = (m' - qn')d$, showing that d is also a divisor of r; therefore d is a common divisor of n and r. Similarly, each common divisor of n and r is also a common divisor of m and n. Therefore the pair m, n has the same set of common divisors as the pair n, r; hence the greatest common divisors of these pairs are equal, that is, $(m, n) = (n, r)$.   □

Applying (3.11) to the divisions in (3.10):

$$(2782, 2249) = (2249, 533) = (533, 117) = (117, 65)$$
$$= (65, 52) = (52, 13) = (13, 0) = 13$$

which justifies the algorithm in this instance. We now treat the general case.

Theorem 3.12   (The Euclidean algorithm). Let m and n be positive integers with $m > n$. Write $r_1 = m$, $r_2 = n$; then the greatest common divisor (m, n) of m and n is equal to the last non-zero r_k in the sequence of divisions:

divide r_1 by r_2 : $r_1 = q_3 r_2 + r_3$ with $0 < r_3 < r_2$

divide r_2 by r_3 : $r_2 = q_4 r_3 + r_4$ with $0 < r_4 < r_3$

$\qquad \vdots \qquad\qquad \vdots \qquad\qquad\quad \vdots$

divide r_{k-2} by r_{k-1} : $r_{k-2} = q_k r_{k-1} + r_k$ with $0 < r_k < r_{k-1}$

divide r_{k-1} by r_k : $r_{k-1} = q_{k+1} r_k$.

Proof   Starting with $r_1 = m$ and $r_2 = n$, the algorithm consists of dividing r_i by r_{i+1} obtaining r_{i+2} as remainder, for $i = 1, 2, 3, \ldots$; these remainders form a decreasing sequence of non-negative integers

$$r_1 > r_2 > r_3 \ldots > r_i > r_{i+1} > \ldots$$

and so eventually a zero remainder is obtained, $r_{k+1} = 0$ say, and the algorithm terminates. By (3.11) we have

$$(m, n) = (r_1, r_2) = (r_2, r_3) = \ldots = (r_{k-1}, r_k) = (r_k, 0) = r_k.$$ □

It is not necessary for the remainders r_i in the Euclidean algorithm to be non-negative, and a useful modification is obtained if the remainder r, on dividing by n, is arranged to satisfy $-n/2 < r \leqslant n/2$ (the reader should check using (3.6) that this is always possible). For instance, the determination of the g.c.d.

(144, 89) using non-negative remainders, as in (3.12), requires ten divisions; if negative remainders are allowed, then the number of steps is practically halved:

$$144 = 2 \times 89 - 34, \qquad 89 = 3 \times 34 - 13,$$
$$34 = 3 \times 13 - 5, \qquad 13 = 3 \times 5 - 2,$$
$$5 = 2 \times 2 + 1, \qquad 2 = 2 \times 1.$$

An algorithm typically requires a simple operation to be performed a large (but finite) number of times until 'something' happens; when several procedures of this kind are to be carried out (by hand, or more likely, by computer), the efficiency of the algorithm becomes an important consideration: how long does it take to execute? Are there quicker ways of arriving at the answers? We cannot go into these questions here, except to say that although the Euclidean algorithm has in a sense been improved upon in recent years (a method of finding g.c.d.s using binary numbers is now known, which involves more (but simpler) steps and takes less time on some machines), nevertheless it is regarded as a very effective procedure and has been called the world's most famous algorithm!

Definition 3.13

An integer $p \neq \pm 1$ is called **prime** if ± 1 and $\pm p$ are the *only* divisors of p.

The sequence of positive primes begins:

$$2, 3, 5, 7, 11, 13, 17, 19, 23, \ldots.$$

Notice that 1 is not prime, and that the natural number $p > 1$ is prime if and only if p cannot be factorized into a product of smaller natural numbers.

Lemma 3.14

Let d, m, n, p be integers with p prime.

(a) If $d \mid mn$ and $(d, m) = 1$, then $d \mid n$.
(b) If $p \mid mn$, then either $p \mid m$ or $p \mid n$ (or both).

Proof

(a) As $(d, m) = 1$, there are integers s, t such that $sd + tm = 1$ by (3.9); as $d \mid mn$, there is an integer z with $zd = mn$. Therefore

$$n = 1 \times n = (sd + tm)n - (sn + tz)d$$

and so $d \mid n$ as $sn + tz \in \mathbb{Z}$.

(b) There are only two possibilities for (p, m): either $(p, m) = p$ or $(p, m) = 1$, since (p, m) is a positive divisor of the prime p.

If $(p, m) = p$, then $p \mid m$ by the definition of g.c.d.; if $(p, m) = 1$, by the paragraph above we see $p \mid n$. ▢

The last theorem of this section will be already familiar to the reader, for it is a well-known fact that every natural number (except 1) is either prime or expressible as a product of primes in one and only one way.

Theorem 3.15

(The fundamental theorem of arithmetic). Each natural number $n > 1$ can be expressed

$$n = p_1^{e_1} p_2^{e_2} \ldots p_k^{e_k}$$

where $p_1, p_2, \ldots, p_k$ are distinct positive primes and the exponents $e_1, e_2, \ldots, e_k$ are natural numbers; further, the above factorization is unique apart from the ordering of the factors.

Proof

Let $P(m)$ be the statement: all natural numbers n with $1 < n \leqslant m$ have the above property. Then $P(1)$ is true as there are no natural numbers n with $1 < n \leqslant 1$ (the reader may think that this is cheating (it isn't!) and prefer to start the induction by verifying $P(2)$). Suppose that $P(m-1)$ is true where $m > 1$. We deduce $P(m)$ from $P(m-1)$ in two stages, showing first that m is either prime or a product of positive primes: if m is not prime, then $m = m_1 m_2$ where $1 < m_1, m_2 < m$. By $P(m-1)$, both m_1 and m_2 are either prime or products of positive primes; hence $m = m_1 m_2$ is a product of positive primes. Secondly, we prove the uniqueness of the factorization: suppose

$$m = p_1^{e_1} p_2^{e_2} \ldots p_k^{e_k} = q_1^{f_1} q_2^{f_2} \ldots q_l^{f_l}$$

where $p_1, p_2, \ldots, p_k$ are distinct positive primes, $q_1, q_2, \ldots, q_l$ are distinct positive primes and the exponents e_i and f_j are positive integers. Now $p_1 \mid q_1^{f_1} q_2^{f_2} \ldots q_l^{f_l}$, and so the prime p_1 divides at least one of the factors in the product $q_1^{f_1} q_2^{f_2} \ldots q_l^{f_l}$ by (3.14)(b) (strictly, there is another induction proof here). We choose the notation so that $p_1 \mid q_1$; as q_1 is prime we conclude $p_1 = q_1$, and so cancellation produces

$$p_1^{e_1-1} p_2^{e_2} \ldots p_k^{e_k} = q_1^{f_1-1} q_2^{f_2} \ldots q_l^{f_l}$$

which are two factorizations of m/p_1; as $m/p_1 \leqslant m - 1$, by $P(m-1)$, these factorizations are identical apart from the ordering of the factors, that is, $k = l$ and (on renumbering $q_2, q_3, \ldots, q_l$ if necessary) $p_2 = q_2, \ldots, p_k = q_k$, and $e_1 - 1 = f_1 - 1$, $e_2 = f_2, \ldots, e_k = f_k$. Hence the original two factorizations of m

are identical, except for the ordering of the factors. The inductive step is now complete for we have shown that the truth of $P(m-1)$ implies the truth of $P(m)$. By the principle of induction (3.5), we conclude that $P(m)$ *is* true for all natural numbers m. $\square$

No efficient algorithm is known for factorizing a given natural number n into primes, or for deciding whether or not n is itself prime; these can be difficult tasks for large n, although recent estimates of the amount of computer time required indicate that such problems may prove to be tractable.

Exercises 3.2

1. (a) Find the ternary representations (representations in the scale of 3) of

   40, 80, 82, 3280.

   (b) Find the binary representations of 1000, 63, 1063, 937, and 63000. Hence write down the representations of these numbers in the scale of 4.

   (c) Show that every positive integer n can be expressed in **balanced ternary form**, that is

   $$n = 3^k + a_{k-1}3^{k-1} + \ldots + a_1 3 + a_0$$

   where each $a_i = -1, 0, 1$.

   (d) Express 547 in balanced ternary form. Explain how 547 grams of sugar can be weighed using a two-pan balance (weights may be placed in either pan) and seven weights of 1, 3, 9, 27, 81, 243, 729 grams. What is the least number of grams which cannot be weighed by this apparatus?

2. Use the Euclidean algorithm to determine the following g.c.d.s (m, n) and in each case find integers s and t satisfying $sm + tn = (m, n)$.
   (a) $(57, 51)$ (b) $(125, 81)$ (c) $(233, 144)$
   (d) $(7497, 5474)$.

3. (a) Let m and n be integers, not both zero. Write $m' = m/(m, n)$, $n' = n/(m, n)$. Use (3.9) to show that $(m', n') = 1$.

   (b) Let m_1, m_2, and n be integers such that $(m_1, m_2) = 1$ with $m_1 \mid n$ and $m_2 \mid n$. Prove that $m_1 m_2 \mid n$.

4. (a) The integers m and n satisfy $m^2 = 2n^2$. Show that $m = n = 0$. (Hint: suppose m and n are not both zero as in question 3(a).) Deduce that $\sqrt{2}$ is **irrational** (that is $\sqrt{2} \notin \mathbb{Q}$).

   (b) Let $\mathbb{Q}(\sqrt{2})$ denote the set of all real numbers of the form $x + y\sqrt{2}$ where x and y are rational numbers. Assuming that $\mathbb{Q}(\sqrt{2})$ is a commutative ring (using the usual addition and multiplication), prove that this system is a field. Express each of the following elements of $\mathbb{Q}(\sqrt{2})$ in the above form:

$$(\tfrac{1}{2} - \tfrac{1}{3}\sqrt{2})^2, \qquad (2 + 3\sqrt{2})^{-1}, \qquad (\tfrac{1}{10} + \tfrac{2}{5}\sqrt{2})^{-1}.$$

(c) Prove that $\sqrt{p}$ is irrational for all positive primes p. More generally, prove that $\sqrt{n}$ is rational, where $n > 1$ is a natural number, if and only if all the exponents $e_1, e_2, \ldots, e_k$ in the factorization (3.15) of n are even.

5. The integer z is called a **common multiple** of the integers m and n if $m \mid z$ and $n \mid z$.

(a) If m and n are non-zero integers, show that $[m, n] = |mn/(m, n)|$ is the **least** common multiple (l.c.m.) of m and n, that is,

(i) $[m, n]$ is a positive common multiple of m and n and
(ii) $[m, n]$ is a divisor of all common multiples of m and n.

(b) Determine the l.c.m.s $[91, 77]$, $[777, 370]$.

(c) If l, m, n are non-zero integers, prove that

$$[(l, n), (m, n)] = ([l, m], n).$$

(Hint: first let l, m, n be non-negative powers of the prime p; then use (3.15) to treat the general case.)

6. (a) If $n > 1$ is not prime, show that n has a positive prime divisor p with $p^2 \leq n$. Show that 263 is prime. Is 391 prime?

(b) Find the g.c.d. of 111111 and 9731; hence factorize these integers into primes.

(c) Let $p_1, p_2, \ldots, p_k$ be distinct positive primes. List the positive divisors of

(i) 100   (ii) $p_1^2 p_2^2$ (iii) 30   (iv) $p_1 p_2 p_3$.

Show that $p_1^{e_1}$ has $e_1 + 1$ positive divisors and find a formula for the number of positive divisors of $n = p_1^{e_1} p_2^{e_2} \ldots p_k^{e_k}$.

(d) (Euclid) Show that there are an infinite number of positive primes. (Hint: Suppose not, and apply (3.15) to $n = 1 + p_1 p_2 \ldots p_k$ where $p_1, p_2, \ldots, p_k$ are all the positive primes.)

Congruence properties

The integral domain $\mathbb{Z}$ is the progenitor (Big Daddy) of a family of finite commutative rings $\mathbb{Z}_n$, there being one ring in this family for each natural number n. We now discus these rings, while urging the reader to keep in mind the underlying idea of the construction: $\mathbb{Z}_n$ is formed from $\mathbb{Z}$ by *discarding* all integer multiples of n; indeed we shall see that the set of integer multiples of n is the 0-element of $\mathbb{Z}_n$.

**Definition
3.16**

Let n be a given natural number. The integers x and y are called **congruent modulo** n (or **congruent** to the **modulus** n) and we write

$$x \equiv y \pmod{n}$$

if n is a divisor of their difference $x - y$.

Therefore $37 \equiv 13 \pmod{12}$, $-7 \equiv 29 \pmod{12}$, $48 \equiv 0 \pmod{12}$ are examples of congruence modulo 12. We show next, as the notation suggests, that congruence modulo n is an equivalence relation on $\mathbb{Z}$, and also that it is compatible with integer addition and multiplication.

**Proposition
3.17**

Let n be a natural number. Congruence modulo n is an equivalence relation on $\mathbb{Z}$. If $x \equiv x' \pmod{n}$ and $y \equiv y' \pmod{n}$, then

$$x + y \equiv x' + y' \pmod{n} \text{ and}$$

$$xy \equiv x'y' \pmod{n}$$

where $x, x', y, y' \in \mathbb{Z}$.

Proof

We verify that congruence modulo n satisfies the conditions of (1.25). Reflexive law: $x \equiv x \pmod{n}$ for all integers x, as $n \mid 0$, that is, $n \mid (x - x)$. Symmetric law: suppose $x \equiv x' \pmod{n}$; then $qn = x - x'$ for some integer q. As $-q$ is an integer and $(-q)n = x' - x$, we see that $n \mid (x' - x)$ and so $x' \equiv x \pmod{n}$. Transitive law: suppose $x \equiv x' \pmod{n}$ and $x' \equiv x'' \pmod{n}$; then there are integers q and q' with $qn = x - x'$ and $q'n = x' - x''$. Adding these equations: $(q + q')n = x - x''$; as $q + q'$ is an integer we see $n \mid (x - x'')$ and so $x \equiv x'' \pmod{n}$. Therefore, by (1.25), congruence modulo n is an equivalence relation on $\mathbb{Z}$.

Suppose $x \equiv x' \pmod{n}$ and $y \equiv y' \pmod{n}$; this means that there are integers p and q with $pn = x - x'$ and $qn = y - y'$. Adding these equations: $(p + q)n = (x + x') - (y + y')$ which shows $x + y \equiv x' + y' \pmod{n}$. Multiplying the equations $pn + x' = x$ and $qn + y' = y$, and rearranging the result: $(pqn + py' + qx')n = xy - x'y'$ which shows $xy \equiv x'y' \pmod{n}$ as $pqn + py' + qx'$ is an integer. □

In this context, the equivalence class $\bar{x}$ of the integer x is called the **residue** (or **congruence**) **class** $(\bmod\, n)$ of x. For example the residue class $(\bmod\, 3)$ of 2 is

$$\bar{2} = \{\ldots, -7, -4, -1, 2, 5, 8, 11, 14, \ldots\}$$

as $\bar{2}$ is the set of those integers which differ from 2 by a multiple of 3, that is, those integers which leave remainder 2 on division by 3. There are only three residue classes (mod 3), since there are three possible remainders on dividing an arbitrary integer by 3, namely $\bar{2}$ above together with

$$\bar{1} = \{\ldots, -8, -5, -2, 1, 4, 7, 10, 13, \ldots\}$$

and

$$\bar{0} = \{\ldots, -9, -6, -3, 0, 3, 6, 9, 12, \ldots\}.$$

More generally, there are exactly n residue classes (mod n): for dividing the integer x by n gives $x = qn + r$ where $0 \leqslant r < n$; this equation can be rewritten as $qn = x - r$ showing that $x \equiv r$ (mod n). Therefore x belongs to exactly one of the following residue classes (mod n):

$$\bar{0}, \bar{1}, \bar{2}, \ldots, \overline{n-1},$$

and so these n residue classes partition $\mathbb{Z}$.

Notation   The set of residue classes (mod n) is denoted by $\mathbb{Z}_n$ (or $\mathbb{Z}/(n)$).

Taking $n = 2$, we obtain $\mathbb{Z}_2 = \{\bar{0}, \bar{1}\}$, as before in (1.32), where $\bar{0}$ is the set of all even integers, $\bar{1}$ is the set of all odd integers; taking $n = 3$, we have $\mathbb{Z}_3 = \{\bar{0}, \bar{1}, \bar{2}\}$ and generally

$$\mathbb{Z}_n = \{\bar{0}, \bar{1}, \bar{2}, \ldots, \overline{n-1}\}.$$

We show now that $\mathbb{Z}_n$ inherits binary operations of addition and multiplication from its 'ancestor' $\mathbb{Z}$: for let $\bar{x}$ and $\bar{y}$ belong to $\mathbb{Z}_n$. The sum $\bar{x} + \bar{y}$ is defined to be the residue class (mod n) of $x + y$, that is,

$$\bar{x} + \bar{y} = \overline{x + y}$$

though the reader should realize that this definition makes sense only because (3.17) assures us that $\bar{x} = \overline{x'}$ and $\bar{y} = \overline{y'}$ imply $\overline{x + y} = \overline{x' + y'}$. Similarly, the product $(\bar{x})(\bar{y})$ is defined to be the residue class (mod n) of xy, that is,

$$(\bar{x})(\bar{y}) = \overline{xy}$$

which is unambiguous as $\bar{x} = \overline{x'}$ and $\bar{y} = \overline{y'}$ imply $\overline{xy} = \overline{x'y'}$ by (3.17). When we refer to $\mathbb{Z}_n$ we shall mean the system $(\mathbb{Z}_n, +, \times$ with addition and multiplication as above.

As an illustration, let $n = 10$ and consider the two element $\{\ldots, -13, -3, 7, 17, \ldots\}$ and $\{\ldots, -16, -6, 4, 14, \ldots\}$ of $\mathbb{Z}_{10}$ Pick any integer from the first set and any integer from the secon set: for instance pick -13 and 4; the sum of these elements o

$\mathbb{Z}_{10}$ is $\overline{-13+4} = \{\ldots, -19, -9, 1, 11, \ldots\}$ and the product of these elements of $\mathbb{Z}_{10}$ is

$$\overline{-13 \times 4} = \{\ldots, -12, -2, 8, 18, \ldots\}.$$

By (3.17) we know that, no matter which integers are picked, their sum will belong to $\overline{-13+4}$ and their product will belong to $\overline{-13 \times 4}$. Test this for yourself!

The addition and multiplication tables of $\mathbb{Z}_4$ are set out below:

+	$\bar{0}$	$\bar{1}$	$\bar{2}$	$\bar{3}$		×	$\bar{0}$	$\bar{1}$	$\bar{2}$	$\bar{3}$
$\bar{0}$	$\bar{0}$	$\bar{1}$	$\bar{2}$	$\bar{3}$		$\bar{0}$	$\bar{0}$	$\bar{0}$	$\bar{0}$	$\bar{0}$
$\bar{1}$	$\bar{1}$	$\bar{2}$	$\bar{3}$	$\bar{0}$		$\bar{1}$	$\bar{0}$	$\bar{1}$	$\bar{2}$	$\bar{3}$
$\bar{2}$	$\bar{2}$	$\bar{3}$	$\bar{0}$	$\bar{1}$		$\bar{2}$	$\bar{0}$	$\bar{2}$	$\bar{0}$	$\bar{2}$
$\bar{3}$	$\bar{3}$	$\bar{0}$	$\bar{1}$	$\bar{2}$		$\bar{3}$	$\bar{0}$	$\bar{3}$	$\bar{2}$	$\bar{1}$

To perform addition or multiplication in $\mathbb{Z}_4$, do the operation in the usual way and adjust the result, if necessary, by a suitable multiple of 4.

Our next theorem deals with the algebraic structure of the system $\mathbb{Z}_n$.

Theorem 3.18   The system $\mathbb{Z}_n$ is a commutative ring for each natural number n. Further, $\mathbb{Z}_n$ is a field if and only if n is prime.

Proof   We start by verifying that the distributive law (law 5 of (2.2)) holds in $\mathbb{Z}_n$, and so let $\bar{x}, \bar{y}, \bar{z} \in \mathbb{Z}_n$. As x, y, and z are integers and the left distributive law holds in $\mathbb{Z}$, we have

$$(\bar{x})(\bar{y} + \bar{z}) = (\bar{x})(\overline{y + x}) = \overline{x(y + z)} = \overline{xy + xz}$$
$$= \overline{xy} + \overline{xz} = (\bar{x})(\bar{y}) + (\bar{x})(\bar{z})$$

using the definition of addition and multiplication of congruence classes; the above equation shows that the left distributive law holds in $\mathbb{Z}_n$. In a similar way, each commutative ring law holding in $\mathbb{Z}$ implies that this law holds also in $\mathbb{Z}_n$; we omit the details, but note that the set $\bar{0}$, of all integer multiples of n, is the 0-element of $\mathbb{Z}_n$, the set $\bar{1} = \{qn + 1 : q \in \mathbb{Z}\}$ is the 1-element of $\mathbb{Z}_n$, and $-(\bar{x}) = \overline{(-x)}$ for all $x \in \mathbb{Z}$. Incidentally, as $\mathbb{Z}_n$ has exactly n elements we see that $\mathbb{Z}_n$ is a non-trivial ring for $n > 1$.

Suppose now that n is prime and write $n = p$. Let $\bar{x}$ be a non-zero element of $\mathbb{Z}_p$; by (2.18) we must show that $\bar{x}$ has an inverse in $\mathbb{Z}_p$. As $\bar{x} \neq \bar{0}$, by (1.28) we have $x \not\equiv 0 \pmod{p}$, that is, p is not a divisor of x. Therefore g.c.d. $(p, x) = 1$ and so by (3.9) there are integers s and t with $sp + tx = 1$. So $1 - tx = sp$, that is,

$1 \equiv tx \pmod{p}$, which can be rewritten as $\bar{1} = \overline{tx} = (\bar{t})(\bar{x})$, showing that $\bar{x}$ has inverse $\bar{t}$ in $\mathbb{Z}_p$. Therefore $\mathbb{Z}_p$ is a field.

Suppose, on the other hand, that n is not prime. If $n = 1$ then $\mathbb{Z}_n$ is trivial and so not a field. If $n > 1$, then $n = n_1 n_2$ where n_1 and n_2 are integers with $1 < n_1, n_2 < n$. As n is not a divisor of either n_1 or n_2, we see $\bar{n}_1 \neq \bar{0}$ and $\bar{n}_2 \neq \bar{0}$, but $(\bar{n}_1)(\bar{n}_2) = \overline{n_1 n_2} = \bar{n} = \bar{0}$, showing that $\mathbb{Z}_n$ has zero-divisors (2.15). Therefore $\mathbb{Z}_n$ is not an integral domain, and so $\mathbb{Z}_n$ is certainly not a field, by (2.19). □

Definition 3.19

The rings $\mathbb{Z}_n$ are called **residue class rings**. The fields $\mathbb{Z}_p$ are known as the **finite prime fields**.

The term prime field is applied to every field which does not contain a smaller field; we shall see that the fields $\mathbb{Z}_p$ have this property and that, apart from the rational field $\mathbb{Q}$, they are the only such fields.

Suppose we wish to find the inverse of $\overline{79}$ in the field $\mathbb{Z}_{199}$ (the reader may verify that 199 is prime). As 1 is the g.c.d. of 199 and 79 we use the Euclidean algorithm to find integers s and t with $199s + 79t = 1$, although in this instance we require only t.

$$199 = 2 \times 79 + 41, \qquad 79 = 2 \times 41 - 3, \qquad 41 = 14 \times 3 - 1.$$

Tracing the algorithm backwards gives:

$$1 = 14 \times 3 - 41 = 14(2 \times 41 - 79) - 41 = 27 \times 41 - 14 \times 79$$
$$= 27(199 - 2 \times 79) - 14 \times 79 = 27 \times 199 - 68 \times 79.$$

So $-68 \times 79 \equiv 1 \pmod{199}$ and so the inverse of $\overline{79}$ in $\mathbb{Z}_{199}$ is $\overline{-68} = \overline{199 - 68} = \overline{131}$.

The above technique can be used to find the inverse (if it exists) of an element $\bar{x}$ in the residue class ring $\mathbb{Z}_n$; in fact $\bar{x}$ has an inverse in $\mathbb{Z}_n$ if and only if the g.c.d. (n, x) equals 1.

We now discuss linear congruences, which are, for all practical purposes, linear equations between elements of $\mathbb{Z}_n$. Once again it is the Euclidean algorithm which provides us with an efficient method of solution; for it tells us whether or not solutions exist, and when they do it tells us how to find them all! First we deal with cancellation in congruences, for this can often shorten calculations.

Definition 3.20

The integers m and n are called **coprime (relatively prime)** if their g.c.d. (m, n) equals 1.

So two integers are coprime if they have no prime factors in common; for instance 20 and 9 are coprime. Let m and n be integers which are not both zero; then $m' = m/(m, n)$ and $n' = n/(m, n)$ are coprime: for on cancelling the common factor (m, n) from the equation $sm + tn = (m, n)$ of (3.9), we obtain

$$sm' + tn' = 1$$

which tells us that 1 is the g.c.d. of m' and n' (every common divisor of m' and n' is a divisor of $sm' + tn'$ and hence of 1). For example, we know from (3.10) that $(2782, 2249) = 13$; therefore $2782/13 = 214$ and $2249/13 = 173$ are coprime.

Suppose we wish to find all integers x satisfying $24x \equiv 21 \pmod{45}$, which is an example of a linear congruence. We may begin by cancelling the common factor 3 (not forgetting to cancel 3 from the modulus 45) obtaining $8x \equiv 7 \pmod{15}$; this congruence can be rewritten as $8x \equiv -8 \pmod{15}$, from which the factor 8 may be cancelled (*without* changing the modulus, as 15 and 8 are coprime) to give $x \equiv -1 \pmod{15}$. So the solutions of the original congruence are the integers $x = -1 + 15y$, where y is any integer, that is, the solutions are

$$\ldots, -16, -1, 14, 29, 44, \ldots.$$

We next establish the general rule of cancellation in congruences.

Lemma 3.21

Let x, y, z, n be integers with z non-zero and n positive. Then $xz \equiv yz \pmod{n}$ if and only if $x \equiv y \pmod{n'}$ where $n' = n/(n, z)$.

Proof

We first do the 'uphill' part. Suppose $xz \equiv yz \pmod{n}$; there is an integer q with $xz - yz = qn$; writing $z' = z/(n, z)$, we obtain $(x - y)z' = qn'$ on cancelling (n, z). Therefore $z' \mid qn'$ and so $z' \mid q$ by (3.14)(a) as $(n', z') = 1$; so $q = q'z'$ where q' is an integer. Cancelling z' gives $x - y = q'n'$, showing $x \equiv y \pmod{n'}$.

Now suppose $x \equiv y \pmod{n'}$; there is an integer q' with $x - y = q'n'$. Multiplying by z produces $xz - yz = q'n'z = q'z'n$, where as before $z' = z/(n, z)$; therefore $xz \equiv yz \pmod{n}$. □

So we see that the non-zero factor z may be cancelled from a congruence provided that the modulus n is divided by (n, z).

Let a, b, n be given integers with n positive; the general **linear congruence** is of the form

$$ax \equiv b \pmod{n}$$

and we are required to describe the set of all integer solutions x in terms of a, b, and n. The following proposition provides us with an exact description of these solutions.

Proposition 3.22   Let a, b, and n be given integers with a and n positive. The linear congruence

$$ax \equiv b \pmod{n}$$

has an integer solution if and only if (n, a) is a divisor of b; when this is the case, the general solution is

$$x = (tb + yn)/(n, a)$$

where y is any integer and $(n, a) = sn + ta$ with s and t integers.

Proof   Suppose first that $ax \equiv b \pmod{n}$ has an integer solution x; then $ax - b = qn$ where q is an integer. Writing $a' = a/(n, a)$ and $n' = n/(n, a)$, we obtain $(a'x - qn')(n, a) = b$, showing $(n, a) \mid b$ as $a'x - qn' \in \mathbb{Z}$.

Now suppose $(n, a) \mid b$ and write $b' = b/(n, a)$. By (3.9), there are integers s and t with $sn + ta = (n, a)$; multiplying this equation by b' and rearranging gives $atb' - b = (-sb')n$, that is, $a(tb') \equiv b \pmod{n}$ as $-sb' \in \mathbb{Z}$. Therefore we have found an explicit solution, because:

> $tb/(n, a)$ is a particular solution of the congruence $ax \equiv b \pmod{n}$ if $(n, a) \mid b$.

To find the general solution, let x be any integer satisfying $ax \equiv b \pmod{n}$; subtracting $a(tb') \equiv b \pmod{n}$ produces $a(x - tb') \equiv 0 \pmod{n}$. Using (3.21) to cancel the factor a gives $x - tb' \equiv 0 \pmod{n'}$, and so $x - tb' = yn'$ where $y \in \mathbb{Z}$; therefore $x = tb' + yn' = (tb + yn)/(n, a)$ is the general solution of $ax \equiv b \pmod{n}$, for it is straightforward to verify that all integers of the form $tb' + yn'$ satisfy $ax \equiv b \pmod{n}$.   □

Therefore the first step towards finding the integer solutions x of a linear congruence $ax \equiv b \pmod{n}$ is to find the g.c.d. (n, a), using the Euclidean algorithm if necessary. If (n, a) is not a divisor of b, there are no integer solutions x to be found. If (n, a) is a divisor of b, next find integers s and t such that $sn + ta = (n, a)$, by using the Euclidean algorithm in reverse (actually only t is required); the general solution can now be written down in terms of the arbitrary integer y using the formula of (3.22).

For example, consider $7x \equiv 3 \pmod{12}$; as $(12, 7) = 1$ which is a divisor of 3, this congruence has integer solutions. As $3 \times 12 - 5 \times 7 = 1$, we see that one such solution is $tb/(n, a) = (-5) \times 3/(12, 7) = -15$, and the general solution is $x = -15 + 12y$ where $y \in \mathbb{Z}$. With small numbers, as in this example, there is no need for the Euclidean algorithm.

The congruence $6x \equiv 14 \pmod{45}$ has no integer solutions as $(45, 6) = 3$, and 3 is not a divisor of 14.

Consider now the congruence $2249x \equiv 182 \pmod{2782}$; in this case we use the Euclidean algorithm to determine $(2782, 2249) = 13$ (see (3.10)). Now $182 = 14 \times 13$ and so there are integer solutions; from (3.10) we have the equation $38 \times 2782 - 47 \times 2249 = 13$, and so a particular solution is $-47 \times 14 = -658$, and the general solution is $x = -658 + (2782/13)y = -658 + 214y$ where $y \in \mathbb{Z}$. Taking $y = 3$ we see that -16 is a solution; the smallest of the positive solutions is 198 given by $y = 4$.

Simultaneous congruences involving a single unknown integer can be solved by an *iteration* of the above technique. To get the idea, consider the simultaneous congruences

$$13x \equiv 40 \pmod{77}, \qquad 11x \equiv 29 \pmod{63}.$$

Solving the first congruence: as $77 = 6 \times 13 - 1$, we may take $t = 6$; so a particular solution is $6 \times 40 = 240$ and the general solution is $x = 240 + 77y$ where $y \in \mathbb{Z}$. It is a good policy to simplify wherever possible; here we may replace y by $y - 3$ and write the general solution of the first congruence as $x = 9 + 77y$, where $y \in \mathbb{Z}$. Substituting this expression of x in the second congruence gives $11(9 + 77y) \equiv 29 \pmod{63}$, which simplifies to $28y \equiv -7 \pmod{63}$ on discarding integer multiples of 63; we may simplify further to $4y \equiv -1 \pmod{9}$. The general solution of this congruence is $y = 2 + 9z$, where $z \in \mathbb{Z}$. Substituting back for y now gives $x = 9 + 77(2 + 9z) = 163 + 693z$, which is the general solution of the original pair of congruences, z being any integer.

This method can be used to find the general solution of any finite number of simultaneous linear congruences in the integer x; if an insoluble congruence arises in this process, then there is no integer satisfying all the original congruences.

Special cases of our next theorem were recorded long ago by mathematicians in China—hence its title. We shall see that this result on simultaneous congruences to coprime moduli is particularly significant, for it forms the heart of the theory of decomposition of many algebraic structures.

Theorem   (The Chinese remainder theorem). Let n_1 and n_2 be coprime
3.23   natural numbers and let r_1 and r_2 be integers. Then there is an
                 integer x satisfying both

$$x \equiv r_1 \pmod{n_1}, \qquad x \equiv r_2 \pmod{n_2}$$

and any two such integers x are congruent modulo $n_1 n_2$.

Proof   We find the general solution of the above pair of congruences.
As n_1 and n_2 are coprime, there are integers t_1 and t_2 with
$t_1 n_1 + t_2 n_2 = 1$ by (3.9). The first congruence $x \equiv r_1 \pmod{n_1}$ has
general solution $x = r_1 + y n_1$ where $y \in \mathbb{Z}$. Substituting for x in the
second congruence $x \equiv r_2 \pmod{n_2}$ produces $n_1 y \equiv r_2 - r_1 \pmod{n_2}$,
which has general solution $y = (r_2 - r_1)t_1 + z n_2$ where z is an
arbitrary integer by (3.22). Substituting back for y, we see that
every solution of the original pair of congruences has the form
$x = r_1 + (r_2 - r_1)t_1 n_1 + z n_1 n_2$ where $z \in \mathbb{Z}$; conversely every integer x
of this type satisfies both the given congruences and so we have
found the general solution which can be expressed symmetrically as

$$x = r_2 t_1 n_1 + r_1 t_2 n_2 + z n_1 n_2$$

where $\quad t_1 n_1 + t_2 n_2 = 1 \quad$ and $\quad z \in \mathbb{Z}$.

In particular $r_2 t_1 n_1 + r_1 t_2 n_2$ satisfies both congruences and every
other solution is congruent modulo $n_1 n_2$ to this solution. $\qquad \square$

The above proof shows that, if n_1 and n_2 are coprime, then the
pair   of   simultaneous   congruences $x \equiv r_1 \pmod{n_1}$ and
$x \equiv r_2 \pmod{n_2}$ have the same solutions as the single congruence

$$x \equiv r_2 t_1 n_1 + r_1 t_2 n_2 \pmod{n_1 n_2}.$$

As a numerical illustration of the Chinese remainder theorem,
consider   the   simultaneous   congruences $x \equiv 41 \pmod{81}$,
$x \equiv 42 \pmod{64}$. The moduli 81 and 64 are coprime and by the
Euclidean algorithm:

$$81 = 64 + 17, \qquad 64 = 4 \times 17 - 4, \qquad 17 = 4 \times 4 + 1$$

from which we deduce $-15 \times 81 + 19 \times 64 = 1$. Therefore $x =$
$r_2 t_1 n_1 + r_1 t_2 n_2 = 42 \times (-15) \times 81 + 41 \times 19 \times 64 = -1174$ is a solution of both congruences. The set of all solutions is the congruence
class of -1174 modulo 81×64; so the smallest positive solution is
$-1174 + 81 \times 64 = 4010$.

An intimation of the theoretical significance of the Chinese
remainder theorem can be acquired by considering the smallest
case, that is, $n_1 = 2$, $n_2 = 3$. Each residue class (mod 6) corresponds

Table 3.1

$\mathbb{Z}_6$	$\mathbb{Z}_2 \times \mathbb{Z}_3$
$\bar{r}$	$(\bar{r})\alpha = (\bar{r}, \bar{r})$
$\bar{0}$	$(\bar{0})\alpha = (\bar{0}, \bar{0})$
$\bar{1}$	$(\bar{1})\alpha = (\bar{1}, \bar{1})$
$\bar{2}$	$(\bar{2})\alpha = (\bar{0}, \bar{2})$
$\bar{3}$	$(\bar{3})\alpha = (\bar{1}, \bar{0})$
$\bar{4}$	$(\bar{4})\alpha = (\bar{0}, \bar{1})$
$\bar{5}$	$(\bar{5})\alpha = (\bar{1}, \bar{2})$

to an ordered pair of residue classes, the first being (mod 2) and the second being (mod 3); to get the idea of this correspondence, notice that if r is an integer with $r \equiv 5$ (mod 6), then dividing by 2 and 3 gives $r \equiv 1$ (mod 2) and $r \equiv 2$ (mod 3) and from (3.23) we obtain the converse, namely that if $r \equiv 1$ (mod 2) and $r \equiv 2$ (mod 3), then $r \equiv 5$ (mod 6). There is, in fact, a simple mapping

$$\alpha : \mathbb{Z}_6 \to \mathbb{Z}_2 \times \mathbb{Z}_3$$

from $\mathbb{Z}_6$ to the cartesian product (1.12) of $\mathbb{Z}_2$ and $\mathbb{Z}_3$, namely $\bar{r}$ (in $\mathbb{Z}_6$) maps by α to the pair $(\bar{r}, \bar{r})$ (in $\mathbb{Z}_2 \times \mathbb{Z}_3$); the Chinese remainder theorem tells us that α is bijective, as can be verified from Table 3.1. We shall see that this means that $\mathbb{Z}_6$ can be built up from its simpler components $\mathbb{Z}_2$ and $\mathbb{Z}_3$.

Exercises 3.3

1. Write out the addition and multiplication tables of the field $\mathbb{Z}_3$. Express each of the following elements of $\mathbb{Z}_3$ as $\bar{0}$, $\bar{1}$, or $\bar{2}$;

$$(\bar{2})^{-1}, \quad (\bar{2})^3, \quad \overline{100}, \quad (-\overline{100})^4, \quad (\overline{10})^2 + \overline{10} + \bar{1}.$$

2. Write out the addition and multiplication tables of the field $\mathbb{Z}_5$. Express each of the following elements of $\mathbb{Z}_5$ as $\bar{0}$, $\bar{1}$, $\bar{2}$, $\bar{3}$, or $\bar{4}$:

$$\overline{-1}, \quad \overline{-2}, \quad \overline{-3}, \quad \overline{-4}, \quad (\overline{-1})^{-1}, \quad (\overline{-2})^{-1},$$
$$(\overline{-3})^{-1}, \quad (\overline{-4})^{-1}, \quad (\overline{64})(\overline{93}).$$

Find the elements x of $\mathbb{Z}_5$ which satisfy $x^2 = -\bar{1}$ (write $x = \bar{m}$ where m is an integer with $0 \leqslant m < 5$). Find the elements x of $\mathbb{Z}_5$ satisfying $x^5 - x$.

3. Write out the addition and multiplication tables of the ring $\mathbb{Z}_6$. Is $\mathbb{Z}_6$ a field or an integral domain? Which elements x of $\mathbb{Z}_6$ satisfy $x^2 = x$?

4. Find the inverse of $\overline{63}$ in the field $\mathbb{Z}_{257}$, expressing your answer in the form $\overline{t}$ where $0 < t < 257$. Hence find the smallest positive integer x satisfying $63x \equiv 5 \pmod{257}$.

5. (a) Find the inverse of $\overline{127}$ in the ring $\mathbb{Z}_{256}$. Determine all the elements x in $\mathbb{Z}_{256}$ such that $x^2 = \overline{1}$.
   (b) Let $n = p^e > 4$ where p is prime. Show that $\mathbb{Z}_n$ contains two or four elements x with $x^2 = \overline{1}$ according as p is odd or even.

6. Show that each of the rings $\mathbb{Z}_4$, $\mathbb{Z}_8$, $\mathbb{Z}_9$, $\mathbb{Z}_{12}$, $\mathbb{Z}_{16}$, $\mathbb{Z}_{18}$ contains an element $x \neq \overline{0}$ with $x^2 = \overline{0}$; show that $\mathbb{Z}_n$ ($n > 1$) does not contain such an element x if and only if n is a product of distinct positive primes.

7. Show that the element $\overline{m}$ of the ring $\mathbb{Z}_n$ has an inverse in $\mathbb{Z}_n$ if and only if $(m, n) = 1$.

8. Find the general solution of each of the following congruences; when there are integer solutions, find the solution of least absolute value and the smallest positive solution.

   (i) $81x \equiv 3 \pmod{100}$,   (ii) $36x \equiv 42 \pmod{156}$,
   (iii) $323x \equiv 76 \pmod{437}$,   (iv) $961x \equiv 527 \pmod{1705}$.

9. Find the smallest positive integer having divisor 167 and decimal representation (representation in the scale of 10) terminating with the digits 999.

10. Two motor racing tracks (red and blue) have nothing in common except a short narrow bridge. The red car sets out from the red pits A (distant 3 miles from the bridge) and travels at 60 m.p.h. along the red track (length 17 miles). Simultaneously, the blue car sets out from the blue pits B (distant 5 miles from the bridge) and travels at 70 m.p.h. along the blue track (length 19 miles). For how long can the cars be driven, in the directions shown, before a head-on collision takes place? If this happens, find the number of circuits completed by each car.

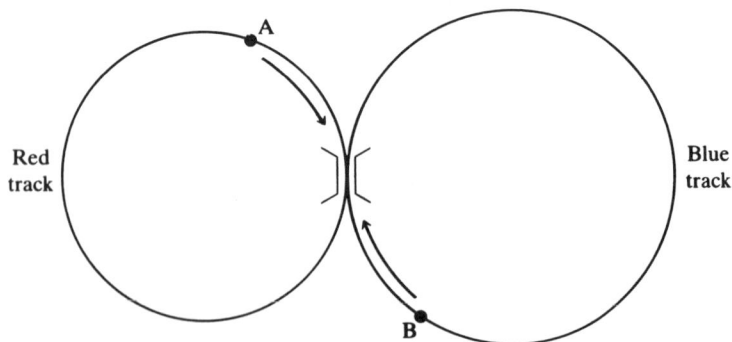

11. (a) Prove by induction:

$$10^n \equiv 1 \pmod 9, \qquad 10^n \equiv (-1)^n \pmod{11},$$

for all natural numbers n.

(b) Let $r_k \ldots r_1 r_0$ be the representation of the natural number m in the scale of 10. Prove

$$m \equiv r_k + \ldots + r_1 + r_0 \pmod 9,$$
$$m \equiv r_k + \ldots + r_1 + r_0 \pmod 3,$$
$$m \equiv (-1)^k r_k + \ldots + r_2 - r_1 + r_0 \pmod{11}.$$

Hence prove that m has divisor 9 if and only if the sum $r_k + \ldots + r_1 + r_0$ has divisor 9; state the corresponding rules for divisors 3 and 11.

(c) Write down the smallest natural number which has divisor 99 and all its digits (in the scale of 10) equal 1.

(d) Let $r_k \ldots r_1 r_0$ be the binary representation of the natural number m. Find a condition on $r_k, \ldots, r_1, r_0$ for m to have divisor 3.

12. Find in each case the general solution and the smallest positive solution of the simultaneous congruences:

(i) $x \equiv 40 \pmod{64}$, $x \equiv 28 \pmod{81}$;
(ii) $x \equiv 25 \pmod{95}$, $x \equiv 4 \pmod{119}$;
(iii) $5x \equiv 51 \pmod{63}$, $6x \equiv 29 \pmod{77}$;
(iv) $x \equiv 5 \pmod 7$, $x \equiv 9 \pmod{11}$, $x \equiv 11 \pmod{13}$.

13. In an unusual university, Professor A begins a course of lectures on the first day of term (a Monday) and continues lecturing every second day. Professor B starts on the second day of term and lectures every third day, Professor C starts on the fourth day of term and lectures every fifth day, while Professor D starts on the fifth day of term and lectures every fourth day.

When will the four professors first find themselves lecturing on

(i) the same day   (ii) a Sunday?

14. Let n be a given natural number. An equivalence relation 'domination by n' may be introduced on $\mathbb{Z}$ as follows: write $x \equiv y \pmod{\text{dom } n}$ if $\lfloor x/n \rfloor = \lfloor y/n \rfloor$, that is, if x/n and y/n have the same integer part $(x, y \in \mathbb{Z})$.

(a) List the integers in the set $\{x \in \mathbb{Z} : x \equiv 0 \pmod{\text{dom } 10}\}$. Show that every equivalence class in 'domination by n' consists of exactly n integers. Describe the partitioning of $\mathbb{Z}$ in 'domination by 2'.

(b) If $n > 1$, show that 'domination by n' is not compatible with either integer addition or multiplication.

(c) Show $x \equiv y \pmod{n}$ and $x \equiv y \pmod{\text{dom } n}$ imply $x = y$.

(d) (Long division in the scale of b.) Let m, b, k be integers with $m \geqslant 1$, $b \geqslant 2$, $k \geqslant 0$, and let $m \equiv 0$ (dom nb^{k+1}). Show that there is a unique integer r_k with $0 \leqslant r_k < b$ such that $m - r_k nb^k \equiv 0$ (dom nb^k). Assuming $r_k \neq 0$, show by iterating the above procedure (replacing m by $m - r_k nb^k$) that $\lfloor m/n \rfloor$ has representation $r_k \ldots r_1 r_0$ in the scale of b.

4 Polynomials

Every field F gives rise to an integral domain $F[x]$, which can be thought of as the smallest integral domain containing F and the symbol x called an **indeterminate** (we deliberately assume as little as possible about x). As $F[x]$ is closed under addition and multiplication, every **polynomial** over F, that is, every expression of the form

$$a_0 + a_1 x + a_2 x^2 + \ldots + a_n x^n \quad (a_0, a_1, a_2, \ldots, a_n \in F)$$

belongs to $F[x]$; further, because of the smallness property, each element in $F[x]$ is of the above form. Polynomials arise in many branches of mathematics, for instance in connection with approximations in analysis and with enumerations in combinatorial theory; we shall see that they play a crucial role in algebra.

This chapter begins with the construction of polynomial rings in general and then takes up the close analogy between $F[x]$ and the integral domain $\mathbb{Z}$: for in the system $F[x]$ it is possible to divide one polynomial by a non-zero polynomial obtaining quotient and remainder polynomials—in fact *all* the division and congruence properties of integers have polynomial counterparts! In particular the analogues of the fields $\mathbb{Z}_p$ provide us with further significant examples of fields.

In Chapter 2 we saw that every complex number w has an n th root z belonging to $\mathbb{C}$; in other words, the polynomial $x^n - w$ over $\mathbb{C}$ has **zero** z (for $z^n - w = 0$). Here we study zeros of polynomials over an arbitrary field, though the complex field takes pride of place; our discussion centres on the famous **fundamental theorem of algebra:**

Every non-constant polynomial over $\mathbb{C}$ has a zero in $\mathbb{C}$.

This was proved by the mathematician Gauss in 1799. This important theorem enables us to describe, theoretically, the factorization of all real and complex polynomials. Also, we see that the process (of adjoining the zero i of $x^2 + 1$) which led us from $\mathbb{R}$ to $\mathbb{C}$ cannot lead us any further, for $\mathbb{C}$ already contains all zeros of all complex polynomials.

Polynomial rings

We begin with a discussion of polynomials over an arbitrary ring.

Definition 4.1 Let R be a ring. A **polynomial** over R in the **indeterminate** x is an expression of the form

$$a_0 + a_1 x + a_2 x^2 + \ldots + a_n x^n \quad \text{where } a_0, a_1, a_2, \ldots, a_n \in R.$$

The ring element a_i is called the **coefficient** of x^i in the polynomial.

So a polynomial in x is made up of a finite number of non-negative powers of x; negative powers of x are not allowed. Although the coefficients a_i belong to R, the indeterminate x should not be regarded as an element of R but rather as a special symbol (satisfying certain rules of manipulation discussed below). For instance, $1 + 3x + 3x^2 + x^3$ is a polynomial over $\mathbb{Z}$ and $1 + (\frac{2}{3})x^3 + (\frac{1}{4})x^4$ is a polynomial over the rational field $\mathbb{Q}$; but $1/x$, $1/(x+1)$, $1 + x + x^2 + \ldots + x^n + \ldots$ are *not* polynomials (a polynomial cannot have an *infinite* number of non-zero coefficients).

Let $\mathbb{N}_0$ denote the set of non-negative integers. Given the polynomial $a_0 + a_1 x + a_2 x^2 + \ldots + a_n x^n$, it is convenient to define $a_i = 0$ for all integers $i > n$ and denote this polynomial by

$$\sum a_i x^i$$

it being understood that i ranges over $\mathbb{N}_0$ in the summation.

Definition 4.2 Two polynomials are called **equal** if corresponding coefficients are equal, that is,

$$\sum a_i x^i = \sum b_i x^i$$

means $a_i = b_i$ for all $i \in \mathbb{N}_0$.

The role of x in the above definition is a pure formality; however, the significance of the indeterminate will become clear when polynomial products are defined. We use f, g, h, etc. to denote polynomials.

Definition 4.3 Let $f = \sum a_i x^i$ be a polynomial over the ring R. If $a_i = 0$ for all i in $\mathbb{N}_0$, then f is called the **zero polynomial** over R and we write $f = 0$. If $f \neq 0$, then the largest integer i with $a_i \neq 0$ is called the **degree** of f and denoted by $\deg f$.

Therefore a polynomial is non-zero if it has at least one non-zero coefficient. The zero polynomial and the polynomials of degree 0 are called **constant** polynomials; these polynomials contain no positive powers of x.

There are two constant polynomials over $\mathbb{Z}_2$, namely 0 and 1 (for simplicity we write 0 and 1 in place of $\bar{0}$ and $\bar{1}$ while not forgetting the exact meaning of these symbols). There are two **linear** (degree 1) polynomials over $\mathbb{Z}_2$, namely x and $x + 1$, and four **quadratic** (degree 2) polynomials over $\mathbb{Z}_2$, namely x^2, $x^2 + 1$, $x^2 + x$, and $x^2 + x + 1$, since the coefficients are either 0 or 1. Similarly, there are eight **cubic** (degree 3) polynomials over $\mathbb{Z}_2$, and, more generally, 2^n polynomials of degree n over $\mathbb{Z}_2$. Note that a polynomial is *over* a ring R if its coefficients belong to R.

To get the idea behind the multiplication rule (4.4) for polynomials, consider 'multiplying out' the product below and 'collecting up' terms in the usual way:

$$(a_0 + a_1 x + a_2 x^2 + \ldots)(b_0 + b_1 x + b_2 x^2 + \ldots)$$
$$= a_0 b_0 + (a_0 b_1 + a_1 b_0)x + (a_0 b_2 + a_1 b_1 + a_2 b_0)x^2 + \ldots$$

the coefficient of x^i in the product being $a_0 b_i + a_1 b_{i-1} + \ldots + a_i b_0$. This operation is legitimate if all the elements involved belong to a ring in which x^i *commutes* with a_j and b_j for all $i, j \in \mathbb{N}_0$. Our aim is to construct a ring having precisely these properties, that is, to form a ring having polynomials as elements in which powers of x commute with constant polynomials.

Definition 4.4 Let $f = \sum a_i x^i$ and $g = \sum b_i x^i$ be polynomials over the ring R. Their **sum** is the polynomial

$$f + g = \sum (a_i + b_i)x^i$$

and their **product** is the polynomial

$$fg = \sum c_i x^i \quad \text{where } c_i = a_0 b_i + a_1 b_{i-1} + \ldots + a_i b_0.$$

The sum and product of polynomials over R are again polynomials over R.

Notation The system consisting of the set of all polynomials in x over the ring R, with addition and multiplication as above, is denoted by $R[x]$.

For instance $\mathbb{Z}_2[x]$ stands for the system of all polynomials in x over $\mathbb{Z}_2$; for $f = 1 + x + x^2$ and $g = 1 + x^2$ belonging to $\mathbb{Z}_2[x]$ we have

$$f + g = 1 + 1 + x + (1 + 1)x^2 = x,$$
$$fg = 1 + x + (1 + 1)x^2 + x^3 + x^4 = 1 + x + x^3 + x^4,$$

using $1 + 1 = 0$.

Applying the product law (4.4) with $f = x^i$ and $g = b_0$ (constant polynomial) gives

$$x^i b_0 = b_0 x^i$$

showing that powers of x commute with constant polynomials. It is usual to identify the constant polynomial $g = b_0$ with the ring element b_0; we adopt this reasonable convention, for sums and products agree and our notation makes no distinction. Therefore R may be regarded as being inside $R[x]$ (in fact R is a **subring** (see (5.6)) of $R[x]$) and the above equation says that powers of x commute with elements of R in the system $R[x]$.

Theorem 4.5   Let R be a ring. Then $R[x]$ is also a ring.

Proof   We begin by showing that the associative laws (laws 1 and 6 of (2.2)) hold in $R[x]$, and so let $f = \sum a_i x^i$, $g = \sum b_i x^i$, and $h = \sum c_i x^i$ belong to $R[x]$; as law 1 holds in R we have $(a_i + b_i) + c_i = a_i + (b_i + c_i)$, that is, the coefficients of x^i in $(f + g) + h$ and $f + (g + h)$ are equal for all $i \in \mathbb{N}_0$. Therefore $(f + g) + h = f + (g + h)$ by (4.2), showing that law 1 holds in $R[x]$.

As for law 6, consider the following triangular array:

$$a_0 b_0 c_i$$
$$a_0 b_1 c_{i-1} \quad a_1 b_0 c_{i-1}$$
$$\vdots \qquad\qquad \vdots$$
$$a_0 b_{i-1} c_1 \quad a_1 b_{i-2} c_1 \quad \cdots \quad a_{i-1} b_0 c_1$$
$$a_0 b_i c_0 \qquad a_1 b_{i-1} c_0 \quad \cdots \quad a_{i-1} b_1 c_0 \quad a_i b_0 c_0$$

Adding the entries in each row, and then adding the row sums together, produces the coefficient of x^i in $(fg)h$. On the other hand, adding the entries in each column, and then adding the column sums together, gives the coefficient of x^i in $f(gh)$. These coefficients are therefore equal, and so we conclude $(fg)h = f(gh)$, showing that law 6 holds in $R[x]$.

We leave the straightforward verification of the remaining ring laws to the reader. Note that the zero polynomial (4.3) is the

0-element of $R[x]$, the constant polynomial 1 is the 1-element of $R[x]$, and the negative of $f = \sum a_i x^i$ is $-f = \sum (-a_i) x^i$.   □

If R is trivial, then so also is $R[x]$, as the zero polynomial is the only polynomial over R; if R is non-trivial, then $R[x]$ contains an infinite number of elements, for example 1, x, x^2, x^3, ..., x^n, ... belong to $R[x]$ and are all different as $1 \neq 0$.

Definition 4.6   Let $f = a_0 + a_1 x + \ldots + a_n x^n$ be a polynomial of degree n over the ring R. Then a_n is called the **leading coefficient** in f. If $a_n = 1$, f is called **monic**.

The leading coefficient is therefore the coefficient of the highest power of x present in the polynomial; for instance $(2 + x)^3 = 8 + 12x + 6x^2 + x^3$ has leading coefficient 1 and so is monic.

Proposition 4.7   Let R be an integral domain. Then $R[x]$ is also an integral domain. Further $\deg fg = \deg f + \deg g$ where $f, g \in R[x]^*$.

Proof   We know that $R[x]$ is a ring by (4.5); as R is non-trivial, so also is $R[x]$. To show $R[x]$ is commutative, consider $f = \sum a_i x^i$ and $g = \sum b_i x^i$ in $R[x]$. As R is commutative, we see $a_0 b_i + a_1 b_{i-1} + \ldots + a_i b_0 = b_0 a_i + b_1 a_{i-1} + \ldots + b_i a_0$, showing that the coefficients of x^i in fg and gf are equal for all i in $\mathbb{N}_0$, and so $fg = gf$.

Suppose now that f and g are non-zero polynomials. We investigate the 'top end' of these polynomials, that is, the terms involving the highest powers of x. Let $m = \deg f$ and $n = \deg g$, and so $a_m \neq 0$ is the leading coefficient in f and $b_n \neq 0$ is the leading coefficient in g. As R is an integral domain and

$$fg = (a_0 + \ldots + a_m x^m)(b_0 + \ldots + b_n x^n)$$
$$= a_0 b_0 + \ldots + a_m b_n x^{m+n}$$

we see that $a_m b_n \neq 0$ is the leading coefficient in fg. Therefore $fg \neq 0$ and so $R[x]$ is an integral domain by (2.16); also

$$\deg fg = m + n = \deg f + \deg g.$$ □

It is customary to decree $-\infty$ to be the degree of the zero polynomial, so that the degree formula of (4.7) is valid for all polynomials f and g, whether zero or not, over an integral domain. We adopt this convenient convention, for it gives every polynomial a degree; the constant polynomials are then the polynomials of non-positive degree. In fact, the degree formula of (4.7) is valid for polynomials f and g over an arbitrary ring R, provided that one of f and g is monic.

Our next proposition should be compared with the division law (3.6) for integers; it is a formalization of the process of dividing the polynomial f by the monic polynomial g, obtaining a unique quotient polynomial q and unique remainder polynomial r of smaller degree than g.

Proposition 4.8 (The polynomial division law.) Let f and g be polynomials in x over the non-trivial ring R and let g be monic. Then there are unique polynomials q and r in $R[x]$ such that

$$f = qg + r \quad \text{where} \quad \deg r < \deg g.$$

Proof We mimic the proof of (3.6): let X denote the set of all polynomials in $R[x]$ of the type $f - hg$ where $h \in R[x]$; by the well-ordering principle, X contains a polynomial r of least degree (either r is the zero polynomial or $\deg r$ is a non-negative integer). So $r = f - qg$ for some $q \in R[x]$. To show $\deg r < \deg g$, suppose to the contrary that $\deg r \geq \deg g$. We write $m = \deg r$ and $n = \deg g$ and let c be the leading coefficient in r; then

$$r' = r - cx^{m-n}g$$

contains no terms involving x^i for $i > m$ and, by construction, the coefficient of x^m in r' is zero also. Therefore $\deg r' < m = \deg r$ which is contrary to the leastness of $\deg r$, for $r' = f - (q + cx^{m-n})g$ belongs to X; so in fact $\deg r < \deg g$, showing that there are polynomials q and r as stated.

To prove the uniqueness of q and r, suppose $f = qg + r$ and $f = q'g + r'$ where r and r' are of smaller degree than g; therefore

$$(q - q')g = r' - r$$

and as g is monic we may use the degree formula to give

$$\deg(q - q') + \deg g = \deg(r' - r) < \deg g$$

showing that $q - q'$ has negative degree. As the zero polynomial is the only polynomial of negative degree, we have $q - q' = 0$; so $q = q'$ and hence $r = r'$. □

When f and g are specific polynomials, then q and r can be found by the polynomial version of long division. For instance, to divide $f = x^4 - 3x^3 + 2x^2 - x + 5$ by $g = x^2 + 2x - 1$ over $\mathbb{Z}$, we proceed as below:

$$
\begin{array}{r}
x^2 - 5x + 13 \\
x^2 + 2x - 1\overline{\smash{\big)}\ x^4 - 3x^3 + 2x^2 - x + 5} \\
\underline{x^4 + 2x^3 - x^2} \\
-5x^3 + 3x^2 - x \\
\underline{-5x^3 - 10x^2 + 5x} \\
13x^2 - 6x + 5 \\
\underline{13x^2 + 26x - 13} \\
-32x + 18
\end{array}
$$

So in this case $q = x^2 - 5x + 13$ and $r = -32x + 18$, giving the equation

$$ x^4 - 3x^3 + 2x^2 - x + 5 = (x^2 - 5x + 13)(x^2 + 2x - 1) - 32x + 18. $$

Notice that division by the quadratic polynomial g produces a remainder r of degree *less* than 2.

Definition 4.9

Let f and g belong to $R[x]$, where R is a commutative ring. If there is q in $R[x]$ with $f = qg$, then g is called a **divisor (factor)** of f and we write $g \mid f$.

Working over $\mathbb{Z}_{12}$ (and omitting bars), we see $x(x + 1) = (x + 4)(x + 9)$ on multiplying out these polynomials; so $x + 4$ and $x + 9$ are divisors of $x^2 + x$ over the ring $\mathbb{Z}_{12}$. Anomalies of this kind can be avoided by restricting the coefficient ring to be a field, which we now proceed to do. If F is a field, it will soon become apparent that there is a great similarity between $F[x]$ and $\mathbb{Z}$; what is more, the techniques of the last chapter (notably the Euclidean algorithm) carry over with only minor changes.

Definition 4.10

Let g and g' be polynomials over the field F. If there is c in F^* with $g = cg'$ we say g and g' are **associate** and write $g \sim g'$.

In other words, two polynomials are associate if they differ by a non-zero constant factor. For instance, the following polynomials are associate over $\mathbb{Q}$:

$$ \tfrac{1}{4}x^2 + \tfrac{2}{3}x - \tfrac{1}{5}, \qquad 15x^2 + 40x - 12, \qquad x^2 + \tfrac{8}{3}x - \tfrac{4}{5}. $$

Let g be a non-zero polynomial over the field F with leading coefficient c. Then $g' = (1/c)g$ is the unique monic polynomial over

F which is associate with *g*; we say *g'* is '*g* **made monic**'. The reader should mentally verify that 'being associate' satisfies (1.25) (here $\sim$ takes the place of $\equiv$) and so is an equivalence relation on $F[x]$; the equivalence classes are called associate classes. Changing a polynomial into an associate polynomial is analogous to changing the sign of an integer; as we shall see, the relevant point is that divisor properties are unaffected by these changes.

Our next result is a direct consequence of the polynomial division law (4.8).

Corollary 4.11

(The division law for polynomials over a field.) Let *f* and *g* be polynomials in *x* over the field *F* with *g* non-zero. Then there are unique polynomials *q* and *r* in $F[x]$ with $f = qg + r$ where $\deg r < \deg g$.

Proof

Let *c* denote the leading coefficient in *g* and let $g' = (1/c)g$. As *g'* is monic, by (4.8) there are unique polynomials *q'* and *r* over *F* with $f = q'g' + r$ and $\deg r < \deg g' = \deg g$; therefore $q = (1/c)q'$ and *r* are the unique polynomials of the statement. □

We shall use the above form of the division law to describe the polynomial version of the Euclidean algorithm.

Suppose now that *g* and *g'* are non-zero polynomials over the field *F* such that each is a divisor of the other, that is, $g' \mid g$ and $g \mid g'$. Then $qg' = g$ and $q'g = g'$ where $q, q' \in F[x]$; therefore $qq'g = g$ and so $qq' = 1$, which gives $\deg q + \deg q' = 0$ on comparing degrees. So $\deg q = \deg q' = 0$, showing that *q* and *q'* are non-zero constant polynomials; writing $q = c$ we obtain $g = cg'$, and so *g* and *g'* are associate.

Definition 4.12

Let *f* and *g* belong to $F[x]$ where *F* is a field. If *d* in $F[x]$ satisfies $d \mid f$ and $d \mid g$, then *d* is called a **common divisor** of *f* and *g*.

A monic polynomial *d* is called the **greatest common divisor** of *f* and *g* if *d* is a common divisor of *f* and *g*, and $d' \mid d$ for all common divisors *d'* of *f* and *g*.

Generally, the integral domain $F[x]$ cannot be ordered in the sense of (3.1), and so inequalities between polynomials (as opposed to inequalities between degrees of polynomials) are meaningless; as (4.12) makes clear, the g.c.d. of polynomials is 'greatest' in terms of divisor properties—there is no other sense in this context!

A given pair of polynomials cannot have more than one g.c.d.: for let d and d_1 be g.c.d.s of f and g in $F[x]$. Then $d \mid d_1$ and $d_1 \mid d$ which imply as above that d and d_1 are associate; as d and d_1 are both monic, we conclude that $d = d_1$. The existence of polynomial g.c.d.s is guaranteed by our next proposition.

Proposition 4.13
Let f and g (not both zero) belong to $F[x]$ where F is a field. Then f and g have a unique greatest common divisor (f, g), and there are polynomials s and t in $F[x]$ with $sf + tg = (f, g)$.

Proof
The existence of (f, g) can be established as in (3.9), which is the corresponding result for integers, and so the following outline should be enough for the reader. Let X denote the set of non-zero polynomials of the form $s'f + t'g$ where s', $t' \in F[x]$. As X is non-empty, we may pick a polynomial in X of least degree, make it monic and call the result (f, g). Then $(f, g) = sf + tg$ for some polynomials s and t in $F[x]$ and so every common divisor of f and g is a divisor of (f, g). As division of every polynomial in X by (f, g) leaves remainder zero, we see that (f, g) is a common divisor of f and g. So (f, g) satisfies (4.12). $\square$

As in the case of integers, we write g.c.d. $(0, 0) = 0$, and so every pair of polynomials in $F[x]$ has a unique g.c.d. The important matter of determining g.c.d.s and polynomials s and t as in (4.13) is taken care of by the Euclidean algorithm.

Example 4.14
Consider the polynomials $r_1 = x^4 - x^3 - 4x^2 - x + 5$ and $r_2 = x^3 - 3x^2 + x + 1$ over the rational field $\mathbb{Q}$. To find the g.c.d. (r_1, r_2), carry out the same sequence of divisions as in the Euclidean algorithm (3.12) for integers until a zero remainder is obtained, that is, divide r_i by r_{i+1} obtaining r_{i+2} as remainder for $i = 1, 2, \ldots$; then (r_1, r_2) is the last non-zero remainder *made monic*. Starting with the given polynomials r_1 and r_2 we obtain:

divide r_1 by r_2: $r_1 = (x + 2)r_2 + r_3$ where $r_3 = x^2 - 4x + 3$,

divide r_2 by r_3: $r_2 = (x + 1)r_3 + \underline{r_4}$ where $r_4 = 2x - 2$,

divide r_3 by r_4: $r_3 = \frac{1}{2}(x - 3)r_4$.

Therefore $(r_1, r_2) = x - 1$ in this case, for $x - 1$ is the last non-zero remainder (underlined above) made monic. Tracing the algorithm backwards, that is, successively substituting for each remainder in terms of previous remainders, determines ultimately polynomials s

and t over $\mathbb{Q}$ with $sr_1 + tr_2 = (r_1, r_2)$. In our case:

$$r_4 = r_2 - (x + 1)r_3 = r_2 - (x + 1)(r_1 - (x + 2)r_2)$$
$$= -(x + 1)r_1 + (x^2 + 3x + 3)r_2$$

giving $s = -\frac{1}{2}(x + 1)$, $t = \frac{1}{2}(x^2 + 3x + 3)$ as $(r_1, r_2) = \frac{1}{2}r_4$.

Theorem 4.15

(The Euclidean algorithm for polynomials.) Let f and g be non-zero polynomials in x over a field F with $\deg f \geqslant \deg g$. Write $r_1 = f$ and $r_2 = g$; if r_{i+1} is not a divisor of r_i, let r_{i+2} be the remainder on dividing r_i by r_{i+1}, for $i = 1, 2, \ldots$. This sequence of divisions terminates in a zero remainder; the last non-zero remainder, r_k say, made monic, is then the greatest common divisor (f, g).

Proof

Because of the close analogy between $\mathbb{Z}$ and $F[x]$, the following outline should be enough. Analogous to (3.11) we see that $(f, g) = (g, r)$ if $f = qg + r$ where $f, g, q, r \in F[x]$. As

$$\deg r_1 \geqslant \deg r_2 > \deg r_3 > \ldots > \deg r_i > \deg r_{i+1} > \ldots$$

form a decreasing sequence of non-negative integers, there cannot be an infinite number of non-zero remainders in the given sequence of divisions (4.11); let r_k be the last non-zero remainder, and so r_k is a divisor of r_{k-1}. As in (3.12) we conclude

$$(f, g) = (r_1, r_2) = (r_2, r_3) = \ldots (r_{k-1}, r_k) = (r_k, 0) = r_k'$$

where r_k' is r_k made monic.   □

The calculations involved in the Euclidean algorithm can often be simplified if the remainders r_i which arise are replaced in the next step by associate polynomials. Thus when working over a field such as $\mathbb{Z}_p$, the remainders r_i can be made monic so that division by r_i is easier. Similarly, when working with polynomials over $\mathbb{Q}$, one may prevent the appearance of polynomials with fractional coefficients if the polynomial to be divided is first multiplied by a suitable non-zero integer. Making these replacements will not change the g.c.d. we are looking for.

Example 4.16

(a) We write 0, 1, 2, 3, 4 for the elements of $\mathbb{Z}_5$ while remembering to calculate modulo 5; so, for instance $2 \times 3 = 1$, $1/2 = 3$, $1/3 = 2$, $1/4 = 4$ in $\mathbb{Z}_5$. Consider the polynomials

$$r_1 = x^4 + 3x^3 + 2, \quad r_2 = x^4 + 2x + 3$$

over $\mathbb{Z}_5$. Applying the Euclidean algorithm:

divide r_1 by r_2: $r_1 = r_2 + r_3$ where $r_3 = 3x^3 + 3x + 4$.

Let $r_3' = 2r_3 = x^3 + x + 3$; then r_3' is r_3 made monic.

Divide r_2 by r_3': $r_2 = xr_3' + r_4$ where $r_4 = 4x^2 + 4x + 3$.

Let $r_4' = 4r_4 = x^2 + x + 2$.

Divide r_3' by r_4': $r_3' = (x+4)r_4'$.

Therefore g.c.d. $(r_1, r_2) = r_4' = x^2 + x + 2$.

(b) Consider the polynomials over $\mathbb{Q}$

$$r_1 = x^4 + 2x^3 + 3x^2 + 2x + 1, \qquad r_2 = x^4 - x^3 - 2x^2 - 3x - 1.$$

Divide r_1 by r_2: $r_1 = r_2 + r_3$ where $r_3 = 3x^3 + 5x^2 + 5x + 2$.

Divide $9r_2$ by r_3: $9r_2 = (3x - 8)r_3 + r_4$ where $r_4 = 7(x^2 + x + 1)$.

Divide r_3 by $\frac{1}{7}r_4$: $r_3 = (3x + 2)\frac{1}{7}r_4$.

Therefore $(r_1, r_2) = x^2 + x + 1$.

Exercises 4.1

1. For each pair f, g below, find q and r in $\mathbb{Q}[x]$ with $f = qg + r$ and $\deg r < \deg g$.

  (i) $f = x^4 + 5x^3 + 2x^2 + 1$, $\quad g = x + 2$;
  (ii) $f = x^4 + 7x^3 + 5x^2$, $\quad g = x^2 + 1$;
  (iii) $f = x^5 + 2x^4 + 3x^3 + 3x^2 + 2x + 1$, $\quad g = x^3 + x^2 + x + 1$.

2. List the eight cubic polynomials over $\mathbb{Z}_2$ (each coefficient is either 0 or 1). Which of these polynomials have divisor $x^2 + x + 1$? Find the quotient and remainder on dividing $x^7 + 1$ by $x^3 + 1$ over $\mathbb{Z}_2$.

3. List the nine monic quadratic polynomials over $\mathbb{Z}_3$ (each coefficient is either 0, 1, or -1). How many monic polynomials of degree n over $\mathbb{Z}_3$ are there? Multiply out the product $(x^2 + 1)(x^2 - x - 1)(x^2 + x - 1)$ in the ring $\mathbb{Z}_3[x]$.

4. Make monic the polynomial $f = 4x^3 + 3x^2 + 2x + 1$ over $\mathbb{Z}_5$ and write down the two further polynomials in $\mathbb{Z}_5[x]$ which are associate with f. How many cubic polynomials over $\mathbb{Z}_5$ are there, and how many associate classes of such polynomials are there?

5. For each pair of polynomials f and g over the rational field $\mathbb{Q}$, find the g.c.d. (f, g) and polynomials s and t over $\mathbb{Q}$ with $sf + tg = (f, g)$.

  (i) $f = x^4 + x^2 + 1$, $\quad g = x^3 + 1$;
  (ii) $f = x^4 + x^3 - 2x^2 - x + 1$, $\quad g = x^3 - 1$;
  (iii) $f = x^6 + x^4 - 1$, $\quad g = x^5 - x$.

6. In each of the following cases, working over the indicated field, find the g.c.d. (f, g) and the polynomials f' and g' such that $f = f'(f, g)$ and $g = g'(f, g)$.

    (i) $f = x^5 + x^4 + x^3 + 1$, $g = x^5 + x^2 + x + 1$ over $\mathbb{Z}_2$;
    (ii) $f = x^4 + x^3 + x^2 - 1$, $g = x^4 + 1$ over $\mathbb{Z}_3$;
    (iii) $f = 2x^3 + x^2 + 3x + 4$, $g = 3x^3 + 4x^2 + 2x + 1$ over $\mathbb{Z}_5$;
    (iv) $f = 2x^3 + x^2 + 4x + 3$, $g = 3x^3 + 4x^2 + 2x + 1$ over $\mathbb{Z}_5$;
    (v) $f = x^{27} - 1$, $g = x^{16} - 1$ over $\mathbb{Q}$.

7. Divide $3x^2 + 2x + 1$ by $x + 2$ over $\mathbb{Z}$ and hence determine the integers n such that $n + 2$ is a divisor of $3n^2 + 2n + 1$.

8. Let R be a ring. Complete the proof of (4.5) by showing that the commutative law of addition and the distributive laws (laws 4 and 5 of (2.2)) hold in $R[x]$.

9. Write out a detailed proof of (4.13) using the given outline.

10. Let f and f' be associate polynomials and let g and g' be associate polynomials over the field F. Show that $f \mid g$ if and only if $f' \mid g'$; hence show that $(f, g) = (f', g')$.

11. Let f and g be non-zero polynomials over $\mathbb{Z}$ and suppose $\deg f = n \geqslant m = \deg g$. If c is the leading coefficient of g, show that there are unique polynomials q and r over $\mathbb{Z}$ with

$$(c^{n-m+1})f = qg + r \quad \text{and} \quad \deg r < \deg g.$$

12. Let m and n be positive integers and let F be a field.
    (a) Show that $x^m - 1$ is a divisor of $x^n - 1$ (as polynomials over F) if and only if m is a divisor of n (as integers).
    (b) Let $d = (m, n)$. Show that $x^d - 1$ is the g.c.d. of the polynomials $x^m - 1$ and $x^n - 1$ over F.
    (c) Let l be an integer with $l > 1$. Show that $l^d - 1$ is the g.c.d. of the integers $l^m - 1$ and $l^n - 1$, where $d = (m, n)$.

Factorization of polynomials

We begin by introducing the polynomial analogue of a prime integer.

Definition 4.17   Let p be a polynomial of positive degree over the field F. Then p is called **irreducible** over F if there is *no* factorization $p = fg$ where f and g are polynomials of smaller degree than p over F; otherwise p is called **reducible** over F.

Let p be an irreducible polynomial over the field F; this means p is not a constant polynomial and the only factorizations $p = fg$, where f and g are polynomials over F, satisfy $\deg f = \deg p$ with $\deg g = 0$ (that is, f and p are associate and g is a non-zero constant polynomial) or vice-versa: $\deg f = 0$ with $\deg g = \deg p$. In other words, irreducible polynomials are like primes, for they cannot be factorized in a non-trivial way.

Every linear polynomial $a_1x + a_0$ ($a_0, a_1 \in F$ with $a_1 \neq 0$) is irreducible over F, for no polynomial of degree 1 is a product of constant polynomials, and this is true for every field F. However, it is usually necessary to specify F when discussing irreducibility: for instance $x^2 + 1$ is *reducible* as a polynomial over the complex field $\mathbb{C}$ because $x^2 + 1 = (x + \mathrm{i})(x - \mathrm{i})$ where $\mathrm{i}^2 = -1$. But $x^2 + 1$ is *irreducible* as a polynomial over the real field $\mathbb{R}$ as we now show: let $x^2 + 1 = fg$ where f and g are polynomials of degree less than 2 over $\mathbb{R}$. As $\deg f + \deg g = 2$, we see $\deg f = \deg g = 1$ and so $f = ax + b$ and $g = cx + d$ ($a, b, c, d \in \mathbb{R}$). Comparing coefficients in the polynomial equation

$$x^2 + 1 = fg = (ax + b)(cx + d) = acx^2 + (ad + bc)x + bd$$

gives $ac = bd = 1$ and $ad + bc = 0$; hence $(ad)^2 = -abcd = -1$ which is impossible in the ordered field $\mathbb{R}$, for $(ad)^2$ is positive and -1 is negative! In fact we have shown that $x^2 + 1$ is irreducible as a polynomial over any ordered field.

We can determine irreducible polynomials of low degree over $\mathbb{Z}_2$ by inspection: as x and $x + 1$ are the only linear polynomials over $\mathbb{Z}_2$, we see that x^2, $x(x + 1)$, $(x + 1)^2$ are the only reducible quadratic polynomials over $\mathbb{Z}_2$ (notice that $x^2 + 1 = (x + 1)^2$ is reducible over $\mathbb{Z}_2$). Therefore $x^2 + x + 1$ is the only irreducible quadratic polynomial over $\mathbb{Z}_2$ (there are only four quadratic polynomials over $\mathbb{Z}_2$). Carrying on in this way, the reader may verify that there are six reducible cubics over $\mathbb{Z}_2$ and hence there are two irreducible cubics over $\mathbb{Z}_2$, namely $x^3 + x + 1$ and $x^3 + x^2 + 1$.

For convenience we restrict ourselves to monic polynomials in our discussion of factorization. Notice therefore that if the monic polynomial f has a factorization $f = gh$, then $f = g'h'$ where g' and h' are g and h made monic: for $1 = ab$ where a and b are the leading coefficients of g and h, and so $f = ((1/a)g)((1/b)h)$. It follows that a monic polynomial $p \neq 1$ is irreducible over the field F if and only if 1 and p are its only monic divisors over F.

Definition 4.18 Let f and g belong to $F[x]$ where F is a field. If the greatest common divisor $(f, g) = 1$, then f and g are called **coprime** (**relatively prime**) polynomials.

For instance the polynomials $x + 1$, $x - 1$ over $\mathbb{R}$ are coprime; more generally any two distinct monic irreducible polynomials over a field are coprime, for 1 is their only monic common divisor.

Let p, p', f be polynomials over the field F, where p and p' are monic and irreducible over F and suppose p is a divisor of $p'f$. If $p = p'$ there is nothing more to be said, and so let us suppose $p \neq p'$. Therefore $(p, p') = 1$. Arguing as in (3.14), by (4.9) and (4.13) there are polynomials q, s, t over F with $qp = p'f$ and $sp + tp' = 1$. Hence $(sf + tq)p = f$ showing that p is a divisor of f.

Suppose now that p is a divisor of $p_1 p_2 \ldots p_k$ where $p, p_1, \ldots, p_k$ are monic irreducible polynomials over F. Using the previous paragraph, either $p = p_1$ or p is a divisor of $p_2 p_3 \ldots p_k$ (or both); in the latter case we may repeat the argument: either $p = p_2$ or p is a divisor of $p_3 p_4 \ldots p_k$ (or both). If p is different from all of $p_1, p_2, \ldots, p_{k-1}$, then ultimately we see that p is a divisor of p_k which means $p = p_k$ as these polynomials are monic and irreducible. The conclusion is that $p = p_i$ for some i $(1 \leq i \leq k)$; this conclusion should really be proved by induction! We leave the reader to reorganize these remarks into an inductive proof of the statement $P(k)$:

$p \mid (p_1 p_2 \ldots p_k)$ implies $p = p_i$ for some i $(1 \leq i \leq k)$ where $p, p_1, p_2, \ldots, p_k$ are monic irreducible polynomials over F.

Our next theorem is the polynomial analogue of (3.15).

Theorem 4.19

Let f be a monic polynomial of positive degree over the field F. Then $f = p_1 p_2 \ldots p_k$ where each p_i is monic and irreducible over F, and, this factorization of f is unique, apart from the ordering of the factors.

Proof

We use induction on $n = \deg f$. If $n = 1$, then $f = x + a_0$ is irreducible; by (4.17), $f = p_1 p_2 \ldots p_k$ implies $k = 1$ and so $f = p_1$. Let $P(n)$ be the statement that (4.19) is true whenever $1 \leq \deg f \leq n$; then, as we have just seen, $P(1)$ is true.

Suppose $n > 1$ and that $P(n - 1)$ is true. To prove that $P(n)$ is true, consider a polynomial f of degree n over F. If f is irreducible, then the argument used in the case $n = 1$ applies. If f is reducible, then $f = f_1 f_2$ where $1 \leq \deg f_1 \leq n - 1$ and $1 \leq \deg f_2 \leq n - 1$. Using the truth of $P(n - 1)$ we obtain $f_1 = p_1 p_2 \ldots p_l$, $f_2 = p_{l+1} p_{l+2} \ldots p_k$, where the polynomials p_i are monic and irreducible over F for $1 \leq i \leq k$; therefore $f = p_1 p_2 \ldots p_l p_{l+1} \ldots p_k$.

To show the uniqueness of the above factorization, suppose also that $f = p'_1 p'_2 \ldots p'_{k'}$ where $p'_1, p'_2, \ldots, p'_{k'}$ are monic irreducible polynomials over F. As

$$p_1 p_2 \ldots p_k = p_1' p_2' \ldots p_{k'}'$$

we see $p_1 \mid p_1' p_2' \cdots p_{k'}'$ and so by the preliminary remarks $p_1 = p_i'$ for some i with $1 \leqslant i \leqslant k'$; we assume by renumbering $p_1', p_2', \ldots, p_{k'}'$ if necessary that $i = 1$, and so $p_1 = p_1'$ may be cancelled to give

$$p_2 \ldots p_k = p_2' \ldots p_{k'}'$$

which are two factorizations of the polynomial f/p_1. As $1 \leqslant \deg (f/p_1) \leqslant n - 1$ we deduce from $P(n - 1)$ that $k = k'$ and $p_i = p_i'$ for all i with $1 \leqslant i \leqslant k$ (a further renumbering of $p_2', \ldots, p_k'$ may be necessary!). Collecting up the pieces, we see that the two factorizations of f into monic irreducible polynomials are identical apart from the order in which the factors appear. Therefore the truth of $P(n)$ follows from the truth of $P(n - 1)$; by the principle of induction $P(n)$ *is* therefore true.　　　　□

The irreducible polynomials in (4.19) might not be distinct; we arrange the notation so that $p_1, p_2, \ldots, p_l$ are distinct and $p_{l+1}, \ldots, p_k$ are repetitions of $p_1, p_2, \ldots, p_l$, and then f is expressed

$$f = p_1^{e_1} p_2^{e_2} \ldots p_l^{e_l}$$

where the exponents e_i are positive integers and the monic irreducible polynomials p_i are pairwise coprime for $1 \leqslant i \leqslant l$ (that is, $(p_i, p_j) = 1$ for all i and j with $1 \leqslant i < j \leqslant l$).

The uniqueness statement in (4.19) tells us that, over any field F, the monic divisors of $x^2(x + 1)^2$ are the obvious ones; they can be arranged in the form of a lattice (Fig. 4.1) which forms the same pattern as the lattice of positive divisors of 36.

Fig. 4.1

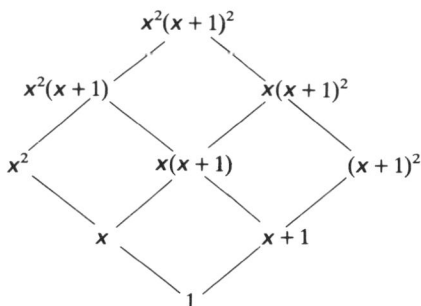

The four quadratic polynomials over $\mathbb{Z}_2$ factorize into irreducible polynomials: x^2, $x(x+1)$, $(x+1)^2$, x^2+x+1. Similarly, the eight cubic polynomials over $\mathbb{Z}_2$ factorize as follows:

$$x^3, (x+1)^3, x^2(x+1), x(x+1)^2, x(x^2+x+1),$$
$$(x+1)(x^2+x+1), x^3+x+1, x^3+x^2+1.$$

The reader may check (by multiplying out) that the above polynomials are all different; for if two were equal we would have a cubic polynomial which factorized in two different ways, contrary to (4.19).

In fact (4.19) can be used to determine the number of monic irreducible quadratics over a finite field F: writing $q = |F|$, there are q quadratics of the form $(x+a_0)^2$, for there are q choices for a_0 in F; there are $q(q-1)/2$ quadratics of the form $(x+a_0)(x+b_0)$ where $a_0 \neq b_0$ ($a_0, b_0 \in F$). This accounts for all the monic *reducible* quadratics over F. By (4.2) there are in all q^2 monic quadratics over F, namely $x^2+a_1x+a_0$, there being q choices for a_0 and q choices for a_1 ($a_0, a_1 \in F$). As $q^2 - q - q(q-1)/2 = q(q-1)/2$, we see that:

There are $q(q-1)/2$ monic irreducible quadratic polynomials over the finite field F with $|F| = q$.

Taking $q = 2$ we see again that there is just one monic irreducible quadratic over $\mathbb{Z}_2$. Taking $q = 3$, the above formula tells us that there are exactly three such polynomials over $\mathbb{Z}_3$; when we have discussed **zeros** of **polynomials** (coming shortly), it will be clear that the three monic irreducible quadratics over $\mathbb{Z}_3$ are

$$x^2+1, \ x^2-x-1, \ x^2+x-1,$$

for none of these expressions become zero on replacing x by any of the elements -1, 0, or 1 of $\mathbb{Z}_3$.

Generally, there is no simple method of determining whether or not a given polynomial is irreducible; however algorithms for factorizing polynomials over $\mathbb{Z}_p$ have been found in recent years (it turns out that row-reduction of matrices (Chapter 8) is useful in this connection), and so polynomial factorization is, in certain cases at least, apparently easier than integer factorization.

We now discuss the polynomial analogue of the rings $\mathbb{Z}_n$; this amounts to selecting a certain polynomial p over a commutative ring R (we are interested mainly in the case $R = F$ (a field) and p irreducible over F) and deriving a new system from the ring $R[x]$ by working **modulo** p, that is, by discarding all polynomials having p as divisor.

Definition 4.20

Let p be a polynomial over the commutative ring R. The polynomials f and g over R are called **congruent modulo p** and we write

$$f \equiv g \pmod{p}$$

if p is a divisor of their difference $f - g$.

For instance $x^5 \equiv x \pmod{x^2 + 1}$ as $x^2 + 1$ is a divisor of $x^5 - x = x(x^2 - 1)(x^2 + 1)$. As (4.20) is the polynomial version of (3.16), corresponding to (3.17) we see that congruence modulo p is an equivalence relation on $R[x]$; the equivalence class $\bar{f}$ of the polynomial f is called the **residue class** $(\mathrm{mod}\, p)$ of f. If p is monic and R non-trivial, then on dividing f by p by (4.8) we obtain polynomials q and r with $f = qp + r$ where $\deg r < \deg p$; so $f \equiv r \pmod{p}$ and hence $\bar{f} = \bar{r}$, showing that every residue class modulo p can be expressed uniquely in the form

$\bar{r}$ where $\deg r < \deg p$.

For instance, with $p = x^2 + x + 1$ over $\mathbb{Z}_2$, the polynomials in $\mathbb{Z}_2[x]$ are partitioned into four residue classes $(\mathrm{mod}\, p)$:

$$\bar{0} = \{0, p, xp, (x + 1)p, x^2 p, \ldots\}$$
$$\bar{1} = \{1, p + 1, xp + 1, (x + 1)p + 1, x^2 p + 1, \ldots\}$$
$$\bar{x} = \{x, p + x, xp + x, (x + 1)p + x, x^2 p + x, \ldots\}$$
$$\overline{x + 1} = \{x + 1, p + x + 1, xp + x + 1,$$
$$(x + 1)p + x + 1, x^2 p + x + 1, \ldots\}$$

according as the remainder on division by p is 0, 1, x, or $x + 1$.

Notation

Let $p \in R[x]$, where R is a commutative ring. The set of residue classes $(\mathrm{mod}\, p)$ of polynomials over R is denoted by $R[x]/(p)$.

Therefore $\mathbb{Z}_2[x]/(x^2 + x + 1) = \{\bar{0}, \bar{1}, \bar{x}, \overline{x + 1}\}$, where the four elements of this set are the four residue classes as above. If our aim was merely to construct a set having four elements, we could scarcely have taken longer or chosen a more complicated notation for it! In fact we are on the point of constructing a *field* with four elements, for (returning to the general case) congruence modulo p is compatible with the operations of addition and multiplication on $R[x]$, as may be shown by mimicking the proof of (3.17). It is therefore legitimate to define addition and multiplication of residue classes $(\mathrm{mod}\, p)$ of polynomials by the rules:

$$\bar{f} + \bar{g} = \overline{f + g}, \quad (\bar{f})(\bar{g}) = \overline{fg} \qquad \text{for } \bar{f}, \bar{g} \in R[x]/(p).$$

We may therefore refer to the system $R[x]/(p)$, meaning the set of residue classes $(\bmod\, p)$ of polynomials with addition and multiplication as defined above; in this system sums and products are formed in the usual way, though the results may be adjusted by any multiple of p.

For instance in the system $\mathbb{Z}_2[x]/(x^2 + x + 1)$ we discard integer multiples of 2 and polynomial multiples of $x^2 + x + 1$; in particular

$$\overline{x} + \overline{x+1} = \overline{2x+1} = \overline{1}$$
$$(\overline{x})(\overline{x}) = \overline{x^2} = \overline{x^2 + 0} = \overline{x^2 + x^2 + x + 1} = \overline{2x^2 + x + 1} = \overline{x+1}$$

showing how addition and multiplication are carried out. As in the case of complex numbers, we simplify the notation by writing $aj + b$ in place of $\overline{ax + b}$ $(a, b \in \mathbb{Z}_2)$; then $j = \overline{x}$ satisfies $j^2 = j + 1$ (this equation plays a similar role to $i^2 = -1$ in $\mathbb{C}$) and the addition and multiplication tables are as shown in Table 4.1.

Table 4.1

+	0	1	j	$j+1$		$\times$	0	1	j	$j+1$
0	0	1	j	$j+1$		0	0	0	0	0
1	1	0	$j+1$	j		1	0	1	j	$j+1$
j	j	$j+1$	0	1		j	0	j	$j+1$	1
$j+1$	$j+1$	j	1	0		$j+1$	0	$j+1$	1	j

As $p = x^2 + x + 1$ is irreducible over $\mathbb{Z}_2$, it follows from our next theorem (the polynomial version of (3.18)) that this system is indeed a field; in fact, apart from notational variations, it is the only field with exactly four elements and is denoted by $\mathbb{F}_4$. Notice that $\mathbb{F}_4$ contains the field $\mathbb{Z}_2$ (just as $\mathbb{C}$ contains $\mathbb{R}$), as the rows and columns headed by 0 and 1 in the above tables (the NW-quarters) form the tables of $\mathbb{Z}_2$; we say that $\mathbb{Z}_2$ is a **subfield** of $\mathbb{F}_4$, or equivalently, that $\mathbb{F}_4$ is an **extension** field of $\mathbb{Z}_2$.

Theorem 4.21

Let p be a monic polynomial over a commutative ring R. Then the system $R[x]/(p)$ is a commutative ring; further, this system is a field if and only if R is a field and p is irreducible over R.

Proof

We mimic the proof of (3.18), giving only an outline. The operations of addition and multiplication on $R[x]/(p)$ are directly derived from its progenitor $R[x]$, which is a commutative ring; therefore $R[x]/(p)$ is a commutative ring, the 0-element being the set $\overline{0}$ of all polynomials in $R[x]$ having p as divisor, and the 1-element being the set $\overline{1}$ of all polynomials in $R[x]$ which leave remainder 1 on division by p.

Notice that $R[x]/(p)$ is trivial if and only if $p = 1$; we therefore assume $\deg p > 0$. Suppose $R[x]/(p)$ is a field. To show that R is a field, let $c \in R^*$; then $\bar{c} \neq \bar{0}$ and so $\bar{c}$ has inverse $\bar{f}$ in $R[x]/(p)$, where $\deg f < \deg p$. Therefore $\overline{fc} = \bar{1}$, that is, $fc - 1 = qp$ for some $q \in R[x]$; comparing degrees gives $q = 0$ and hence $\deg f = 0$. So $f = c'$ is a constant polynomial over R and, as $c'c = 1$, we see that c has inverse c' in R. Therefore R is a field.

Suppose that p is irreducible over the field F and let $\bar{f}$ be a non-zero element of $F[x]/(p)$. As the g.c.d. $(p, f) = 1$, by (4.13) there is a polynomial t in $F[x]$ with $\overline{tf} = \bar{1}$, that is, $\bar{f}$ has inverse $\bar{t}$. Therefore $F[x]/(p)$ is a field. Finally, if p is reducible, then $F[x]/(p)$ has zero divisors and so is not a field. $\qquad\square$

Let us assume now that the monic polynomial p of (4.21) has *positive* degree over R; then the residue classes $\bar{c}$ of constant polynomials c (for all $c \in R$) form a subsystem of $R[x]/(p)$, for the sum and product of residue classes of this type are again of this type; what is more, this subsystem is, for all practical purposes, the same as R (in fact we may identify c with $\bar{c}$, for working modulo p does not affect calculations involving only constants). Therefore whenever we are presented with in irreducible polynomial p over a field F, we can use p as in (4.21) to construct a larger field, namely $F[x]/(p)$, which extends the original field F; the extension $\mathbb{F}_4$ of $\mathbb{Z}_2$ is formed in this way, and we now present further examples of this important construction.

Example 4.22(a) As $p = x^2 + 1$ is irreducible over the real field $\mathbb{R}$, the system $E = \mathbb{R}[x]/(x^2 + 1)$ is an extension field of $\mathbb{R}$. We shall see in a moment that E is a very familiar field! As p is quadratic, each element $\bar{f}$ of E can be expressed uniquely in the form $\bar{f} = \overline{c + dx}$; in fact if $f = \sum a_i x^i$, then $c = a_0 - a_2 + a_4 - a_6 + \ldots$ and $d = a_1 - a_3 + a_5 - a_7 + \ldots$, for x^2 may be replaced by -1 when working modulo $x^2 + 1$. For instance

$$6 + 4x + x^2 + 7x^3 \equiv 5 - 3x \pmod{x^2 + 1}.$$

The elements of E are, effectively, the remainders on division by $x^2 + 1$. As the sum of two remainders is itself a remainder, the rule of addition in E is

$$\overline{(c + dx)} + \overline{(c_1 + d_1 x)} = \overline{(c + c_1) + (d + d_1)x}.$$

The product of two remainders is not, in general, a remainder and so some adjustment is necessary; as

$$cc_1 + (cd_1 + dc_1)x + dd_1x^2$$
$$\equiv (cc_1 - dd_1) + (cd_1 + dc_1)x \pmod{x^2 + 1}$$

we obtain the product rule for elements of E:

$$\overline{(c + dx)(c_1 + d_1x)} = \overline{(cc_1 - dd_1) + (cd_1 + dc_1)x}.$$

The reader should now be in no doubt, for writing $c + di$ in place of $\overline{c + dx}$ we see that $i = \overline{x}$ satisfies $i^2 = -1$ and the above rules become the familiar sum and product rules for complex numbers; therefore, apart from a minor difference in notation, the extension field E is precisely the complex field $\mathbb{C}$.

Example 4.22(b)

Here we use the irreducible cubic $p = x^3 + x + 1$ over $\mathbb{Z}_2$ to construct the extension field $E = \mathbb{Z}_2[x]/(x^3 + x + 1)$ of $\mathbb{Z}_2$. Each residue class $(\bmod\, p)$ of polynomials over $\mathbb{Z}_2$ can be uniquely expressed in the form $\overline{ax^2 + bx + c}$ where a, b, $c \in \mathbb{Z}_2$; so E has exactly 8 elements corresponding to the 8 remainders on division by p. Addition of elements in E is carried out modulo 2; multiplication in E uses the congruence $x^3 \equiv x + 1 \pmod{p}$ to eliminate x^3 and higher powers of x; for instance $(x^2 + x + 1)(x^2 + 1) = x^4 + x^3 + x + 1$ in $\mathbb{Z}_2[x]$ and $x^4 + x^3 + x + 1 \equiv x(x + 1) + x + 1 + x + 1 \pmod{p}$ combine to show $\overline{(x^2 + x + 1)(x^2 + 1)} = \overline{x^2 + x}$ in E. As before, we simplify the notation, replacing $\overline{ax^2 + bx + c}$ by $aj^2 + bj + c$, and so $j = \overline{x}$ satisfies $j^3 = j + 1$. The most illuminating way to write out the multiplication table of E (Table 4.2) is to notice that every non-zero element is a power of j; because of this property j is called a **primitive** element of E. The elements of E^* are $j, j^2, j^3 = j + 1, j^4 = j^2 + j, j^5 = j^2 + j + 1, j^6 = j^2 + 1, j^7 = 1$.

Table 4.2

$\times$	0	1	j	j^2	j^3	j^4	j^5	j^6
0	0	0	0	0	0	0	0	0
1	0	1	j	j^2	j^3	j^4	j^5	j^6
j	0	j	j^2	j^3	j^4	j^5	j^6	1
j^2	0	j^2	j^3	j^4	j^5	j^6	1	j
j^3	0	j^3	j^4	j^5	j^6	1	j	j^2
j^4	0	j^4	j^5	j^6	1	j	j^2	j^3
j^5	0	j^5	j^6	1	j	j^2	j^3	j^4
j^6	0	j^6	1	j	j^2	j^3	j^4	j^5

In this case E is denoted by $\mathbb{F}_8$, for it is, in effect, the only field having exactly eight elements.

The properties of finite fields are beautiful and intricate; although a thorough treatment is beyond the scope of this book, the reader should be aware of the following facts: let p be a positive prime and n a natural number. Then there is at least one monic irreducible polynomial p' of degree n over $\mathbb{Z}_p$; the method of (4.21) can now be used to construct $\mathbb{Z}_p[x]/(p')$, which is a field having exactly p^n elements and is denoted by $\mathbb{F}_{p^n}$ (or by $GF(p^n)$) after the French mathematician Galois). It can be shown that $\mathbb{F}_{p^n}$ has a primitive element, that $\mathbb{F}_{p^n}$ is (apart from variations of notation) the *only* field having exactly p^n elements, and that *every* finite field is of this type. One work of caution: although p' (as above) exists, it may be difficult to find such a polynomial, except in certain special cases; similarly, although it is nice to know that every finite field has a primitive element (for this tells us that its multiplicative structure is particularly simple), one may be reduced to 'trial and error' in order to locate such an element.

Zeros of polynomials

The theory of linear congruences and the Chinese remainder theorem have polynomial analogues, but we shall not deal with them here. Rather, we discuss an important aspect of polynomials for which there is no integer equivalent.

Definition 4.23 Let $f = a_0 + a_1x + a_1x + a_2x^2 + \ldots + a_nx^n$ be a polynomial over the commutative ring R and let $c \in R$. The element $f(c) = a_0 + a_1c + a_2c^2 + \ldots + a_nc^n$ of R is called the **evaluation** of f at c. If $f(c) = 0$, then c is called a **zero** of f.

If $f = x^3 + 9x^2 + 9x + 1$, then $f(10) = 1991$; as $f(-1) = 0$ we see that -1 is a zero of f.

Corollary 4.24 (The remainder theorem.) Let f be a polynomial over the non-trivial commutative ring R and let $c \in R$. Then division of f by $x - c$ leaves remainder $f(c)$.

Proof By (4.8) there are polynomials q and r over R such that $f = (x - c)q + r$ where $\deg r < 1$; so r is a constant polynomial and we write $r = a$ where $a \in R$. Because of the way in which polynomial addition and multiplication is defined (4.4), the polynomials $x - c$, q, and r can be evaluated separately at c, and these evaluations combined to produce the evaluation of f at c; in short, $f(c) = (c - c)q(c) + a$. Therefore $r = a = f(c)$. □

Let f and g be polynomials over the commutative ring R and let $c \in R$. By adding and multiplying out the elements $f(c)$ and $g(c)$ of R and collecting up powers of c we obtain

$$(f + g)(c) = f(c) + g(c), \quad (fg)(c) = f(c)g(c),$$

which combine to show that polynomials over R may be evaluated at c in a *piecemeal* way (as in the proof of (4.24)); in fact the operation of evaluation at c, when carried out on all polynomials over R, is a ring homomorphism (5.1) of $R[x]$ to R.

Corollary 4.25   (The factor theorem.) Let f be a polynomial over the non-trivial commutative ring R and let $c \in R$. Then c is a zero of f if and only if $x - c$ is a factor of f.

Proof   Let $f(c) = 0$; then $f = (x - c)q$ for some q in $R[x]$ by (4.24), that is, $x - c$ is a factor of f. Conversely, using the factorization $f = (x - c)q$ to evaluate f at c, gives $f(c) = (c - c)q(c) = 0$. □

As -1 is a zero of $f = x^3 + 9x^2 + 9x + 1$, by (4.25), $x + 1$ is a factor of f; in fact $f = (x + 1)(x^2 + 8x + 1)$.

We now restrict our attention to polynomials over a field. Let f be an irreducible polynomial over the field F; as f has no factorization into polynomials of lower degree over F, by (4.25) we see that f has no zeros in F if $\deg f > 1$.

There is a *partial* converse to the preceding paragraph: let f be a quadratic or cubic polynomial over F; if f has no zeros in F, then f is irreducible over F. For if f were reducible it would have two linear factors (if $\deg f = 2$), or a linear and a quadratic factor or three linear factors (if $\deg f = 3$); in any case f would have a factor $x - c$ and hence $f(c) = 0$.

For instance, consider $f = x^3 + x + 1$ over $\mathbb{Z}_7$; this field has seven elements only, which may be taken as 0, ± 1, ± 2, ± 3. Now $f(3) = 31$ and $31 \neq 0 \pmod 7$, showing that 3 is not a zero of f; in fact, none of these elements are zeros of f and so we conclude that the cubic $x^3 + x + 1$ is *irreducible* over $\mathbb{Z}_7$.

The same idea can be used to test irreducibility over the rational field. For example consider $f = 3x^3 + 2x^2 + x + 5$ over $\mathbb{Q}$; we show that f has no zeros in $\mathbb{Q}$. So suppose $f(m/n) = 0$ where m and n are coprime integers and n is positive; rearranging this equation to $3m^3 + 2m^2n + mn^2 = -5n^3$ shows that m is a divisor of $-5n^3$, since *m* is clearly a divisor of the left-hand side. By (3.14)(a), we see $m \mid 5$ and so $m = \pm 1$, ± 5; rearranging the above equation so that all the terms involving *n* appear on one side, gives $n \mid 3m^3$ and so $n \mid 3$ as the g.c.d. $(m, n) = 1$. So $n = 1$ or $n = 3$, showing that ± 1, ± 5, $\pm 1/3$, $\pm 5/3$ are the only possible rational zeros of *f*; we next test each of these rationals in turn—in fact none of them are zeros of *f* and so *f* has no rational zeros at all. As *f* is cubic we conclude that $3x^3 + 2x^2 + x + 5$ is *irreducible* over $\mathbb{Q}$.

As we pointed out earlier, it is necessary to specify the field when discussing irreducibility. The polynomial $g = 3x^3 + 2x^2 + x + 5$ over $\mathbb{R}$ is, strictly speaking, different to the polynomial *f* of the previous paragraph; in fact *g* is *reducible* over $\mathbb{R}$, because $g(-2) = -13$ and $g(-1) = 3$ and so the graph of $y = g(x)$ crosses the *x*-axis at a real zero *c* of *g* where $-2 < c < -1$. By (4.25) *g* factorizes $g = (x - c)q$ where *q* is a quadratic polynomial over $\mathbb{R}$.

Notice that the polynomial $h = x^4 + 3x^2 + 2$ over $\mathbb{R}$ has no real zeros, as $c^4 + 3c^2 + 2 \geqslant 2$ for all $c \in \mathbb{R}$; however $h = (x^2 + 1)(x^2 + 2)$ and so *h* is reducible over $\mathbb{R}$. Generally, if $\deg h \geqslant 4$ and *h* has no zeros, then no conclusion can be drawn about the irreducibility of *h*.

We come now to the main theorem of this section.

Theorem 4.26

(The fundamental theorem of algebra.) Every polynomial of positive degree over the complex field $\mathbb{C}$ has a zero in $\mathbb{C}$.

Proof

Let *f* be a polynomial of positive degree *n* over $\mathbb{C}$; we may assume that *f* is monic, for making *f* monic will not change its zeros, and so $f = x^n + a_{n-1}x^{n-1} + \ldots + a_1 x + a_0$ where $a_{n-1}, \ldots, a_1, a_0 \in \mathbb{C}$. We assume $a_0 \neq 0$, for otherwise 0 is a zero of *f*. The idea of the proof is as follows: imagine the complex number *z* to be moving on the circle $|z| = r$ in the Argand diagram, where *r* is a given positive real number. What can be said of $f(z)$ as *z* makes one complete anticlockwise revolution? As *z* describes a closed curve (*z* finishes where it begins), $f(z)$ also describes a closed curve which we denote by C_r. If *r* is sufficiently small, C_r is contained in the disk with centre a_0 and radius $|a_0|/2$ and so $f(z)$, in describing C_r, does not go round the origin at all (Fig. 4.2(a)). If *r* is sufficiently large, we show that $f(z)$ goes *n* times anticlockwise around the origin (Fig. 4.2(b)). We appeal to the reader's geometric intuition for the final step (a rigorous treatment of the analytic concepts involved can be given: see, for instance, chapter 6 of C. T. C. Wall, *A Geometric Introduction to Topology*, Addison–Wesley, 1972). As *r* increases (from the smaller to the larger value), the curve in Fig. 4.2(a) undergoes a

Fig. 4.2

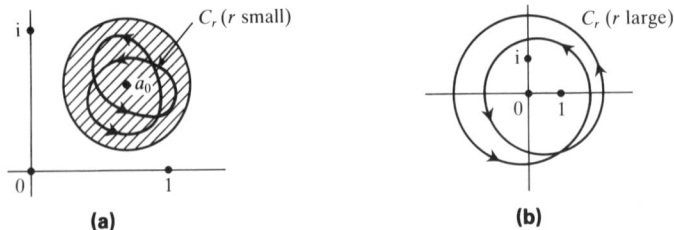

(a)   (b)

continuous deformation (arising from the continuous function $f(z)$) until it becomes the curve of Fig. 4.2(b); at some intermediate stage in the deformation, with $r = r'$ say, $C_{r'}$ must pass through the origin, showing that f has a zero of modulus r'.

That's the idea of the proof! Here are the non-analytic details, beginning with the case 'r small'. Suppose $|z| = r \leqslant 1$; then $|z^j| \leqslant |z|$ for $j = 1, 2, \ldots, n$. Hence

$$|f(z) - a_0| = |z^n + a_{n-1}z^{n-1} + \ldots + a_1 z|$$
$$\leqslant |z|^n + |a_{n-1}z^{n-1}| + \ldots + |a_1 z|$$
$$\leqslant (1 + |a_{n-1}| + \ldots + |a_1|)\,|z|$$

using standard properties of the modulus. So let r denote the minimum of 1 and $(|a_0|/2)(1 + |a_{n-1}| + \ldots + |a_1|)^{-1}$. Then $|f(z) - a_0| \leqslant |a_0|/2$ and therefore $f(z)$ belongs to the disk (shaded in Fig. 4.2(a)) with centre a_0 and radius $|a_0|/2$ if $|z| = r$; so C_r does not go round the origin in this case.

For large r, we compare $f(z)$ with z^n as follows; for $z \neq 0$,

$$f(z) = z^n + a_{n-1}z^{n-1} + \ldots + a_1 z + a_0 = z^n(1 + w)$$

where

$$w = (a_{n-1}/z) + \ldots + (a_1/z^{n-1}) + (a_0/z^n).$$

We arrange for w to be small ($|w| \leqslant \frac{1}{2}$ will be good enough) by taking $|z| = r$ sufficiently large: in fact $|1/z^{n-j}| \leqslant |1/z|$ for $|z| \geqslant 1$ and $j = 0, 1, \ldots, n - 1$. As before

$$|w| \leqslant |a_{n-1}/z| + \ldots + |a_1/z^{n-1}| + |a_0/z^n|$$
$$\leqslant (|a_{n-1}| + \ldots + |a_1| + |a_0|)/|z|$$

and so, taking r to be the maximum of 1 and $2(|a_{n-1}| + \ldots + |a_1| + |a_0|)$ gives $|w| \leqslant \frac{1}{2}$ for $|z| = r$. As $|1 + w - 1| = |w| \leqslant \frac{1}{2}$, we see that $1 + w$ belongs to the disk (shaded in Fig. 4.3(a)) with centre 1 and radius $\frac{1}{2}$; from Fig. 4.3(a) we see that $1 + w$ has argument ϕ where $-\pi/6 \leqslant \phi \leqslant \pi/6$, and $\frac{1}{2} \leqslant |1 + w| \leqslant \frac{3}{2}$. Writing $z = r(\cos \theta + \mathrm{i} \sin \theta)$,

Fig. 4.3

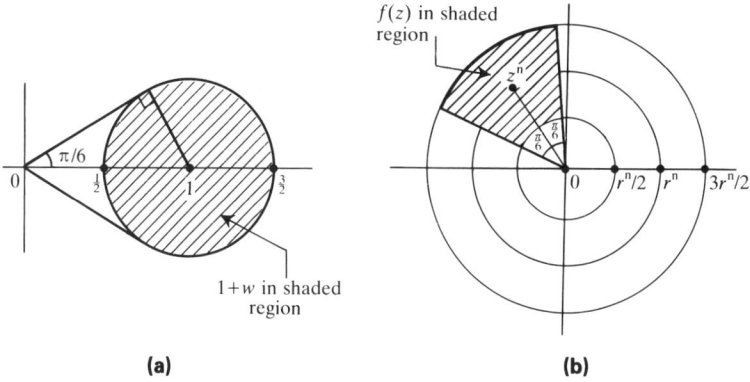

(a)

(b)

we have $z^n = r^n(\cos n\theta + i \sin n\theta)$, and using the formulae for the modulus and argument of the product $z^n(1 + w) = f(z)$ gives

$$|f(z)| = |z^n| \, |1 + w| \quad \text{and so} \quad r^n/2 \leqslant |f(z)| \leqslant 3r^n/2,$$

$\arg(f(z)) = \arg(z^n) + \arg(1 + w)$ and so $f(z)$ has argument ψ where $n\theta - \pi/6 \leqslant \psi \leqslant n\theta + \pi/6$. Therefore $f(z)$ belongs to the sector of the annulus shaded in Fig. 4.3(b). Finally, let θ increase from 0 to 2π; then z makes one anticlockwise lap around the circle $|z| = r$ while z^n describes n complete circuits of the circle with centre 0 and radius r^n; but as $f(z)$ remains 'caught' in the shaded region surrounding z^n throughout the motion, we see that $f(z)$, like z^n, goes n times anticlockwise around 0, that is C_r goes n times around 0 (as in Fig. 4.2(b) where $n = 2$).

Continuity properties now tell us, as mentioned earlier in the proof, that for some r', between the values of r discussed above, $C_{r'}$ passes through 0, showing that f has a zero in $\mathbb{C}$. □

A number of deductions can be made from (4.26). Let p be a monic irreducible polynomial over $\mathbb{C}$; by (4.26) p has a zero c in $\mathbb{C}$, and so p has factor $x - c$ by (4.25). As 1, p are the only monic factors of p, we conclude that $p = x - c$, showing that all irreducible polynomials over $\mathbb{C}$ are linear. We now determine the irreducible polynomials over the real field $\mathbb{R}$.

Proposition 4.27 The monic irreducible polynomials over $\mathbb{R}$ are $x - c$ and $x^2 + ax + b$ where $a^2 < 4b$.

Proof Note first that all polynomials of the above form are irreducible over $\mathbb{R}$: for writing $p = x^2 + ax + b$ we see $p = (x + a/2)^2 + b - a^2/4$

and so

$$p(x) \geq b - a^2/4 > 0 \quad \text{for all } x \in \mathbb{R}$$

showing that p has no zeros in $\mathbb{R}$; being quadratic, p is therefore irreducible over $\mathbb{R}$.

Conversely, let p be a monic irreducible polynomial over $\mathbb{R}$. Regarding p as a polynomial over $\mathbb{C}$, by (4.26) p has a zero c in $\mathbb{C}$; if c is real, then p has the polynomial $x - c$ over $\mathbb{R}$ as factor, and so $p = x - c$ as the only monic factors in $R[x]$ of p are 1 and p. If c is not real, then taking the complex conjugate of the equation $p(c) = 0$ produces $p(c^*) = 0$ (here we use (2.23) and the fact that p has real coefficients); so c and c^* are zeros of p and hence the coprime polynomials $x - c$ and $x - c^*$ are factors of p. Therefore the polynomial with real coefficients $(x - c)(x - c^*) = x^2 - (c + c^*)x + cc^*$ is a factor of p; as the only monic polynomials over $\mathbb{R}$ which are factors of p are 1 and p, we conclude that $p = x^2 - (c + c^*)x + cc^*$. So $p = x^2 + ax + b$ where $a = -(c + c^*)$ and $b = cc^*$; hence $4b - a^2 = 4cc^* - (c + c^*)^2 = -(c - c^*)^2 > 0$ as $c - c^*$ is imaginary. So $a^2 < 4b$. $\qquad\square$

Combining (4.19) and (4.26), we see that every monic polynomial of positive degree over $\mathbb{C}$ factorizes into monic linear polynomials; from (4.19) we deduce that every monic polynomial of positive degree over $\mathbb{R}$ factorizes into linear and quadratic polynomials over $\mathbb{R}$ as described in (4.27). The reader should realize that, in spite of these apparently reassuring results, it is usually difficult to resolve a given polynomial over $\mathbb{R}$ into its irreducible factors. However, there are exceptions, for instance, $f = x^7 + x^6 + x^5 + x^4 + x^3 + x^2 + x + 1$; the trick here is to notice that $(x - 1)f = x^8 - 1$, and so the zeros of f are $w, w^2, \ldots, w^7$ where $w = (1 + i)/\sqrt{2}$, that is, the eighth roots of 1 apart from 1 itself; now $w^4 = -1$ and the remaining zeros of f occur in conjugate pairs, namely $w, w^7; w^2, w^6; w^3, w^5$. So

$$f = (x - w)(x - w^7)(x - w^2)(x - w^6)(x - w^3)(x - w^5)(x - w^4)$$

is the factorization of f into monic linear polynomials over $\mathbb{C}$. Now $(x - w)(x - w^7) = (x - (1 + i)/\sqrt{2})(x - (1 - i)/\sqrt{2}) = x^2 - \sqrt{2}x + 1$ and multiplying all pairs of conjugate factors produces

$$f = (x^2 - \sqrt{2}x + 1)(x^2 + 1)(x^2 + \sqrt{2}x + 1)(x + 1)$$

which is the factorization of f into monic irreducible polynomials over $\mathbb{R}$. As $(x^2 - \sqrt{2}x + 1)(x^2 + \sqrt{2}x + 1) = x^4 + 1$ is the only factorization of $x^4 + 1$ into monic irreducible polynomials over $\mathbb{R}$, we see that $x^4 + 1$ is irreducible over $\mathbb{Q}$ and so

$$f = (x^4 + 1)(x^2 + 1)(x + 1)$$

is the factorization of f into monic irreducible polynomials over $\mathbb{Q}$. We next discuss repeated zeros of polynomials.

Definition 4.28

Let f be a non-zero polynomial over the non-trivial commutative ring R and let f have zero c in R. The largest positive integer m, such that $(x - c)^m$ is a divisor of f, is called the **multiplicity** of the zero c of f.

For instance, $x^3 - 10x^2 + 33x - 36 = (x - 3)^2(x - 4)$ has zero 3 (of multiplicity 2) and zero 4 (of multiplicity 1). By (4.25), every zero of f has multiplicity at least one.

Proposition 4.29

Let f be a polynomial of non-negative degree n over the field F. Then f has at most n zeros in F, each zero being counted according to its multiplicity.

Proof

If $n = 0$, then f is a non-zero constant polynomial and so has no zeros in F. We may suppose that f is monic and of positive degree, for making f monic does not change its zeros or its degree. By (4.19), f can be resolved into monic irreducible polynomials over F; let q denote the product of all the irreducible polynomials p with $\deg p > 1$ in the factorization of f (if there are no such irreducible polynomials, let $q = 1$). On collecting together identical linear factors of f, we obtain

$$f = (x - c_1)^{m_1}(x - c_2)^{m_2} \ldots (x - c_s)^{m_s} q$$

where $c_1, c_2, \ldots, c_s$ are distinct elements of F and $m_1, m_2, \ldots, m_s$ are natural numbers ($s \geq 0$). As q has no zeros in F, we see $f(c) \neq 0$ for all c in F which are distinct from $c_1, c_2, \ldots, c_s$; so f has zeros $c_1, c_2, \ldots, c_s$ of multiplicities $m_1, m_2, \ldots, m_s$ and no further zeros in F. As $\deg q \geq 0$, on comparing degrees in the factorization of f, we obtain $n \geq m_1 + m_2 + \ldots + m_s$. □

Notice that (4.29) becomes false if F is replaced by a commutative ring; for instance, the quadratic polynomial $f = x^2 - x$ over $\mathbb{Z}_6$ has zeros 0, 1, 3, 4, and factorizes in two ways: $f = x(x - 1) = (x - 3)(x - 4)$.

We may restate (4.29) as follows: if f is a polynomial of degree at most n over the field F with $n + 1$ zeros in F, then f is the zero polynomial. We end this chapter with a related result. To get the idea and become familiar with the notation, suppose we construct the cubic polynomial g over $\mathbb{Q}$ such that 1, 2, 3 are zeros of g and $g(4) = 1$; in fact $g = a(x-1)(x-2)(x-3)$ where $1 = a(4-1)(4-2)(4-3)$, that is,

$$g = \left(\frac{x-1}{4-1}\right)\left(\frac{x-2}{4-2}\right)\left(\frac{x-3}{4-3}\right).$$

So g is the product of linear polynomials $(x-j)/(4-j)$ for $j = 1, 2, 3$; we introduce the **product symbol** $\prod$ (the multiplicative analogue of the summation sign Σ) and write

$$g = \prod_{j=1}^{3} \frac{x-j}{4-j}.$$

In our next proposition we build up a polynomial f with specified properties using polynomials g of the above form; think of f as being a cake we want to make, g_i as the ingredients, and the following formula as the recipe!

Proposition 4.30

(Lagrange's interpolation formula.) Let $c_0, c_1, \ldots, c_n$ be distinct elements of the field F and let $d_0, d_1, \ldots, d_n \in F$. Then the polynomial

$$f = d_0 g_0 + d_1 g_1 + \ldots + d_n g_n \quad \text{where} \quad g_i = \prod_{j \neq i} \frac{x - c_j}{c_i - c_j}$$

is the unique polynomial of degree at most n over F satisfying $f(c_i) = d_i$ for $i = 0, 1, \ldots, n$.

Proof

If there were two such polynomials, f_1 and f_2 say, then $f_1 - f_2$ would have degree at most n and zeros $c_0, c_1, \ldots, c_n$. By (4.29), this means $f_1 - f_2 = 0$, that is, $f_1 = f_2$.

Let g_i be the product of the n linear polynomials $(x - c_j)/(c_i - c_j)$ for $j = 0, 1, \ldots, n$ and $j \neq i$, that is $g_i = \prod_{j \neq i} (x - c_j)/(c_i - c_j)$ for $i = 0, 1, \ldots, n$. Then $\deg g_i = n$, $g_i(c_i) = 1$, and $g_i(c_j) = 0$ for $j \neq i$. The polynomial f we are looking for can be built up from $g_0, g_1, \ldots, g_n$; in fact $f = d_0 g_0 + d_1 g_1 + \ldots + d_n g_n$, as $\deg f \leq n$ and $f(c_i) = d_i g_i(c_i) = d_i$. $\square$

To find the quadratic polynomial f over $\mathbb{Q}$ with $f(0) = 3$, $f(1) = 6$, and $f(2) = 7$, let $f = a_2 x^2 + a_1 x + a_0$; then $a_0 = 3$, $a_2 + a_1 + a_0 = 6$, and $4a_2 + 2a_1 + a_0 = 7$, giving $f = -x^2 + 4x + 3$. Alternatively, as in the above proof, we can build up f from $g_0 = \frac{1}{2}(x-1)(x-2)$, $g_1 = -x(x-2)$, $g_2 = \frac{1}{2}x(x-1)$; in this case $f = 3g_0 + 6g_1 + 7g_2$.

Exercises
4.2

1. Determine the zeros of the following polynomials in the indicated field F and resolve each of them into monic irreducible factors over F.

    (i) $x^3 - 3x + 2$ over $\mathbb{Q}$,   (ii) $x^4 + x^3 + x + 1$ over $\mathbb{Z}_2$,
    (iii) $x^3 - 8$ over $\mathbb{R}$,   (iv) $x^3 - 8$ over $\mathbb{C}$.

2. Resolve the following polynomials into linear factors over $\mathbb{C}$ and deduce their factorizations into irreducible polynomials (a) over $\mathbb{R}$, (b) over $\mathbb{Q}$.

    (i) $x^6 - 1$ (ii) $x^6 - 9$ (iii) $x^6 - 8$ (iv) $x^6 + 1$.

3. (a) Find the g.c.d. of $x^4 + 2x^3 + 3x^2 + 2x + 2$ and $x^4 - x^3 - 3x^2 - 4x + 2$ over $\mathbb{Q}$, and hence find their factorizations into irreducible polynomials over $\mathbb{Q}$, $\mathbb{R}$, and $\mathbb{C}$.

    (b) Show that one of $x^4 + 4$, $x^4 + 2$ is reducible over $\mathbb{Q}$ and the other is irreducible over $\mathbb{Q}$.

    (c) Show that one of $2x^3 + x^2 + 5x + 3$, $2x^3 - x^2 + 5x + 3$ has a zero in $\mathbb{Q}$ and the other does not, and hence find their irreducible factorizations over $\mathbb{Q}$.

    (d) Let f be a monic polynomial with integer coefficients. If c is a rational zero of f, show that c is an integer.

4. Let $f = \sum a_i x^i$ be a polynomial over a non-trivial commutative ring. Show that $x^2 - 1$ is a divisor of f if and only if $a_0 + a_2 + a_4 + \ldots = 0$ and $a_1 + a_3 + a_5 + \ldots = 0$. Under what (similar) conditions is $x^2 + 1$ a divisor of f?

    Find the least natural number n such that $x^n + x^{n-1} + \ldots + x + 1$ has divisor $x^4 - 1$ over $\mathbb{Z}_5$.

5. (a) Find the unique reducible polynomial of degree 4 over $\mathbb{Z}_2$ which has no zeros in $\mathbb{Z}_2$. Hence find the three irreducible polynomials of degree 4 over $\mathbb{Z}_2$. How many reducible polynomials of degree 5 over $\mathbb{Z}_2$, having no zeros in $\mathbb{Z}_2$, are there?

    (b) Let F be a finite field. Show that there are $(q+1)q(q-1)/3$ monic irreducible cubic polynomials over F, where $q = |F|$.

6. Find the cubic polynomial f over $\mathbb{Q}$ satisfying $f(0) = 6$, $f(1) = 2$, $f(2) = -2$, $f(3) = 12$. Hence, or otherwise, find the cubic polynomial g over $\mathbb{Q}$ such that $g(0) = 12$, $g(1) = -2$, $g(2) = 2$, $g(3) = 6$.

7. (a) Find the eight zeros of $x^2 - x$ in $\mathbb{Z}_{30}$ and hence factorize $x^2 - x$ into linear polynomials over $\mathbb{Z}_{30}$ in four different ways.

    (b) Factorize $x^7 - x$ into linear factors over $\mathbb{Z}_7$. Let f be a polynomial over $\mathbb{Z}_7$. Show that $x^7 - x$ is a divisor of f if and only if $f(c) = 0$ for all $c \in \mathbb{Z}_7$.

8. (a) Verify that $p = x^2 + 1$ is irreducible over $\mathbb{Z}_3$. Write out the addition and multiplication tables of the field $\mathbb{Z}_3[x]/(p)$ (see (4.21)). (This field $\mathbb{F}_9$ has nine elements $a + ib$ where $i^2 = -1$ and $a, b \in \mathbb{Z}_3$.)

(b) Let $p = x^2 + x$ over $\mathbb{Z}_2$. Write out the addition and multiplication tables of the ring $\mathbb{Z}_2[x]/(p)$. (This ring has four elements.)

(c) As $\sqrt{2}$ is irrational, the polynomial $p = x^2 - 2$ is irreducible over $\mathbb{Q}$. State the addition and multiplication rules for the elements $\overline{ax + b}$ and $\overline{a_1 x + b_1}$ of the field $\mathbb{Q}[x]/(p)$; if a_1 and b_1 are not both zero, find a and b with $(\overline{a_1 x + b_1})^{-1} = \overline{ax + b}$ where $a, b, a_1, b_1 \in \mathbb{Q}$. What is the connection between $\mathbb{Q}[x]/(p)$ and the field $\mathbb{Q}(\sqrt{2})$ of real numbers of the form $a + b\sqrt{2}$, where $a, b \in \mathbb{Q}$?

9. Let $f = a_0 + a_1 x + \ldots + a_n x^n$ be a polynomial over a field F. The polynomial $f' = a_1 + 2a_2 x + \ldots + na_n x^n$ is called the (formal) derivative of f.

(a) Prove $(f + g)' = f' + g'$ and $(fg)' = f'g + fg'$ for all $f, g \in F[x]$.

(b) Let p and f be polynomials over F. If p^2 is a divisor of f, show that p is a divisor of the g.c.d. (f, f'). If p is irreducible over F and $p' \neq 0$, show that $p \mid (f, f') \Rightarrow p^2 \mid f$.

(c) Find the g.c.d. (f, f') if $f = 2x^5 + 3x^4 + 4x^3 + x^2 - 1$ over $\mathbb{Q}$ and hence factorize f into irreducible polynomials over $\mathbb{Q}$.

5 Ring theory*

Here we take stock of the rings, integral domains, and fields we have already met together with the ways in which they arise. What common features are there? Can the constructions be generalized? In answering these questions, we shall discover important new aspects of rings (the notion of an **ideal**, for instance); nevertheless 'consolidation', rather than 'development' is our key-note, and the reader who is so inclined may omit this chapter on a first reading.

In abstract algebra, it is essential to have the means of comparing one system with another. In the case of sets (regarded as systems with no structure), arbitrary mappings do the job. But meaningful comparisons of rings are provided only by mappings which respect the ring operations; such mappings (defined below) form the foundation of our discussion.

Definition 5.1 Let R and R' be rings. A mapping $\alpha : R \to R'$ is called a **ring homomorphism** if $(x + y)\alpha = (x)\alpha + (y)\alpha$ and $(xy)\alpha = ((x)\alpha)((y)\alpha)$ for all x and y in R, and $(1)\alpha = 1$.

The equation $(x + y)\alpha = (x)\alpha + (y)\alpha$ for all x and y in R, is expressed by saying that α **respects addition**; notice that $(x + y)\alpha$ is formed using addition on R, whereas $(x)\alpha + (y)\alpha$ involves addition on R'. Similarly the equation $(xy)\alpha = ((x)\alpha)((y)\alpha)$ for all x and y in R, relates multiplication on R with multiplication on R', and says that α **respects multiplication**. The equation $(1)\alpha = 1$ means α maps the 1-element of R to the 1-element of R'.

The natural mapping $\eta : \mathbb{Z} \to \mathbb{Z}_n$, defined by $(x)\eta = \bar{x}$ for all x in $\mathbb{Z}$, a ring homomorphism because

$$\left. \begin{array}{l} (x + y)\eta = \overline{x + y} = \bar{x} + \bar{y} = (x)\eta + (y)\eta \\ (xy)\eta = \overline{xy} = (\bar{x})(\bar{y}) = ((x)\eta)((y)\eta) \end{array} \right\} \text{ for all } x, y \in \mathbb{Z}$$

and $(1)\eta = \bar{1}$ is the 1-element of $\mathbb{Z}_n$ (see (1.31), (3.17), (3.18)). We shall see that η tells us concisely all about the relationship of $\mathbb{Z}$ to $\mathbb{Z}_n$; as η is surjective, $\mathbb{Z}_n$ owes all it has to its progenitor $\mathbb{Z}$.

The polynomial analogue of the above paragraph is covered by (4.20), (4.21): let p be a monic polynomial over the non-trivial commutative ring R; then the natural mapping $\eta : R[x] \to R[x]/(p)$

is a ring homomorphism. In other words, η, which maps every polynomial f over R to its residue class $\bar{f}$ modulo p, satisfies (5.1); as above, this is a direct consequence of the way addition and multiplication of residue classes of polynomials is introduced.

Let c be an element of the commutative ring R, and let $\sigma_c : R[x] \to R$ be defined by $(f)\sigma_c = f(c)$ for all $f \in R[x]$; so σ_c evaluates every polynomial f at c (see (4.23)). Because of the way polynomial addition and multiplication is defined, σ_c is a ring homomorphism.

Lemma 5.2   Let $\alpha : R \to R'$ be a ring homomorphism. Then $(0)\alpha = 0$ and $(-x)\alpha = -(x)\alpha$ for all $x \in R$. If α is bijective, then $\alpha^{-1} : R' \to R$ is a ring homomorphism.

Proof   As α respects addition, we obtain the following equation relating elements of R':

$$1 + (0)\alpha = (1)\alpha + (0)\alpha = (1 + 0)\alpha = (1)\alpha = 1.$$

Adding -1 to the above equation gives $(0)\alpha = 0$, that is, α maps the 0-element of R to the 0-element of R'. Applying α to $-x + x = 0$ now produces $(-x)\alpha + (x)\alpha = (0)\alpha = 0$; therefore $(-x)\alpha$ is the negative of $(x)\alpha$, that is, $(-x)\alpha = -(x)\alpha$.

Suppose that α is bijective. Let x' and y' be elements of R' and write $x = (x')\alpha^{-1}$, $y = (y')\alpha^{-1}$. As α respects addition, we have $(x + y)\alpha = (x)\alpha + (y)\alpha = x' + y'$. Applying α^{-1} we obtain $(x' + y')\alpha^{-1} = x + y = (x')\alpha^{-1} + (y')\alpha^{-1}$, showing that α^{-1} respects addition. Similarly, α^{-1} respects multiplication and $(1)\alpha^{-1} = 1$; therefore α^{-1} is a ring homomorphism by (5.1).   □

As we have seen, essentially the same ring can appear in different contexts and in different notations; for instance, in (4.22)(a) the field $E = \mathbb{R}[x]/(x^2 + 1)$ amounts to a thinly-disguised version of the complex field $\mathbb{C}$. Whether the disguise is thick or thin, our next definition gives a precise meaning to 'essentially the same'.

Definition 5.3   A bijective ring homomorphism is called a **ring isomorphism**. Two rings are called **isomorphic** if there is a ring isomorphism between them.

Isomorphic rings are essentially the same; they are abstractly identical and the only possible difference between them is notational—nevertheless it can be difficult to decide whether or not two given rings are isomorphic.

Notation   Write $\alpha : R \cong R'$ if $\alpha : R \to R'$ is a ring isomorphism.

As mentioned above, the field E of (4.22)(a) is isomorphic to $\mathbb{C}$; in fact $\alpha : E \cong \mathbb{C}$, where $(\overline{c + dx})\alpha = c + di$ for all $c, d \in \mathbb{R}$.

As another illustration, consider the Boolean ring $P(U)$, where $U = \{a\}$ has only one element (see (2.7)). The addition and multiplication tables of $P(U)$ are:

+	$\varnothing$	U		$\times$	$\varnothing$	U
$\varnothing$	$\varnothing$	U		$\varnothing$	$\varnothing$	$\varnothing$
U	U	$\varnothing$		U	$\varnothing$	U

The same patterns occur in the addition and multiplication tables of the field $\mathbb{Z}_2$ (see (1.32)), that is, $\varnothing$ corresponds to $\bar{0}$, U corresponds to $\bar{1}$; in other words, the mapping $\alpha : P(U) \to \mathbb{Z}_2$, defined by $(\varnothing)\alpha = \bar{0}$, $(U)\alpha = \bar{1}$, is a ring isomorphism and so $\alpha : P(U) \cong \mathbb{Z}_2$.

The reader may verify that the composition of two compatible ring homomorphisms is itself a ring homomorphism; it follows that $R \cong R'$ and $R' \cong R''$ imply $R \cong R''$. In fact the symbol $\cong$ satisfies the three laws (1.25) of an equivalence relation; we therefore refer to the **isomorphism class** of the ring R, meaning all the rings isomorphic to R. Of particular interest is the *number* of isomorphism classes of rings. In this connection we note that certain properties are *preserved* by isomorphisms: for example, suppose $\alpha : R \cong R'$ and let R have zero-divisors. So there are x and y in R^* with $xy = 0$; then $x' = (x)\alpha$ and $y' = (y)\alpha$ are in R'^* and $x'y' = 0$, showing that R' has zero-divisors. In other words, a ring with zero-divisors cannot be isomorphic to a ring without zero-divisors; in particular $\mathbb{Z}_4$ and the field $\mathbb{F}_4$ are not isomorphic. Similarly a ring which is isomorphic to an integral domain (or field) is itself an integral domain (or field).

Example 5.4   Consider the irreducible polynomial $p = x^2 + x + 1$ over the real field $\mathbb{R}$. We show that the field $E = \mathbb{R}[x]/(p)$ of (4.21) is isomorphic to the complex field $\mathbb{C}$. For all practical purposes, the elements of E are of the form $c + dj$ where $j^2 + j + 1 = 0$ and $c, d \in \mathbb{R}$. There are, in fact, many isomorphisms between E and $\mathbb{C}$; we settle for the one which leaves real numbers unaffected and replaces j by $(-1 + i\sqrt{3})/2$: let $\alpha : E \to \mathbb{C}$ be defined by

$$(c + dj)\alpha = c + d(-1 + i\sqrt{3})/2 \quad \text{for all } c, d \in \mathbb{R}.$$

We leave the reader to verify that α satisfies (5.1). As α is bijective $((x + iy)\alpha^{-1} = x + y(1 + 2j)/\sqrt{3}$ for $x, y \in \mathbb{R})$, we have $\alpha : E \cong \mathbb{C}$, for α is a ring isomorphism.

The argument of (5.4) can be generalized to show that the fields $\mathbb{R}[x]/(p)$ are isomorphic to $\mathbb{C}$ for all irreducible quadratic polynomials p over $\mathbb{R}$, that is, all these fields belong to the same isomorphism class. On the other hand there are two isomorphism classes of rings $\mathbb{R}[x]/(p)$ where p is a *reducible* quadratic polynomial over $\mathbb{R}$, depending on whether p has distinct zeros (e.g. $p = x(x-1)$) or p has a repeated zero (e.g. $p = x^2$).

Let D be an ordered integral domain as in (3.4) with D_+ well-ordered; then D is isomorphic to the ordered integral domain $\mathbb{Z}$ of integers. For consider the mapping $\chi : \mathbb{Z} \to D$, defined by $(m)\chi = me$ for all m in $\mathbb{Z}$, where e denotes the 1-element of D. By the laws of indices (2.12), χ is a ring homomorphism. Now $m > m'$ implies $(m - m')e \in D_+$, that is, $me > m'e$, and so χ is order-preserving and hence injective; by (3.4)(c), χ is surjective. Therefore $\chi : \mathbb{Z} \cong D$, showing that all ordered integral domains D, with D_+ well-ordered, are isomorphic to $\mathbb{Z}$.

Definition 5.5   A ring isomorphism $\alpha : R \cong R$ is called an **automorphism** of the ring R.

Complex conjugation is an automorphism of the complex field $\mathbb{C}$; for by (2.23), the mapping $\alpha : \mathbb{C} \to \mathbb{C}$, defined by $(x + iy)\alpha = x - iy$ for all x and y in $\mathbb{R}$, respects addition and multiplication, is bijective $(\alpha^{-1} = \alpha)$, and leaves all real numbers (1 in particular) unchanged. Apart from the identity mapping of $\mathbb{C}$ (which is also an automorphism of $\mathbb{C}$), conjugation is the *only* automorphism of $\mathbb{C}$ leaving all real numbers unaffected.

Consider once again the integral domain $\mathbb{Z}[\sqrt{2}]$ of (2.17). Incidentally, the elements $m + n\sqrt{2}$ $(m, n \in \mathbb{Z})$ of this integral domain are precisely the real numbers which arise on evaluating polynomials in $\mathbb{Z}[x]$ at $\sqrt{2}$. The reader may verify, as in (2.23), that an automorphism α of $\mathbb{Z}[\sqrt{2}]$ is defined by

$$(m + n\sqrt{2})\alpha = m - n\sqrt{2} \quad \text{for all } m \text{ and } n \text{ in } \mathbb{Z}.$$

Whereas complex conjugation leaves real numbers unchanged and interchanges the zeros i and $-i$ of $x^2 + 1$, the above automorphism fixes every integer and interchanges the zeros $\sqrt{2}$ and $-\sqrt{2}$ of $x^2 - 2$. Notice that $\mathbb{Z}[\sqrt{2}]$ inherits an ordering from the real field $\mathbb{R}$, that is, $\mathbb{Z}[\sqrt{2}]$ is ordered in the sense of (3.1), where 'positive' has its usual meaning. However, the automorphism α can be used to introduce an *unusual* ordering on $\mathbb{Z}[\sqrt{2}]$: if $m + n\sqrt{2}$ is positive, call $(m + n\sqrt{2})\alpha$ **u-positive**. As α respects addition and multiplication, the set of all u-positive elements in $\mathbb{Z}[\sqrt{2}]$ satisfies (3.1)(i), as α is bijective and $(0)\alpha = 0$, we see that (3.1)(ii) is satisfied, and so the set of u-positive elements (which includes $-\sqrt{2}$) can be used in the role of D_+. The upshot is that $\mathbb{Z}[\sqrt{2}]$ can be ordered in two ways.

We have seen that some rings occur within other rings. For instance, $\mathbb{Z}[\sqrt{2}]$ is best thought of as part of the real field $\mathbb{R}$; not only in $\mathbb{Z}[\sqrt{2}]$ a subset of $\mathbb{R}$, but sums and products in the system $\mathbb{Z}[\sqrt{2}]$ are worked out using the operations of addition and multiplication on the ambient (or parent) field $\mathbb{R}$. Our next definition spells out this type of relationship between rings.

Definition 5.6

Let S be a subset of R. Then the ring $(S, \oplus, \otimes)$ is called a **subring** of the ring $(R, +, \times)$ if the **inclusion mapping** $\iota : S \to R$ $((x)\iota = x$ for all x in $S)$ is a ring homomorphism.

The inclusion mapping ι is a convenient concept (having the same status as the identity mapping (1.24) which it becomes when $R = S$) which helps us express concisely what we are trying to say. Supposing ι (as above) to be a ring homomorphism we obtain

$$x \oplus y = (x \oplus y)\iota = (x)\iota + (y)\iota = x + y \quad \text{for all } x, y \in S.$$

Therefore the operation $\oplus$ on S is no more than the **restriction** of the operation $+$ on R; so S is closed under $+$, and $\oplus$ is said to be **inherited** from $+$. As ι respects multiplication, in an analogous way we obtain

$$x \otimes y = (x \otimes y)\iota = (x)\iota \times (y)\iota = x \times y \quad \text{for all } x, y \in S.$$

Therefore the operation $\otimes$ on S is merely the restriction of the operation $\times$ on R; so S is closed under $\times$, and $\otimes$ is said to be inherited from $\times$. The condition $(1)\iota = 1$ tells us that the 1-element of R belongs to S.

As the operations on a subring are inherited from the parent ring, we may test a subset S of a ring for being (the set of elements of) a subring as below.

Proposition 5.7

(Criterion for a subring.) Let S be a subset of the ring R. Then S is a subring of R if and only if the following conditions are satisfied:

    (i) S is closed under addition and multiplication,
    (ii) S is closed under negation, that is, $-x \in S$ for all $x \in S$,
    (iii) S contains the 1-element of R.

Proof

Suppose first that S is a subring of R. The preceding discussion shows that conditions (i) and (iii) are satisfied. As S is a ring in its own right by (5.6), with inherited addition, law 3 of (2.2) tells us that S contains the negative of each of its elements, showing that condition (ii) is satisfied.

Now suppose that S satisfies conditions (i), (ii), (iii). By (i), it makes sense to refer to the operations of addition and multiplication on S inherited from those on R. What is more, conditions (ii) and (iii) ensure that S, with its inheritance, is a ring, that is, laws 1–7 of (2.2) hold in S as we now show.

By condition (iii), $1 \in S$ and so the 1-element of R is also the 1-element of S. By condition (ii), $-1 \in S$ and hence $0 = -1 + 1 \in S$, as S is closed under addition; so the 0-element of R is also the 0-element of S. By condition (ii), we see that each element of S has a negative in S. Therefore laws 2, 3, and 7 of (2.2) hold in S. Laws 1, 4, 5, and 6 of (2.2) hold in S simply because they hold in the parent ring R, and so S is itself a ring. Finally, the inclusion mapping $\iota : S \to R$ is a ring homomorphism because the ring operations on S are the restrictions of those on R and because $1 \in S$.
□

It is usually easy to see whether or not conditions (i), (ii), (iii) of (5.7) are satisfied in any given situation. For instance, taking S to be the set of even integers and $R = \mathbb{Z}$, we see that conditions (i) and (ii) are satisfied, but condition (iii) is not, showing that the even integers do not form a subring of $\mathbb{Z}$. In fact the only subring of $\mathbb{Z}$ is $\mathbb{Z}$ itself: for let S be a subring of $\mathbb{Z}$. As $1 \in S$ and S is closed under addition, we see (by induction) that S contains all positive integers; as S is closed under negation, S contains all negative integers. Hence $S = \mathbb{Z}$ as $0 \in S$.

A subring S of a ring R is called **proper** if $S \neq R$. The above argument shows that $\mathbb{Z}$ has no proper subrings; similarly the residue class rings $\mathbb{Z}_n$ have no proper subrings.

Notice that if R is an integral domain and S is a subset of R satisfying the conditions of (5.7), then S is also an integral domain; we call S a **subdomain** of R. For instance, $\mathbb{Z}[\sqrt{2}]$ is a subdomain of $\mathbb{R}$; the set of all rational numbers expressible in the form m/n with n odd, is a subdomain of $\mathbb{Q}$.

Definition 5.8

Let S be a subring of the field F. If S is a field, then S is called a **subfield** of F and F is called an **extension field** of S.

For example $\mathbb{Q}$ and $\mathbb{R}$ are subfields of $\mathbb{C}$, and $\mathbb{R}$ and $\mathbb{C}$ are extension fields of $\mathbb{Q}$. However the prime field $\mathbb{Z}_p$ contains no proper subfields, for it contains no proper subrings. Our next proposition is a modification of (5.7).

Proposition 5.9 (Criterion for a subfield.) Let S be a subset of the field F. Then S is a subfield of F if and only if conditions (i), (ii), (iii) of (5.7) are satisfied together with:

(iv) S^* is closed under inversion, that is, $x^{-1} \in S^*$ for all $x \in S^*$.

Proof We have already done the hard work in (5.7). Let S be a subfield of F. Then (5.7)(i), (ii), (iii) are satisfied, as S is a subring of F. By (2.18) each element of S^* has an inverse in S^*; as multiplication on S is inherited, this means condition (iv) above is satisfied.

Conversely, suppose S satisfies conditions (i), (ii), (iii) of (5.7) and condition (iv) above. By (5.7), S is a subring of F. As 0 and 1 are distinct elements of S, we see that S is non-trivial. As F is commutative, so also is S. Finally, condition (iv) tells us that every non-zero element of S has an inverse in S. By (2.18), S is a field and so S is a subfield of F. $\qquad\square$

Let S be a subfield of the rational field $\mathbb{Q}$. As S is a subring, we see (as in the case of subrings of $\mathbb{Z}$) that S contains all integers; therefore $m, n \in S$ for all integers m and n with $n \neq 0$. By condition (iv) of (5.9), we have $1/n \in S$, and as S is closed under multiplication we deduce $m/n \in S$. Therefore $S = \mathbb{Q}$, showing that $\mathbb{Q}$ is a prime field, that is, $\mathbb{Q}$ has no proper subfields.

Definition 5.10 Let $\alpha : R \to R'$ be a ring homomorphism. The **image** of α is the subset $\operatorname{im} \alpha = \{x' \in R' : x' = (x)\alpha \text{ for some } x \in R\}$ of R'.

So the image of α, denoted by $\operatorname{im} \alpha$, consists of the images by α of all the elements in R. For instance let $\alpha : \mathbb{Z}[x] \to \mathbb{R}$ be the ring homomorphism defined by $(f)\alpha = f(\sqrt{2})$ for all $f \in \mathbb{Z}[x]$; then $\operatorname{im} \alpha = \mathbb{Z}[\sqrt{2}]$, for $m + n\sqrt{2} = (m + nx)\alpha$, showing $\mathbb{Z}[\sqrt{2}] \subseteq \operatorname{im} \alpha$, and

$$(a_0 + a_1 x + a_2 x^2 + a_3 x^3 + \ldots)\alpha = m + n\sqrt{2}$$

where

$$m = a_0 + 2a_2 + 4a_4 + 8a_6 + \ldots + 2^i a_{2i} + \ldots \text{ and}$$
$$n = a_1 + 2a_3 + 4a_5 + 8a_7 + \ldots + 2^i a_{2i+1} + \ldots$$

showing $\operatorname{im} \alpha \subseteq \mathbb{Z}[\sqrt{2}]$.

Notice that $\operatorname{im} \alpha = R'$ if and only if $\alpha : R \to R'$ is surjective.

Corollary 5.11 Let $\alpha : R \to R'$ be a ring homomorphism. Then $\operatorname{im} \alpha$ is a subring of R'.

Proof   We apply the subring criterion (5.7) with im α in place of S, and R' in place of R. Let $x', y' \in$ im α; then $x' = (x)\alpha$ and $y' = (y)\alpha$ for some $x, y \in R$. So $x' + y' = (x + y)\alpha$ and $x'y' = (xy)\alpha$; as $x + y$ and xy belong to R we see that $x' + y'$ and $x'y'$ belong to im α, showing that im α is closed under addition and multiplication. Also $-x' = (-x)\alpha$ by (5.2), showing that im α is closed under negation. As $(1)\alpha = 1$, we see that the 1-element of R' belongs to im α. Therefore im α satisfies the conditions of (5.7) and so is a subring of R'.   □

A given ring R can beget further rings in a number of ways; we have seen that R is parent to all its subrings (one-parent families are common in mathematics!), but that $\mathbb{Z}$, having no proper subrings, is impotent using this method. On the other hand, there is a more fruitful procedure, typified by $\mathbb{Z}$ giving rise to all the residue class rings $\mathbb{Z}_n$, as we now explain.

Definition   Let R be a ring. The rings of the form im α, for some ring
5.12   homomorphism $\alpha : R \to R'$, are called **homomorphic images** of R.

As the natural homomorphism $\eta : \mathbb{Z} \to \mathbb{Z}_n$ is surjective, we see that $\mathbb{Z}_n = $ im η is a homomorphic image of $\mathbb{Z}$; we shall show that the rings $\mathbb{Z}_n$, for n a natural number, together with $\mathbb{Z}$ itself, are the *only* homomorphic images of $\mathbb{Z}$, that is, all homomorphic images of $\mathbb{Z}$ are isomorphic either to $\mathbb{Z}_n$ or to $\mathbb{Z}$. Now $\mathbb{Z}_n$ is formed from $\mathbb{Z}$ by 'losing' all integer multiples of n; in other words, the integer multiples of n are precisely the integers which are mapped by η to the 0-element of $\mathbb{Z}_n$. We now deal with the general case.

Definition   Let $\alpha : R \to R'$ be a ring homomorphism. The **kernel** of α is the
5.13   subset ker $\alpha = \{x \in R : (x)\alpha = 0\}$ of R.

Therefore the kernel of a ring homomorphism is the set of elements (of the first ring) mapped to the 0-element (of the second ring), that is, ker α is the set of elements 'lost' by α (Fig. 5.1). In the case of the natural homomorphism $\eta : \mathbb{Z} \to \mathbb{Z}_2$, we see ker η is the set of even integers. The evaluation homomorphism $\sigma_0 : \mathbb{Z}[x] \to \mathbb{Z}$ maps every polynomial f over $\mathbb{Z}$ to its constant term a_0, that is, $(f)\sigma_0 = a_0$ where $f = \Sigma\, a_i x^i$; therefore ker σ_0 consists of all polynomials over $\mathbb{Z}$ having zero constant term. The kernel of the composite homomorphism $\sigma_0\eta : \mathbb{Z}[x] \to \mathbb{Z}_2$ is the set of all polynomials over $\mathbb{Z}$ having *even* constant term.

Let $\alpha : R \to R'$ be a ring homomorphism and let us suppose that

Fig. 5.1

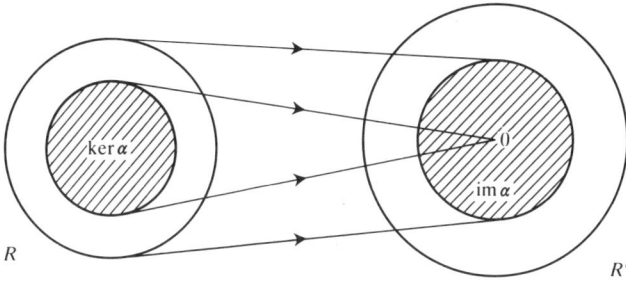

kernel and image of $\alpha : R \to R'$

ker α is a subring of R. Can this happen? We show that the answer is: almost never! For $1 \in \ker \alpha$ in this case and so $(1)\alpha = 0$; as $(1)\alpha = 1$, we see $1 = 0$ in R', showing that R' is trivial by (2.10). Conversely, if R' is trivial, α maps every element of R to the 0-element of R', because $R' = \{0\}$; therefore ker $\alpha = R$ which is a subring of R. If kernels are not usually subrings, what then is the relationship between them and their parent rings? Our next definition provides the tailor-made answer.

Definition 5.14

Let K be a subset of the ring R. Then K is called an **ideal** of R if

(i) K is closed under addition and negation, and $0 \in K$,
(ii) xk and kx belong to K for all $x \in R$ and $k \in K$.

The internal structure of the ideal K is described by condition (i) above. The special relationship between an ideal and its parent ring is set out in condition (ii): not only is K closed under multiplication, but the product (in either order) of each ring element x with each ideal element k is itself an ideal element.

The set of even integers is an ideal of $\mathbb{Z}$; the set of polynomials over $\mathbb{Z}$ with even constant term is an ideal of $\mathbb{Z}[x]$. Both these sets are kernels of ring homomorphisms; we see next that this is no coincidence.

Lemma 5.15

Let $\alpha : R \to R'$ be a ring homomorphism. Then ker α is an ideal of R.

Proof

Write $K = \ker \alpha$ and let $k, l \in K$. Then $(k)\alpha = (l)\alpha = 0$ by (5.13), and so $(k + l)\alpha = (k)\alpha + (l)\alpha = 0 + 0 = 0$ and $(-k)\alpha = -(k)\alpha = 0$. As $(0)\alpha = 0$, we see $0 \in K$ and so K satisfies (5.14)(i).

Now let $x \in R$ and $k \in K$. Then $(xk)\alpha = ((x)\alpha)((k)\alpha) = ((x)\alpha)0 = 0$, showing that $xk \in K$. Similarly $kx \in K$ and so K satisfies (5.14)(ii). Therefore K is an ideal of R. □

The integer multiples of the given integer n form an ideal of $\mathbb{Z}$ which is denoted by (n); thus (2) denotes the ideal of all even integers. Similarly, if R is a commutative ring, the polynomials in $R[x]$, having the given polynomial p over R as factor, form an ideal of $R[x]$ denoted by (p). The reader should keep these examples of ideals in mind during the following theory, which generalizes the construction of the rings $\mathbb{Z}_n$ (also denoted $\mathbb{Z}/(n)$) and $R[x]/(p)$ in (3.18) and (4.21).

Let K be an ideal of the ring R. Using K, we introduce an equivalence relation on R as follows: write

$$x \equiv x' \pmod{K} \quad \text{if} \quad x - x' \in K, \quad \text{where } x, x' \in R.$$

The reader may verify that, as K satisfies (5.14)(i), the symbol $\equiv$ satisfies the laws (1.25) of an equivalence relation. If $x \equiv x'$ $\pmod{K}$ we say that x and x' are **congruent modulo** K; so two elements of R are congruent modulo K if their difference belongs to K. In this context, the congruence classes have a special name and a simple notation.

Definition 5.16

Let K be an ideal of the ring R. A subset of R of the form

$$K + x_0 = \{k + x_0 : k \in K\} \quad \text{where } x_0 \in R$$

is called a **coset** of K in R. Each element $k + x_0$ is called a **representative** of $K + x_0$.

We refer to *the* coset of K in R having representative x_0, for this coset is exactly the congruence class $\pmod{K}$ of x_0:

$$\bar{x}_0 = \{x \in R : x \equiv x_0 \pmod{K}\}$$
$$= \{x \in R : x - x_0 = k \in K\}$$
$$= \{k + x_0 : k \in K\} = K + x_0.$$

For example, the element $\bar{r}$ of $\mathbb{Z}_n$ is a coset of the ideal (n) in $\mathbb{Z}$; in fact $\bar{r} = (n) + r$, that is, $\bar{r}$ consists of all integers of the form $qn + r$. Returning to the general case of an ideal K of the ring R, we show that congruence modulo K is compatible with addition and multiplication. To do this, suppose $x \equiv x' \pmod{K}$ and $y \equiv y' \pmod{K}$ therefore $x = k + x'$ and $y = l + y'$ where $k, l \in K$. Adding and multiplying these equations, as in (3.17), gives

$$(x + y) - (x' + y') = k + l \in K \quad \text{by (5.14)(i)},$$

$$xy - x'y' = kl + ky' + x'l \in K \quad \text{by (5.14)(i) and (ii)},$$

that is, $x + y \equiv x' + y' \pmod{K}$ and $xy \equiv x'y' \pmod{K}$. It therefore makes sense to define addition and multiplication of cosets in terms of representatives:

$$\left.\begin{array}{l} (K + x) + (K + y) = K + (x + y) \\ (K + x)(K + y) = K + xy \end{array}\right\} \quad \text{for all } x, y \in R.$$

Effectively, coset sums and products are carried out as in R, except that elements of K are discarded; the system consisting of the set of all cosets of K in R, together with the above operations of coset addition and multiplication, is denoted by R/K and called the result of **factoring** R by K. For instance, $\mathbb{Z}_2$ results from factoring $\mathbb{Z}$ by the ideal (2); from (4.22)(a) we see that the complex field $\mathbb{C}$ results from factoring $\mathbb{R}[x]$ by the ideal $(x^2 + 1)$. Our next theorem should be compared to (3.18) and (4.21).

Theorem 5.17   Let K be an ideal of the ring R. Then the system R/K is a ring, called the factor (quotient) ring of R by K. The natural mapping $\eta : R \to R/K$ is a ring homomorphism and $K = \ker \eta$.

Proof   We leave the reader to show that R/K is a ring by modifying the proof of (3.18); this involves writing $K + x$ in place of $\bar{x}$, $K + y$ in place of $\bar{y}$, etc. The 0-element and 1-element of R/K are K $(= K + 0)$ and $K + 1$ respectively, the negative of $K + x$ is $K - x$. Now $\eta : R \to R/K$ is defined by $(x)\eta = K + x$ for all $x \in R$ by (1.31); because of the way coset addition and multiplication is defined, η is a ring homomorphism. Also, $\ker \eta = \{x \in R : (x)\eta = K\}$ as K is the 0-element of R/K; but $(x)\eta = K \Leftrightarrow K + x = K \Leftrightarrow x \in K$, showing that $K = \ker \eta$.   □

As a further example of a factor ring, let K be the ideal of $\mathbb{Z}[x]$ consisting of polynomials $k = \sum a_i x^i$ with a_0 and a_1 even. There are four possibilities for the parities of b_0 and b_1 where $f = \sum b_i x^i$ is a general polynomial over $\mathbb{Z}$; hence there are four cosets of K in $\mathbb{Z}[x]$, namely K, $K + 1$, $K + x$, $K + x + 1$ (for instance $K + x$ consists of all polynomials $\sum b_i x^i$ over $\mathbb{Z}$ with b_0 even and b_1 odd). The ring $\mathbb{Z}[x]/K$ has these four cosets as its elements, with addition and multiplication tables as in Table 5.1.

Table 5.1

+	K	$K+1$	$K+x$	$K+x+1$
K	K	$K+1$	$K+x$	$K+x+1$
$K+1$	$K+1$	K	$K+x+1$	$K+x$
$K+x$	$K+x$	$K+x+1$	K	$K+1$
$K+x+1$	$K+x+1$	$K+x$	$K+1$	K

$\times$	K	$K+1$	$K+x$	$K+x+1$
K	K	K	K	K
$K+1$	K	$K+1$	$K+x$	$K+x+1$
$K+x$	K	$K+x$	K	$K+x$
$K+x+1$	K	$K+x+1$	$K+x$	$K+1$

Combining (5.15) and (5.17) we see that the concepts 'ideal of a ring' and 'kernel of a ring homomorphism' coincide, that is, each ideal is the kernel of a suitable ring homomorphism, and the kernel of each ring homomorphism is an ideal. Our next theorem tells us that factor rings and homomorphic images amount to the same thing also.

Theorem 5.18

(The first isomorphism theorem.) Let $\alpha : R \to R'$ be a ring homomorphism. Then $\bar{\alpha} : R/\ker \alpha \cong \operatorname{im} \alpha$, where $(\bar{x})\bar{\alpha} = (x)\alpha$ for all $x \in R$.

Proof

By (5.15), $\ker \alpha$ is an ideal of R, and so the factor ring $R/\ker \alpha$ of (5.17) can be formed. By (5.11), $\operatorname{im} \alpha$ is a subring of R' and so $\operatorname{im} \alpha$ is a ring in its own right by (5.6). We must show that the rings $R/\ker \alpha$ and $\operatorname{im} \alpha$ are isomorphic, the ring isomorphism $\bar{\alpha}$ involved being closely related to the given ring homomorphism α; here $\bar{x}$ denotes the congruence class (mod $\ker \alpha$) of x. Note that $(x)\alpha = (x')\alpha \Leftrightarrow (x - x')\alpha = 0 \Leftrightarrow x - x' \in \ker \alpha \Leftrightarrow x \equiv x' \pmod{\ker \alpha} \Leftrightarrow \bar{x} = \bar{x'}$. This sequence of implications shows not only that it makes sense to define $\bar{\alpha}$ as above by $(\bar{x})\bar{\alpha} = (x)\alpha$ (for α maps every element of $\bar{x}$ to the same element of $\operatorname{im} \alpha$), but also that $\bar{\alpha}$ is injective (for α maps elements in different congruence classes (mod $\ker \alpha$) to different elements of $\operatorname{im} \alpha$). Also $\bar{\alpha}$ is surjective by definition (5.10) of $\operatorname{im} \alpha$; so $\bar{\alpha}$ is bijective.

To show that $\bar{\alpha}$ is a ring homomorphism, consider $\bar{x}, \bar{y} \in R/\ker \alpha$. Then $(\bar{x} + \bar{y})\bar{\alpha} = (\overline{x + y})\bar{\alpha} = (x + y)\alpha = (x)\alpha + (y)\alpha = (\bar{x})\bar{\alpha} + (\bar{y})\bar{\alpha}$ showing that $\bar{\alpha}$ respects addition. Similarly $\bar{\alpha}$ respects multiplication, and $(\bar{1})\bar{\alpha} = (1)\alpha = 1$. By (5.3), $\bar{\alpha}$ is a ring isomorphism between $R/\ker \alpha$ and $\operatorname{im} \alpha$. $\qquad\square$

So each ring homomorphism gives rise to a ring isomorphism; this isomorphism can be used to recognize factor rings as 'known' rings. For example, the evaluation homomorphism $\sigma_0 : \mathbb{Z}[x] \to \mathbb{Z}$ is surjective, that is, im $\sigma_0 = \mathbb{Z}$, and ker $\sigma_0 = (x)$, the ideal of all polynomials over $\mathbb{Z}$ with divisor x; by (5.18), $\mathbb{Z}[x]/(x) \cong \mathbb{Z}$ (this isomorphism should come as no surprise, for $\mathbb{Z}[x]/(x)$ is merely the algebraists' way of losing x from $\mathbb{Z}[x]$, the result being simply $\mathbb{Z}$). Similarly, the composite homomorphism $\sigma_0\eta : \mathbb{Z}[x] \to \mathbb{Z}_2$ is surjective and its kernel is the ideal $(2, x)$ of all polynomials over $\mathbb{Z}$ having even constant term; by (5.18) we see that $\mathbb{Z}[x]/(2, x) \cong \mathbb{Z}_2$ (again, this is no surprise, for the factor ring $\mathbb{Z}[x]/(2, x)$ has only two elements, namely $(2, x)$ itself and the coset $(2, x)+1$ of all polynomials over $\mathbb{Z}$ having odd constant term).

Let us apply (5.18) to the natural mapping $\eta : \mathbb{Z} \to \mathbb{Z}_n$: the mapping $\bar{\eta} : \mathbb{Z}/\ker \eta \cong \text{im } \eta$ is given by $(\bar{x})\bar{\eta} = (x)\eta = \bar{x}$ for all $x \in \mathbb{Z}$, that is, $\bar{\eta}$ is no more than the identity mapping! As ker $\eta = (n)$ and im $\eta = \mathbb{Z}_n$, we obtain $\mathbb{Z}/(n) = \mathbb{Z}_n$ (which accounts for the alternative notation for $\mathbb{Z}_n$). Similarly applying (5.18) to $\eta : R[x] \to R[x]/(p)$, where p is a monic polynomial over the non-trivial commutative ring R, produces the equation (of doubtful value) $\bar{\eta} : R[x]/(p) = R[x]/(p)$, $\bar{\eta}$ again being the identity mapping; however the reader should now realize why the notation $R[x]/(p)$ is adopted in (4.21), namely, to conform with the notation of a general factor ring.

From (5.18) we see that every homomorphic image of the ring R is obtained (up to isomorphism) on factoring R by an ideal K; therefore, if all ideals of R are known, then all homomorphic images of R can be found. We now discuss a method of constructing ideals of a commutative ring; this will enable us to describe all the ideals of $\mathbb{Z}$ and all the ideals of $F[x]$, where F is a field.

Let $k_1, k_2, \ldots, k_n$ be given elements of the commutative ring R and let

$$K = \{x_1 k_1 + x_2 k_2 + \ldots + x_n k_n : x_1, x_2, \ldots, x_n \in R\},$$

that is, K consists of all **linear combinations** of $k_1, k_2, \ldots, k_n$. It is straightforward to check that K satisfies the conditions of (5.14) (in particular, taking $x_1 = x_2 = \ldots = x_n = 0$ we see $0 \in K$). So K is an ideal of R.

Definition 5.19

The above set K is called the ideal of R **generated** by $k_1, k_2, \ldots, k_n$ and we write $K = (k_1, k_2, \ldots, k_n)$.

Let us take $R = \mathbb{Z}[x]$, $k_1 = 2$, $k_2 = x$. Then $K = (2, x)$ consists of all polynomials of the form $2f + xg$, where $f, g \in \mathbb{Z}[x]$. Now $2f + xg$

has even constant term; conversely every polynomial $\sum a_i x^i$ over $\mathbb{Z}$ with even constant term $a_0 = 2b_0$ can be expressed $2b_0 + x(\sum a_{i+1} x^i)$ and so belongs to K. Therefore $(2, x)$ is *exactly* the ideal of $\mathbb{Z}$ consisting of polynomials with even constant term (we have already used the notation $(2, x)$ for this ideal).

Definition 5.20

Let K be an ideal of the commutative ring R. If there is an element k in R such that $K = (k)$, then K is called a **principal ideal** of R.

So an ideal is principal if it is generated by a single element. Consider the ideal $K = (12, 18)$ of $\mathbb{Z}$ (the notation should suggest to the reader that g.c.d.s are involved). As K consists of all integers of the form $12x + 18y$ where $x, y \in \mathbb{Z}$, we see that 6 is a divisor of every integer in K and also $6 = -12 + 18$ is in K, showing $K = (6)$. More generally, by (3.9) the ideal (m, n) of $\mathbb{Z}$ is principal, being generated by the g.c.d. $(m, n) = sm + tn$, and this is true for all integers m and n (including the case $m = n = 0$).

The set $K = \{0\}$, consisting solely of the 0-element of the ring R, satisfies (5.14) and is called the **zero ideal** of R.

Theorem 5.21

Every non-zero ideal of $\mathbb{Z}$ is principal and has a unique positive generator.

Proof

Let K be a non-zero ideal of $\mathbb{Z}$. Then K contains a non-zero integer k; as $k, -k \in K$ by (5.14), we see that K contains a positive integer. By the well-ordering principle, we may refer to n, the *least* of the positive integers in K (notice the similarity to the proof of (3.9)). As $n \in K$, by (5.14)(ii) we see $xn \in K$ for all $x \in \mathbb{Z}$, that is $(n) \subseteq K$. On the other hand, dividing an arbitrary integer k in K by n gives $k = qn + r$, where $0 \le r < n$. But as $k, n \in K$ we see $r = k - qn$ belongs to K also. By the 'leastness' of n, r cannot be positive; therefore $r = 0$ and so $k = qn$. This means $K \subseteq (n)$ and so $K = (n)$, showing that K is principal with positive generator n.

Now $(n) \subseteq (n')$ if and only if $n' \mid n$ $(n, n' \in \mathbb{Z})$. Therefore if n and n' are both positive and $(n) = (n')$, we must have $n \mid n'$ and $n' \mid n$; hence $n = n'$, showing that K has a unique positive generator. $\square$

The polynomial analogue of (5.21) states that every non-zero ideal of $F[x]$ (F a field) is principal and has a unique monic generator. As the zero ideals of $\mathbb{Z}$ and $F[x]$ are clearly principal (being generated by 0), we see that these rings are examples of

principal ideal domains, that is, integral domains such that *all* their ideals are principal.

Corollary 5.22	Every homomorphic image of $\mathbb{Z}$ is isomorphic either to $\mathbb{Z}$ or to $\mathbb{Z}_n$ for some natural number n.
Proof	Let R be a homomorphic image of $\mathbb{Z}$. By (5.12) there is a surjective ring homomorphism $\alpha : \mathbb{Z} \to R$; let $K = \ker \alpha$. By (5.18), $\bar{\alpha} : \mathbb{Z}/K \cong R$. Suppose $K = \{0\}$; in this case $x \equiv x' \pmod{K} \Leftrightarrow x = x'$, that is, congruence modulo the zero ideal is simply equality. Therefore $\mathbb{Z}/\{0\} \cong \mathbb{Z}$ and so $\mathbb{Z} \cong R$. Suppose $K \neq \{0\}$. By (5.21) there is a unique natural number n with $K = (n)$; therefore $\mathbb{Z}_n = \mathbb{Z}/(n) \cong R$. $\qquad\square$

Let F be a field. The polynomial analogue of (5.22) states that the homomorphic images of $F[x]$ are isomorphic either to $F[x]$ itself or to a factor ring of the form $F[x]/(k)$, where k is a monic polynomial over F.

We close this section with an application of (5.21) to an arbitrary ring R. There is a unique ring homomorphism $\chi : \mathbb{Z} \to R$, for necessarily $(m)\chi = me$ for all integers m, where e is the 1-element of R. In this case im χ consists of all integer multiples of e; in fact im χ is the *smallest* subring of R, for it is contained in every subring of R. As $\ker \chi$ is an ideal of $\mathbb{Z}$, by (5.21) $\ker \chi$ is principal and has a unique non-negative generator.

Definition 5.23	Let R be a ring. The non-negative generator of $\ker \chi$ is called the **characteristic** of R and denoted by $\chi(R)$.

Suppose first that R has characteristic zero, that is, $\chi(R) = 0$; this means that $me = 0$ *only* if $m = 0$. For instance $\mathbb{Z}, \mathbb{Z}[\sqrt{2}], \mathbb{Q}, \mathbb{R}, \mathbb{C}$ all have characteristic zero. In this case $\ker \chi = \{0\}$ and applying (5.18) to χ produces $\bar{\chi} : \mathbb{Z}/(0) \cong$ im χ, that is, in a ring of characteristic zero, the integer multiples of its 1-element form a subring isomorphic to $\mathbb{Z}$.

Secondly, suppose that R has positive characteristic n; therefore $me = 0$ if and only if n is a divisor of m. The rings $\mathbb{Z}_n$, $\mathbb{Z}_n[x]$, $\mathbb{Z}_n[x]/(k)$ where k is a monic polynomial of positive degree over $\mathbb{Z}_n$, are of characteristic n. In this case $\ker \chi = (n)$ and by (5.18) we obtain $\mathbb{Z}_n \cong$ im χ. So in a ring of characteristic n, the integer multiples of its 1-element form a subring isomorphic to $\mathbb{Z}_n$. By (2.10) R is trivial if and only if R has characteristic 1.

Let R be an integral domain of positive characteristic n. Then

$n > 1$ as R is non-trivial; further n is prime, for otherwise $\mathbb{Z}_n$, and hence R, would have zero-divisors. Writing $n = p$, in this case the integer multiples of the 1-element form a field isomorphic to $\mathbb{Z}_p$.

The characteristic $\chi(F)$ of a field F is a significant number, for it completely describes the smallest subfield P of F. If $\chi(F) = 0$, then P is isomorphic to the rational field $\mathbb{Q}$ (the rational number m/m' corresponds to the element $(me)(m'e)^{-1}$ of P). If $\chi(F) = p > 0$, then p is prime and $\mathbb{Z}_p \cong \text{im } \chi = P$. Summing up:

> Every field F contains a prime subfield P
> and either $P \cong \mathbb{Q}$ or $P \cong \mathbb{Z}_p$ for some
> positive prime p.

For instance, the fields $\mathbb{Z}_2$, $\mathbb{F}_4$, $\mathbb{F}_8$ have characteristic 2, the prime subfield in each case being isomorphic to $\mathbb{Z}_2$.

Exercises 5.1

1. Let $\alpha : R \to R'$ and $\beta : R' \to R''$ be ring homomorphisms. Show that $\alpha\beta : R \to R''$ is a ring homomorphism.

2. Unusual operations of addition and multiplication are defined on $\mathbb{Z}$ by:

$$x \oplus y = x + y + 1, \qquad x \otimes y = xy + x + y,$$

for all $x, y \in \mathbb{Z}$. Show that α, defined by $(x)\alpha = x + 1$ for all $x \in \mathbb{Z}$, is an isomorphism of $(\mathbb{Z}, \oplus, \otimes)$ to $\mathbb{Z}$ (with the usual addition and multiplication).

3. Let $\alpha : (R, +, \times) \cong (R, \oplus, \otimes)$ be a ring isomorphism. If $(x)\alpha = 1 - x$ for all $x \in R$, express $x \oplus y$ and $x \otimes y$ in terms of x and y using the operations $+$ and $\times$.

4. Show that there is a unique ring homomorphism $\alpha : \mathbb{Z}_6 \to \mathbb{Z}_3$ and list the elements in (i) im α (ii) ker α. Show that $\mathbb{Z}_6$ has exactly four ideals. How many ideals has $\mathbb{Z}_3$?

5. Decide whether or not the factor ring $\mathbb{Z}_2[x]/(x(x + 1))$ is isomorphic to the Boolean ring $P(\{a, b\})$, $a \neq b$.

6. (a) Let R be a ring. Use (5.7) to show that the intersection of any number of subrings of R is itself a subring of R.

(b) Show that the set S of rational numbers of the form $m/2^n$ ($m \in \mathbb{Z}$, $n \in \mathbb{N}$) is a subring of $\mathbb{Q}$. Show that S is the smallest subring of $\mathbb{Q}$ with $\frac{1}{2} \in S$.

(c) Describe the smallest subring S of $\mathbb{Q}$ with $\frac{1}{2}, \frac{1}{3} \in S$.

7. Show that the rings $\mathbb{Z}_2[x]/(x^2)$ and $\mathbb{Z}_2[x]/((x+1)^2)$ are isomorphic. Find four rings R with $|R| = 4$ such that no two are isomorphic.

8. (a) Let S be a subring of the Boolean ring $R = P(U)$ where $U = \{1, 2, 3, 4\}$ (see (2.7)). If S contains $\{1, 2\}$ and $\{1, 3\}$, show that $S = R$. Show that R has a unique proper subring containing both $\{1, 2\}$ and $\{1, 2, 3\}$.

(b) Let U be a finite set and let S be a subring of $P(U)$. Show that S gives rise to an equivalence relation on U as follows: $u_1 \equiv u_2$ if there is no X in S with $u_1 \in X$, $u_2 \notin X$. Show further that $\bar{u} \in S$ for all $u \in U$ and that S is isomorphic to $P(\bar{U})$.

Prove that subrings of $P(U)$ correspond to partitions (1.29) of U. Find the number of subrings of $P(U)$ if $|U| = 4$.

9. Let $\alpha : R \to R'$ be a ring homomorphism.

(a) Show that α is injective if and only if $\ker \alpha = \{0\}$.
(b) Let S be a subring of R. Use (5.9) to show that $S' = \{(x)\alpha : x \in S\}$ is a subring of R'.
(c) Let K' be an ideal of R'. Show that $K = \{x \in R : (x)\alpha \in K'\}$ is an ideal of R.

10. (a) Let K_1 and K_2 be ideals of the ring R. Show that $K_1 \cap K_2$ and $K_1 + K_2 = \{k_1 + k_2 : k_1 \in K_1, k_2 \in K_2\}$ are ideals of R. Find a generator of $K_1 \cap K_2$ and a generator of $K_1 + K_2$ in the case $R = \mathbb{Z}$, $K_1 = (16)$, $K_2 = (24)$.

(b) Use the division law (4.11) to establish the polynomial analogue of (5.21): every non-zero ideal of $F[x]$, where F is a field, is principal with a unique monic generator. Find monic generators of $K_1 \cap K_2$ and $K_1 + K_2$ in the case $F = \mathbb{Q}$, $K_1 = (x^3 - x^2 - x - 2)$, $K_2 = (x^3 - 3x^2 + 4)$.

11. Let K be an ideal and S a subring of the ring R. Prove that $K \cap S$ is an ideal of S and $K + S = \{k + s : k \in K, s \in S\}$ is a subring of R. Let $\alpha : S \to (K + S)/K$ be defined by $(s)\alpha = K + s$ for all $s \in S$. Show that α is a ring homomorphism and find $\ker \alpha$ and $\operatorname{im} \alpha$. Use (5.18) to prove the **second isomorphism theorem**: $\bar{\alpha} : S/(K \cap S) \cong (K + S)/K$.

Taking $R = \mathbb{Z}[x]$, $S = \mathbb{Z}[x^2]$, $K = (x^3)$, describe, in terms of their coefficients, the polynomials in (i) $K \cap S$, (ii) $K + S$. Using representatives of lowest degree, state the rules of coset multiplication in the factor rings $S/(K \cap S)$ and $(K + S)/K$. What is the form of $\bar{\alpha}$ in terms of these representatives?

12. Let K be an ideal of the ring R and let K contain a unit (2.11) of R. Show that $K = R$.

Let F be a field. Show that $\{0\}$ and F are the only ideals of F, and deduce, as in (5.22), the form of the homomorphic images of F.

13. (a) Let V be a subset of the set U. Show that $\alpha : P(U) \rightarrow P(V)$, defined by $(X)\alpha = X \cap V$ for all $X \subseteq U$, is a homomorphism of Boolean rings. Find im α and ker α; explain the meaning of congruence (mod ker α) in this case and describe the isomorphism $\bar{\alpha}$ obtained, as in (5.18), from α.

(b) Let K consist of all the finite subsets of the set U; show that K is an ideal of $P(U)$. Let S denote the set of all subsets X of U such that either X or X' (the complement of X in U) is finite; show that S is a subring of $P(U)$.

(c) If U is finite, show that every ideal of $P(U)$ is principal. Is the condition 'finite' necessary?

14. (a) Let α be an automorphism of $\mathbb{Z}[\sqrt{2}]$ (see (2.17)). Show that $(m)\alpha = m$ for all integers m. If $f(c) = 0$ where $c \in \mathbb{Z}[\sqrt{2}]$ and $f \in \mathbb{Z}[x]$, show that $f((c)\alpha) = 0$. Taking $f = x^2 - 2$, deduce that $\mathbb{Z}[\sqrt{2}]$ has exactly two automorphisms.

(b) Let α be an automorphism of the field $\mathbb{Q}(\sqrt{2}) = \{q + \sqrt{2}\, q' : q, q' \in \mathbb{Q}\}$. If α is not the identity mapping, show that $(q + \sqrt{2}\, q')\alpha = q - \sqrt{2}\, q'$ for all $q, q' \in \mathbb{Q}$. Deduce that $\mathbb{Q}(\sqrt{2})$ can be ordered in two different ways.

15. Let α be an automorphism of the real field $\mathbb{R}$. Show $(q)\alpha = q$ for all $q \in \mathbb{Q}$; show $x \leqslant y \Leftrightarrow (x)\alpha \leqslant (y)\alpha$ for $x, y \in \mathbb{R}$. Deduce that α is the identity mapping. (Hint: every positive real number has a real square root; there is a rational number between any two different real numbers.)

16. Show that the fields $\mathbb{Q}(\sqrt{2})$ and $\mathbb{Q}(\sqrt{3}) = \{q + \sqrt{3}\, q' : q, q' \in \mathbb{Q}\}$ are not isomorphic.

17. Show that $\mathbb{Z}_2[x]/(x^3 + x + 1)$ and $\mathbb{Z}_2[x]/(x^3 + x^2 + 1)$ are isomorphic fields.

18. Let p be a quadratic polynomial over the real field $\mathbb{R}$. Show that there are three isomorphism classes of rings $\mathbb{R}[x]/(p)$. Is this result valid if $\mathbb{R}$ is replaced by (i) $\mathbb{Z}_2$, (ii) $\mathbb{Q}$?

Constructions

A general technique of mathematics consists of decomposing a given system into simpler systems in order to understand its structure; the simpler systems then recombine in a straightforward way to produce the original system. Here we deal with the direct sum construction in which two rings combine together to give a larger ring.

Let R and R' be rings. We form the set $R \times R'$ of all ordered pairs (x, x') where $x \in R$, $x' \in R'$. Addition and multiplication of ordered pairs is defined in a *componentwise* way as follows:

$$\left.\begin{aligned}(x, x') + (y, y') &= (x + y, x' + y') \\ (x, x')(y, y') &= (xy, x'y')\end{aligned}\right\} \quad \text{for all } x, y \in R \text{ and } x', y' \in R'.$$

Definition 5.24

The above system is denoted by $R \oplus R'$ and called the **direct sum** of the rings R and R'.

The reader may verify that $R \oplus R'$ is itself a ring. The 0-element of $R \oplus R'$ is the pair $(0, 0')$ made up of the 0-element of R and the 0-element of R'; similarly the pair $(1, 1')$ of 1-elements of R and R' is the 1-element of $R \oplus R'$. The negative of (x, x') is $(-x, -x')$, showing that negation is carried out componentwise, as are all operations in $R \oplus R'$. Indeed 'componentwise' is the key-word in forming the direct sums of all types of system, for, as we shall see, the same idea is used in every case.

The direct sum $\mathbb{Z}_2 \oplus \mathbb{Z}_2$ is the ring with elements $(0, 0)$, $(1, 0)$, $(0, 1)$, and $(1, 1)$, where $1 + 1 = 0$ and $1 \neq 0$. In this ring, $(1, 0) + (0, 1) = (1, 1)$ and $(1, 0)(0, 1) = (0, 0)$; as $(0, 0)$ is the 0-element, we see that $(1, 0)$ and $(0, 1)$ are non-zero elements having product zero, showing that $\mathbb{Z}_2 \oplus \mathbb{Z}_2$ has zero-divisors. Have we met this ring before? In fact it is isomorphic to the Boolean ring $\mathrm{P}(\{a, b\})$ of (2.5), for $\alpha : \mathbb{Z}_2 \oplus \mathbb{Z}_2 \cong \mathrm{P}(\{a, b\})$ defined by $(0, 0)\alpha = \varnothing$, $(1, 0)\alpha = \{a\}$, $(0, 1)\alpha = \{b\}$, $(1, 1)\alpha = \{a, b\}$, is a ring isomorphism (the reader should show that α matches the addition and multiplication tables of $\mathbb{Z}_2 \oplus \mathbb{Z}_2$ with those of $\mathrm{P}(\{a, b\})$). So $\mathrm{P}(\{a, b\})$ is (isomorphic to) the direct sum of two copies of $\mathbb{Z}_2$.

Notice that $R \oplus R'$ is trivial if and only if both R and R' are trivial; also $R \oplus R'$ is commutative if and only if both R and R' are commutative. However, $R \oplus R'$ has zero-divisors if R and R' are non-trivial, for $(1, 0)(0, 1') = (0, 0')$; in particular, the direct sum of two integral domains is *not* an integral domain. The units (invertible elements) of $R \oplus R'$ are easily described in terms of the units of the component rings R and R': for (x, x') is a unit of $R \oplus R'$ if and only if x is a unit of R and x' is a unit of R'.

It is a straightforward matter to define the direct sum $R_1 \oplus R_2 \oplus \ldots \oplus R_n$ of a finite number of rings R_i $(1 \leq i \leq n)$: the elements of this direct sum are n-tuples $(x_1, x_2, \ldots, x_n)$ where $x_i \in R_i$, addition and multiplication being carried out componentwise, that is,

$$\left.\begin{aligned}(x_1, \ldots, x_n) + (y_1, \ldots, y_n) &= (x_1 + y_1, \ldots, x_n + y_n) \\ (x_1, \ldots, x_n)(y_1, \ldots, y_n) &= (x_1 y_1, \ldots, x_n y_n)\end{aligned}\right\} \quad \begin{aligned}&\text{for all} \\ &x_i, y_i \in R_i.\end{aligned}$$

As before $R_1 \oplus R_2 \oplus \ldots \oplus R_n$ is itself a ring.

We now use the Chinese remainder theorem (3.23) to decompose $\mathbb{Z}_n$, as mentioned at the close of Chapter 3. Let n_1 and n_2 be natural numbers and let $x, x' \in \mathbb{Z}$. As $x \equiv x' \pmod{n_1 n_2}$ implies $x \equiv x' \pmod{n_1}$ and $x \equiv x' \pmod{n_2}$, a mapping $\alpha : \mathbb{Z}_{n_1 n_2} \to \mathbb{Z}_{n_1} \oplus \mathbb{Z}_{n_2}$ is unambiguously defined by

$$(\bar{x})\alpha = (\bar{x}, \bar{x}) \quad \text{for all } x \in \mathbb{Z}.$$

congruence class   congruence class   congruence class
$(\mod n_1 n_2)$ of x $(\mod n_1)$ of x $(\mod n_2)$ of x

This mapping is written out in full at the end of Chapter 3 in the case $n_1 = 2$, $n_2 = 3$; the general case is covered by the following theorem.

Theorem 5.25

Let n_1 and n_2 be coprime natural numbers. Then the above mapping $\alpha : \mathbb{Z}_{n_1 n_2} \cong \mathbb{Z}_{n_1} \oplus \mathbb{Z}_{n_2}$ is a ring isomorphism.

Proof

As the operations in the residue class rings $\mathbb{Z}_n$ derive from integer addition and multiplication, α is a ring homomorphism: specifically

$$(\bar{x} + \bar{y})\alpha = \overline{(x+y)}\alpha = (\overline{x+y}, \overline{x+y}) = (\bar{x} + \bar{y}, \bar{x} + \bar{y})$$
$$= (\bar{x}, \bar{x}) + (\bar{y}, \bar{y}) = (\bar{x})\alpha + (\bar{y})\alpha \quad \text{for all } x, y \in \mathbb{Z}.$$

Similarly $((\bar{x})(\bar{y}))\alpha = ((\bar{x})\alpha)((\bar{y})\alpha)$ and $(\bar{1})\alpha = (\bar{1}, \bar{1})$, showing that α satisfies the conditions of (5.1).

Let $(\bar{r}_1, \bar{r}_2)$ belong to $\mathbb{Z}_{n_1} \oplus \mathbb{Z}_{n_2}$. The integer x of (3.23) is such that $(\bar{x})\alpha = (\bar{r}_1, \bar{r}_2)$, and so α is surjective; as any two such integers x are congruent $(\mod n_1 n_2)$, we see that α is injective. $\quad\square$

For instance, (5.25) tells us $\mathbb{Z}_{60} \cong \mathbb{Z}_4 \oplus \mathbb{Z}_{15}$ and $\mathbb{Z}_{15} \cong \mathbb{Z}_3 \oplus \mathbb{Z}_5$; we may combine these isomorphisms to give $\mathbb{Z}_{60} \cong \mathbb{Z}_4 \oplus \mathbb{Z}_3 \oplus \mathbb{Z}_5$, which is the decomposition of $\mathbb{Z}_{60}$ corresponding to the factorization of $60 = 4 \times 3 \times 5$ into prime powers. More generally, any factorization of n into mutually coprime factors leads, by (5.25), to a direct sum decomposition of $\mathbb{Z}_n$; in particular if $n = q_1 q_2 \ldots q_k$ is the factorization (3.15) of n (q_i is a power of the positive prime p_i and $p_1, p_2, \ldots, p_k$ are distinct), then

$$\mathbb{Z}_n \cong \mathbb{Z}_{q_1} \oplus \mathbb{Z}_{q_2} \oplus \ldots \oplus \mathbb{Z}_{q_k}.$$

What is more, each $\mathbb{Z}_{q_i}$ is **indecomposable**, that is, it *cannot* be expressed as a direct sum of two non-trivial rings (see Exercises 5.2, Question 5); so the above decomposition of $\mathbb{Z}_n$ cannot be improved upon.

We now derive a formula for the number of units of $\mathbb{Z}_n$ in terms of the prime factorization (3.15) of the natural number n. The elements of $\mathbb{Z}_n$ are of the form $\bar{r}$ for $1 \leqslant r \leqslant n$. Suppose first that $\bar{r}$ is a unit of $\mathbb{Z}_n$; then $\bar{1} = \bar{s}\bar{r}$ for some integer s, and so $1 \equiv sr \pmod{n}$ which means $1 = sr + tn$ for some integer t. Therefore the g.c.d. $(r, n) = 1$, that is, r is coprime to n. Conversely, if r is coprime to n, the above steps may be reversed using (3.9) to show that $\bar{r}$ is a unit of $\mathbb{Z}_n$.

Definition 5.26

Let n be a positive integer. The *number* of integers r in the range $1 \leqslant r \leqslant n$ with r coprime to n is denoted by $\phi(n)$, and $\phi : \mathbb{N} \to \mathbb{N}$ is called **Euler's function**.

Euler's function plays an important role in number theory. From the above discussion, the ring $\mathbb{Z}_n$ has exactly $\phi(n)$ units. There are four integers r with $1 \leqslant r \leqslant 8$ and $(r, 8) = 1$, namely 1, 3, 5, and 7, and so $\phi(8) = 4$; there are two integers r with $1 \leqslant r \leqslant 6$ and $(r, 6) = 1$, namely 1 and 5, and so $\phi(6) = 2$. Notice that $\phi(p) = p - 1$ for p prime, as $r = p$ is the only integer with $1 \leqslant r \leqslant p$ which is *not* coprime to p. More generally $\phi(p^e) = p^e - p^{e-1}$ if p prime, because the only integers r with $1 \leqslant r \leqslant p^e$ which are not coprime to p^e are the p^{e-1} integer multiples of p.

The isomorphism (5.25) can be used to determine $\phi(n)$. For $\mathbb{Z}_{n_1 n_2}$ has $\phi(n_1 n_2)$ units, while $\mathbb{Z}_{n_1} \oplus \mathbb{Z}_{n_2}$ has $\phi(n_1)\phi(n_2)$ units (x_1, x_2), there being $\phi(n_1)$ choices for x_1 (any unit of $\mathbb{Z}_{n_1}$) and $\phi(n_2)$ choices for x_2 (any unit of $\mathbb{Z}_{n_2}$). From (5.25) we obtain the **multiplicative property** of ϕ:

$$\phi(n_1 n_2) = \phi(n_1)\phi(n_2) \text{ for } n_1 \text{ and } n_2 \text{ coprime}$$

as the isomorphic rings $\mathbb{Z}_{n_1 n_2}$ and $\mathbb{Z}_{n_1} \oplus \mathbb{Z}_{n_2}$ each contain the same number of units. Therefore $\phi(60) = \phi(4)\phi(15)$, $\phi(15) = \phi(3)\phi(5)$ and hence $\phi(60) = \phi(4)\phi(3)\phi(5) = 2 \times 2 \times 4 = 16$. More generally, if $n = q_1 q_2 \ldots q_k$ where $q_i = p_i^{e_i}$ $(p_1, p_2, \ldots, p_k$ distinct positive primes), then $\phi(n) = \phi(q_1)\phi(q_2) \ldots \phi(q_k)$ by the above multiplicative property of ϕ. As $\phi(q_i) = p_i^{e_i} - p_i^{e_i - 1}$, we obtain:

$$\phi(n) = (p_1^{e_1} - p_1^{e_1 - 1})(p_2^{e_2} - p_2^{e_2 - 1}) \ldots (p_k^{e_k} - p_k^{e_k - 1})$$

where $n = p_1^{e_1} p_2^{e_2} \ldots p_k^{e_k}$. This is the formula for the number $\phi(n)$ of units of $\mathbb{Z}_n$.

We turn now to a construction of a different kind: every integral domain D gives rise to a field F (called the field of fractions of D) which can be thought of as the smallest field containing D. For instance, the smallest field containing the integral domain $\mathbb{Z}$ is the rational field $\mathbb{Q}$; the reader should keep this case in mind through-out, for it is the model on which the general construction is based.

Let D be an integral domain. We form $X = \{(x, y) : x \in D, y \in D^*\}$, the idea being to treat the ordered pair (x, y) as if it were the fraction x/y; so the entries in (x, y) are destined to become a numerator x and a (non-zero) denominator y. Using this idea, operations of addition and multiplication on X are introduced:

$$\left.\begin{array}{l} (x, y) + (x_1, y_1) = (xy_1 + yx_1, yy_1) \\ (x, y)(x_1, y_1) = (xx_1, yy_1) \end{array}\right\} \quad \text{for } x, x_1 \in D \text{ and } y, y_1 \in D^*.$$

Notice that X is closed under the above operations, for D has no zero-divisors and so the product yy_1 of the non-zero 'denominators' y and y_1 is itself non-zero; what is more, the above operations are suggested by the familiar rules: $(x/y) + (x_1/y_1) = (xy_1 + yx_1)/(yy_1)$ and $(x/y)(x_1/y_1) = (xx_1)/(yy_1)$. So, without assuming the existence of fractions (that would be jumping the gun!), we have nevertheless imbued the ordered pairs in X with some of the properties of fractions.

We now arrange for equality of fractions to 'happen' in our system, being guided by the familiar cross-multiplication rule: $x/y = x'/y' \Leftrightarrow xy' = yx'$. Formally, we write

$$(x, y) \equiv (x', y') \text{ if } xy' = yx', \quad \text{where } x, x' \in D \text{ and } y, y' \in D^*.$$

In other words, two ordered pairs in X are equivalent if and only if they correspond to the same fraction.

Lemma 5.27   Let D be an integral domain. Using the above notation, $\equiv$ is an equivalence relation on X which is compatible with addition and multiplication.

Proof   As D is commutative and has no zero-divisors, it is straight-forward to verify that $\equiv$ satisfies the conditions of (1.25); so $\equiv$ is an equivalence relation on X.

Let $(x, y) \equiv (x', y')$ and $(x_1, y_1) \equiv (x_1', y_1')$; so $xy' = yx'$ and $x_1y_1' = y_1x_1'$. Multiplying the first of these equations by y_1y_1', the second by yy', and adding, gives $(xy_1 + yx_1)(y'y_1') = (yy_1)(x'y_1' + y'x_1')$; therefore $(x, y) + (x_1, y_1) \equiv (x', y') + (x_1', y_1')$ showing that $\equiv$ is compatible with addition. Multiplying the original equations together produces $(xx_1)(y'y_1') = (yy_1)(x'x_1')$; therefore $(x, y)(x_1, y_1) \equiv (x', y')(x_1', y_1')$, and so $\equiv$ is compatible with multi-plication. $\quad\square$

The system F that we are aiming for can now be formed: the elements of F are the equivalence classes $\overline{(x, y)}$ of ordered pairs in X; addition and multiplication of elements in F is carried out according to the rules:

$$\left.\begin{array}{l} \overline{(x, y)} + \overline{(x_1, y_1)} = \overline{(xy_1 + yx_1, yy_1)} \\ \overline{(x, y)}\ \overline{(x_1, y_1)} = \overline{(xx_1, yy_1)} \end{array}\right\} \quad \text{for } x, x_1 \in D \text{ and } y, y_1 \in D^*.$$

Notice that these rules make sense by (5.27). For example, taking $D = \mathbb{Z}$ we see $\overline{(1, 2)} = \{\ldots, (-1, -2), (1, 2), (2, 4), (3, 6), \ldots\}$ ($\frac{1}{2}$ is the practical notation for this set!), and $\overline{(2, -3)} = \{\ldots, (-4, 6), (-2, 3), (2, -3), (4, -6), \ldots\}$ (which is an impartial, but impractical, way of denoting $-\frac{2}{3}$). Using the above rules: $\overline{(1, 2)} + \overline{(2, -3)} = \overline{(1, -6)}$ and $\overline{(1, 2)}\ \overline{(2, -3)} = \overline{(2, -6)}$ (that is, in usual notation: $\frac{1}{2} - \frac{2}{3} = -\frac{1}{6}$ and $\frac{1}{2}(-\frac{2}{3}) = -\frac{2}{6}$).

Theorem 5.28

(Properties of the field of fractions.) Let D be an integral domain. Then the above system F is a field. Further $S = \{\overline{(x, 1)} : x \in D\}$ is a subdomain of F which is isomorphic to D. The only subfield of F containing S is F itself.

Proof

The reader may verify that all the laws of (2.2) hold in the underlying system X, except possibly law 3 (the existence of negatives) and law 5 (the distributive law); however, laws 3 and 5 hold with equality replaced by equivalence, and hence F is a commutative ring. In particular $\overline{(0, 1)} = \{(0, y) : y \in D^*\}$ is the 0-element of F and $\overline{(-x, y)}$ is the negative of $\overline{(x, y)}$. The 1-element of F is $\overline{(1, 1)} = \{(y, y) : y \in D^*\}$ which is different from $\overline{(0, 1)}$ as $(1, 1) \neq (0, 1)$; so F is non-trivial. The element $\overline{(x, y)}$ of F is non-zero if and only if x is non-zero, in which case its inverse is $\overline{(y, x)}$. Therefore F is a field.

The mapping $\alpha : D \to F$, defined by $(x)\alpha = \overline{(x, 1)}$, is a ring homomorphism. As $\operatorname{im} \alpha = S$, we see that S is a subdomain of F by (5.11); now α is injective: for $(x)\alpha = (x_1)\alpha$ means $\overline{(x, 1)} = \overline{(x_1, 1)}$, that is, $(x, 1) \equiv (x_1, 1)$ and so $x = x_1$. Therefore S is isomorphic to D.

Finally let F' be a subfield of F with $S \subseteq F'$. Now $\overline{(x, 1)}$ and $\overline{(y, 1)}$ belong to S, and hence to F', for all $x \in D$, $y \in D^*$. Therefore

$$\overline{(x, y)} = \overline{(x, 1)}\overline{(1, y)} = \overline{(x, 1)}\overline{(y, 1)}^{-1} \in F'$$

by (5.9). So F' contains every element in F, that is, $F' = F$. $\quad\square$

It is customary to regard the above injection α as identifying the element x of D with the element $\overline{(x, 1)}$ of S; writing $\overline{(x, 1)} = x$, for all x in D, produces $\overline{(x, y)} = xy^{-1}$ as the typical element of F. The conclusion is that fractions with numerators and denominators from an integral domain can be manipulated in the usual way; at the same time we see that every integral domain can be extended to its field of fractions.

Exercises 5.2

1. Write out the addition and multiplication tables of the ring $\mathbb{Z}_2 \oplus \mathbb{Z}_2$. Show that this ring has two automorphisms (5.5).

2. (a) Let R be a ring. Show that $\delta : R \to R \oplus R$, defined by $(x)\delta = (x, x)$ for all $x \in R$, is a ring homomorphism. Show that $\tau : R \oplus R \to R \oplus R$ defined by $(x, y)\tau = (y, x)$ for all $x, y \in R$, is an automorphism.
   (b) Show that R and R' are homomorphic images of $R \oplus R'$. Show that $\{(x, 0) : x \in R\}$ and $\{(0, x') : x' \in R'\}$ are ideals of $R \oplus R'$.

3. (a) Let $\alpha : \mathbb{Z}_{12} \cong \mathbb{Z}_4 \oplus \mathbb{Z}_3$ be the isomorphism of (5.25). List the images by α of the elements of $\mathbb{Z}_{12}$. Verify that $\bar{r}$ is a unit (invertible element) of $\mathbb{Z}_{12}$ if and only if $(\bar{r})\alpha$ is a unit of $\mathbb{Z}_4 \oplus \mathbb{Z}_3$.
   (b) By considering $\alpha : \mathbb{Z}_{90} \cong \mathbb{Z}_2 \oplus \mathbb{Z}_9 \oplus \mathbb{Z}_5$ determine the number of zeros of $x^2 - 1$ over $\mathbb{Z}_{90}$ and hence find these zeros. Find the zeros of $x^2 - x$ over $\mathbb{Z}_{90}$.

4. Verify $\phi(3)\phi(5) = \phi(15)$, where ϕ denotes Euler's function (5.26). Evaluate $\phi(272)$ and $\phi(480)$. Determine all the natural numbers n such that $\phi(n) = 2^7$.

5. The element x of the ring R is called (i) an **idempotent** if $x^2 = x$, (ii) **central** if $xy = yx$ for all $y \in R$.
   (a) Show that 0 and 1 are the only idempotents of $\mathbb{Z}_{p^e}$ (p prime).
   (b) Use (5.25) to show that $\mathbb{Z}_n$ contains exactly 2^k idempotents, where k is the number of distinct positive prime divisors of the natural number n.

   (c) A ring is **decomposable** if it is isomorphic to a direct sum $R \oplus R'$ of non-trivial rings (otherwise it is **indecomposable**). Show that integral domains and fields are indecomposable.
   By considering $(1, 0)$ and $(0, 1')$ in $R \oplus R'$, show that a decomposable ring contains non-zero central idempotents x and x' with $xx' = 0$ and $x + x' = 1$. Deduce that $\mathbb{Z}_{p^e}$ is indecomposable (p prime).
   (d) Let the ring R_1 contain non-zero central idempotents x and x' such that $xx' = 0$ and $x + x' = 1$. Show that R_1 is decomposable. (Hint: consider $K = \{xy : y \in R_1\}$ and $K' = \{x'y : y \in R_1\}$.)
   (e) Show that the ring R_1 is decomposable if and only if R_1 contains non-zero ideals K and K' with $K \cap K' = \{0\}$ and $K + K' = R_1$.

6. (a) Let p_1 and p_2 be monic coprime polynomials over the field F. Show that the rings $F[x]/(p_1p_2)$ and $(F[x]/(p_1)) \oplus (F[x]/(p_2))$ are isomorphic by establishing the analogue of (5.25).

(b) Express $\mathbb{Z}_2[x]/(p_1p_2)$ as a direct sum of fields, where $p_1 = x$ and $p_2 = x^2 + x + 1$.

(c) Express $\mathbb{R}[x]/(p_1p_2)$ as a direct sum of indecomposable rings, where $p_1 = x^2$ and $p_2 = x^2 + 1$.

7. (a) Let D be a subdomain of the field E. Show that $F' = \{xy^{-1} : x \in D, y \in D^*\}$ is a subfield of E; show that F' is isomorphic to the field F of fractions of D.

(b) For each of the following integral domains D, describe its field F of fractions.

(i) $D = \{m/n : m, n \in \mathbb{Z} \text{ with } n \text{ odd}\}$,
(ii) $D = \{m/2^n : m \in \mathbb{Z}, n \in \mathbb{N}\}$,
(iii) $D = \mathbb{Z}[\sqrt{2}] = \{m + n\sqrt{2} : m, n \in \mathbb{Z}\}$,
(iv) $D = \mathbb{Q}(\sqrt{2}) = \{q + q'\sqrt{2} : q, q' \in \mathbb{Q}\}$.

(c) Let F and F' be the fields of fractions of the integral domains D and D'. If $\beta : D \to D'$ is an injective ring homomorphism, show that $\bar{\beta} : F \to F'$ is unambiguously defined by $\overline{(x, y)\bar{\beta}} = \overline{((x)\beta, (y)\beta)}$ for all $x \in D$, $y \in D^*$, and that $\bar{\beta}$ is also an injective ring homomorphism. Hence show that isomorphic integral domains have isomorphic fields of fractions.

Part II: Linear algebra

6 Vector spaces

Here we make a fresh start, assuming little more than a nodding acquaintance with the concept of a field (see (2.18)). The reader, who is familiar with $\mathbb{Z}_2$ (see (1.32)) and the rational and real fields $\mathbb{Q}$ and $\mathbb{R}$, is ready to begin the study of **vector spaces** which will occupy us throughout the remaining chapters. Vector spaces arise in a wide variety of situations—they provide the framework within which many 'real world' problems can be formulated and solved—and yet their structure is simple enough to be easily analysed.

We begin with a review of cartesian 3-dimensional space $\mathbb{R}^3$, for the idea of a vector space and the terminology used stems from this familiar example. The elements of $\mathbb{R}^3$ are ordered triples (x, y, z) of real numbers; in this context (x, y, z) is called a **vector**, for it can be thought of as the position vector of the point P with co-ordinates x, y, z (and represented by an arrow drawn from the origin 0 to P (Fig. 6.1)). We write $V = \mathbb{R}^3$, and so V is the set of vectors $v = (x, y, z)$ under consideration. In this context and in contrast to vectors, individual real numbers are called **scalars**. So $v = (\sqrt{2}, 0, -\frac{1}{4})$ is a vector, whereas $a = \pi/2$ is a scalar.

Fig. 6.1

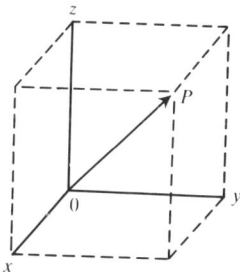

Our immediate concern is *not* with the geometric aspects of $\mathbb{R}^3$ (the concepts of angle, distance, area, etc. will be discussed later) but rather with the operations of addition and scalar multiplication of vectors. In the case of $V = \mathbb{R}^3$, these operations are carried out *componentwise*: for instance, if $u = (3, 2, 4)$ and $v = (4, 7, 1)$, then $u + v = (7, 9, 5)$ and $\sqrt{2}v = (4\sqrt{2}, 7\sqrt{2}, \sqrt{2})$; notice that u and v (and hence $u + v$, $\sqrt{2}v$) are the position vectors of points in the

Fig. 6.2

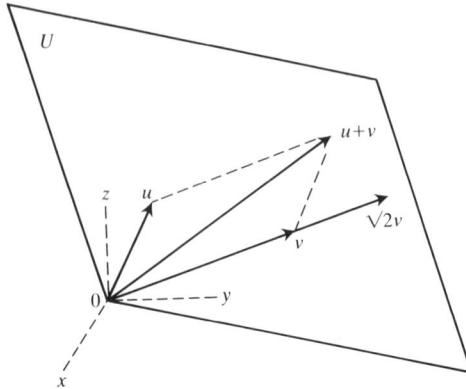

plane U with equation $2x - y - z = 0$ (Fig. 6.2). The reader who is already accustomed to vectors will know that the sum of vectors in $\mathbb{R}^3$ can be found by the parallelogram rule—as in the case of complex numbers (see (2.26)); also the arrows representing av and v are in the same (opposite) direction if a is positive (negative), their lengths being in the ratio $|a| : 1$. However, what is important from our point of view is that the operations of addition and scalar multiplication of vectors in $V = \mathbb{R}^3$ satisfy seven laws, similar to the ring laws (2.2); this is what is meant by saying that V is a **vector space over the field** $\mathbb{R}$.

This chapter is devoted to the basic theory of vector spaces over an *arbitrary* field F. The substitution of F in place of $\mathbb{R}$ involves no extra work and has several advantages: the foundations of the subject are laid once and for all, and the theory is thrown into sharper relief; what is more, many of the applications (to coding theory for instance) involve fields other than $\mathbb{R}$.

Elementary properties of vector spaces

Throughout F will denote a field, its elements being called **scalars**. Let V be a set and let $\alpha : V \times V \to V$ and $\mu : F \times V \to V$ be mappings. The elements of V are called **vectors** and the mappings α and μ are interpreted as operations of **vector addition** and **scalar multiplication** of vectors by writing

$$u + v = (u, v)\alpha, \quad av = (a, v)\mu \quad \text{for all } u, v \in V \text{ and } a \in F.$$

Therefore the sum of vectors is itself a vector, that is, V is closed under vector addition; the product of each scalar and vector is itself a vector, that is, V is closed under scalar multiplication. By the system $(V, F, +, \times)$ we mean the set V together with the field F and the mappings α and μ interpreted as above.

The reader should compare the above preliminaries and our next definition with (2.2).

Definition 6.1   The system $(V, F, +, \times)$, or simply V, is called a **vector space over the field** F if the following laws hold:

1. $(u + v) + w = u + (v + w)$ for all $u, v, w \in V$.
2. There is a vector 0 in V with $0 + v = v$ for all $v \in V$.
3. For each $v \in V$ there is $-v \in V$ satisfying $-v + v = 0$.
4. $u + v = v + u$ $\qquad\qquad$ for all $u, v \in V$.
5. $\left.\begin{array}{l} a(u + v) = au + av \\ (a + b)v = av + bv \end{array}\right\}$ $\quad$ for all $u, v \in V$ and $a, b \in F$.
6. $(ab)v = a(bv)$ $\qquad\quad$ for all $v \in V$ and $a, b \in F$.
7. The 1-element of F satisfies $1v = v$ for all $v \in V$.

The laws of a vector space should remind the reader of the laws (2.2) of a ring; certainly the two concepts have much in common: being unique, the vector 0 of law 2 is called the **zero vector** of V and each vector v has a unique negative $-v$. However there are differences: no vector product is postulated in (6.1) (we shall introduce the scalar product of vectors later) but instead we have the lop-sided product (scalar) $\times$ (vector) = vector; vectors av are called **scalar multiples** of v. As to the content of (6.1), it tells us that all the laws which one might reasonably expect to hold in this context, do in fact hold! In other words, a vector space is a forum in which the business of manipulating vectors can be carried out subject to the 'usual' laws.

We have already mentioned that $\mathbb{R}^3$ has the structure of a vector space over $\mathbb{R}$, the operations of vector addition and scalar multiplication being performed componentwise. This example can be generalized: for let F be an arbitrary field and n a natural number. We take the n-fold cartesian product F^n (see (1.13)) to be our set V of vectors; so each vector is a **row vector**, that is, an n-tuple $(a_1, a_2, \ldots, a_n)$ of elements a_i belonging to F. Vector addition and scalar multiplication of row vectors are defined:

$$(a_1, a_2, \ldots, a_n) + (b_1, b_2, \ldots, b_n)$$
$$= (a_1 + b_1, a_2 + b_2, \ldots, a_n + b_n)$$
$$a(a_1, a_2, \ldots, a_n) = (aa_1, aa_2, \ldots, aa_n) \quad \text{for all } a, a_i, b_i \in F.$$

Proposition 6.2 Let F be a field and n a natural number. Then F^n, with operations of vector addition and scalar multiplication as above, is a vector space over F.

Proof We should verify that laws 1–7 of (6.1) hold with $V = F^n$. To verify $a(u + v) = au + av$, write $u = (a_1, a_2, \ldots, a_n)$, $v = (b_1, b_2, \ldots, b_n)$. As a, a_i, b_i belong to F and the distributive law (law 5 of (2.2)) holds in F, we see $a(a_i + b_i) = aa_i + ab_i$, showing that the ith entries in the n-tuples $a(u + v)$ and $au + av$ agree for $1 \leq i \leq n$. Therefore these n-tuples are identical, that is, $a(u+v) = au + av$.

The remaining laws of (6.1) may be verified similarly. We omit the details, noting that the zero vector of F^n is $(0, 0, \ldots, 0)$, that is, the n-tuple with the 0-element of F in each place; also $-(a_1, a_2, \ldots, a_n) = (-a_1, -a_2, \ldots, -a_n)$. $\qquad\square$

In particular, the cartesian plane $\mathbb{R}^2$ and cartesian 3-dimensional space $\mathbb{R}^3$ are vector spaces over $\mathbb{R}$. The vector space $\mathbb{Z}_2^3$ consists of the eight vectors $(0, 0, 0)$, $(1, 0, 0)$, $(0, 1, 0)$, $(0, 0, 1)$, $(0, 1, 1)$, $(1, 0, 1)$, $(1, 1, 0)$, $(1, 1, 1)$, where 0 and 1 stand for the scalars in $\mathbb{Z}_2$, and so $1 + 1 = 0$; in $\mathbb{Z}_2^3$ each vector is equal to its negative and $(1, 1, 0) + (1, 1, 1) = (0, 0, 1)$, $(1, 1, 0) + (0, 1, 1) = (1, 0, 1)$ etc. More generally, if F is a finite field with $|F| = q$, then F^n consists of exactly q^n vectors, there being q choices for each of the n entries in an n-tuple belonging to F^n. So $\mathbb{Z}_2^{10}$ is a vector space over $\mathbb{Z}_2$ containing exactly $2^{10} = 1024$ vectors.

The vector spaces F^n play a central role in the ensuing theory. However, the following examples should convince the reader that vector spaces can arise in different forms, some of which do not rely on co-ordinates.

Example 6.3 (a) The Boolean ring $P(U)$ of subsets of U (2.7) can be thought of as a vector space over $\mathbb{Z}_2$. In this case the subsets X of U are regarded as vectors, symmetric difference (2.4) being the vector sum $X + Y$. We capitalize on the equations $mX = \emptyset$ (m even), $mX = X$ (m odd) and define scalar multiplication by the rules: $0 \times X = \emptyset$, $1 \times X = X$ where $\mathbb{Z}_2 = \{0, 1\}$. The laws of (6.1) follow from (2.6) and (2.12)(a), the empty set $\emptyset$ being the zero vector. As the product XY plays no part here, we see that the ring $P(U)$ becomes a vector space over $\mathbb{Z}_2$ on *ignoring* the ring multiplication.

(b) The integral domain $F[x]$ of all polynomials over F may be regarded as a vector space over F. In this case polynomials are vectors, polynomial addition being vector addition and multiplication by constant polynomials being scalar multiplication; the laws (6.1) are a direct consequence of (4.5). As in the preceding example, a vector space structure remains on discounting part of the ring structure: $F[x]$ becomes a vector space over F on ignoring the product of polynomials of positive degree.

The same idea can be used to think of the real field $\mathbb{R}$ as being a vector space over its subfield $\mathbb{Q}$; here $\mathbb{R}$ is the set of vectors, while the smaller field $\mathbb{Q}$ is the field of scalars. On ignoring the product of irrational numbers, $\mathbb{R}$ becomes a vector space over $\mathbb{Q}$. Similarly, on ignoring the product of non-real complex numbers, $\mathbb{C}$ is a vector space over $\mathbb{R}$ (in fact 1 and i form a **basis** (6.18) of this vector space, which is therefore 2-dimensional, as every complex number can be expressed uniquely $x \times 1 + y \times i$ where $x, y \in \mathbb{R}$). More generally, every extension field E of F becomes a vector space over F on ignoring the product of elements not in F.

(c) Vector spaces of the following type are the subject of functional analysis. Let X be a non-empty set and let V denote the set of all functions $f : X \to F$ (in this context the image of x by f is denoted $f(x)$); so vectors are functions of X to the field F. Vector addition (sum of functions) and scalar multiplication (constant multiples of functions) have their usual meanings, that is,

$$\left. \begin{array}{l} (f+g)(x) = f(x) + g(x) \\ (af)(x) = af(x) \end{array} \right\} \quad \text{for all } f, g \in V, a \in F, x \in X.$$

It is straightforward to verify that V satisfies the laws of (6.1) and so is a vector space over F. Taking $X = F = \mathbb{R}$, we obtain the vector space of all real-valued functions of a real variable.

Our next proposition is the vector space counterpart of (2.8).

Proposition 6.4 Let V be a vector space over F.

(a) The zero vector of V is unique.
(b) Each vector in V has a unique negative.
(c) Given $u, v \in V$, there is a unique $w \in V$ with $w + u = v$.
(d) $0v = 0 = a0$ for all $v \in V$ and $a \in F$.
(e) $\left. \begin{array}{l} (-a)v = -(av) = a(-v) \\ (-a)(-v) = av \end{array} \right\}$ for all $v \in V$ and $a \in F$.
(f) $av = 0$ implies either $a = 0$ or $v = 0$.

Proof The proofs of (a)–(e) follow closely the proofs of (2.8)(a)–(c), and we shall say no more, except to point out that the same notation is being used for the scalar zero and the zero vector: therefore (d) reads:

(scalar $0)v$ = zero vector = a(zero vector).

To prove (f), suppose $av = 0$. If $a = 0$ there is nothing to do. If $a \neq 0$, multiply $av = 0$ by $a^{-1} \in F$ and use laws 6 and 7 of (6.1) obtaining

$$v = 1v = (a^{-1}a)v = a^{-1}(av) = a^{-1}0 = 0.$$

So either $a = 0$ or $v = 0$.   □

By law 2 of (6.1), each vector space V contains at least one vector, namely 0. If $V = \{0\}$, that is, V contains no non-zero vectors, then V is called **trivial**.

Most vector spaces occur not in isolation, but as part of a 'parent' space such as F^n. We now investigate this important source of vector spaces, for, as one might expect, dull parents sometimes produce interesting children! Throughout the following general discussion the reader may keep in mind the case $V = \mathbb{R}^3$, $F = \mathbb{R}$, and $U = \{(x, y, z) \in \mathbb{R}^3; \ 2x - y - z = 0\}$, that is, U consists of position vectors of points in the plane with equation $2x - y - z = 0$ (Fig. 6.2).

Let V be a vector space over the field F and let U be a subset of V; so U consists of some of the vectors in V. We assume that U is closed under the operation of vector addition, that is, $u + v \in U$ for all $u, v \in U$: the sum of each pair of vectors in U is itself a vector in U. Therefore the subset U acquires, from V, an operation of vector addition; we say U **inherits** this operation from V. We assume also that U is closed under the operation of scalar multiplication of vectors, that is, $au \in U$ for all $a \in F$ and $u \in U$: scalar multiples of vectors in U are themselves in U. Therefore U acquires, from V, an operation of scalar multiplication of vectors; as above, we say U **inherits** this operation from V.

Definition 6.5   Let V be a vector space over the field F and let U be a subset of V. Then U is called a **subspace** of V if (i) U is closed under addition and scalar multiplication of vectors, and (ii) U, together with its inherited operations, forms a vector space over F.

So a subspace U is a vector space in its own right, although it owes everything to its parent V. It is straightforward to verify that $U = \{(x, y, z) \in \mathbb{R} : 2x - y - z = 0\}$ is a subspace of $\mathbb{R}^3$; for the sum and scalar (constant) multiples of vectors (x, y, z) satisfying

$2x - y - z = 0$ are again vectors satisfying this equation, and further, these vectors satisfy the seven laws (6.1). Luckily, the tedious verification of the vector space laws required by (6.5)(ii) can be avoided—this is the message of our next proposition.

Proposition 6.6

(Criterion for a subspace.) Let V be a vector space over F and let U be a subset of V. Then U is a subspace of V if and only if U is closed under addition and scalar multiplication of vectors and U contains the zero vector of V.

Proof

Suppose first that U is a subspace of V. As (6.5)(i) is satisfied, we must show $0 \in U$. As U is a vector space, by law 2 of (6.1), U contains a vector $0'$ such that $0' + u = u$ for all $u \in U$; in particular $0' + 0' = 0'$. Comparing this equation with $0 + 0' = 0'$, from (6.4)(c) we deduce $0 = 0' \in U$. Therefore the zero vector 0 of V belongs to U.

Secondly suppose U is closed under addition and scalar multiplication of vectors and $0 \in U$. As (6.5)(i) is satisfied, we must show that laws 1–7 of (6.1) hold in U. However all these laws, except law 2 and law 3, hold in U simply because they hold in the parent space V; for instance, $u + v = v + u$ for all $u, v \in U$ because $u + v = v + u$ for all $u, v \in V$. As $0 \in U$, the zero vector of V is also the zero vector of U; so law 2 holds in U. Finally, the negative of each vector in U belongs to U: for $u \in U$ implies $(-1)u \in U$, as U is closed under scalar multiplication, and $-u = (-1)u$ by (6.4)(e). Therefore $-u \in U$ for all $u \in U$ and so law 3 holds in U. So (6.5)(ii) is satisfied, showing that U is a subspace of V. □

Let V be the vector space over $\mathbb{R}$ consisting of all functions $f : \mathbb{R} \to \mathbb{R}$ (6.3)(c). Let U be the subset of *continuous* functions $f : \mathbb{R} \to \mathbb{R}$. The sum and constant multiples of continuous functions are again continuous and the zero function (the function which takes the value 0 at every real number) is continuous; applying (6.6) we see that the continuous functions form a subspace U of V. Similarly, the *differentiable* functions $f : \mathbb{R} \to \mathbb{R}$ form a subspace of V. The solutions f of a given linear homogeneous differential equation, such as $x^2(d^2f/dx^2) - 2x(df/dx) + 2f = 0$, form a subspace of V; for the sum and constant multiples of solutions are again solutions and the zero function is a solution. In fact $f(x) = ax + bx^2$, where a and b are real constants, is the general solution in this case.

Notice that each vector space V has two extreme subspaces: the smallest subspace is $\{0\}$, the trivial subspace consisting solely of the zero vector of V, while the largest subspace of V is V itself.

The subspaces of the vector spaces F^n (6.2) will be of special concern to us. For example, let

$$U = \{(x_1, x_2, x_3, x_4) \in \mathbb{Z}_2^4 : x_1 + x_2 + x_3 + x_4 = 0\},$$

that is, U consists of all solutions of the linear homogeneous equation $x_1 + x_2 + x_3 + x_4 = 0$, working over the field $\mathbb{Z}_2$. As the sum and scalar multiples of solutions are again solutions and the zero vector $(0, 0, 0, 0)$ of $\mathbb{Z}_2^4$ is a solution, U is a subspace of $\mathbb{Z}_2^4$ by (6.6). In fact U consists of the eight vectors $(1, 1, 1, 1)$, $(1, 1, 0, 0)$, $(1, 0, 1, 0)$, $(1, 0, 0, 1)$, $(0, 1, 1, 0)$, $(0, 1, 0, 1)$, $(0, 0, 1, 1)$, $(0, 0, 0, 0)$.

Corollary 6.7

Let U and W be subspaces of the vector space V over the field F. Then their intersection $U \cap W$ is a subspace of V.

Proof

We apply (6.6) to $U \cap W$. Let $u, v \in U \cap W$. Then $u, v \in U$ and $u, v \in W$. Therefore $u + v \in U$, since U, being a subspace, is closed under vector addition; similarly $u + v \in W$, for W is closed under vector addition. So $u + v \in U \cap W$, showing that $U \cap W$ is closed under vector addition.

Let $u \in U \cap W$ and $a \in F$. Then $u \in U$ and hence $au \in U$, as U, being a subspace, is closed under scalar multiplication of vectors; as $u \in W$, we see $au \in W$ for the same reason. Therefore $au \in U \cap W$, showing that $U \cap W$ is closed under scalar multiplication of vectors.

As $0 \in U$ and $0 \in W$, we see $0 \in U \cap W$. Therefore $U \cap W$ is a subspace of V by (6.6). $\qquad\square$

For instance, let $U = \{(x, y, z) \in \mathbb{R}^3 : 2x + y = 0\}$ and let $W = \{(x, y, z) \in \mathbb{R}^3 : x - \frac{1}{2}y + z = 0\}$. Then U and W are subspaces of $\mathbb{R}^3$ and may be pictured as two planes through the origin (Fig. 6.3).

Fig. 6.3

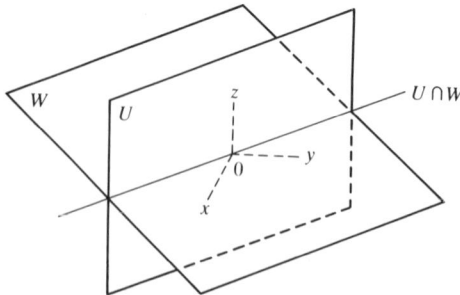

The subspace $U \cap W$ is the line of intersection of these planes. In algebraic terms, $U \cap W$ consists of the solutions (x, y, z) in $\mathbb{R}^3$ of the simultaneous equations $2x + y = 0$, $x - \frac{1}{2}y + z = 0$. Expressing y and z in terms of x, the general solution is

$$(x, y, z) = (x, -2x, -2x) = x(1, -2, -2) \quad \text{where} \quad x \in \mathbb{R}.$$

So $U \cap W$ is the subspace of all scalar multiples of $(1, -2, -2)$.

Definition 6.8

Let $v_1, v_2, \ldots, v_m$ belong to the vector space V over F. Vectors of the form

$$a_1 v_1 + a_2 v_2 + \ldots + a_m v_m \quad \text{where} \quad a_1, a_2, \ldots, a_m \in F$$

are called **linear combinations** of $v_1, v_2, \ldots, v_m$. The subset of V consisting of all such vectors is denoted by $\langle v_1, v_2, \ldots, v_m \rangle$.

Therefore $\langle v_1 \rangle$ is the set of all scalar multiples of the single vector v_1; if v_1 is a non-zero vector in $\mathbb{R}^3$, then $\langle v_1 \rangle$ is the line through the origin and the point v_1. A typical vector in the plane W with equation $x - \frac{1}{2}y + z = 0$ is $(x, y, z) = (x, 2x + 2z, z) = x(1, 2, 0) + z(0, 2, 1)$ on eliminating y, and so $W = \langle (1, 2, 0), (0, 2, 1) \rangle$.

Corollary 6.9

Let $v_1, v_2, \ldots, v_m$ belong to the vector space V over F. Then $\langle v_1, v_2, \ldots, v_m \rangle$ is a subspace of V.

Proof

We write $U = \langle v_1, v_2, \ldots, v_m \rangle$ and apply (6.6). Let $u, v \in U$; then $u = a_1 v_1 + \ldots + a_m v_m$, $v = b_1 v_1 + \ldots + b_m v_m$, where $a_i, b_i \in F$. Using the laws of (6.1) we obtain: $u + v = (a_1 + b_1)v_1 + \ldots + (a_m + b_m)v_m$ and $au = (aa_1)v_1 + \ldots + (aa_m)v_m$ for $a \in F$. Therefore $u + v$ and au are linear combinations of $v_1, v_2, \ldots, v_m$, that is, $u + v, au \in U$; so U is closed under vector addition and scalar multiplication. As $0 = 0v_1 + \ldots + 0v_m$, we see $0 \in U$. Therefore $U = \langle v_1, v_2, \ldots, v_m \rangle$ is a subspace of V by (6.6). □

As $v_1 = 1v_1 + 0v_2 + \ldots + 0v_m$, we see $v_1 \in \langle v_1, v_2, \ldots, v_m \rangle$; similarly $v_1, v_2, \ldots, v_m$ belong to $\langle v_1, v_2, \ldots, v_m \rangle$. Let U be a subspace of V containing $v_1, v_2, \ldots, v_m$; as U is closed under scalar multiplication and addition of vectors, we see that $a_1 v_1 + a_2 v_2 + \ldots + a_m v_m \in U$ for all scalars $a_1, a_2, \ldots, a_m$. Therefore

$$\langle v_1, v_2, \ldots, v_m \rangle \subseteq U$$

showing that $\langle v_1, v_2, \ldots, v_m \rangle$ is the *smallest* subspace containing the vectors $v_1, v_2, \ldots, v_m$.

Definition 6.10   Let $v_1, v_2, \ldots, v_m$ belong to the vector space V. Then $\langle v_1, v_2, \ldots, v_m \rangle$ is called the subspace of V **spanned (generated)** by $v_1, v_2, \ldots, v_m$.

Each non-zero vector v_1 of $\mathbb{R}^3$ spans a line through the origin, namely $\langle v_1 \rangle$; the vectors in $\langle v_1 \rangle$ are proportional to v_1. The non-proportional vectors v_1 and v_2 of $\mathbb{R}^3$ span the plane $\langle v_1, v_2 \rangle$ through the origin (Fig. 6.4). Notice that each plane through the origin is spanned by any pair of its vectors which are non-proportional. For instance

$$\langle (1, 2, 0), (0, 2, 1) \rangle = \langle (1, 2, 0), (1, 4, 1) \rangle$$
$$= \langle (2, 6, 1), (1, 4, 1) \rangle \text{ etc.}$$

since $\langle v_1, v_2 \rangle = \langle v_1, v_1 + v_2 \rangle = \langle 2v_1 + v_2, v_1 + v_2 \rangle$. This point will be taken up when we study **bases** of vector spaces later in the chapter.

Fig. 6.4

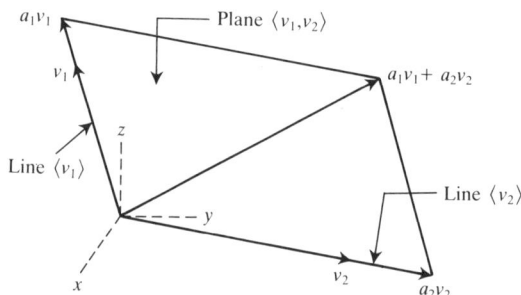

We saw in (6.7) that the intersection of subspaces is again a subspace. However, the union $U \cup W$ of the subspaces U and W is *not* in general a subspace. For example, consider the subspaces $U = \langle (1, 0) \rangle$ and $W = \langle (0, 1) \rangle$ of the cartesian plane $\mathbb{R}^2$; U is the x-axis and W is the y-axis of $\mathbb{R}^2$, and so $U \cup W$ is the 'cross' consisting of points on at least one of these axes. Now $(1, 0)$ and $(0, 1)$ both belong to $U \cup W$, but $(1, 0) + (0, 1) = (1, 1) \notin U \cup W$, showing that $U \cup W$ is not closed under vector addition. Therefore $U \cup W$ is not a subspace of $\mathbb{R}^2$.

We next introduce the concept of subspace sum, which steps into the role that subspace union refuses to play.

Corollary 6.11   Let U and W be subspaces of the vector space V over F. Then their sum

$$U + W = \{u + w : u \in U, w \in W\}$$

is also a subspace of V.

Proof

We use (6.6) once again. Let $v, v' \in U + W$; then $v = u + w$ and $v' = u' + w'$ where $u, u' \in U$ and $w, w' \in W$. Therefore $v + v' = u + w + u' + w' = (u + u') + (w + w') \in U + W$ as $u + u' \in U$ and $w + w' \in W$. Also $av = a(u + w) = au + aw \in U + W$, as $au \in U$ and $aw \in W$ where $a \in F$. As U and W both contain 0, we see $0 = 0 + 0 \in U + W$. So $U + W$ is a subspace of V by (6.6). $\square$

The sum $U + W$ consists of those vectors which are expressible as the sum of two vectors, one from U and the other from W. Consider again the subspaces $U = \langle (1, 0) \rangle$ and $W = \langle (0, 1) \rangle$ of $\mathbb{R}^2$. Now $u = (x, 0) = x(1, 0)$ belongs to U and $w = (0, y) = y(0, 1)$ belongs to W. As every vector (x, y) in $\mathbb{R}^2$ can be decomposed:

$$(x, y) = (x, 0) + (0, y) = u + w \in U + W,$$

we see that $U + W = \mathbb{R}^2$. Notice that $U \cap W$ is the trivial subspace of $\mathbb{R}^2$, that is, $U \cap W = \{(0, 0)\}$.

Let U and W be subspaces of the vector space V and let S be any subspace of V with $U \subseteq S$ and $W \subseteq S$. Therefore $u \in S$ for all $u \in U$, and $w \in S$ for all $w \in W$. As S is closed under vector addition, we see $u + w \in S$ for all $u \in U$ and $w \in W$; so $U + W \subseteq S$. Therefore every subspace containing U and W must contain $U + W$, and so $U + W$ is the *smallest* subspace which contains U and W.

Exercises
6.1

1. Let $u = (1, 2, 4, \frac{1}{2})$, $v = (\frac{1}{2}, 2, 3, -\frac{1}{3})$.

(a) Express as 4-tuples of rational numbers: $u + v$, $u - v$, $6v$, $2u + 3v$, $\frac{1}{3}u + \frac{1}{2}v$.

(b) Find the rational numbers a, b, c, d such that $au + bv = (c, 1, 1, d)$.

2. An attempt is made to turn the set $\mathbb{Z}^3$ of triples of integers into a vector space over the real field $\mathbb{R}$ by defining:

$$(l, m, n) + (l', m', n') = (l + l', m + m', n + n')$$

$$x(l, m, n) = (\lfloor x \rfloor l, \lfloor x \rfloor m, \lfloor x \rfloor n)$$

where $\lfloor x \rfloor$ is the integer part of x. Which of the vector space laws (6.1) hold in this system?

3. Which of the following subsets of $\mathbb{R}^3$ are (i) closed under vector addition, (ii) closed under multiplication by real scalars, (iii) subspaces of $\mathbb{R}^3$?

(a) $\{(x_1, x_2, x_3) : x_i \geqslant 0 \text{ for } i = 1, 2, 3\}$,
(b) $\{(x_1, x_2, 0) : |x_1| \geqslant |x_2|\}$,
(c) $\{(x_1, x_2, x_3) : x_i \in \mathbb{Q} \text{ for } i = 1, 2, 3\}$,
(d) $\{(x, 2x, 3x) : x \in \mathbb{R}\}$,
(e) $\{(x, y, xy) : x, y \in \mathbb{R}\}$,
(f) $\{(x + y, x - y, -x) : x, y \in \mathbb{R}\}$,
(g) $\{(x + y, x - y, z + 1) : x, y, z \in \mathbb{R}\}$.

4. Let $U = \{(x, y, z) \in \mathbb{R}^3 : 2x + 3y + z = 0\}$ and $W = \{(x, y, z) \in \mathbb{R}^3 : x + 2y + 3z = 0\}$. Find vectors v_1, v_2, v_3, v_4, v_5 in $\mathbb{R}^3$ such that $U = \langle v_1, v_2 \rangle$ and $W = \langle v_3, v_4 \rangle$, $U \cap W = \langle v_5 \rangle$. Make a geometric sketch of these subspaces and find vectors u_1, u_2, u_3 such that $U = \langle u_1, u_2 \rangle$ and $W = \langle u_2, u_3 \rangle$.

5. (a) Show that $(1, 0, -3)$ and $(0, 1, -4)$ belong to the plane $U = \langle (1, -1, 1), (4, -5, 8) \rangle$ in $\mathbb{R}^3$, and hence determine the equation of U.
   (b) List the eight vectors in the subspace $U = \langle (1, 1, 1, 0), (1, 1, 0, 1), (1, 0, 1, 1) \rangle$ of $\mathbb{Z}_2^4$. Show that U consists of all solutions of a certain linear homogeneous equation.

6. Let v_1 and v_2 belong to the vector space V over the field F.

   (a) Show that $\langle v_1 + v_2, v_2 \rangle = \langle v_1, v_2 \rangle$
   (b) Show that $\langle v_1 + v_2, v_1 - v_2 \rangle = \langle v_1, v_2 \rangle$ if $F = \mathbb{Q}$.
   (c) If $\langle v_1 + v_2, v_1 - v_2 \rangle \neq \langle v_1, v_2 \rangle$, show that $1 + 1 = 0$ in F.

7. (a) Let U be a *non-empty* subset of the vector space V. Use (6.6) to show that U is a subspace of V if and only if U is closed under addition and scalar multiplication of vectors.
   (b) Let U and W be subspaces of the vector space V. Show that $U \cup W$ is a subspace of V if and only if either $U \subseteq W$ or $W \subseteq U$. (Hint: suppose not.)

8. Which of the following are subspaces of $F[x]$ (regarded as a vector space over the field F (6.3)(b))?

   (i) The constant polynomials over F.
   (ii) The polynomials f over F with $f(1) = 0$.
   (iii) The polynomials $f = \sum a_i x^i$ over F with $a_i = 0$ for all even i.
   (iv) The polynomials of even degree over F.
   (v) The polynomials of degree at most n over F, where n is a given natural number.

9. Let U_1, U_2, U_3 be subspaces of the vector space V.

   (a) Show that subspace sum is not distributive over intersection, that is,

   $$(U_1 \cap U_2) + (U_1 \cap U_3) \neq U_1 \cap (U_2 + U_3)$$

   by considering three subspaces of $\mathbb{R}^2$.

(b) Establish the **modular law**:

$$(U_1 \cap U_2) + (U_1 \cap U_3) = U_1 \cap ((U_1 \cap U_2) + U_3).$$

Is the dual modular law (obtained by interchanging the symbols $+$ and $\cap$ in the above equation) true?

Bases and dimension

We now concentrate on **finite-dimensional** vector spaces, that is, spaces which are spanned by a finite number of their vectors (6.10). The world we live in can be regarded, for most purposes, as being 3-dimensional, and the reader may therefore wonder if higher-dimensional vector spaces are of any practical use. The answer is bound up with the fact that the entries in an n-tuple might *not* arise from the spatial co-ordinates of a point, as in the geometric interpretation of $\mathbb{R}^3$, but from a completely different source. For instance, consider a supermarket which stocks 1000 $(= n$ say) types of food items numbered type 1, type 2, ..., type n. Type 1 items might be cartons of milk, type 2 items bags of sugar, etc. Let a_i be the number of items of type i on display (ready to be purchased); then $v = (a_1, a_2, \ldots, a_n)$ specifies the total goods on display and varies with time in a 1000-dimensional vector space (customers will cause the entries in v to decrease, while assistants re-stocking the shelves will make these entries increase). The point is that every-day activities require a large number of variables to specify them, which means that vector spaces of large dimension do indeed have their uses.

Definition 6.12

Let V be a vector space over the field F. If V contains vectors $v_1, v_2, \ldots, v_m$ such that $V = \langle v_1, v_2, \ldots, v_m \rangle$, then V is called **finite-dimensional**.

So a vector space V is finite-dimensional if it is spanned by a finite number of its vectors $v_1, v_2, \ldots, v_m$, that is, every vector in V is a linear combination of $v_1, v_2, \ldots, v_m$.

The vectors $e_1 = (1, 0, 0)$, $e_2 = (0, 1, 0)$, $e_3 = (0, 0, 1)$ span $\mathbb{R}^3$ as every vector $v = (a_1, a_2, a_3)$ in $\mathbb{R}^3$ is expressible

$$v = (a_1, a_2, a_3) = (a_1, 0, 0) + (0, a_2, 0) + (0, 0, a_3)$$
$$= a_1 e_1 + a_2 e_2 + a_3 e_3.$$

So $\mathbb{R}^3$ is finite-dimensional being spanned by 3 vectors.

More generally, the vector space F^n of (6.2) is finite-dimensional. For let $e_i = (0, \ldots, 0, 1, 0, \ldots, 0)$ be the n-tuple having i th entry 1 (the 1-element of F) and all other entries 0 (the 0-element of F). Each element $v = (a_1, a_2, \ldots, a_n)$ of F^n can be expressed:

$$v = a_1 e_1 + a_2 e_2 + \ldots + a_n e_n,$$

and so $F^n = \langle e_1, e_2, \ldots, e_n \rangle$.

On the other hand, $F[x]$ is not a finite-dimensional vector space (6.3)(b). For suppose it possible that $F[x]$ is spanned by a finite number of polynomials $f_1, f_2, \ldots, f_m$ over F. Let d be the maximum of the degrees of $f_1, f_2, \ldots, f_m$. Then $\deg(a_1 f_1 + a_2 f_2 + \ldots + a_m f_m) \leq d$ for all $a_1, a_2, \ldots, a_m \in F$, and so x^{d+1} is not a linear combination of $f_1, f_2, \ldots, f_m$. This contradiction shows that $F[x]$ is not spanned by any finite number of polynomials.

A finite-dimensional vector space V is spanned by a finite number of its vectors. We now prepare to answer the important question: how can the *most economical* ways of spanning V be found? In other words, how can vectors $v_1, v_2, \ldots, v_m$ spanning V be found so that m is as *small* as possible?

Definition 6.13

Let V be a vector space over the field F. The vectors $v_1, v_2, \ldots, v_m$ of V are called **linearly dependent** if there are scalars $a_1, a_2, \ldots, a_m$ in F, *not all zero*, such that

$$a_1 v_1 + a_2 v_2 + \ldots + a_m v_m = 0.$$

An equation of the above form, with at least one non-zero coefficient a_i, is called a **linear dependence relation** between $v_1, v_2, \ldots, v_m$.

The vectors $v_1 = (-1, 3, 2)$, $v_2 = (2, -1, 1)$, and $v_3 = (1, 1, 2)$ are linearly dependent as $3v_1 + 4v_2 - 5v_3 = 0$. This linear dependence relation can be used to express each of v_1, v_2, v_3 as a linear combination of the other two; we opt to express v_3 in terms of its predecessors v_1 and v_2 by $v_3 = \frac{3}{5} v_1 + \frac{4}{5} v_2$ and so $v_3 \in \langle v_1, v_2 \rangle$. Notice that v_3 is **redundant** (provided that v_1 and v_2 remain employed), meaning

$$\langle v_1, v_2, v_3 \rangle = \langle v_1, v_2 \rangle,$$

since $x_1 v_1 + x_2 v_2 + x_3 v_3 = (x_1 + \frac{3}{5} x_3) v_1 + (x_2 + \frac{4}{5} x_3) v_2$ shows that every linear combination of v_1, v_2, and v_3 is a linear combination of v_1 and v_2 ($x_1, x_2, x_3 \in \mathbb{Q}$).

Linear dependence is treated systematically in our next proposition. It is convenient to regard the empty set $\varnothing$ of vectors as spanning the trivial vector space, that is, we adopt the convention $\langle \varnothing \rangle = \{0\}$.

Proposition 6.14

Let V be a vector space over the field F. Then the vectors $v_1, v_2, \ldots, v_m$ of V are linearly dependent if and only if there is an integer i with $1 \le i \le m$ such that $v_i \in \langle v_1, v_2, \ldots, v_{i-1} \rangle$.

Proof

Suppose first that $v_1, v_2, \ldots, v_m$ are linearly dependent. By (6.13), there is a linear dependence relation $a_1 v_1 + a_2 v_2 + \ldots + a_m v_m = 0$; let i be the *largest* suffix with $a_i \ne 0$. On omitting the last $m - i$ terms (each of which is zero as the scalar coefficient is zero), the above equation becomes $a_1 v_1 + a_2 v_2 + \ldots + a_i v_i = 0$, which rearranges to

$$v_i = -a_i^{-1}(a_1 v_1 + a_2 v_2 + \ldots + a_{i-1} v_{i-1}) \in \langle v_1, v_2, \ldots, v_{i-1} \rangle.$$

(Notice that $i = 1$ gives $v_1 = 0 \in \langle \varnothing \rangle$.)

Secondly, suppose $v_i \in \langle v_1, v_2, \ldots, v_{i-1} \rangle$ for some i with $1 \le i \le m$. Then there are scalars $a_1, a_2, \ldots, a_{i-1}$ such that $v_i = a_1 v_1 + a_2 v_2 + \ldots + a_{i-1} v_{i-1}$; this equation can be expressed

$$a_1 v_1 + a_2 v_2 + \ldots + a_{i-1} v_{i-1} + (-1)v_i + 0v_{i+1} + \ldots + 0v_m = 0$$

which is a linear dependence relation between $v_1, v_2, \ldots, v_m$ as -1 (the coefficient of v_i) is non-zero. Therefore $v_1, v_2, \ldots, v_m$ are linearly dependent. $\qquad\square$

So the vectors $v_1, v_2, \ldots, v_m$ are linearly dependent if and only if either $v_1 = 0$ or there is a vector v_i $(1 < i \le m)$ which is a linear combination of its predecessors $v_1, v_2, \ldots, v_{i-1}$.

In the case of $v_1 = (1, -2, 3)$ and $v_2 = (-2, 4, -6)$ in $\mathbb{Q}^3$, we see $v_2 = -2v_1 \in \langle v_1 \rangle$ and so v_1 and v_2 are linearly dependent by (6.14). More generally, two vectors are linearly dependent if and only if they are proportional (one is a scalar multiple of the other).

Consider $v_1 = (1, 2, 3)$, $v_2 = (2, -1, 1)$, $v_3 = (1, 3, 3)$ in $\mathbb{Q}^3$. One can see at a glance that $v_1 \ne 0$ and $v_2 \notin \langle v_1 \rangle$. Let us suppose $v_3 \in \langle v_1, v_2 \rangle$. Then there are rational numbers x_1 and x_2 with $x_1 v_1 + x_2 v_2 = v_3$, that is,

$$(x_1 + 2x_2, 2x_1 - x_2, 3x_1 + x_2) = (1, 3, 3).$$

Comparing the first and second entries gives: $x_1 + 2x_2 = 1$ and $2x_1 - x_2 = 3$ from which we obtain $x_1 = \frac{7}{5}$ and $x_2 = -\frac{1}{5}$. However, these numbers do *not* satisfy $3x_1 + x_2 = 3$ and so the third entries in the above vectors are not equal. Therefore there are no rational numbers x_1 and x_2 with $x_1 v_1 + x_2 v_2 = v_3$, which means $v_3 \notin \langle v_1, v_2 \rangle$. By (6.14), we conclude that v_1, v_2, v_3 are not linearly dependent.

On the other hand, $v_1 = (1, 2, 3)$, $v_2 = (2, -1, 1)$, $v_3 = (1, 3, 4)$ in $\mathbb{Q}^3$ are linearly dependent: for arguing as in the previous paragraph produces $v_3 = \frac{7}{5} v_1 - \frac{1}{5} v_2 \in \langle v_1, v_2 \rangle$, and so $7v_1 - v_2 - 5v_3 = 0$.

We deal next with the effect (or rather, the lack of effect) of deleting a vector v_i, as in (6.14), which is a linear combination of its predecessors $v_1, v_2, \ldots, v_{i-1}$. We use $\langle v_1, \ldots, \not{v_i}, \ldots, v_m \rangle$ to denote the subspace spanned by $v_1, \ldots, v_{i-1}, v_{i+1}, \ldots, v_m$.

Lemma 6.15

Let $v_1, v_2, \ldots, v_m$ belong to the vector space V over the field F and suppose $v_i \in \langle v_1, v_2, \ldots, v_{i-1} \rangle$ for some integer i with $1 \leqslant i \leqslant m$. Then

$$\langle v_1, \ldots, v_i, \ldots, v_m \rangle = \langle v_1, \ldots, \not{v_i}, \ldots, v_m \rangle.$$

Proof

We simplify the notation by writing $U = \langle v_1, \ldots, v_i, \ldots, v_m \rangle$ and $W = \langle v_1, \ldots, \not{v_i}, \ldots, v_m \rangle$. As $v_1, \ldots, v_{i-1}, v_{i+1}, \ldots, v_m$ belong to the subspace U, all linear combinations of these vectors belong to U, that is, $W \subseteq U$. As $v_1, \ldots, v_{i-1}$ belong to the subspace W, all linear combinations of these vectors belong to W; in particular $v_i \in W$. So in fact $v_1, \ldots, v_i, \ldots, v_m$ all belong to W, and hence all linear combinations of these vectors belong to W, that is, $U \subseteq W$. Therefore $U = W$. $\qquad\square$

Suppose $V = \langle v_1, v_2, \ldots, v_m \rangle$, where m is as small as possible. Could $v_1, v_2, \ldots, v_m$ be linearly dependent? If so, there would be at least one vector v_i with $v_i \in \langle v_1, v_2, \ldots, v_{i-1} \rangle$. Such vectors v_i are redundant, for $V = \langle v_1, \ldots, \not{v_i}, \ldots, v_m \rangle$ by (6.15). So V would be spanned by $m - 1$ vectors. This contradiction shows that $v_1, v_2, \ldots, v_m$ are *not* linearly dependent; we now introduce the appropriate terminology for vectors with this important property.

Definition 6.16

Let V be a vector space over the field F. The vectors $v_1, v_2, \ldots, v_m$ of V are called **linearly independent** if they are not linearly dependent.

From (6.13) we see that $v_1, v_2, \ldots, v_m$ are linearly independent if and only if

$$a_1 v_1 + a_2 v_2 + \ldots + a_m v_m = 0 \quad \text{with} \quad a_1, a_2, \ldots, a_m \in F$$

holds *only* in the case $a_1 = a_2 = \ldots = a_m = 0$.

Consider $v_1 = (1, 1, 1)$, $v_2 = (0, 1, 1)$, $v_3 = (0, 0, 1)$ in $\mathbb{Q}^3$ and suppose $a_1 v_1 + a_2 v_2 + a_3 v_3 = 0$ for some rational numbers a_1, a_2, a_3. Then

$$(a_1, a_1 + a_2, a_1 + a_2 + a_3) = (0, 0, 0)$$

and so $a_1 = a_1 + a_2 = a_1 + a_2 + a_3 = 0$; the only solution of these equations is $a_1 = a_2 = a_3 = 0$ showing that v_1, v_2, v_3 are linearly independent. Similarly, we see that $e_1 = (1, 0, 0)$, $e_2 = (0, 1, 0)$, $e_3 = (0, 0, 1)$ in F^3 are linearly independent.

Our next corollary is a restatement of (6.14).

Corollary 6.17

Let V be a vector space over the field F. Then the vectors $v_1, v_2, \ldots, v_m$ of V are linearly independent if and only if $v_i \notin \langle v_1, v_2, \ldots, v_{i-1} \rangle$ for all integers i with $1 \leqslant i \leqslant m$.

Proof

Let $v_1, v_2, \ldots, v_m$ be linearly independent. If there was an integer i $(1 \leqslant i \leqslant m)$ with $v_i \in \langle v_1, v_2, \ldots, v_{i-1} \rangle$, then $v_1, v_2, \ldots, v_m$ would be linearly dependent by (6.14); therefore there is no such integer, that is, $v_i \notin \langle v_1, v_2, \ldots, v_{i-1} \rangle$ for all integers i with $1 \leqslant i \leqslant m$.

Suppose $v_i \notin \langle v_1, v_2, \ldots, v_{i-1} \rangle$ for all i $(1 \leqslant i \leqslant m)$. Then $v_1, v_2, \ldots, v_m$ cannot be linearly dependent; for if this were so, by (6.14) there would be an integer i $(1 \leqslant i \leqslant m)$ with $v_i \in \langle v_1, v_2, \ldots, v_{i-1} \rangle$. Therefore $v_1, v_2, \ldots, v_m$ are linearly independent. $\qquad \square$

Note that $v_i \notin \langle v_1, v_2, \ldots, v_{i-1} \rangle$ becomes $v_1 \notin \langle \varnothing \rangle$ when $i = 1$, that is, $v_1 \neq 0$. Therefore $v_1, v_2, \ldots, v_m$ are linearly independent if and only if $v_1 \neq 0$, $v_2 \notin \langle v_1 \rangle$, $v_3 \notin \langle v_1, v_2 \rangle$, $\ldots$, $v_m \notin \langle v_1, v_2, \ldots, v_{m-1} \rangle$. Following (6.14) we showed that $v_1 = (1, 2, 3)$, $v_2 = (2, -1, 1)$, $v_3 = (1, 3, 3)$ in $\mathbb{Q}^3$ are linearly independent by verifying $v_1 \neq 0$, $v_2 \notin \langle v_1 \rangle$, and $v_3 \notin \langle v_1, v_2 \rangle$. It will become apparent, especially in Chapter 8, that this recursive (vector by vector) approach to linear independence is a very practical one.

The linear independence of v_1, v_2, v_3 in $\mathbb{R}^3$ has a simple geometric interpretation. The non-zero vector v_1 spans the line $\langle v_1 \rangle$ through 0. As $v_2 \notin \langle v_1 \rangle$, the lines $\langle v_1 \rangle$ and $\langle v_2 \rangle$ are distinct. As $v_3 \notin \langle v_1, v_2 \rangle$, we see that v_3 does not lie in the plane $\langle v_1, v_2 \rangle$. Let v be any vector of $\mathbb{R}^3$. We may construct the parallelepiped (drunken box) having v as diagonal and edges parallel to v_1, v_2, v_3 (Fig. 6.5). Using the parallelogram construction for the vector sum, we obtain $v = x_1 v_1 + x_2 v_2 + x_3 v_3$ where $x_1 v_1$, $x_2 v_2$, $x_3 v_3$ are the vectors along the edges emanating from 0. The upshot is that every

Fig. 6.5

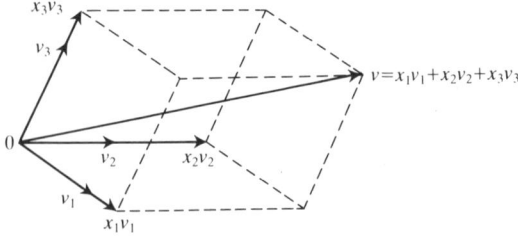

vector in $\mathbb{R}^3$ is a linear combination of v_1, v_2, v_3, that is, $\mathbb{R}^3 = \langle v_1, v_2, v_3 \rangle$. So the vectors v_1, v_2, v_3 give rise to a (generally oblique) system of co-ordinates x_1, x_2, x_3. In later chapters we shall meet a range of problems, notably those involving characteristic vectors (eigenvectors), which are best solved by reference to a co-ordinate system of this type.

It is time to meet the fundamental concept in vector space theory.

Definition 6.18

The n-tuple $(v_1, v_2, \ldots, v_n)$ of vectors belonging to the vector space V is called a **basis** of V if $v_1, v_2, \ldots, v_n$ are linearly independent and span V.

In other words, a basis of V is an ordered set (or sequence) of linearly independent vectors of V which also span. We adopt the usual practice of omitting the round brackets in this context, referring to $(v_1, v_2, \ldots, v_n)$ as the basis $v_1, v_2, \ldots, v_n$.

For example, consider again $v_1 = (1, 1, 1)$, $v_2 = (0, 1, 1)$, $v_3 = (0, 0, 1)$ in $\mathbb{Q}^3$. Following (6.16), we proved that v_1, v_2, v_3 are linearly independent. Let $v = (a_1, a_2, a_3) \in \mathbb{Q}^3$. Then $v = a_1 v_1 + (a_2 - a_1) v_2 + (a_3 - a_2) v_3$, showing that $\mathbb{Q}^3 = \langle v_1, v_2, v_3 \rangle$. Therefore v_1, v_2, v_3 form a basis of $\mathbb{Q}^3$. Interchanging v_1 and v_2 we obtain the different basis v_2, v_1, v_3 of $\mathbb{Q}^3$.

Following (6.12), we saw that the vectors $e_1, e_2, \ldots, e_n$ span F^n. To show that these vectors are linearly independent, suppose $a_1 e_1 + a_2 e_2 + \ldots + a_n e_n = 0$ where $a_1, a_2, \ldots, a_n \in F$. This equality of n-tuples is $(a_1, a_2, \ldots, a_n) = (0, 0, \ldots, 0)$, and so $a_1 = a_2 = \ldots = a_n = 0$ on comparing entries.

Definition 6.19

Let F be a field and n a natural number. Let e_i denote the n-tuple in F^n with i th entry 1 and all other entries 0. Then $e_1, e_2, \ldots, e_n$ (in that order) is called the **standard basis** of F^n.

The standard basis of $\mathbb{R}^3$ is therefore $e_1 = (1, 0, 0)$, $e_2 = (0, 1, 0)$, $e_3 = (0, 0, 1)$.

Suppose V has a basis $v_1, v_2, \ldots, v_n$. Then $V = \langle v_1, v_2, \ldots, v_n \rangle$ and so V is finite-dimensional by (6.12). We show next that every finite-dimensional vector space V has a basis (we adopt the convention that the trivial vector space $\{0\}$ has basis $\varnothing$).

Proposition 6.20   Let $v_1, v_2, \ldots, v_m$ span the vector space V over the field F and let $u_1, u_2, \ldots, u_r$ be linearly independent vectors of V. Then

(a) there are vectors forming a basis of V included among $v_1, v_2, \ldots, v_m$,

(b) there is a basis of V which includes $u_1, u_2, \ldots, u_r$.

Proof   (a) We use the following technique:

From an ordered set of vectors which span V, delete those vectors which are linear combinations of their predecessors; the remaining vectors form a basis of V.

So starting with $v_1, v_2, \ldots, v_m$, we delete all the vectors v_i such that $v_i \in \langle v_1, v_2, \ldots, v_{i-1} \rangle$ (v_1 is deleted $\Leftrightarrow v_1 = 0$). The vectors which remain are linearly independent by (6.17); as $V = \langle v_1, v_2, \ldots, v_m \rangle$, the vectors which remain span V by (6.15). Therefore the remaining vectors form a basis of V and are included among $v_1, v_2, \ldots, v_m$.

(b) We apply the same technique to the ordered set

$$u_1, u_2, \ldots, u_r, v_1, v_2, \ldots, v_m$$

in which the u's precede the v's. As $v_1, v_2, \ldots, v_m$ span V, so do the vectors in the above set. By (6.17), none of $u_1, u_2, \ldots, u_r$ are deleted as $u_i \notin \langle u_1, u_2, \ldots, u_{i-1} \rangle$ for all integers i $(1 \leq i \leq r)$. Therefore the vectors which remain form a basis of V which includes $u_1, u_2, \ldots, u_r$.   □

Consider $v_1 = (1, 1, 0)$, $v_2 = (1, -1, 0)$, $v_3 = (4, -2, 0)$, $v_4 = (1, 1, 1)$, $v_5 = (3, 2, 1)$ in $\mathbb{Q}^3$ and let $V = \langle v_1, v_2, v_3, v_4, v_5 \rangle$. Applying the above technique: retain v_1 (as $v_1 \neq 0$), retain v_2 (as v_2 is not a scalar multiple of v_1), delete v_3 (as $v_3 = v_1 + 3v_2$), retain v_4 (comparing last entries gives $v_4 \notin \langle v_1, v_2 \rangle = \langle v_1, v_2, v_3 \rangle$), and delete v_5 (as $v_5 = \frac{3}{2}v_1 + \frac{1}{2}v_2 + v_4$). We conclude that v_1, v_2, v_4 is a basis of V.

Let $u_1 = (1, 1, 0)$ and $u_2 = (1, -1, 0)$ in $\mathbb{Q}^3$. Then u_1 and u_2 are linearly independent, and we may apply the technique of (6.20)(b) to u_1, u_2, e_1, e_2, e_3 where e_1, e_2, e_3 is the standard basis of $\mathbb{Q}^3$: u_1

and u_2 are retained, e_1 and e_2 are deleted (as $e_1 = \frac{1}{2}(u_1 + u_2)$, $e_2 = \frac{1}{2}(u_1 - u_2)$), and e_3 is retained (as $e_3 \notin \langle u_1, u_2 \rangle$). Therefore u_1, u_2, e_3 form a basis of $\mathbb{Q}^3$.

It is usual to paraphrase (6.20)(b) as follows.

The linearly independent vectors $u_1, u_2, \ldots, u_r$ of the finite-dimensional vector space V can be extended to a basis of V.

We shall make frequent use of this fact in the ensuing theory. In particular the above vectors u_1 and u_2 of $\mathbb{Q}^3$ extend to the basis u_1, u_2, e_3 of $\mathbb{Q}^3$.

In Chapter 8 we develop a method of determining bases of subspaces of F^n which uses the technique of (6.20); this method is direct, systematic, and efficient (just like the Euclidean algorithm), and will allow us to treat examples which are beyond our means at present.

Our next theorem is crucial to the theory of vector spaces. It provides a simple comparison between two sets of vectors belonging to the same vector space V, the vectors of the first set being linearly independent, while the vectors in the second set span V: there cannot be more vectors in the first set than in the second!

Theorem 6.21

(The exchange theorem.) Let $u_1, u_2, \ldots, u_r$ be linearly independent vectors of the vector space V over the field F, and let $v_1, v_2, \ldots, v_m$ span V. Then $r \leq m$.

Proof

The idea of the proof is as follows: starting with $v_1, v_2, \ldots, v_m$ which span V, one v_i is exchanged for u_1, the vector v_i being chosen so that the resulting m vectors still span V. Next, another vector v_i is exchanged for u_2 so that again the resulting m vectors span V. If $m < r$, then m exchanges of this kind can be made, the outcome being that $u_1, u_2, \ldots, u_m$ span V (all the v's have been exchanged for some of the u's). So there remains $u_{m+1} \in V = \langle u_1, u_2, \ldots, u_m \rangle$, which is contrary to the linear independence of $u_1, u_2, \ldots, u_r$ by (6.17). This contradiction shows that the assumption $m < r$ is false; therefore $r \leq m$.

The proof itself is a finite induction. Suppose $m < r$ and let t be an integer with $0 \leq t \leq m$. Let $P(t)$ be the statement:

there are $m - t$ vectors among $v_1, v_2, \ldots, v_m$ which together with $u_1, u_2, \ldots, u_t$ span V.

The induction starts at $t = 0$. As $P(0)$ says that all of the v's together with none of the u's span V, we see that $P(0)$ is true, for $v_1, v_2, \ldots, v_m$ span V by hypothesis. Suppose now that $P(t-1)$ is true where $t \geq 1$; this means that there are $m - t + 1$ vectors (for convenience of notation we take them to be $v_t, v_{t+1}, \ldots, v_m$) among $v_1, v_2, \ldots, v_m$ such that $u_1, u_2, \ldots, u_{t-1}, v_t, v_{t+1}, \ldots, v_m$ span V. In particular u_t is a linear combination of these vectors, and so $u_1, u_2, \ldots, u_{t-1}, u_t, v_t, v_{t+1}, \ldots, v_m$ span V and are linearly dependent. By (6.14) one of these vectors is a linear combination of its predecessors; as none of $u_1, u_2, \ldots, u_t$ have this property by (6.17), the vector in question must be one of $v_t, v_{t+1}, \ldots, v_m$ (for convenience of notation we take it to be v_t). By (6.15) v_t is redundant, that is, $u_1, u_2, \ldots, u_{t-1}, u_t, v_{t+1}, \ldots, v_m$ span V. We have exchanged v_t for u_t and shown that $P(t)$ is true provided $P(t-1)$ is true, for $1 \leq t \leq m$.

Therefore $P(m)$ *is* true, with no strings attached! So $u_1, u_2, \ldots, u_m$, together with no v's, span V. In other words $u_1, u_2, \ldots, u_m$ span V. This leads to a contradiction (see the first paragraph of the proof), showing that $r \leq m$. □

The exchange theorem tells us that a finite-dimensional vector space cannot contain more linearly independent vectors than the number of vectors needed to span it. For example, the standard basis vectors e_1, e_2, e_3 span $\mathbb{Q}^3$; therefore every four vectors in $\mathbb{Q}^3$ must be linearly dependent (for their linear independence would imply $4 \leq 3$ by (6.21)). The reader may carry out a spot check on the theory by writing down at random four vectors in $\mathbb{Q}^3$ (or three vectors in $\mathbb{Q}^2$) and determining a linear dependence relation between them. As e_1, e_2, e_3 are linearly independent vectors of $\mathbb{Q}^3$, no two vectors can span $\mathbb{Q}^3$ (for if they did, (6.21) would say $3 \leq 2$).

Corollary 6.22

Let V be a finite-dimensional vector space over the field F. Then there is a non-negative integer n such that every basis of V consists of exactly n vectors.

Proof

If $V = \{0\}$ is trivial, then $\varnothing$ is the only basis of V, and so $n = 0$ in this case. So suppose V is non-trivial and consider the bases $u_1, u_2, \ldots, u_m$ and $v_1, v_2, \ldots, v_n$ of V (by (6.20) V does have a basis). As the u's are linearly independent and the v's span V, we obtain $m \leq n$ from (6.21). As the v's are linearly independent and the u's span V, we obtain $n \leq m$ from (6.21). Therefore $m = n$. □

From (6.15), we know that bases provide the most economical way of spanning a finite-dimensional vector space V. Now we have the complete answer, for (6.22) tells us that *any* basis of V will do!

In particular, the basis of V obtained by the technique of (6.20) consists of as few vectors as possible.

Definition 6.23

Let the vector space V over the field F have a basis consisting of exactly n vectors. Then n is called the **dimension** of V. We write $n = \dim V$ and refer to the n-**dimensional** vector space V.

So a trivial vector space $\{0\}$ has dimension 0. Every 1-dimensional vector space V consists of all scalar multiples of a given non-zero vector, and so $V = \langle v_1 \rangle$ where $v_1 \neq 0$. The 1-dimensional subspaces of $\mathbb{R}^3$ are the lines through the origin of $\mathbb{R}^3$; for instance the x_1-axis (earlier called the x-axis) is $\langle e_1 \rangle = \{(x_1, 0, 0) : x_1 \in \mathbb{R}\}$. Each non-zero vector of a 1-dimensional vector space V is a basis of V.

A 2-dimensional vector space V consists of all linear combinations of two given linearly independent vectors, and so $V = \langle v_1, v_2 \rangle$ where $v_1 \neq 0$ and $v_2 \notin \langle v_1 \rangle$. The cartesian plane $\mathbb{R}^2$ has basis $e_1 = (1, 0)$ and $e_2 = (0, 1)$ and so $2 = \dim \mathbb{R}^2$. The planes through the origin in $\mathbb{R}^3$ are precisely the 2-dimensional subspaces of $\mathbb{R}^3$.

The reader should be relieved to know that cartesian 3-dimensional space $\mathbb{R}^3$ *is* 3-dimensional in the sense of (6.23), as every basis of $\mathbb{R}^3$ consists of 3 vectors. More generally F^n is n-dimensional, as the standard basis (6.19) of F^n (and hence every basis of F^n) consists of exactly n vectors, and so $n = \dim F^n$.

The reader should be a little wary of our next corollary, which is not an *unqualified* declaration that 'spanning' is the same as 'being linearly independent', for it is necessary that the number of vectors under consideration be equal to the dimension of the space containing them.

Corollary 6.24

Let $v_1, v_2, \ldots, v_n$ belong to the n-dimensional vector space V. Then $v_1, v_2, \ldots, v_n$ span V if and only if $v_1, v_2, \ldots, v_n$ are linearly independent.

Proof

Let $v_1, v_2, \ldots, v_n$ span V. By (6.20)(a) there is a basis of V included among these n vectors. As $n = \dim V$, every basis of V consists of n vectors. Therefore $v_1, v_2, \ldots, v_n$ must form a basis of V, and so these vectors are linearly independent.

Let $v_1, v_2, \ldots, v_n$ be linearly independent. By (6.20)(b) these n vectors are included in some basis of V. But every basis of V consists of n vectors and so $v_1, v_2, \ldots, v_n$ themselves form a basis and so span V. $\square$

Following (6.17), we saw that three linearly independent vectors of $\mathbb{R}^3$ necessarily span $\mathbb{R}^3$; this is a special case of (6.24), which we use in the following way: to prove that $v_1, v_2, \ldots, v_n \in F^n$ form a basis of F^n, it is enough to show *either* that these vectors span F^n *or* that they are linearly independent. For instance, following (6.14) we showed that $v_1 = (1, 2, 3)$, $v_2 = (2, -1, 1)$, $v_3 = (1, 3, 3)$ are linearly independent vectors of $\mathbb{Q}^3$; from (6.24) v_1, v_2, v_3 must span $\mathbb{Q}^3$ and so they form a basis of $\mathbb{Q}^3$.

Lemma 6.25

Let the vector space V over the field F have basis $v_1, v_2, \ldots, v_n$. Then for each vector v in V there are unique scalars $a_1, a_2, \ldots, a_n$ in F such that $v = a_1 v_1 + a_2 v_2 + \ldots + a_n v_n$.

Proof

As $v_1, v_2, \ldots, v_n$ span V, there are $a_1, a_2, \ldots, a_n \in F$ with $v = a_1 v_1 + a_2 v_2 + \ldots + a_n v_n$. Suppose also $v = b_1 v_1 + b_2 v_2 + \ldots + b_n v_n$ where $b_1, b_2, \ldots, b_n \in F$. Subtracting:

$$0 = v - v = (a_1 - b_1)v_1 + (a_2 - b_2)v_2 + \ldots + (a_n - b_n)v_n.$$

As $v_1, v_2, \ldots, v_n$ are linearly independent, the scalar coefficients of these vectors, in the above equation, are all zero: $a_1 - b_1 = a_2 - b_2 = \ldots = a_n - b_n = 0$. Therefore $a_1 = b_1$, $a_2 = b_2, \ldots, a_n = b_n$, showing that v can be expressed in one and only one way as a linear combination of $v_1, v_2, \ldots, v_n$. $\qquad\square$

Using the basis $v_1, v_2, \ldots, v_n$ of V, a bijection of V to F^n may be constructed which respects vector addition and scalar multiplication of vectors (v maps to $(a_1, a_2, \ldots, a_n)$ where $a_1, a_2, \ldots, a_n$ are the scalars of (6.25)). We shall investigate this bijection, which co-ordinatizes V, in the next chapter (see (7.24)); in fact V and F^n are **isomorphic** spaces (they are abstractly identical, and their only differences are notational), the above bijection being an **isomorphism** between them. The reader should not be deterred by these high-sounding terms: for we shall see in Chapter 7 that the abstract structure of V is completely specified by its ground field F and the single integer $n = \dim V$, an isomorphism being little more than a certain type of square matrix.

Let us put $F = \mathbb{Z}_2$ in (6.25). As 0 and 1 are the only elements of $\mathbb{Z}_2$, there are two possibilities for each of $a_1, a_2, \ldots, a_n$ and so 2^n possibilities in all; by (6.25) there are exactly 2^n vectors in the n-dimensional vector space V over the field $\mathbb{Z}_2$. For instance, $V = \{0, v_1, v_2, v_1 + v_2\}$ for $n = 2$, while $V = \{0, v_1, v_2, v_3, v_2 + v_3, v_1 + v_3, v_1 + v_2, v_1 + v_2 + v_3\}$ for $n = 3$; as v_1, v_2, v_3 are linearly independent, no two vectors in the same list are equal.

More generally, let F be a finite field with $|F| = q$. There are q possible values for each of $a_1, a_2, \ldots, a_n \in F$, and hence q^n possibilities in all. By (6.25):

$|V| = q^n$ where V is an n-dimensional
vector space over a finite field F and $q = |F|$.

For instance, every 5-dimensional space over $\mathbb{Z}_3$ consists of exactly $3^5 = 243$ vectors, as $|\mathbb{Z}_3| = 3$. Notice that the above formula is valid for $n = 0$, for every 0-dimensional vector space is trivial and so contains just $q^0 = 1$ vector.

The vector space $V = \mathbb{Z}_2^3$ consists of eight vectors. How many bases does V have? By (6.17) and (6.24) a basis of V is an ordered triple v_1, v_2, v_3 of vectors in V such that $v_1 \neq 0$, $v_2 \notin \langle v_1 \rangle$, $v_3 \notin \langle v_1, v_2 \rangle$. There are seven choices for v_1 (any of the $8 - 1$ vectors in V which are not the zero vector). There are six remaining choices for v_2 (any of the $8 - 2$ vectors in V which do not belong to $\langle v_1 \rangle = \{0, v_1\}$). The vectors v_1, v_2 are linearly independent and so $\langle v_1, v_2 \rangle$ consists of four vectors: $0, v_1, v_2, v_1 + v_2$. Therefore four choices remain for v_3 (any of the $8 - 4$ vectors in V which are not in $\langle v_1, v_2 \rangle$). So $\mathbb{Z}_2^3$ has $7 \times 6 \times 4 = 168$ bases.

More generally, let V be an n-dimensional vector space over a finite field F. Using the method of the preceding paragraph, we now find a formula for the number of bases of V in terms of n and $q = |F|$; this number is of importance in group theory, which is introduced in Chapter 9. In how many ways can a basis $v_1, v_2, \ldots, v_n$ of V be constructed, vector by vector? Let us suppose that $v_1, v_2, \ldots, v_{i-1}$ have been chosen ($1 \leqslant i \leqslant n$); as these vectors are linearly independent, they span an $(i - 1)$-dimensional vector space over F, which therefore contains exactly q^{i-1} vectors, that is, $|\langle v_1, v_2, \ldots, v_{i-1} \rangle| = q^{i-1}$. By (6.17) v_i can be any vector of V with $v_i \notin \langle v_1, v_2, \ldots, v_{i-1} \rangle$; as $|V| = q^n$, there are exactly $q^n - q^{i-1}$ choices for v_i. Taking $i = 1, 2, \ldots, n$ successively, we see that:

Every n-dimensional vector space V over the finite field F has exactly $(q^n - 1)(q^n - q)(q^n - q^2) \ldots (q^n - q^{n-1})$ bases, where $q = |F|$.

In particular, a 3-dimensional vector space over $\mathbb{Z}_5$ has $(5^3 - 1) \times (5^3 - 5)(5^3 - 5^2) = 1,488,000$ bases, as $q = 5$ in this case.

Exercises
6.2

1. In each case find a linear dependence relation between the given vectors of $\mathbb{Q}^3$.

    (a) $(1, 0, 1)$, $(1, 1, 0)$, $(0, 1, -1)$;
    (b) $(1, -1, 2)$, $(2, 1, 2)$, $(1, 5, -2)$;
    (c) $(1, 1, 0)$, $(1, -1, 0)$, $(1, 1, 1)$, $(3, 2, 1)$.

2. In each case show that the given vectors of $\mathbb{Q}^3$ are linearly independent and express $(1, 0, 0)$ as a linear combination of them.

    (a) $(1, 1, 1)$, $(1, -1, 1)$, $(1, 1, -1)$;
    (b) $(1, 0, 1)$, $(0, 1, 2)$, $(1, 2, 3)$;
    (c) $(1, 7, 8)$, $(0, 1, 9)$, $(0, 0, 1)$.

3. (a) The vectors $(1, x, 3)$ and $(x + y, y, 1)$ of $\mathbb{Q}^3$ are linearly dependent. Find x and y.

    (b) The vectors $(1, 2, 1)$, $(1, 1, 2)$, $(x, 1, 1)$ do not span $\mathbb{Q}^3$. Find the rational number x.

    (c) Find the rational number x such that $(1 + x, 2x, 3x)$ belongs to $\langle (2, 1, 1), (1, 2, 1) \rangle$.

4. (a) In each case find a basis included among the given vectors.

    (i) $(1, 2, 1)$, $(1, 1, -1)$, $(0, 2, 4)$, $(2, 3, 0)$, $(2, 3, 1)$ in $\mathbb{Q}^3$.
    (ii) $(1, 2, 1, 2)$, $(1, 1, 2, 2)$, $(1, 3, 0, 2)$, $(2, 1, 1, 2)$, $(1, 2, 2, 1)$, $(1, 1, 1, 1)$ in $\mathbb{Q}^4$.
    (iii) $(1, 1, 1, 1)$, $(1, 1, 0, 0)$, $(0, 0, 1, 1)$, $(1, 0, 0, 1)$, $(0, 1, 1, 0)$, $(1, 0, 1, 0)$ in $\mathbb{Q}^4$.

    (b) Show that there are 36 bases of $\mathbb{Q}^3$ consisting of vectors selected from (i) above. How many bases of $\mathbb{Q}^4$ can be formed using the vectors of (ii) above, and how many using the vectors of (iii) above?

5. Let V be a vector space.

    (a) Suppose $v_1, v_2, \ldots, v_m$ are linearly dependent vectors of V such that $v_1, v_2, \ldots, v_{m-1}$ are linearly independent. Prove, directly from (6.13) and (6.16), that $v_m \in \langle v_1, v_2, \ldots, v_{m-1} \rangle$.

    (b) Let $v_1, v_2, \ldots, v_m$ span V. Show that $V = \langle v_1, \ldots, \not{v_i} \ldots, v_m \rangle$ if and only if v_i has a non-zero coefficient in some linear dependence relation between $v_1, v_2, \ldots v_m$.

    (c) Let $u_1, u_2, \ldots, u_r$ be linearly independent vectors of V and let $0 \neq u \in V$. Show that $u_i \in \langle u, u_1, u_2, \ldots, u_{i-1} \rangle$ for some integer i $(1 \leqslant i \leqslant r)$ if and only if $u \in \langle u_1, u_2, \ldots, u_r \rangle$.

6. Let u_0 and v_0 be linearly independent vectors of the vector space V over the field F, and let $u_1 = au_0 + bv_0$ and $v_1 = cu_0 + dv_0$ $(a, b, c, d \in F)$. Show that u_1 and v_1 are linearly independent if and only if $ad - bc \neq 0$.

7. (a) Show that 1, $\sqrt{2}$, $\sqrt{3}$ are linearly independent vectors of $\mathbb{R}$ regarded as a vector space over the rational field $\mathbb{Q}$ (see (6.3)(b)).

(b) The complex numbers $\mathbb{C}$ form a vector space over the real field $\mathbb{R}$ ((6.3)(b)). Show that this vector space has basis 1, i and deduce that 1, z, z^2 are linearly dependent over $\mathbb{R}$, for all $z \in \mathbb{C}$. Find a linear dependence relation over $\mathbb{R}$ between 1, $\sqrt{2} + i\sqrt{3}$, $(\sqrt{2} + i\sqrt{3})^2$.

8. Here the set $P(U)$, of all subsets of the set U, is regarded as a vector space over $\mathbb{Z}_2$ (6.3)(a).

(a) Let $U = \{1, 2, 3, 4, 5\}$. Show that $v_1 = \{1, 2, 3\}$, $v_2 = \{1, 3, 4\}$, $v_3 = \{1, 2, 5\}$, $v_4 = \{1, 4, 5\}$ are linearly dependent, but that any three of these vectors are linearly independent. Find a basis of $P(U)$ which includes $u_1 = \{1, 2, 3\}$, $u_2 = \{1, 2, 3, 4\}$, $u_3 = \{1, 2, 3, 4, 5\}$.

(b) If U is a finite set, show that the $|U|$ subsets $\{x\}$ ($x \in U$) form a basis of $P(U)$. Deduce that $|U| = \dim P(U)$.

(c) Show that $P(U)$ is finite-dimensional if and only if U is finite.

Complementary subspaces

A finite-dimensional vector space V can be decomposed (broken down into components) in a variety of ways, and we shall see in the following chapters that this property is extremely useful: problems which arise in connection with V can often be solved by reference to a suitable decomposition of V. Here we deal with the decomposition of V into two complementary subspaces U and W, which is symbolized by the equation

$$V = U \oplus W$$

the essential fact being that each vector v in V is *uniquely* expressible in the form

$$v = u + w \qquad (u \in U, w \in W).$$

We show first that the dimension of a subspace cannot be greater than the dimension of its parent space.

Proposition 6.26 Let U be a subspace of the finite-dimensional vector space V. Then U is finite-dimensional and $\dim U \leqslant \dim V$.

Proof To show that U is finite-dimensional, we must find a finite number of vectors *belonging to* U which span U. (Although the conclusion of (6.26) is not surprising, nevertheless there is something to be proved!) Let $n = \dim V$ and let $u_1, u_2, \ldots, u_r$ be linearly independent vectors of U. As $u_1, u_2, \ldots, u_r$ belong also to V, by the exchange theorem we see $r \leq n$. Suppose now that r is chosen as *large* as possible ($r = 0 \Leftrightarrow U$ is trivial). Then $u_1, u_2, \ldots, u_r$ span U: for otherwise there would be u_{r+1} in U with $u_{r+1} \notin \langle u_1, u_2, \ldots, u_r \rangle$ giving, by (6.17), $r + 1$ linearly independent vectors $u_1, u_2, \ldots, u_r, u_{r+1}$ in U, contrary to the choice of r. So $u_1, u_2, \ldots, u_r$ form a basis of U, which is therefore finite-dimensional, and dim $U = r \leq n = $ dim V. $\qquad\square$

Notice that an n-dimensional vector space V has only one n-dimensional subspace, namely V itself: for if U is a subspace of V with dim $U = $ dim V, any basis of U spans V by (6.24), and so $U = V$.

The subspaces of $\mathbb{R}^3$ are: the origin (the unique 0-dimensional subspace), lines through the origin (the 1-dimensional subspaces), planes through the origin (the 2-dimensional subspaces), $\mathbb{R}^3$ itself (the unique 3-dimensional subspace).

Definition 6.27

Let U be a subspace of the vector space V. A subspace W of V is called a **complement** of U in V if

$$U + W = V \quad \text{and} \quad U \cap W = 0.$$

Let W be a complement of U in V; then every vector in V is expressible in the form $u + w$ ($u \in U$, $w \in W$), for this is the meaning of the equation $U + W = V$ by (6.11), and the zero vector is the only vector common to U and W (we denote the trivial subspace by 0 rather than $\{0\}$). By the symmetry of (6.27), U is a complement of W in V, and we therefore refer to the **complementary pair** of subspaces U and W.

Fig. 6.6

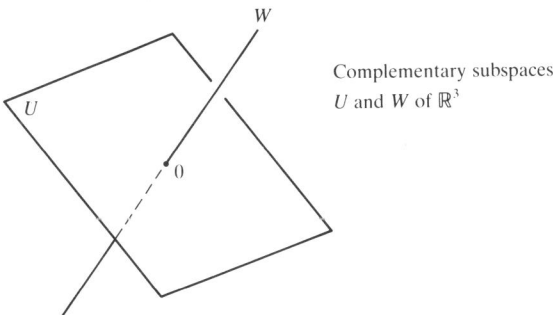

Complementary subspaces
U and W of $\mathbb{R}^3$

For example, let $V = \mathbb{R}^3$ and let U denote the plane in $\mathbb{R}^3$ with equation $2x - y + 2z = 0$; so $U = \langle (1, 2, 0), (0, 2, 1) \rangle$. Let W denote the line $\langle (1, 1, 1) \rangle$ (Fig. 6.6). Then U has a basis $v_1 = (1, 2, 0)$, $v_2 = (0, 2, 1)$ and W has a basis $v_3 = (1, 1, 1)$; taken together we obtain the basis v_1, v_2, v_3 of $\mathbb{R}^3$. As v_1, v_2, v_3 span

$\mathbb{R}^3$, we see $U + W = \mathbb{R}^3$; as v_1, v_2, v_3 are linearly independent, we obtain $U \cap W = 0$, for $v_3 \notin \langle v_1, v_2 \rangle$ tells us that non-zero multiples of v_3 do not belong to U. So U and W are a complementary pair of subspaces of $\mathbb{R}^3$. More generally, any plane U through the origin in $\mathbb{R}^3$ and any line W through the origin in $\mathbb{R}^3$ are complementary subspaces of $\mathbb{R}^3$ provided that W does not lie in U.

Let U and W be a complementary pair of subspaces of V and suppose $u + w = u' + w'$ where $u, u' \in U$ and $w, w' \in W$. We rearrange this equation so that vectors in U appear on one side and vectors in W appear on the other side:

$$u - u' = w' - w.$$

As $u - u' \in U$ and $w' - w \in W$, the above equation tells us that these vectors, being equal, belong to U *and* W, that is, to $U \cap W$. As $U \cap W = 0$ we conclude $u - u' = w' - w = 0$, that is, $u = u'$ and $w = w'$. Therefore every v in V is *uniquely* expressible $v = u + w$, where $u \in U$ and $w \in W$. (In fact, we may regard V as being built up from its complementary 'component' subspaces U, W and use the ordered pair notation (u, w) instead of $u + w$ for a typical vector of V.)

Definition 6.28

Let U and W be complementary subspaces of the vector space V. Then V is said to be the (**internal**) **direct sum** of U and W, and we write

$$V = U \oplus W.$$

So the equation $V = U \oplus W$ is shorthand for: each v in V is uniquely expressible as $v = u + w$ where $u \in U$ and $w \in W$. (The reader who is familiar with Chapter 5, should compare (6.28) with (5.24) and become reconciled to the present usage of the symbol $\oplus$.)

For instance $\mathbb{R}^2 = \langle e_1 \rangle \oplus \langle e_2 \rangle$ shows that the cartesian plane $\mathbb{R}^2$ is the direct sum of the x_1-axis and the x_2-axis. Let V be the vector space of all functions $f : \mathbb{R} \to \mathbb{R}$ (see (6.3)(c)), let U denote the 1-dimensional subspace of all constant functions, and let W be the subspace of all functions g in V satisfying $g(0) = 0$; then $V = U \oplus W$ $(f(x) = f(0) + (f(x) - f(0)))$ is the only way of expressing f as the sum of two functions, one in U and one in W).

Extra care is required when using $\oplus$ in this context: $U \oplus W$ makes sense *only* for subspaces U and W of a vector space V with $U \cap W = 0$ (in which case U and W are complementary subspaces of $U + W$, and so $U \oplus W = U + W$).

We come to grips with the concept of direct sum in our next proposition, for, in the finite-dimensional case, it is closely related to bases.

Proposition 6.29 Let U and W be subspaces of the vector space V over the field F. Let U have basis $u_1, u_2, \ldots, u_r$ and let W have basis $w_1, w_2, \ldots, w_s$. Then $V = U \oplus W$ if and only if $u_1, u_2, \ldots, u_r, w_1, \ldots, w_s$ is a basis of V.

Proof Let ℓ denote the basis $u_1, u_2, \ldots, u_r$ of U and let ℓ' denote the basis $w_1, w_2, \ldots, w_s$ of W. We write $\ell \cup \ell'$ for the ordered set of vectors $u_1, u_2, \ldots, u_r, w_1, w_2, \ldots, w_s$.

Suppose first that $V = U \oplus W$ and consider $v \in V$. There are unique vectors $u \in U$, $w \in W$ with $v = u + w$. Expressing u and w as linear combinations of the vectors in ℓ and ℓ' respectively, we see from (6.25) that there are unique scalars $a_1, a_2, \ldots, a_r, b_1, b_2, \ldots, b_s$ in F with

$$v = a_1 u_1 + a_2 u_2 + \ldots + a_r u_r + b_1 w_1 + b_2 w_2 + \ldots + b_s w_s.$$

In particular, if $v = 0$ then all these scalars are zero. Therefore the vectors in $\ell \cup \ell'$ are linearly independent and span V; so $\ell \cup \ell'$ is a basis of V.

Secondly, suppose that $\ell \cup \ell'$ is a basis of V. Reversing the steps in the foregoing argument, we see that for each v in V there are unique vectors $u \in U$ and $w \in W$ with $v = u + w$. Therefore $V = U + W$; taking $v = 0$ we see $u = w = 0$, that is, the *only* way of expressing 0 as a sum of two vectors, one in U and one in W, is $0 = 0 + 0$. Let $u_0 = w_0$ belong to $U \cap W$, and so $0 = u_0 + (-w_0)$. Therefore $u_0 = -w_0 = 0$ as $u_0 \in U$ and $-w_0 \in W$. The conclusion is $U \cap W = 0$, and so $V = U \oplus W$. □

It is worth noting that if $u_1, u_2, \ldots, u_r$ are linearly independent vectors of V and if $w_1, w_2, \ldots, w_s$ are also linearly independent vectors of V, then it is generally *false* that taken together $u_1, u_2, \ldots, u_r, w_1, w_2, \ldots, w_s$ are linearly independent. For instance in $\mathbb{Q}^3$, $(1, 0, 0)$ and $(0, 1, 0)$ are linearly independent and $(0, 0, 1)$ and $(1, 1, 1)$ are also linearly independent, but $(1, 0, 0)$, $(0, 1, 0)$, $(0, 0, 1)$, $(1, 1, 1)$ are linearly dependent.

Corollary 6.30 Let U be a subspace of the finite-dimensional vector space V. Then U has at least one complement W in V, and

$$\dim U + \dim W = \dim V.$$

Proof   Let $n = \dim V$. By (6.26), $r = \dim U$ where $0 \leqslant r \leqslant n$. Let $u_1, u_2, \ldots, u_r$ form a basis of U. By (6.20)(b), this basis of U can be extended to a basis $u_1, u_2, \ldots, u_r, u_{r+1}, \ldots, u_n$ of V. By (6.29), $W = \langle u_{r+1}, \ldots, u_n \rangle$ is a complement of U in V. As the $n - r$ vectors $u_{r+1}, \ldots, u_n$ form a basis of W, we see $n - r = \dim W$; therefore $\dim U + \dim W = \dim V$.   □

For instance, let $U = \langle (1, 2, 3) \rangle$ and $V = \mathbb{Q}^3$. To find a complement W of U in V, extend the basis $(1, 2, 3)$ of U to a basis of V; one way to extend is: $(1, 2, 3)$, $(1, 0, 0)$, $(0, 1, 0)$. By (6.30), we may take $W = \langle (1, 0, 0), (0, 1, 0) \rangle$.

The trivial subspace 0 has a unique complement in V, namely V itself; similarly V has a unique complement in V, namely 0. Generally, a subspace U will have several complements in V.

For example, consider the 1-dimensional subspace $U = \langle u_1 \rangle$ of $V = \mathbb{Z}_2^3$. How many complements W does U have in V? As $3 = \dim V$, from (6.30) we see that each W is 2-dimensional; in fact, $W = \langle u_2, u_3 \rangle$ where u_1, u_2, u_3 form a basis of V. Here we require the number of ways of extending u_1 to a basis u_1, u_2, u_3 of V; there are 6 choices for u_2, and 4 remaining choices for u_3, giving 24 bases of V beginning with the vector u_1. So the subspaces W have, between themselves, 24 different bases. However each W is a 2-dimensional vector space over $\mathbb{Z}_2$ and so has 6 bases. Therefore U has $24/6 = 4$ complements W in V.

The dimension of a vector space (over a given field) is a measure of its size. We end this chapter with an application of complements, proving the analogue, for finite-dimensional vector spaces, of the equation $|X_1 \cup X_2| + |X_1 \cap X_2| = |X_1| + |X_2|$ relating the sizes of (the number of elements in) the finite sets $X_1, X_2, X_1 \cup X_2, X_1 \cap X_2$.

Proposition 6.31   Let U_1 and U_2 be finite-dimensional subspaces of the vector space V. Then $U_1 + U_2$ is finite-dimensional and

$$\dim(U_1 + U_2) + \dim(U_1 \cap U_2) = \dim U_1 + \dim U_2.$$

Proof   Write $U = U_1 \cap U_2$. As U_1 is finite-dimensional, U has a complement W_1 in U_1 by (6.30).

We show that W_1 and U_2 are complementary subspaces of $U_1 + U_2$. As $U \subseteq U_2$, we have $U_2 = U + U_2$ and so

$$W_1 + U_2 = W_1 + (U + U_2) = (W_1 + U) + U_2 = U_1 + U_2$$

since $U_1 = W_1 + U$. Also

$$W_1 \cap U_2 = (W_1 \cap U_1) \cap U_2 = W_1 \cap (U_1 \cap U_2)$$
$$= W_1 \cap U = 0.$$

Therefore $U_1 + U_2 = W_1 \oplus U_2$; by (6.29) $U_1 + U_2$ is finite-dimensional and $\dim(U_1 + U_2) = \dim W_1 + \dim U_2$. From $U_1 = W_1 \oplus U$ we obtain $\dim(U_1 \cap U_2) = \dim U = \dim U_1 - \dim W_1$, and so $\dim(U_1 + U_2) + \dim(U_1 \cap U_2) = \dim U_1 + \dim U_2$. $\qquad\square$

The reader who is unhappy with (or even unconvinced by) the proof of (6.31) should verify the set equations involved from first principles (e.g. $W_1 + U_2 = U_1 + U_2$ rests on subspace sum introduced in (6.11)).

The reader may be unhappy with the lack of symmetry in the proof of (6.31); this can be remedied by letting W_2 be a complement of U in U_2. Then $U_1 + U_2 = W_1 \oplus U_2 = W_1 \oplus (U \oplus W_2)$; the bracket may be omitted ($\oplus$ is commutative and associative) to give

$$U_1 + U_2 = W_1 \oplus U \oplus W_2.$$

So $U_1 + U_2$ is the direct sum of three subspaces (just as there are three regions inside the Venn diagram of $X_1 \cup X_2$).

As a numerical illustration of (6.31), let $v_1 = (1, 1, 0)$, $v_2 = (0, 1, 1)$, $v_3 = (1, 2, 2)$, $v_4 = (2, 1, 2)$ in $\mathbb{R}^3$, and take $U_1 = \langle v_1, v_2 \rangle$ and $U_2 = \langle v_3, v_4 \rangle$. Then $\dim U_1 = \dim U_2 = 2$. Now $U_1 + U_2$ is spanned by v_1, v_2, v_3, v_4, that is, $U_1 + U_2 = \langle v_1, v_2, v_3, v_4 \rangle$. The reader may verify that v_1, v_2, v_3 are linearly independent and so $3 \leqslant \dim(U_1 + U_2)$; but $U_1 + U_2$ is a subspace of $\mathbb{R}^3$, and so $3 = \dim(U_1 + U_2)$ which means $U_1 + U_2 = \mathbb{R}^3$. The vectors v_1, v_2, v_3, v_4 are linearly dependent (being 4 vectors in a 3-dimensional space); in fact $v_1 + 4v_2 - 3v_3 + v_4 = 0$. We rearrange this equation so that vectors in U_1 appear on one side and vectors in U_2 appear on the other (and so both sides belong to $U_1 \cap U_2$):

$$v_1 + 4v_2 = 3v_3 - v_4 = (1, 5, 4) \in U_1 \cap U_2.$$

Knowing the dimensions of U_1, U_2, $U_1 + U_2$ to be 2, 2, 3 respectively, from (6.31) we deduce $\dim(U_1 \cap U_2) = 1$. Therefore $U_1 \cap U_2 = \langle (1, 5, 4) \rangle$. Geometrically, U_1 and U_2 are planes through the origin of $\mathbb{R}^3$, their intersection being the line $\langle (1, 5, 4) \rangle$. In Chapter 8 we shall find a systematic way of finding a basis of $U_1 \cap U_2$ starting from given bases of U_1 and U_2.

One final comment: nowhere in this chapter have we made use of the commutativity of multiplication in F, and so the theory remains valid if F is a **skew field** (a non-trivial ring within which each non-zero element has an inverse). However, if F is genuinely skew, the system V of (6.1) is called a **left** vector space over F, as in this case it is important to draw attention to the fact that scalars appear on the left of vectors.

Exercises 6.3

1. Show that the following pairs U and W are complementary subspaces of $\mathbb{Q}^3$. In each case find the unique vectors $u \in U$ and $w \in W$ with $u + w = (1, 2, 3)$.

(a) $U = \langle (2, 1, 1), (2, 1, 2) \rangle$, $W = \langle (1, 0, 0) \rangle$;
(b) $U = \langle (1, 1, 1) \rangle$, $W = \langle (2, 1, 0), (1, 2, 0) \rangle$;
(c) $U = \langle (1, 1, 0), (1, 2, -1), (2, 1, 1) \rangle$, $W = \langle (0, 1, 1) \rangle$.

2. Let $v_1 = (1, 1, 1, 1)$, $v_2 = (1, 1, 0, 0)$, $v_3 = (0, 0, 1, 1)$, $v_4 = (0, 1, 1, 0)$, $v_5 = (1, 0, 0, 1)$, $v_6 = (1, 0, 1, 0)$ in $\mathbb{Q}^4$. In which of the following cases is the use of $\oplus$ legitimate (as in (6.28))?

(a) $\langle v_1, v_2 \rangle \oplus \langle v_4, v_6 \rangle$,
(b) $\langle v_2, v_4 \rangle \oplus \langle v_3, v_5 \rangle$,
(c) $\langle v_1, v_2, v_3 \rangle \oplus \langle v_5 \rangle \oplus \langle v_6 \rangle$,
(d) $\langle v_2, v_3 \rangle \oplus \langle v_4 \rangle \oplus \langle v_5 \rangle$.

3. Let $v_1 = (1, 2, 2)$, $v_2 = (1, 1, 2)$, $v_3 = (2, 4, 3)$, $v_4 = (1, 1, 1)$ in $\mathbb{Q}^3$.

(a) Show $\langle v_1, v_2, v_3, v_4 \rangle = \mathbb{Q}^3$.
(b) Find a linear dependence relation between v_1, v_2, v_3, v_4.
(c) Use (6.31) to determine the dimension of $\langle v_1, v_2 \rangle \cap \langle v_3, v_4 \rangle$, and find a basis of this subspace.
(d) Find a basis of $\langle v_1, v_3 \rangle \cap \langle v_2, v_4 \rangle$.

4. (a) Let v_1 and v_2 form a basis of the 2-dimensional vector space V over the field F. Show that each complement of $U = \langle v_1 \rangle$ in V has the form $W_a = \langle a v_1 + v_2 \rangle$ for a unique scalar $a \in F$.

(b) Let U be a subspace of the finite-dimensional vector space V. If $0 \neq U \neq V$, show that U has at least two complements in V.

5. (a) Let u, v, w form a basis of the 3-dimensional vector space V. Show that $\langle v + w \rangle$ is a complement in V of both $\langle u, v \rangle$ and $\langle u, w \rangle$.

(b) Prove that any two r-dimensional subspaces of the n-dimensional vector space V have a common complement in V.

6. (a) Let U_1 and U_2 be subspaces of the vector space V. If $3 = \dim U_1$, $4 = \dim U_2$, $6 = \dim V$, show that $U_1 \cap U_2$ contains a non-zero vector. If $2 = \dim U_1$, $4 = \dim U_2$, $6 = \dim V$, show that $U_1 + U_2 = V$ if and only if $U_1 \cap U_2 = 0$.

(b) Let U_1 and U_2 be distinct $(n-1)$-dimensional subspaces of the n-dimensional vector space V. Show that $n - 2 = \dim(U_1 \cap U_2)$. If $U_1, U_2, \ldots, U_t$ are $(n-1)$-dimensional subspaces of V, prove

$$n - t \leqslant \dim(U_1 \cap U_2 \cap \ldots \cap U_t).$$

7. (a) List the seven 1-dimensional subspaces of $\mathbb{Z}_2^3$. How many 2-dimensional subspaces does $\mathbb{Z}_2^3$ have? How many complements in $\mathbb{Z}_2^3$ does each 2-dimensional subspace of $\mathbb{Z}_2^3$ have?

(b) Show that every non-zero vector of F^3 is proportional to exactly one of $(1, a, b)$, $(0, 1, a)$, $(0, 0, 1)$, where $a, b \in F$. Hence show that the number of 1-dimensional subspaces of $\mathbb{Z}_p^3$ is $p^2 + p + 1$ (p prime).

(c) Let V be an n-dimensional vector space over the finite field F, and let r be an integer with $1 \leqslant r \leqslant n$. Show that the number of bases of r-dimensional subspaces of V is $(q^n - 1)(q^n - q^{n-1}) \ldots (q^n - q^{r-1})$, where $q = |F|$. Hence show that the number of r-dimensional subspaces of V is

$$(q^n - 1)(q^n - q^{n-1}) \ldots (q^n - q^{r-1})/(q^r - 1)(q^r - q) \ldots (q^r - q^{r-1}).$$

Show that each r-dimensional subspace of V has exactly $q^{r(n-r)}$ complements in V.

7 Matrices and linear mappings

Here we introduce the reader to **matrices**, which are nothing more than inanimate rectangular arrays of scalars. Our aim, however, is to give life and meaning to matrices by associating them with mappings of a certain type—linear mappings of vector spaces—operations on matrices being tailor-made for the purpose; in particular, *multiplication* of matrices corresponds to *composition* of mappings. Specifically, suppose we carry out the linear substitutions

$$x_1 = ay_1 + by_2 \qquad\qquad y_1 = a'z_1 + b'z_2$$
$$x_2 = cy_1 + dy_2 \qquad\qquad y_2 = c'z_1 + d'z_2$$

in succession. The first pair of equations expresses the variables x_1 and x_2 in terms of the variables y_1 and y_2; the second pair expresses y_1 and y_2 in terms of z_1 and z_2. Suppressing these variables, we are left with the corresponding matrices:

$$\begin{pmatrix} a & b \\ c & d \end{pmatrix} \qquad\qquad \begin{pmatrix} a' & b' \\ c' & d' \end{pmatrix}$$

that is, the arrays of constants which link the variables. On eliminating the intermediate variables y_1 and y_2 (that is, substituting for y_1 and y_2), we obtain the composite substitution

$$x_1 = (aa' + bc')z_1 + (ab' + bd')z_2$$
$$x_2 = (ca' + dc')z_1 + (cb' + dd')z_2$$

which expresses x_1 and x_2 directly in terms of z_1 and z_2. Suppressing the variables in this substitution gives **the rule of matrix multiplication**:

$$\begin{pmatrix} a & b \\ c & d \end{pmatrix}\begin{pmatrix} a' & b' \\ c' & d' \end{pmatrix} = \begin{pmatrix} aa' + bc' & ab' + bd' \\ ca' + dc' & cb' + dd' \end{pmatrix}$$

for the product of the individual matrices is defined (7.6) to be the matrix of the composite substitution.

A vector space is a system in which operations of vector addition and scalar multiplication of vectors can be carried out (subject to the laws (6.1)). The theory and practice of vector spaces requires the parallel study of **linear mappings**, that is, mappings of vector spaces which respect these operations. We shall say more later, but it turns out that, just as F^n is the co-ordinate form of an abstract n-dimensional vector space over F, so each linear mapping (of non-trivial finite-dimensional vector spaces) can be represented in concrete form by a suitable matrix.

Matrices

We discuss the properties of matrices having entries from a field F.

Definition 7.1

Let F be a field. Any rectangular array

$$\begin{pmatrix} a_{11} & a_{12} & \cdots & a_{1n} \\ a_{21} & a_{22} & \cdots & a_{2n} \\ \vdots & \vdots & & \vdots \\ a_{m1} & a_{m2} & \cdots & a_{mn} \end{pmatrix}$$

of elements a_{ij} belonging to F is called a **matrix** over F. The element a_{ij} in row i and column j is called the (i, j)-**entry** in the matrix.

As indicated in the above matrix, rows are numbered from top to bottom using the first suffix i, and columns are numbered from left to right using the second suffix j. A matrix having m rows and n columns is called an $m \times n$ matrix. Generally, capital letters $A, B, C, \ldots$ will denote matrices, and we write $A = [a_{ij}]$ for the matrix with typical entry a_{ij}. For example $A = \begin{pmatrix} \frac{1}{2} & 0 & \frac{5}{4} \\ -\frac{1}{5} & \frac{4}{3} & 7 \end{pmatrix}$ is a 2×3 matrix over $\mathbb{Q}$; the $(1, 3)$-entry in A is $\frac{5}{4}$ and the $(2, 2)$-entry is $\frac{4}{3}$. The second row of A is $[-\frac{1}{5} \quad \frac{4}{3} \quad 7]$ and the third column of A is $\begin{pmatrix} \frac{5}{4} \\ 7 \end{pmatrix}$.

Definition 7.2

The matrices $A = [a_{ij}]$ and $B = [b_{ij}]$ are called **equal** if they are $m \times n$ matrices for some natural numbers m and n, and $a_{ij} = b_{ij}$ for all i and j $(1 \leq i \leq m, 1 \leq j \leq n)$.

So the matrix equation $A = B$ means that A and B are equal in *every* respect: they have the same number of rows, the same number of columns and corresponding entries agree. The number of 2×2 matrices over $\mathbb{Z}_2$ is 16, as there are two choices (either 0 or 1) for each of the four entries in such matrices; in particular

$$\begin{pmatrix} 1 & 0 \\ 0 & 1 \end{pmatrix} \neq \begin{pmatrix} 1 & 0 \\ 1 & 1 \end{pmatrix}$$

as the $(2, 1)$-entries disagree. More generally, if F is a finite field and $q = |F|$, then the number of $m \times n$ matrices over F is q^{mn}, as there are q choices for each of the mn entries.

**Definition
7.3**

Let $A = [a_{ij}]$ and $B = [b_{ij}]$ be $m \times n$ matrices over the field F. The $m \times n$ matrix $A + B = [a_{ij} + b_{ij}]$ is called the **sum** of A and B.

Notice that $A + B$ makes sense only for matrices A and B of the same **shape** (having the same number of rows and the same number of columns) and having entries from the same field F; the sum of such matrices is formed by adding corresponding entries. For example

$$\begin{pmatrix} 1 & 2 & 3 \\ 4 & 5 & 6 \end{pmatrix} + \begin{pmatrix} 10 & 11 & 12 \\ 13 & 14 & 15 \end{pmatrix} = \begin{pmatrix} 11 & 13 & 15 \\ 17 & 19 & 21 \end{pmatrix}.$$

**Definition
7.4**

Let $A = [a_{ij}]$ be an $m \times n$ matrix over the field F. The $m \times n$ matrices $aA = [aa_{ij}]$, for $a \in F$, are called **scalar multiples** of A.

The matrix aA is formed by multiplying *all* the entries in the matrix A by the scalar a. For instance

$$\tfrac{1}{6}\begin{pmatrix} 1 & 2 & 3 \\ 4 & 5 & 6 \end{pmatrix} = \begin{pmatrix} \frac{1}{6} & \frac{1}{3} & \frac{1}{2} \\ \frac{2}{3} & \frac{5}{6} & 1 \end{pmatrix}.$$

Notation

The set of all $m \times n$ matrices over the field F is denoted by ${}^mF^n$.

As F is closed under addition and multiplication, the sum and scalar multiples of $m \times n$ matrices over F are again $m \times n$ matrices over F, that is, the set ${}^mF^n$ is closed under the operations of matrix addition (7.3) and scalar multiplication of matrices (7.4).

**Proposition
7.5**

Let F be a field. Then ${}^mF^n$, together with the above operations, forms a vector space of dimension mn over F.

Proof

To show that ${}^mF^n$ is a vector space over F, we should verify laws 1–7 of (6.1), regarding each $m \times n$ matrix over F as a vector. However, this can be done by mimicking (6.2), and so a spot check on law 4 together with a few comments should be enough.

Let $A = [a_{ij}]$ and $B = [b_{ij}]$ be $m \times n$ matrices over F. As the commutative law of addition (law 4 of (2.2)) holds in F, we see $a_{ij} + b_{ij} = b_{ij} + a_{ij}$ for all i and j $(1 \leqslant i \leqslant m, \; 1 \leqslant j \leqslant n)$; so the (i, j)-entries in the $m \times n$ matrices $A + B$ and $B + A$ are equal, showing $A + B = B + A$ by (7.2). Therefore law 4 of (6.1) holds in ${}^mF^n$.

The $m \times n$ matrix with the 0-element of F in *every* position is called the **zero matrix** and denoted by 0; this matrix is the zero vector of $^mF^n$, as $0 + A = A$ for all $A \in {}^mF^n$. The **negative** of $A = [a_{ij}]$ is $-A = [-a_{ij}]$ as $(-A) + A = 0$ (to form $-A$ from A, replace *each* entry a_{ij} by $-a_{ij}$).

To show that the dimension of $^mF^n$ is mn, we generalize the notion of standard basis (6.19): for each pair of integers i and j ($1 \leq i \leq m$, $1 \leq j \leq n$), let E_{ij} denote the $m \times n$ matrix over F having the 1-element of F as (i, j)-entry, all other entries being zero (the 0-element of F). Let $A = [a_{ij}]$ belong to $^mF^n$. As each E_{ij} has only one non-zero entry, there is one, and only one, way of expressing A as a linear combination of the mn matrices E_{ij}, namely

$$A = \sum_{i,j} a_{ij} E_{ij}$$

for, comparing entries, the coefficient of E_{ij} must be the (i, j)-entry a_{ij} in A. We have shown, at a stroke, that the mn matrices E_{ij} span $^mF^n$ (for every matrix A in $^mF^n$ is a linear combination of them) and that the mn matrices E_{ij} are linearly independent (for $A = 0$ (the zero matrix) in the above equation only if *all* the coefficients of the E_{ij} are zero). Therefore the mn matrices E_{ij} (in any order we care to choose) form a basis of $^mF^n$, and so $mn = \dim(^mF^n)$. $\square$

For instance, the vector space $^2\mathbb{Q}^3$ of 2×3 matrices over $\mathbb{Q}$ has dimension 6, as the matrices

$$E_{11} = \begin{pmatrix} 1 & 0 & 0 \\ 0 & 0 & 0 \end{pmatrix}, \ E_{12} = \begin{pmatrix} 0 & 1 & 0 \\ 0 & 0 & 0 \end{pmatrix}, \ E_{13} = \begin{pmatrix} 0 & 0 & 1 \\ 0 & 0 & 0 \end{pmatrix},$$

$$E_{21} = \begin{pmatrix} 0 & 0 & 0 \\ 1 & 0 & 0 \end{pmatrix}, \ E_{22} = \begin{pmatrix} 0 & 0 & 0 \\ 0 & 1 & 0 \end{pmatrix}, \ E_{23} = \begin{pmatrix} 0 & 0 & 0 \\ 0 & 0 & 1 \end{pmatrix}$$

form a basis. A typical calculation in $^2\mathbb{Q}^3$ is:

$$\tfrac{2}{3}\begin{pmatrix} 1 & 2 & 3 \\ 4 & 5 & 6 \end{pmatrix} + \tfrac{1}{3}\begin{pmatrix} \tfrac{1}{2} & 0 & -\tfrac{1}{2} \\ 1 & -1 & 0 \end{pmatrix} = \begin{pmatrix} \tfrac{5}{6} & \tfrac{4}{3} & \tfrac{11}{6} \\ 3 & 3 & 4 \end{pmatrix}.$$

Notice that

$$-\begin{pmatrix} \tfrac{1}{6} & \tfrac{1}{3} & \tfrac{1}{2} \\ \tfrac{2}{3} & \tfrac{5}{6} & 1 \end{pmatrix} = \begin{pmatrix} -\tfrac{1}{6} & -\tfrac{1}{3} & -\tfrac{1}{2} \\ -\tfrac{2}{3} & -\tfrac{5}{6} & -1 \end{pmatrix}$$

since a sign in front of a matrix applies to *all* its entries.

The vector space $^mF^n$ is no more than a simple generalization of F^n; throughout, the $1 \times n$ matrix $[x_1 \ \ x_2 \ \ \ldots \ \ x_n]$ is identified with the n-tuple $(x_1, x_2, \ldots, x_n)$, and so $^1F^n = F^n$ as addition and scalar multiplication of $1 \times n$ matrices over F agree with addition and scalar multiplication in F^n (6.2). Similarly, it is convenient (and consistent with (7.2), (7.3), (7.4), and (7.6)) to make no distinction between 1×1 matrices and scalars, regarding the 1×1 matrix $[a]$ as being identical to the scalar a; in other words, $^mF^n$ reduces to F when $m = n = 1$. We use ^mF (rather than $^mF^1$) to denote the m-dimensional vector space of all $m \times 1$ matrices over F (**column vectors** with m entries from F).

We now prepare for the important concept of matrix multiplication, dealing first with a significant special case, namely the product rc of a row vector $r \in F^m$ and a column vector $c \in {}^mF$. Let a_j be the $(1, j)$-entry in r, and let b_j be the $(j, 1)$-entry in c. Then

$$rc = (a_1, a_2, \ldots, a_m) \begin{pmatrix} b_1 \\ b_2 \\ \vdots \\ b_m \end{pmatrix} = a_1 b_1 + a_2 b_2 + \ldots + a_m b_m$$

that is, rc is *defined* to be the scalar obtained by adding up the products of corresponding entries in r and c. For instance

$$(2, 3, 5) \begin{pmatrix} 4 \\ 6 \\ 7 \end{pmatrix} = 2 \times 4 + 3 \times 6 + 5 \times 7 = 61.$$

We refer to rc as the result of multiplying the row r into the column c.

More generally, let A and B be matrices over F. The product AB can be formed only if A has the same number of columns as B has rows, that is, A is an $l \times m$ matrix and B is an $m \times n$ matrix for some natural numbers l, m, n. Let us suppose that $A = [a_{ij}]$ and $B = [b_{jk}]$ are such matrices and so i, j, k are integers in the ranges $1 \leqslant i \leqslant l$, $1 \leqslant j \leqslant m$, $1 \leqslant k \leqslant n$; notice that j serves as column suffix in A and as row suffix in B. We are now ready for the general case.

Definition 7.6   Let $A = [a_{ij}]$ be an $l \times m$ matrix and $B = [b_{jk}]$ be an $m \times n$ matrix over the field F. Then their **product** AB is the $l \times n$ matrix with (i, k)-entry $a_{i1}b_{1k} + a_{i2}b_{2k} + \ldots + a_{im}b_{mk}$.

To grasp the meaning of (7.6), imagine A to be partitioned horizontally into rows $r_1, r_2, \ldots, r_l$ and B to be partitioned vertically into columns $c_1, c_2, \ldots, c_n$. As r_i and c_k each have m entries, we may multiply r_i (row i of A) into c_k (column k of B) to get

$$r_i c_k = a_{i1}b_{1k} + a_{i2}b_{2k} + \ldots + a_{im}b_{mk}$$

and so (7.6) tells us:

The (i, k)-entry in AB is the result of
multiplying row i of A into column k of B.

Therefore the product AB is formed using the scheme:

$$AB = \begin{pmatrix} \leftarrow r_1 \rightarrow \\ \leftarrow r_2 \rightarrow \\ \vdots \\ \leftarrow r_r \rightarrow \end{pmatrix} \begin{pmatrix} \uparrow & \uparrow & & \uparrow \\ c_1 & c_2 & \dots & c_n \\ \downarrow & \downarrow & & \downarrow \end{pmatrix} = \begin{pmatrix} r_1 c_1 & r_1 c_2 & \dots & r_1 c_n \\ r_2 c_1 & r_2 c_2 & \dots & r_2 c_n \\ \vdots & \vdots & & \vdots \\ r_l c_1 & r_l c_2 & \dots & r_l c_n \end{pmatrix}.$$

Here are some numerical examples:

Let $A = \begin{pmatrix} 1 & 3 & 5 \\ 3 & 1 & 4 \end{pmatrix}$ and so $r_1 = (1, 3, 5)$ and $r_2 = (3, 1, 4)$;

let $B = \begin{pmatrix} 6 & 1 & 2 \\ 4 & 1 & 2 \\ 2 & 1 & 1 \end{pmatrix}$ and so $c_1 = \begin{pmatrix} 6 \\ 4 \\ 2 \end{pmatrix}$, $c_2 = \begin{pmatrix} 1 \\ 1 \\ 1 \end{pmatrix}$, $c_3 = \begin{pmatrix} 2 \\ 2 \\ 1 \end{pmatrix}$.

Then $AB = \begin{pmatrix} r_1 \\ r_2 \end{pmatrix} (c_1, c_2, c_3) = \begin{pmatrix} r_1 c_1 & r_1 c_2 & r_1 c_3 \\ r_2 c_1 & r_2 c_2 & r_2 c_3 \end{pmatrix} = \begin{pmatrix} 28 & 9 & 13 \\ 30 & 8 & 12 \end{pmatrix}.$

After a little practice, the reader should be able to calculate matrix products directly: *multiply rows of the first matrix into columns of the second matrix.* For instance:

$$\begin{pmatrix} 2 & 3 & 1 \\ 1 & 3 & 2 \end{pmatrix} \begin{pmatrix} 4 & -1 \\ -1 & 0 \\ 3 & 5 \end{pmatrix} = \begin{pmatrix} 8 & 3 \\ 7 & 9 \end{pmatrix}, \qquad \begin{pmatrix} 4 & 3 \\ 5 & 7 \\ 2 & 2 \end{pmatrix} \begin{pmatrix} -1 & 2 \\ 1 & -1 \end{pmatrix} = \begin{pmatrix} -1 & 5 \\ 2 & 3 \\ 0 & 2 \end{pmatrix}.$$

Let us briefly review the introduction to this chapter: the reader should check that the matrix product carried out there is obtained by multiplying rows of the first matrix into columns of the second, and so conforms with (7.6). What is more, the definition of matrix multiplication may be complicated, but there is a reason for it: whenever we calculate a matrix product, we are, in effect, composing two mappings.

In Chapter 8 we discuss systems of linear equations; these can be compactly expressed using (7.2) and (7.6). For instance, the system

$$2x_1 + 3x_2 + x_3 = 5$$
$$4x_1 \qquad + 7x_3 = 1$$
$$5x_1 - x_2 + 9x_3 = 3$$

can be rewritten as the column equation:

$$\begin{pmatrix} 2 & 3 & 1 \\ 4 & 0 & 7 \\ 5 & -1 & 9 \end{pmatrix} \begin{pmatrix} x_1 \\ x_2 \\ x_3 \end{pmatrix} = \begin{pmatrix} 5 \\ 1 \\ 3 \end{pmatrix}$$

for multiplying out the left-hand side and comparing entries produces the scalar equations of the system.

In the following theory we shall often be required to manipulate matrices, and so it is important to know the "do's and don'ts" of matrix addition and multiplication; we show now that it is the laws of a ring (laws 1–7 of (2.2)) which hold in this context—these laws and only these laws!

Let $A = \begin{pmatrix} 2 & 1 \\ 3 & 2 \end{pmatrix}$ and $B = \begin{pmatrix} 4 & 1 \\ 2 & 3 \end{pmatrix}$.

Then $AB = \begin{pmatrix} 10 & 5 \\ 16 & 9 \end{pmatrix}$,

but multiplying rows of B into columns of A gives

$$BA = \begin{pmatrix} 11 & 6 \\ 13 & 8 \end{pmatrix}.$$

Therefore $AB \neq BA$ in this case and so:

Matrix multiplication is not commutative.

Of course, it is possible to find certain pairs A and B of matrices which *do* commute, but this is the exception rather than the rule.

Let $A = \begin{pmatrix} 1 & -2 \\ -2 & 4 \end{pmatrix}$ and $B = \begin{pmatrix} 6 & 8 \\ 3 & 4 \end{pmatrix}$.

Then $AB = \begin{pmatrix} 0 & 0 \\ 0 & 0 \end{pmatrix}$,

that is, $AB = 0$ although $A \neq 0$ and $B \neq 0$. The product of non-zero matrices can be the zero matrix and so:

Matrix multiplication has zero-divisors.

Therefore the cancellation of a non-zero matrix factor from an equation is *not* generally valid (see (2.14)); for example,

$$X = \begin{pmatrix} 1 & 1 \\ 1 & 1 \end{pmatrix} \text{ and } Y = \begin{pmatrix} 7 & 9 \\ 4 & 5 \end{pmatrix} \text{ satisfy } AX = AY,$$

where $A = \begin{pmatrix} 1 & -2 \\ -2 & 4 \end{pmatrix}$, but $X \neq Y$.

We show in the next proposition that the distributive laws and the associative law of multiplication hold for matrices. As preparation, let $A = [a_{ij}]$ and $B = [b_{jk}]$ be $l \times m$ and $m \times n$ matrices over F; by (7.6) the (i, k)-entry in AB is the sum

$$a_{i1}b_{1k} + a_{i2}b_{2k} + \ldots + a_{im}b_{mk} = \sum_j a_{ij}b_{jk}$$

the range of summation being: $1 \leqslant j \leqslant m$.

Proposition 7.7 Let A, B, C be matrices over a field F. Then
  (a) $A(B + C) = AB + AC$ and $(A + B)C = AC + BC$,
  (b) $(AB)C = A(BC)$.

Proof The above equations mean: if one side can be formed, then so can the other and the two sides are equal.

(a) For either side of the equation $A(B + C) = AB + AC$ to make sense, $A = [a_{ij}]$ must be an $l \times m$ matrix while $B = [b_{jk}]$ and $C = [c_{jk}]$ are $m \times n$ matrices, for some natural numbers l, m, n. In this case, using the distributive law (law 5 of (2.2)) in F, we obtain

$$\sum_{j=1}^{m} a_{ij}(b_{jk} + c_{jk}) = \sum_{j=1}^{m} a_{ij}b_{jk} + \sum_{j=1}^{m} a_{ij}c_{jk}.$$

By (7.3) and (7.6), the above equation tells us that the (i, k)-entries in the $l \times n$ matrices $A(B + C)$ and $AB + AC$ are equal ($1 \leqslant i \leqslant l$, $1 \leqslant k \leqslant n$). By (7.2) we conclude $A(B + C) = AB + AC$. The second (right) distributive law can be proved in the same way.

(b) The condition for either side of $(AB)C = A(BC)$ to make sense is that $A = [a_{ij}]$, $B = [b_{jk}]$, $C = [c_{ks}]$ are $l \times m$, $m \times n$, $n \times t$ matrices for some natural numbers l, m, n, t. Supposing this to be the case, let i and s be given integers with $1 \leqslant i \leqslant l$ and $1 \leqslant s \leqslant t$. In order to compare (i, s)-entries in the $l \times t$ matrices $(AB)C$ and $A(BC)$, we form the $m \times n$ matrix $D = [d_{jk}]$ where $d_{jk} = a_{ij}b_{jk}c_{ks}$. The sum of the mn entries in D can be obtained by adding up the entries in column k of D (this produces $\sum_j d_{jk}$), and then adding up these column sums—the grand total being $\sum_{k=1}^{n}(\sum_{j=1}^{m} d_{jk})$; but equally, the sum of the entries in D can be arrived at by adding the entries in row j of D (producing $\sum_k d_{jk}$) and then adding these row sums together (giving $\sum_j(\sum_k d_{jk})$). We have proved the formula

$$\sum_k \left(\sum_j d_{jk} \right) = \sum_j \left(\sum_k d_{jk} \right) \tag{$*$}$$

for interchanging the order of summation of the entries in *any* matrix D over F. However, in this case $\sum_j d_{jk} = \sum_j a_{ij}b_{jk}c_{ks} = (\sum_j a_{ij}b_{jk})c_{ks} = ((i, k)\text{-entry}$ in $AB) \times ((k, s)\text{-entry}$ in $C)$; so $\sum_k (\sum_j d_{jk})$, being the result of multiplying row i of AB into column s of C, is the (i, s)-entry in $(AB)C$. Similarly,

$$\sum_j \left(\sum_k d_{jk} \right) = \sum_j \left(\sum_k a_{ij}b_{jk}c_{ks} \right) = \sum_j a_{ij}\left(\sum_k b_{jk}c_{ks} \right)$$

$$= \sum_j ((i, j)\text{-entry in } A) \times ((j, s)\text{-entry in } BC),$$

which is the (i, s)-entry in $A(BC)$. The equation $(*)$ above and (7.2) now tell us that $(AB)C = A(BC)$, as the (i, s)-entries in these matrices are equal.   □

So some, at least, of the usual laws of algebra hold for matrices. As an illustration of the associative law (7.7)(b) of matrix multiplication,

let $A = \begin{pmatrix} 2 & 3 & -2 \\ 1 & 0 & 1 \end{pmatrix}$, $B = \begin{pmatrix} 2 & 3 \\ 1 & 0 \\ 2 & -1 \end{pmatrix}$, $C = \begin{pmatrix} 1 & 3 \\ 2 & -4 \end{pmatrix}$.

Then $AB = \begin{pmatrix} 3 & 8 \\ 4 & 2 \end{pmatrix}$ and so $(AB)C = \begin{pmatrix} 19 & -23 \\ 8 & 4 \end{pmatrix}$.

The same matrix can be arrived at by the alternative route:

$$BC = \begin{pmatrix} 8 & -6 \\ 1 & 3 \\ 0 & 10 \end{pmatrix} \text{ and so } A(BC) = \begin{pmatrix} 19 & -23 \\ 8 & 4 \end{pmatrix}.$$

Definition 7.8   Let F be a field and n a natural number. The $n \times n$ matrix $I = [\delta_{ij}]$, where $\delta_{ii} = 1$ and $\delta_{ij} = 0$ for $i \neq j$, is called the $n \times n$ **identity matrix** over F.

For instance the 3×3 identity matrix over $\mathbb{Q}$ is

$$I = \begin{pmatrix} 1 & 0 & 0 \\ 0 & 1 & 0 \\ 0 & 0 & 1 \end{pmatrix}.$$

All the (i, i)-entries in I are 1 (the 1-element of F) and all the (i, j)-entries in I, for $i \neq j$, are 0 (the 0-element of F); the notation δ_{ij} (called the **Kronecker delta**) is a convenient way of specifying the entries in I. Notice that the rows of the $n \times n$ identity matrix I are the vectors $e_1, e_2, \ldots, e_n$ in the standard basis (6.19) of F^n.

Let I be the $n \times n$ identity matrix over F and let $A = [a_{jk}]$ be any $n \times t$ matrix over F. As $\delta_{ii} = 1$ is the only non-zero entry in row i of I, we obtain

$$\sum_{j=1}^{n} \delta_{ij} a_{jk} = \delta_{i1} a_{1k} + \delta_{i2} a_{2k} + \ldots + \delta_{in} a_{nk}$$

$$= \delta_{ii} a_{ik} = a_{ik}$$

showing that the (i, k)-entries in the $n \times t$ matrices IA and A agree, that is, $IA = A$. In particular we obtain the useful rule

$$e_i A = \text{row } i \text{ of } A.$$

Similarly $AI = A$ for all $m \times n$ matrices A over F. Let us denote column j of I by e_j^T (this notation is explained in (7.10)); then

$$Ae_j^T = \text{column } j \text{ of } A.$$

For example, let $A = \begin{pmatrix} 1 & 2 & 3 \\ 4 & 5 & 6 \\ 7 & 8 & 9 \end{pmatrix}$. Then $Ae_2^T = A \begin{pmatrix} 0 \\ 1 \\ 0 \end{pmatrix} = \begin{pmatrix} 2 \\ 5 \\ 8 \end{pmatrix}$, $e_3 A = (0, 0, 1)A = (7, 8, 9)$, $e_3 A e_2^T = 8$ (the $(3, 2)$-entry in A), and more generally

$$e_i A e_j^T = (i, j)\text{-entry in } A.$$

As multiplication by the identity matrix leaves every matrix unchanged, we see $IA = A = AI$ for all $n \times n$ matrices A over F.

Notation

The set of all $n \times n$ matrices over the field F is denoted by $M_n(F)$ (rather than $^n F^n$).

So $M_2(\mathbb{Q})$ stands for the set of all 2×2 matrices having rational entries. Our next corollary deals with structure of $M_n(F)$.

Corollary 7.9

Let F be a field and n a natural number. Then $M_n(F)$, together with the operations of matrix addition and matrix multiplication, is a non-trivial ring.

Proof

Notice first that, by (7.3) and (7.6), the sum and product of $n \times n$ matrices over F are again $n \times n$ matrices over F; therefore $M_n(F)$ is closed under matrix addition and matrix multiplication, and so we may refer to the system $M_n(F)$. From (7.5), laws 1–4 of (2.2) hold in $M_n(F)$ (for these laws coincide with laws 1–4 of (6.1) in this context). Laws 5 and 6 of (2.2) hold in $M_n(F)$ (they are special cases of (7.7)), and from the preceding discussion, the

$n \times n$ identity matrix I is the 1-element of $M_n(F)$. So laws 1–7 of (2.2) hold in $M_n(F)$, which is therefore a ring; as $I \neq 0$, this ring is non-trivial. $\square$

We have been heading towards (7.9) since the introduction of rings in (2.2), and so, having arrived, let us take stock of where we are. Notice first that the ring of 1×1 matrices over F is nothing new, for it is simply F itself, that is, $M_1(F) = F$. However as $1 \neq 0$ we see

$$\begin{pmatrix} 1 & 0 \\ 0 & 0 \end{pmatrix}\begin{pmatrix} 0 & 1 \\ 0 & 0 \end{pmatrix} \neq \begin{pmatrix} 0 & 1 \\ 0 & 0 \end{pmatrix}\begin{pmatrix} 1 & 0 \\ 0 & 0 \end{pmatrix}$$

showing that $M_2(F)$ is a non-commutative ring for every field F. More generally, $M_n(F)$ is non-commutative for $n \geq 2$ (the above example of non-commuting matrices can be enlarged to $n \times n$ matrices by adjoining extra rows and columns of zeros). Also $M_n(F)$ has zero-divisors for $n \geq 2$ (consider the product on the right-hand side above). A practical summary is provided by the rule of thumb:

Matrix manipulation is governed by the ring laws.

In fact, this rule is of wider application than (7.9) suggests, since matrices having entries from a non-trivial ring R feature in more advanced aspects of algebra (the cases $R = \mathbb{Z}$ and $R = F[x]$ are particularly important). Addition and multiplication of such matrices is defined by (7.3) and (7.6) with no modification (beyond replacing F by R); what is more, (7.9) remains valid for matrices over R (the proof goes through unchanged) to give:

The system $M_n(R)$, of all $n \times n$ matrices over the non-trivial ring R, is itself a non-trivial ring.

So the 2×2 matrices having integer entries form the ring $M_2(\mathbb{Z})$; however, for simplicity and because they are sufficient for our present purposes, we restrict our attention to matrices over a field.

Let A be a **square matrix** over F (that is, A is an $n \times n$ matrix over F for some natural number n). As A is an element of the ring $M_n(F)$, we may form the positive powers:

$$A, A^2, A^3, A^4, \ldots$$

all of which belong to $M_n(F)$.

For instance, consider $A = \begin{pmatrix} \frac{1}{2} & \frac{5}{4} \\ -1 & -\frac{1}{2} \end{pmatrix}$ in $M_2(\mathbb{Q})$. Then

$$A^2 = \begin{pmatrix} -1 & 0 \\ 0 & -1 \end{pmatrix} = -I, \qquad A^3 = -A = \begin{pmatrix} -\frac{1}{2} & -\frac{5}{4} \\ 1 & \frac{1}{2} \end{pmatrix}, \qquad A^4 = I.$$

In this case, higher powers of A give repetitions of A, A^2, A^3, A^4; from $1987 = 4 \times 496 + 3$, we see $A^{1987} = (A^4)^{496}A^3 = A^3 = -A$. Now let $B = \begin{pmatrix} \frac{1}{2} & 1 \\ -\frac{1}{4} & -\frac{1}{2} \end{pmatrix}$. Then $B^2 = \begin{pmatrix} 0 & 0 \\ 0 & 0 \end{pmatrix}$, and so all higher powers of B are equal to the zero matrix also, that is, $0 = B^2 = B^3 = B^4 = \dots$. Notice that the matrix A has an inverse (2.11) in the ring $M_2(\mathbb{Q})$: for the equation $A^4 = I$ can be rewritten as $A^3 A = I = A A^3$, showing that $A^{-1} = A^3$ (the product of a matrix and its inverse (in either order) is I); so the negative integer powers of A may be formed: $A^{-1}, A^{-2}, \dots, A^{-n}, \dots$ where $A^{-n} = (A^{-1})^n$. On the other hand, B has no inverse B^{-1} in $M_2(\mathbb{Q})$: for if $B^{-1}B = I$, then $B = IB = (B^{-1}B)B = B^{-1}B^2 = B^{-1}0 = 0$, showing B to be the zero matrix, which is not true; therefore negative integer powers of B make no sense and cannot be formed. We shall study matrix inverses in the next chapter.

The reader should guard against the unwitting use of the commutative law of multiplication when carrying out matrix calculations. For example, $(A + B)^2$ can be expanded, but $A^2 + 2AB + B^2$ is *not* usually the correct expansion! Rather one must use the distributive law (7.7)(a):

$$(A + B)^2 = (A + B)(A + B) = A(A + B) + B(A + B)$$
$$= A^2 + AB + BA + B^2.$$

In fact the 'usual' expansion formula $(A + B)^2 = A^2 + 2AB + B^2$ is valid if and only if $AB = BA$.

The rows of a matrix have the same status as the columns, and it is convenient to introduce next the idea of transposing a matrix, that is, of forming a second matrix by interchanging the rows and columns of the first matrix.

Definition 7.10 Let A be the $m \times n$ matrix with (i, j)-entry a_{ij}. The $n \times m$ matrix A^T having (j, i)-entry a_{ij} is called the **transpose** of A.

For example, if $A = \begin{pmatrix} 1 & 2 & 3 \\ 4 & 5 & 6 \end{pmatrix}$, then $A^T = \begin{pmatrix} 1 & 4 \\ 2 & 5 \\ 3 & 6 \end{pmatrix}$.

More generally, if

$$A = \begin{pmatrix} a_{11} & a_{12} & a_{1n} \\ \vdots & \vdots & \vdots \\ a_{m1} & a_{m2} \dots a_{mn} \end{pmatrix}, \text{ then } A^T = \begin{pmatrix} a_{11} \dots a_{m1} \\ a_{12} \dots a_{m2} \\ \vdots & \vdots \\ a_{1n} \dots a_{mn} \end{pmatrix}.$$

Notice that column i of A^T is the transpose of row i of A; also row j of A^T is the transpose of column j of A. On transposing twice, the original matrix is recovered, that is,

$(A^T)^T = A$.

Our next proposition establishes the properties of matrix transposition.

Proposition 7.11   Let A and B be matrices over the field F. Then

(a) $(A + B)^T = A^T + B^T$,   (b) $(AB)^T = B^T A^T$.

Proof   (a) We suppose $A = [a_{ij}]$ and $B = [b_{ij}]$ are $m \times n$ matrices for some natural numbers m and n (otherwise neither side of the above equation can be formed). The (j, i)-entry in the $n \times m$ matrix $(A + B)^T$ is $a_{ij} + b_{ij}$, which is also the (j, i)-entry in the $n \times m$ matrix $A^T + B^T$; therefore $(A + B)^T = A^T + B^T$.

(b) We suppose $A = [a_{ij}]$ and $B = [b_{jk}]$ are respectively $l \times m$ and $m \times n$ matrices (otherwise there is nothing to be proved). The (k, i)-entry in the $n \times l$ matrix $(AB)^T$ is the (i, k)-entry in AB, namely $\sum_{j=1}^{m} a_{ij} b_{jk}$. The (k, i)-entry in the $n \times l$ matrix $B^T A^T$ is the result of multiplying row k of B^T into column i of A^T, that is,

$$b_{1k} a_{i1} + b_{2k} a_{i2} + \ldots + b_{mk} a_{im} = \sum_j b_{jk} a_{ij}.$$

As multiplication in F is commutative, $b_{jk} a_{ij} = a_{ij} b_{jk}$ for $1 \leq j \leq m$; therefore $\sum_j a_{ij} b_{jk} = \sum_j b_{jk} a_{ij}$, showing that the (k, i)-entries in $(AB)^T$ and $B^T A^T$ are equal, and so $(AB)^T = B^T A^T$.   □

From (7.11)(a) we see that the transpose of a sum of matrices is simply the sum of the individual transposes; however (7.11)(b) is more surprising, for it tells us that the transpose of a product of matrices is the product, in the *opposite* order, of the individual transposes. For instance, transposing

$$\begin{pmatrix} 0 & 2 \\ 3 & 1 \end{pmatrix} \begin{pmatrix} 4 & 5 & 6 \\ 7 & 8 & 9 \end{pmatrix} = \begin{pmatrix} 14 & 16 & 18 \\ 19 & 23 & 27 \end{pmatrix}$$

gives

$$\begin{pmatrix} 4 & 7 \\ 5 & 8 \\ 6 & 9 \end{pmatrix} \begin{pmatrix} 0 & 3 \\ 2 & 1 \end{pmatrix} = \begin{pmatrix} 14 & 19 \\ 16 & 23 \\ 18 & 27 \end{pmatrix}.$$

The reader should realize that the theory of finite-dimensional vector spaces can be expressed concretely either in terms of row vectors or in terms of column vectors; it is often merely a matter of taste as to whether a problem (in simultaneous differential equations, for example) is formulated using rows or columns. Generally we favour rows (for this is in accord with scalars being on the left (as in av) and mappings being on the right (as in $(v)\alpha$)), although the reader will find many texts which are biased towards columns. The point is that *the two approaches are linked by matrix transposition*, and (7.11) tells us how to express the details of the one in terms of the other. Here is an illustration: a linear substitution from the variables x_1 and x_2 to the variables y_1 and y_2 can be set up either as a column equation or as a row equation because the simultaneous scalar equations $x_1 = ay_1 + by_2$ and $x_2 = cy_1 + dy_2$ are expressed by

$$\begin{pmatrix} x_1 \\ x_2 \end{pmatrix} = \begin{pmatrix} a & b \\ c & d \end{pmatrix} \begin{pmatrix} y_1 \\ y_2 \end{pmatrix} \quad \text{and} \quad [x_1 \ x_2] = [y_1 \ y_2] \begin{pmatrix} a & c \\ b & d \end{pmatrix},$$

each being the transpose of the other.

Exercises 7.1

Assume that all matrices are over $\mathbb{Q}$ unless otherwise stated.

1. Express AB, A^2, B^2, BA, $(A+B)^2$, and $(A-B)^2$ as arrays of rational numbers in the following cases:

(a) $A = \begin{pmatrix} 1 & 2 \\ 0 & 1 \end{pmatrix}$, $B = \begin{pmatrix} 0 & 1 \\ 1 & 0 \end{pmatrix}$.

(b) $A = \begin{pmatrix} 1 & 2 \\ 0 & 1 \end{pmatrix}$, $B = \begin{pmatrix} 1 & 3 \\ 0 & 1 \end{pmatrix}$.

(c) $A = \begin{pmatrix} 1 & 0 & 0 \\ 2 & 1 & 0 \\ 3 & 4 & 1 \end{pmatrix}$, $B = \begin{pmatrix} 1 & 1 & 1 \\ 0 & 2 & 1 \\ 0 & 0 & 3 \end{pmatrix}$.

2. Find the various missing entries in the matrix equations:

(a) $\begin{pmatrix} 1 & * \\ 2 & * \end{pmatrix}\begin{pmatrix} * & 3 \\ 2 & * \end{pmatrix} = \begin{pmatrix} 3 & 4 \\ 4 & 6 \end{pmatrix}$, (b) $\begin{pmatrix} 1 & 2 \\ * & * \end{pmatrix}^2 = \begin{pmatrix} * & 3 \\ 3 & * \end{pmatrix}$.

3. Calculate A^2 and A^3 where $A = \begin{pmatrix} -2 & 1 & 1 \\ -2 & 1 & 2 \\ -3 & 1 & 1 \end{pmatrix}$,

and hence find A^6, A^{-1}, and A^{2000}.

4. Calculate A^2 and A^3 where $A = \begin{pmatrix} 2 & 1 & 3 \\ 1 & 1 & 2 \\ -2 & -1 & -3 \end{pmatrix}$.

Use the distributive law (7.7)(a) to multiply out $(I - A)(I + A + A^2)$ and hence find $(I - A)^{-1}$ and $(I + A + A^2)^{-1}$.

5. Let a_i and b_j be either 0 or 1 $(1 \leqslant i, j \leqslant 3)$. Find the smallest positive integer *not* expressible in the form

$$[a_1 \ a_2 \ a_3]A[b_1 \ b_2 \ b_3]^{\mathrm{T}} \quad \text{where} \quad A = \begin{pmatrix} 1 & 2 & 3 \\ 4 & 5 & 6 \\ 7 & 8 & 9 \end{pmatrix}.$$

6. Let $A = \begin{pmatrix} a & b \\ c & d \end{pmatrix}$ where a, b, c, d are elements of a commutative ring. Verify that $A^2 - (a + d)A + (ad - bc)I = 0$.

7. If $A = \begin{pmatrix} 1 & 1 \\ 0 & 1 \end{pmatrix}$, find A^2 and A^3. Postulate the form of A^n (n a positive integer) and verify your assertion by induction. Find A^n for each of the following matrices A:

$$\begin{pmatrix} 1 & \frac{1}{2} \\ 0 & 1 \end{pmatrix}, \quad \begin{pmatrix} 2 & 3 \\ 0 & 2 \end{pmatrix}, \quad \begin{pmatrix} 4 & 3 \\ 3 & -4 \end{pmatrix}.$$

8. Let $A = [a_{ij}]$ be an $m \times m$ matrix over a field F such that $a_{ij} = 0$ for all $i \geqslant j$. Prove, by induction on n, that the (i, j)-entry in A^n is zero for $i + n - 1 \geqslant j$. Deduce that $A^m = 0$.

9. The matrix A is called **symmetric** if $A^{\mathrm{T}} = A$.

(a) Let $B = \begin{pmatrix} 1 & 2 & 3 \\ 4 & 5 & 6 \\ 7 & 8 & 9 \end{pmatrix}$ and $C = \begin{pmatrix} 3 & 1 & 4 \\ 2 & 5 & 1 \end{pmatrix}$.

Calculate the matrices $B + B^{\mathrm{T}}$, CC^{T}, $C^{\mathrm{T}}C$ and verify that they are symmetric.

(b) Let B be a square matrix over a field. Use (7.11)(a) to show that $B + B^{\mathrm{T}}$ is symmetric.

(c) Let C be a matrix over a field. Use (7.11)(b) to show that CC^{T} is symmetric.

(d) Let A and B be symmetric $n \times n$ matrices over the field F. Show that $A + B$ is symmetric. Show that AB is symmetric if and only if $AB = BA$.

10. Let F be a field and let U denote the set of all 2×2 matrices over F of the form $\begin{pmatrix} a & b \\ 0 & c \end{pmatrix}$. Use (6.6) to show that U is a subspace of the vector space of all 2×2 matrices over F. Write down a basis of U and find $\dim U$. Use (5.7) to show that U is a subring of $\mathrm{M}_2(F)$. Show that U is a non-commutative ring having zero-divisors.

(i) Write out the multiplication table of U in the case $F = \mathbb{Z}_2$.

(ii) Taking $F = \mathbb{Z}_3$, find the eight matrices A in U satisfying $A^2 = A$.

11. The matrix $A = [a_{ij}]$ is called **diagonal** if $a_{ij} = 0$ for all $i \neq j$.

(a) Show that the 2×2 matrix A over $\mathbb{Q}$ commutes with $B = \begin{pmatrix} 1 & 0 \\ 0 & -1 \end{pmatrix}$ (that is, $AB = BA$), if and only if A is diagonal.

(b) Show that the 2×2 matrix A over the field F commutes with $\begin{pmatrix} 1 & 0 \\ 0 & 0 \end{pmatrix}$ if and only if A is diagonal. Determine the form of the 2×2 matrices over F which commute with $\begin{pmatrix} 0 & 1 \\ 0 & 0 \end{pmatrix}$. Deduce that A commutes with all 2×2 matrices over F if and only if A is **scalar**, that is, $A = aI$ for some $a \in F$ (A is a scalar multiple of the identity matrix I).

(c) Show that the 2×2 matrix A over the field F commutes with $B = \begin{pmatrix} 0 & 1 \\ 1 & 0 \end{pmatrix}$ if and only if $A = aI + bB$ $(a, b \in F)$.

(d) Show that the 3×3 matrix A over F commutes with

$$B = \begin{pmatrix} 0 & 1 & 0 \\ 0 & 0 & 1 \\ 0 & 0 & 0 \end{pmatrix}$$

if and only if $A = aI + bB + cB^2$ $(a, b, c \in F)$.

(e) Determine the form of the 3×3 matrices over F which commute with

$$B = \begin{pmatrix} 1 & 0 & 0 \\ 0 & 1 & 0 \\ 0 & 0 & 0 \end{pmatrix}.$$

(f) Let $A = [a_{ij}]$ be a diagonal $n \times n$ matrix over F with n distinct diagonal entries a_{ii}. Show that the $n \times n$ matrix B over F commutes with A if and only if B is diagonal.

12. Let $A = [a_{ij}]$ be an $n \times n$ matrix over the field F. The scalar $a_{11} + a_{22} + \ldots + a_{nn}$ is called the **trace** of A.

(a) Find the trace of $B^{\mathrm{T}}B$ where $B = (1, 2, 3)$.

(b) Show trace $(A + B) = $ trace $A + $ trace B, trace $(AB) = $ trace (BA) where A and B are $n \times n$ matrices over F. If B has inverse B^{-1}, deduce trace $(BAB^{-1}) = $ trace A.

(c) Let $A \in M_2(\mathbb{Q})$ satisfy $A^2 = -I$. Prove that trace A is zero and show that there are an infinite number of such matrices A. Are any of them symmetric?

Linear mappings

A mapping of one vector space to another (over the same field) is called **linear** if it respects the vector space operations. Here we study such mappings and establish their close connection with matrices.

Linear mappings, like vector spaces, have a simple structure and yet are versatile enough to have a wide range of applications. In analysis, for example, the local behaviour of a function can often be determined by means of an approximation which is linear. From the point of view of abstract algebra, linear mappings not only provide comparisons between vector spaces, but they form systems which are interesting in their own right—groups (see Chapter 9) as well as non-commutative rings arise from linear mappings of a given vector space. On a more practical level, we shall see that many problems can be expressed in an unprejudiced way by means of linear mappings, and further, the solution of such problems can often be found simply by viewing the linear mappings from the 'correct' vantage point.

Definition 7.12 Let V and V' be vector spaces over the field F. The mapping $\alpha : V \to V'$ is called **linear** if

(i) $(u + v)\alpha = (u)\alpha + (v)\alpha$ for all $u, v \in V$

and

(ii) $(av)\alpha = a((v)\alpha)$ for all $a \in F$ and $v \in V$.

Condition (7.12)(i) tells us that the image by α of a sum of vectors is the sum of the individual images, that is, α **respects vector addition**. Condition (7.12)(ii) says that α **respects scalar multiplication of vectors**: the image by α of a scalar multiple of a vector is that scalar multiple of the image vector. For instance, suppose $\alpha : \mathbb{Q}^2 \to \mathbb{Q}^3$ is linear and such that

$$(1, 2)\alpha = (3, -1, 4), \qquad (3, 5)\alpha = (2, 1, 1).$$

Then the above equations can be added to give $(4, 7)\alpha = (5, 0, 5)$ by (7.12)(i); by (7.12)(ii) any of these equations may be scalar-multiplied to produce: $(\frac{1}{3}, \frac{2}{3})\alpha = (1, -\frac{1}{3}, \frac{4}{3})$, $(\frac{4}{5}, \frac{7}{5})\alpha = (1, 0, 1)$ etc.

We show next how each matrix over a field gives rise to a linear mapping; further, distinct matrices give rise to distinct linear mappings, and—most important of all—products of matrices correspond to compositions of linear mappings.

Proposition Let A be an $m \times n$ matrix over the field F. The mapping
7.13
$$\mu_A : F^m \to F^n, \text{ defined by } (x)\mu_A = xA \text{ for all } x \in F^m,$$

is linear. Further $\mu_A = \mu_B$ implies $A = B$ and $\mu_{AB} = \mu_A \mu_B$ where B is an $n \times t$ matrix over F.

Proof We regard x in F^m as a $1 \times m$ matrix over F, and so xA is a $1 \times n$ matrix over F, that is $(x)\mu_A$ belongs to F^n. Therefore μ_A is a mapping of F^m to F^n. To show that μ_A is linear, let x, $y \in F^m$ and let $a \in F$. Treating scalars as 1×1 matrices, by (7.7)

$$(x + y)\mu_A = (x + y)A = xA + yA = (x)\mu_A + (y)\mu_A$$
$$(ax)\mu_A = (ax)A = a(xA) = a((x)\mu_A)$$

showing that μ_A is a linear mapping.

Suppose $\mu_A = \mu_B$. Then μ_B is also a mapping of F^m to F^n and $(x)\mu_A = (x)\mu_B$ for all x in F^m by (1.17). Therefore B is an $m \times n$ matrix over F. Taking $x = e_i$ (where $e_1, e_2, \ldots, e_m$ form the standard basis of F^m) we obtain

$$(\text{row } i \text{ of } A) = e_i A = (e_i)\mu_A = (e_i)\mu_B$$
$$= e_i B = (\text{row } i \text{ of } B)$$

and so $A = B$, as the above equation says A and B have the same row i for $1 \leqslant i \leqslant m$.

Notice that μ_{AB} and $\mu_A \mu_B$ both map F^m to F^t. By the associative law (7.7)(b),

$$(x)\mu_A \mu_B = ((x)\mu_A)\mu_B = (xA)B = x(AB) = (x)\mu_{AB}$$

for all $x \in F^m$. Therefore $\mu_{AB} = \mu_A \mu_B$ by (1.17). $\qquad\square$

The linear mapping μ_A **postmultiplies** (multiplies on the right) each row vector x by the matrix A. We show in (7.17) that *every* linear mapping of F^m to F^n is of the form μ_A, and so linear mappings (in this context) are nothing to be afraid of, for they are little more than matrices.

Example Linear mappings of the cartesian plane $\mathbb{R}^2$ can be pictured geo-
7.14 metrically. Consider $\mu_A : \mathbb{R}^2 \to \mathbb{R}^2$ where $A = \begin{pmatrix} 0 & 1 \\ 1 & 0 \end{pmatrix}$. Then

$$(x_1, x_2)\mu_A = (x_1, x_2)A = (x_2, x_1)$$

showing that μ_A interchanges x_1 and x_2. In geometric terms: $(x)\mu_A$ is the image of the point $x = (x_1, x_2)$ in the 'mirror' $x_1 = x_2$, that is,

Fig. 7.1

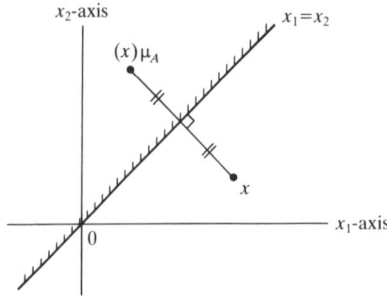

μ_A is **reflection** in the line $x_1 = x_2$ (Fig. 7.1). Notice that $(x)\mu_I = xI = x$ and so μ_I is the identity mapping. Reflecting twice in the line $x_1 = x_2$ (which is silvered on both sides!) we obtain the identity mapping of $\mathbb{R}^2$, that is, $\mu_A^2 = \mu_A \mu_A = \mu_{A^2} = \mu_I$, the corresponding matrix equation being

$$A^2 = \begin{pmatrix} 0 & 1 \\ 1 & 0 \end{pmatrix}\begin{pmatrix} 0 & 1 \\ 1 & 0 \end{pmatrix} = \begin{pmatrix} 1 & 0 \\ 0 & 1 \end{pmatrix} = I.$$

In a similar way, the linear mapping $\mu_B : \mathbb{R}^2 \to \mathbb{R}^2$, where $B = \begin{pmatrix} 1 & 0 \\ 0 & -1 \end{pmatrix}$, is reflection in the x_1-axis. As before $\mu_B^2 = \mu_I$, since reflecting twice in the same mirror is the identity mapping, and this is expressed by the matrix equation $B^2 = I$.

Let $\rho_\phi : \mathbb{R}^2 \to \mathbb{R}^2$ denote rotation through the angle ϕ (see (2.29)). So the point x with polar co-ordinates (r, θ) is mapped to the point $(x)\rho_\phi$ with polar co-ordinates $(r, \theta + \phi)$ (Fig. 7.2). Let us express ρ_ϕ in cartesian co-ordinates: writing $x = (x_1, x_2)$ and $(x)\rho_\phi = (y_1, y_2)$, resolving horizontally and vertically gives $x_1 = r \cos \theta$ and $x_2 = r \sin \theta$,

$$y_1 = r \cos(\theta + \phi) = r \cos \theta \cos \phi - r \sin \theta \sin \phi$$
$$= x_1 \cos \phi - x_2 \sin \phi,$$
$$y_2 = r \sin(\theta + \phi) = r \sin \theta \cos \phi + r \cos \theta \sin \phi$$
$$= x_1 \sin \phi + x_2 \cos \phi,$$

using the trigonometric expansion formulae (see before (2.30)). As ϕ is constant, we see that y_1 and y_2 depend linearly on x_1 and x_2; in fact the above equations combine to give

$$(x_1, x_2)\rho_\phi = (x_1, x_2)\begin{pmatrix} \cos \phi & \sin \phi \\ -\sin \phi & \cos \phi \end{pmatrix} \quad \text{for all } x_1, x_2 \in \mathbb{R}.$$

The above 2×2 matrix is called a **rotation matrix**, and as the

Fig. 7.2

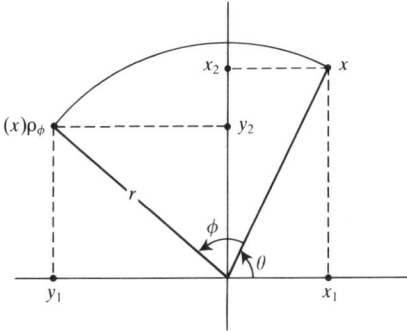

mapping ρ_ϕ amounts to postmultiplication by this matrix, ρ_ϕ is linear by (7.13).

Notice that composing the reflections μ_A and μ_B produces rotations: in fact, $\mu_A\mu_B = \rho_{-\pi/2}$ and $\mu_B\mu_A = \rho_{\pi/2}$, since

$$AB = \begin{pmatrix} 0 & -1 \\ 1 & 0 \end{pmatrix} \quad \text{and} \quad BA = \begin{pmatrix} 0 & 1 \\ -1 & 0 \end{pmatrix}$$

are the rotation matrices with $\phi = -\pi/2$ and $\phi = \pi/2$ respectively.

We now derive some elementary properties of linear mappings.

Proposition 7.15 Let V and V' be vector spaces over the field F and let $\alpha : V \to V'$ be a linear mapping. Then

(a) $(0)\alpha = 0$ (α maps the zero vector of V to the zero vector of V').
(b) $(-v)\alpha = -((v)\alpha)$ (α respects vector negation).
(c) $(a_1 v_1 + \ldots + a_m v_m)\alpha = a_1((v_1)\alpha) + \ldots + a_m((v_m)\alpha)$ (α respects linear combinations of vectors).
(d) $\alpha^{-1} : V' \to V$ is linear if α is bijective.

Proof

(a) Applying α to the vector equation $0 + 0 = 0$ in V produces $(0)\alpha + (0)\alpha = (0)\alpha$ by (7.12)(i). Comparing with $0 + (0)\alpha = (0)\alpha$ gives $(0)\alpha = 0$ by (6.4)(c).

(b) For v in V, we apply α to $-v + v = 0$ obtaining $(-v)\alpha + (v)\alpha = (0)\alpha = 0$ by (7.12)(i) and part (a) above. Therefore $(-v)\alpha$ is the negative of $(v)\alpha$, that is, $(-v)\alpha = -((v)\alpha)$.

(c) We use induction on m, the case $m = 1$ being condition (7.12)(ii). So take $m > 1$ and suppose inductively that

$$(u)\alpha = \sum_{i=1}^{m-1} a_i((v_i)\alpha) \quad \text{where} \quad u = \sum_{i=1}^{m-1} a_i v_i.$$

Then

$$\left(\sum_{i=1}^{m} a_i v_i\right)\alpha = (u + a_m v_m)\alpha = (u)\alpha + (a_m v_m)\alpha$$

$$= (u)\alpha + a_m((v_m)\alpha) = \sum_{i=1}^{m} a_i((v_i)\alpha)$$

using (7.12) and the inductive hypothesis; the induction is now complete, which finishes the proof by (3.5).

(d) Let u' and v' belong to V'. As α is bijective there are unique vectors u and v in V with $(u)\alpha = u'$ and $(v)\alpha = v'$, and so $u = (u')\alpha^{-1}$ and $v = (v')\alpha^{-1}$ by (1.23). As α is linear: $(u + v)\alpha = (u)\alpha + (v)\alpha = u' + v'$ which, on applying α^{-1}, gives $(u')\alpha^{-1} + (v')\alpha^{-1} = u + v = (u' + v')\alpha^{-1}$. So α^{-1} satisfies (7.12)(i). Applying α^{-1} to $(av)\alpha = a((v)\alpha) = av'$ gives $a((v')\alpha^{-1}) = av = (av')\alpha^{-1}$, showing that α^{-1} satisfies (7.12)(ii). Therefore α^{-1} is linear. ☐

Property (7.15)(c) is particularly useful when $v_1, v_2, \ldots, v_m$ form a basis of V. For example, let $v_1 = (1, 2)$ and $v_2 = (3, 5)$ and consider once again a linear mapping $\alpha : \mathbb{Q}^2 \to \mathbb{Q}^3$ such that $(v_1)\alpha = (3, -1, 4)$ and $(v_2)\alpha = (2, 1, 1)$; in fact there is one and only one such mapping α as we now demonstrate. Notice that v_1 and v_2 form a basis of $\mathbb{Q}^2$ and the vectors e_1 and e_2 of the standard basis of $\mathbb{Q}^2$ are related to v_1 and v_2 by

$$e_1 = (1, 0) = -5v_1 + 2v_2, \qquad e_2 = (0, 1) = 3v_1 - v_2.$$

From (7.15)(c) we obtain

$$(e_1)\alpha = -5((v_1)\alpha) + 2((v_2)\alpha) = -5(3, -1, 4) + 2(2, 1, 1)$$
$$= (-11, 7, -18),$$
$$(e_2)\alpha = 3((v_1)\alpha) - ((v_2))\alpha = 3(3, -1, 4) - (2, 1, 1) = (7, -4, 11).$$

Knowing the effect of α on e_1 and e_2, it is easy to see what happens to each vector (x_1, x_2) of $\mathbb{Q}^2$: for $(x_1, x_2) = x_1 e_1 + x_2 e_2$ and so

$$(x_1, x_2)\alpha = x_1((e_1)\alpha) + x_2((e_2)\alpha) = x_1(-11, 7, -18) + x_2(7, -4, 11)$$

$$= (x_1, x_2)\begin{pmatrix} -11 & 7 & -18 \\ 7 & -4 & 11 \end{pmatrix}$$

using (7.15)(c) and matrix multiplication. Denoting the above 2×3 matrix by A, we see $(x)\alpha = xA$ for all $x = (x_1, x_2)$ in $\mathbb{Q}^2$, that is, α is the linear mapping μ_A of (7.13).

We discover next how to specify, as economically as possible, a linear mapping $\alpha : V \to V'$ where V is finite-dimensional: the images of the vectors of a basis of V may be chosen at will (*any* vectors in V' will do!) and any choice determines α uniquely.

Theorem 7.16

Let V and V' be vector spaces over the field F. Let V have basis $v_1, v_2, \ldots, v_m$ and let $v'_1, v'_2, \ldots, v'_m$ be any vectors in V'. Then there is a unique linear mapping $\alpha : V \to V'$ such that

$$(v_1)\alpha = v'_1, \quad (v_2)\alpha = v'_2, \quad \ldots, \quad (v_m)\alpha = v'_m.$$

Further, α is injective if and only if $v'_1, v'_2, \ldots, v'_m$ are linearly independent, and α is surjective if and only if $v'_1, v'_2, \ldots, v'_m$ span V'.

Proof

For each v belonging to V, there are unique scalars a_i $(1 \leq i \leq m)$ such that $v = \sum_{i=1}^{m} a_i v_i$ by (6.25). If α is a linear mapping satisfying the hypothesis, then by (7.15)(c)

$$(v)\alpha = \left(\sum_{i=1}^{m} a_i v_i \right)\alpha = \sum_{i=1}^{m} a_i ((v_i)\alpha) = \sum_{i=1}^{m} a_i v'_i$$

showing that α is completely specified (whatever scalars a_i are required to express v as a linear combination of $v_1, v_2, \ldots, v_m$, the *same* scalars a_i express $(v)\alpha$ as a linear combination of $v'_1, v'_2, \ldots, v'_m$). So there is at most one linear mapping α as in (7.16). What is more, the above equation tells us how to construct the only mapping which has a chance of doing the job (the job is to map each v_j to v'_j $(1 \leq j \leq m)$ and be linear).

Let $\alpha : V \to V'$ be defined by $(v)\alpha = \sum_{i=1}^{m} a_i v'_i$ where $v = \sum_{i=1}^{m} a_i v_i$. It is routine (and is generously left as an exercise for the reader) to verify that α, as defined above, is linear. Using the Kronecker delta (7.8), from $v_j = \sum_{i=1}^{m} \delta_{ji} v_i$ we see

$$(v_j)\alpha = \sum_{i=1}^{m} \delta_{ji} v'_i = v'_j \ (1 \leq j \leq m).$$

Therefore α is the unique linear mapping satisfying the hypothesis of (7.16).

Suppose $v'_1, v'_2, \ldots, v'_m$ are linearly independent. To show that α is injective, consider u and v in V with $(u)\alpha = (v)\alpha$. There are unique scalars a_i with $u - v = \sum_{i=1}^{m} a_i v_i$; however the equation

$$0 = (u)\alpha - (v)\alpha = (u - v)\alpha = \sum_{i=1}^{m} a_i v'_i$$

implies that each a_i is zero by (6.16). Therefore $u - v = 0$, that is, $u = v$ and α is injective.

Conversely suppose α to be injective. To prove that $v'_1, v'_2, \ldots, v'_m$ are linearly independent, suppose $\sum_{i=1}^m a_i v'_i = 0$ where $a_i \in F$. This equation can be rewritten as $(\sum_{i=1}^m a_i v_i)\alpha = (0)\alpha$, which means $\sum_{i=1}^m a_i v_i = 0$ by (1.20). As $v_1, v_2, \ldots, v_m$ are linearly independent, each a_i is zero; therefore $v'_1, v'_2, \ldots, v'_m$ are linearly independent.

Finally, as $(v)\alpha = \sum_{i=1}^m a_i v'_i$, we see that $\alpha : V \to V'$ is surjective (1.21) if and only if each vector in V' is of the form $\sum_{i=1}^m a_i v'_i$, that is by (6.10), if and only if $v'_1, v'_2, \ldots, v'_m$ span V'. $\qquad\square$

From (7.16) we obtain the following useful fact.

If two linear mappings agree on a basis,
then they are equal.

The next corollary tells us all there is to know about the form of linear mappings of F^m to F^n.

Corollary 7.17

Let $\alpha : F^m \to F^n$ be a linear mapping. Then there is a unique $m \times n$ matrix A over the field F such that $\alpha = \mu_A$.

Proof

We use the standard basis $e_1, e_2, \ldots, e_m$ of F^m. Each $(e_i)\alpha$ belongs to F^n and so an $m \times n$ matrix A over F can be constructed by taking $(e_i)\alpha$ as row i of A $(1 \leq i \leq m)$. As

$$(e_i)\alpha = (\text{row } i \text{ of } A) = e_i A = (e_i)\mu_A$$

for each i $(1 \leq i \leq m)$, the linear mappings α and μ_A agree on the standard basis of F^m. From (7.16) (with e_i in place of v_i and $e_i A$ in place of v'_i) we deduce $\alpha = \mu_A$. The matrix A is unique by (7.13). $\qquad\square$

We now turn our attention to linear mappings which are bijective, or as we shall call them (in view of (7.15)(d)) **invertible**, for the inverse of a bijective linear mapping is also linear. The reader should not be surprised to learn that there is a close connection between invertible matrices (matrices having inverses in the ring $M_n(F)$ of (7.9)) and invertible linear mappings. The reflections and rotations of $\mathbb{R}^2$ discussed in (7.14) are invertible: in fact reflections are self-inverse, for the equations $\mu_A^2 = \mu_B^2 = \mu_I$ can be rewritten as $\mu_A^{-1} = \mu_A$ and $\mu_B^{-1} = \mu_B$ (the corresponding matrix equations are $A^{-1} = A$ and $B^{-1} = B$). The inverse of ρ_ϕ is $\rho_{-\phi}$, that is $\rho_\phi^{-1} = \rho_{-\phi}$, for rotating $\mathbb{R}^2$ through $-\phi$ about the origin has the reverse effect to ρ_ϕ; in terms of matrices, the reader may check that the product of

$$\begin{pmatrix} \cos(-\phi) & \sin(-\phi) \\ -\sin(-\phi) & \cos(-\phi) \end{pmatrix} = \begin{pmatrix} \cos\phi & -\sin\phi \\ \sin\phi & \cos\phi \end{pmatrix} \text{ and } \begin{pmatrix} \cos\phi & \sin\phi \\ -\sin\phi & \cos\phi \end{pmatrix},$$

in either order, produces the identity matrix, and so these matrices are inverses of each other. Incidentally the only rotation matrices which are self-inverse are I and the 'about-turn' matrix

$$-I = \begin{pmatrix} -1 & 0 \\ 0 & -1 \end{pmatrix}$$

corresponding to rotation through π.

Invertible linear mappings play a fundamental role in the theory of abstract vector spaces — hence the alternative terminology below.

Definition 7.18   Let V and V' be vector spaces over the same field. Any invertible linear mapping $\alpha : V \to V'$ is called a **(vector space) isomorphism** and denoted by $\alpha : V \cong V'$. If there is such a mapping, V and V' are called **isomorphic**.

Isomorphic vector spaces are abstractly identical and any isomorphism between them matches up the operations of vector addition and scalar multiplication of vectors on the one vector space with the corresponding operations on the other space.

For example, let V be the vector space consisting of the eight subsets of $\{1, 2, 3\}$; V is a vector space over $\mathbb{Z}_2$ (see (6.3)(a)), vector addition being symmetric difference of subsets. Let $\alpha : V \to \mathbb{Z}_2^3$ be defined as follows: for $X \subseteq \{1, 2, 3\}$, let $(X)\alpha = (a_1, a_2, a_3)$ where $a_i = 0$ or 1 according as $i \notin X$ or $i \in X$ ($i = 1, 2, 3$). From Table 1 we see that α is bijective. Let $(X)\alpha = (a_1, a_2, a_3)$ and $(Y)\alpha = (b_1, b_2, b_3)$. Now

$a_i + b_i = 1 \Leftrightarrow$ exactly one of a_i and b_i is 1

$\qquad\qquad \Leftrightarrow i$ belongs to exactly one of X and Y

$\qquad\qquad \Leftrightarrow i$ bclongs to $X + Y$ by (2.4).

Therefore $(X + Y)\alpha = (a_1 + b_1, a_2 + b_2, a_3 + b_3) = (X)\alpha + (Y)\alpha$ for

Table 1

V $\alpha\downarrow$ $\mathbb{Z}_2^3$	$\varnothing$	$\{1\}$	$\{2\}$	$\{3\}$	$\{2, 3\}$	$\{1, 3\}$	$\{1, 2\}$	$\{1, 2, 3\}$
	$(0, 0, 0)$	$(1, 0, 0)$	$(0, 1, 0)$	$(0, 0, 1)$	$(0, 1, 1)$	$(1, 0, 1)$	$(1, 1, 0)$	$(1, 1, 1)$

all subsets X and Y of $\{1, 2, 3\}$, showing that α matches up symmetric difference in V with componentwise addition in $\mathbb{Z}_2^3$, that is, α respects vector addition. As

$$(0X)\alpha = (\varnothing)\alpha = (0, 0, 0) = 0((X)\alpha)$$
$$(1X)\alpha = (X)\alpha = 1((X)\alpha)$$

we see that α respects scalar multiplication of vectors. Using (7.18), all the properties of α are expressed in one breath by writing

$$\alpha : V \cong \mathbb{Z}_2^3$$

for α is an isomorphism between the vector spaces V and $\mathbb{Z}_2^3$.

The reader should compare (7.18) with the analogous concept (5.3) in ring theory. Although it is often difficult to decide whether or not two rings are isomorphic, the corresponding problem for finite-dimensional vector spaces is easily solved.

Theorem 7.19   Let V and V' be finite-dimensional vector spaces over the field F. Then V and V' are isomorphic if and only if $\dim V = \dim V'$.

Proof   As V is finite-dimensional, by (6.20) V has a basis $v_1, v_2, \ldots, v_n$ (if V is trivial, then the basis is $\varnothing$, that is, $n = 0$).

Suppose first that V and V' are isomorphic; let $\alpha : V \cong V'$ be an isomorphism between them. As α is bijective, by (7.16) the vectors $(v_1)\alpha, (v_2)\alpha, \ldots, (v_n)\alpha$ span V' and are linearly independent; these vectors therefore form a basis of V'. Comparing dimensions: $\dim V = n = \dim V'$ by (6.23).

Secondly suppose $\dim V = \dim V'$. By (6.20) V' has a basis $v_1', v_2', \ldots, v_n'$. By (7.16) there is a unique linear mapping $\alpha : V \to V'$ such that $(v_i)\alpha = v_i'$ $(1 \leq i \leq n)$, and α is bijective because $v_1', v_2', \ldots, v_n'$ span V' and are linearly independent. Therefore $\alpha : V \cong V'$ is an isomorphism; so V and V' are isomorphic.   □

As for abstract structure, (7.19) tells us that $\dim V$ is the one and only significant number associated with the finite-dimensional vector space V over the field F; in other words, knowing F and the non-negative integer $\dim V$ is sufficient information to construct an isomorphic copy of V. For instance, if V has dimension 100 over the rational field $\mathbb{Q}$, then V is isomorphic to $\mathbb{Q}^{100}$. However, much of the interest in vector spaces stems not from their abstract properties but from the context in which they arise: the dimension of a subspace *is* important, but so too is its location within the parent space.

Exercises   1. For each of the following linear mappings $\alpha : \mathbb{Q}^2 \to \mathbb{Q}^2$, find $(1, 0)\alpha$
7.2   and $(0, 1)\alpha$ and hence determine the 2×2 matrix A such that $(x)\alpha = xA$ for all $x = (x_1, x_2)$ in $\mathbb{Q}^2$ (that is, $\alpha = \mu_A$ (7.13)).

(a) $(1, 1)\alpha = (2, 1)$, $(1, -1)\alpha = (4, 3)$;
(b) $(2, 0)\alpha = (3, -4)$, $(2, 1)\alpha = (-6, 8)$;
(c) $(3, 5)\alpha = (6, 10)$, $(4, 7)\alpha = (-4, -7)$;
(d) $(7, 8)\alpha = (1, 0)$, $(9, 10)\alpha = (0, 1)$.

Denoting the above linear mappings by α_a, α_b, α_c, α_d, express the mappings below in the form μ_A:

$$\alpha_a \alpha_b, \qquad \alpha_b \alpha_a, \qquad \alpha_a^2 \alpha_b, \qquad \alpha_a \alpha_b \alpha_c, \qquad \alpha_d^{-1}.$$

Use (7.16) to determine which of α_a, α_b, α_c are invertible. Which of the composite mappings above are invertible?

2. In each case decide whether or not there is a linear mapping $\alpha : \mathbb{Q}^m \to \mathbb{Q}^n$ as stated, and if so, find an $m \times n$ matrix A with $\alpha = \mu_A$.

(a) $m = n = 2$; $(2, 1)\alpha = (1, 2)$, $(1, 3)\alpha = (1, 1)$, $(2, -9)\alpha = (-1, 2)$.
(b) $m = n = 2$; $(1, -1)\alpha = (1, 2)$, $(1, 4)\alpha = (2, 3)$, $(1, 1)\alpha = (\frac{7}{5}, \frac{11}{5})$.
(c) $m = 2$, $n = 3$; $(1, 1)\alpha = (1, -2, 1)$, $(1, -1)\alpha = (1, 0, 1)$, $(3, 1)\alpha = (3, -4, 3)$.
(d) $m = 3$, $n = 2$; $(1, 2, 1)\alpha = (1, 0)$, $(2, 1, -1)\alpha = (0, 1)$, $(1, -1, 0)\alpha = (1, 2)$.

3. For each matrix A below, interpret geometrically, as in (7.14), the linear mappings μ_A, μ_A^2, μ_A^{-1} of $\mathbb{R}^2$.

(a) $\begin{pmatrix} -1 & 0 \\ 0 & 1 \end{pmatrix}$,   (b) $\begin{pmatrix} 0 & -1 \\ -1 & 0 \end{pmatrix}$,   (c) $\begin{pmatrix} 2 & 0 \\ 0 & 2 \end{pmatrix}$,

(d) $(1/\sqrt{2})\begin{pmatrix} 1 & -1 \\ 1 & 1 \end{pmatrix}$,   (e) $(1/\sqrt{2})\begin{pmatrix} 1 & 1 \\ 1 & -1 \end{pmatrix}$.

4. (a) Let A_θ denote the rotation matrix $\begin{pmatrix} \cos\theta & \sin\theta \\ -\sin\theta & \cos\theta \end{pmatrix}$. Verify that $A_\theta A_\phi = A_{\theta+\phi}$, $A_\theta^{-1} = A_{-\theta}$. Determine the values of θ such that

(i) $A_\theta^4 = I$,   (ii) $A_\theta^4 = -I$.

(b) Let $\alpha : \mathbb{R}^2 \to \mathbb{R}^2$ be defined by

$$(x_1, x_2)\alpha = (x_1, x_2)\begin{pmatrix} \cos\phi & \sin\phi \\ \sin\phi & -\cos\phi \end{pmatrix} \text{ for all } x_1, x_2 \in \mathbb{R}.$$

Show that α is reflection in the line through the origin of gradient $\tan\frac{1}{2}\phi$.

(c) Let $A = \begin{pmatrix} a & b \\ c & d \end{pmatrix}$ be   a   matrix   over $\mathbb{R}$ which   is

orthogonal, that is, $AA^T = I$. By comparing entries in this matrix equation, show that $(ad - bc)^2 = 1$. Prove that A is a rotation matrix or reflection matrix (as in (b) above) according as $ad - bc = 1$ or $ad - bc = -1$.

5. At Uniform State University, x_1 professors and x_2 assistants together fulfil y_1 hours of teaching each week, their combined annual salary being y_2 thousands of dollars. The mapping

$$\alpha : \mathbb{Q}^2 \to \mathbb{Q}^2 \quad \text{where} \quad (x_1, x_2)\alpha = (y_1, y_2)$$

is known to be linear. Given that $(4, 8)\alpha = (402, 216)$ and $(6, 5)\alpha = (253, 275)$, find each professor's and each assistant's weekly teaching load and annual salary.

6. (a) Let $\alpha : V \to V'$ be linear and let $v_1, v_2, \ldots, v_n \in V$. If $(v_1)\alpha, (v_2)\alpha, \ldots, (v_n)\alpha$ are linearly independent, show that $v_1, v_2, \ldots, v_n$ are linearly independent. If $(v_1)\alpha, (v_2)\alpha, \ldots, (v_n)\alpha$ span V', show by example that $v_1, v_2, \ldots, v_n$ need not span V.
 (b) Let $\alpha : V \to V'$ and $\beta : V' \to V''$ be linear mappings. Show that their composition $\alpha\beta : V \to V''$ is also linear.
 (c) Let V have basis $v_1, v_2, \ldots, v_m$ and let $v_1', v_2', \ldots, v_m'$ belong to V'. Complete (7.16) by verifying the linearity of $\alpha : V \to V'$ defined by

$$(v)\alpha = \sum_{i=1}^{m} a_i v_i' \quad \text{where} \quad v = \sum_{i=1}^{m} a_i v_i.$$

 (d) Let V and V' be vector spaces of the same finite dimension over the same field, and let $\alpha : V \to V'$ be linear. Use (6.24) and (7.16) to show that α is injective if and only if α is surjective.

Representation of linear mappings*

Our theme remains the connection between matrices and linear mappings, but with a different emphasis: instead of starting with a matrix and obtaining a linear mapping from it as in (7.13), we begin with a linear mapping $\alpha : V \to V'$ and find, using bases b and b' of V and V' a matrix A which **represents** α, which means that the abstract linear mapping α is completely specified by b, b', and A; in fact α is no more than the mapping μ_A of (7.13) in terms of co-ordinates defined by the bases b and b'. However, the issues raised by expressing α in co-ordinate form will occupy us through several chapters—we are about to set out on a crusade, punctuated by the incidental slaying of problems en route, its main object being an understanding of the relationship between α and A.

One further comment: the material of this section is rather

abstract, for we are, in effect, setting up our future programme. The reader who gets lost should have no qualms either about turning directly to the next chapter (where there is more 'doing' and less 'theorizing') or about returning here as many times as is necessary once more experience has been gained — after a few fights have been fought (and won), it should be easier to grasp what the contest is all about!

We begin by introducing sums and scalar multiples of linear mappings.

Definition 7.20

Let α, $\beta : V \to V'$ be linear mappings where V and V' are vector spaces over F, and let $a \in F$. The **sum** of α and β is the mapping $\alpha + \beta : V \to V'$ where $(v)(\alpha + \beta) = (v)\alpha + (v)\beta$ for all $v \in V$. The **scalar product** of a and α is the mapping $a\alpha : V \to V'$ where $(v)(a\alpha) = a((v)\alpha)$ for all $v \in V$.

It is routine to verify that $\alpha + \beta$ and $a\alpha$ are also linear mappings; note that the commutative law of scalar multiplication is required for the linearity of $a\alpha$.

Notation

Let V and V' be vector spaces over the field F. The set of all linear mappings of V to V' is denoted by $\operatorname{Hom}(V, V')$.

As linear mappings respect the vector space structure, they are also known as **vector space homomorphisms**—hence the notation above. As sums and scalar multiples of linear mappings of V to V' are again such mappings, $\operatorname{Hom}(V, V')$ is closed under addition and scalar multiplication of mappings.

Proposition 7.21

Let V and V' be vector spaces over F. Then $\operatorname{Hom}(V, V')$ is a vector space over F.

Proof

To show that the left distributive law (part one of law 5 of (6.1)) holds in $\operatorname{Hom}(V, V')$, consider α, $\beta \in \operatorname{Hom}(V, V')$ and $a \in F$. As this law holds in V', we obtain

$$a((v)\alpha + (v)\beta) = a((v)\alpha) + a((v)\beta) \quad \text{for all } v \in V$$

which means $a(\alpha + \beta) = a\alpha + a\beta$ using (7.20) and (1.17). Similarly, the remaining laws of (6.1) hold in $\operatorname{Hom}(V, V')$; in particular, the zero of $\operatorname{Hom}(V, V')$ is the mapping which maps every vector of V to the zero vector of V'. $\qquad\square$

Let us look at $\operatorname{Hom}(V, V')$ in the familiar case $V = F^m$, $V' = F^n$.

By (7.17) every mapping in $\text{Hom}(F^m, F^n)$ is uniquely expressible in the form μ_A where A is an $m \times n$ matrix over F. If B is also an $m \times n$ matrix over F, then

$$(x)(\mu_A + \mu_B) = (x)\mu_A + (x)\mu_B = xA + xB = x(A + B)$$
$$= (x)\mu_{A+B} \quad \text{for all } x \in F^m$$

showing that $\mu_A + \mu_B = \mu_{A+B}$; in other words, the sum (7.20) of linear mappings corresponds to the sum (7.3) of matrices. Similarly $a\mu_A = \mu_{aA}$ shows that scalar multiples of linear mappings correspond to scalar multiples of matrices. Summarizing, $\text{Hom}(F^m, F^n)$ is essentially the same as the vector space $^mF^n$ of all $m \times n$ matrices over F; more exactly, the correspondence

$$\text{Hom}(F^m, F^n) \cong {}^mF^n$$

$$\mu_A \mapsto A$$

is a vector space isomorphism, because this correspondence is bijective by (7.13) and (7.17), and because the above discussion shows that the vector space operations are respected.

We prepare now to discuss a more general version of this correspondence, where F^m and F^n are replaced by *abstract* vector spaces V and V' of dimensions m and n with bases b and b'.

Definition 7.22 Let V and V' be vector spaces over the field F and let $\alpha : V \to V'$ be a linear mapping. Let V have basis b consisting of $v_1, v_2, \ldots, v_m$ and let V' have basis b' consisting of $v'_1, v'_2, \ldots, v'_n$. As $(v_i)\alpha$ belongs to V', there are unique scalars a_{ij} such that

$$(v_i)\alpha = a_{i1}v'_1 + a_{i2}v'_2 + \ldots + a_{in}v'_n \quad (1 \leqslant i \leqslant m).$$

The $m \times n$ matrix $A = [a_{ij}]$ is called **the matrix of α relative to the bases b and b'.**

So the matrix of α relative to two given bases b and b' is formed by expressing the image by α of each vector in b as a linear combination of the vectors in b'; the scalars appearing in these linear combinations form the rows of the matrix of α, which is therefore the array $[a_{ij}]$ of coefficients on the right-hand side of the equations

$$(v_1)\alpha = a_{11}v'_1 + a_{12}v'_2 + \ldots + a_{1n}v'_n$$
$$(v_2)\alpha = a_{21}v'_1 + a_{22}v'_2 + \ldots + a_{2n}v'_n$$
$$\vdots \qquad \vdots \qquad \vdots \qquad \qquad \vdots$$
$$(v_m)\alpha = a_{m1}v'_1 + a_{m2}v'_2 + \ldots + a_{mn}v'_n.$$

What is the point of this elaborate set-up? We next work an example which provides some clues to the answer.

Example 7.23

Suppose we wish to analyse the linear mapping $\alpha = \mu_A$ of $\mathbb{Q}^4$ to $\mathbb{Q}^3$ where

$$A = \begin{pmatrix} 5 & 4 & 4 \\ 7 & 6 & 5 \\ 8 & 6 & 7 \\ 4 & 2 & 5 \end{pmatrix}.$$

Therefore $(x)\alpha = xA$ for all $x = (x_1, x_2, x_3, x_4)$ in $\mathbb{Q}^4$. Let ℓ_s denote the standard basis e_1, e_2, e_3, e_4 of $\mathbb{Q}^4$, and let ℓ'_s denote the standard basis e_1, e_2, e_3 of $\mathbb{Q}^3$ (the number of entries in e_i should be clear from the context). As

$$(e_1)\alpha = e_1A = (5, 4, 4) = 5e_1 + 4e_2 + 4e_3$$
$$(e_2)\alpha = e_2A = (7, 6, 5) = 7e_1 + 6e_2 + 5e_3$$
$$(e_3)\alpha = e_3A = (8, 6, 7) = 8e_1 + 6e_2 + 7e_3$$
$$(e_4)\alpha = e_4A = (4, 2, 5) = 4e_1 + 2e_2 + 5e_3$$

from (7.22) we see that α has matrix A relative to ℓ_s and ℓ'_s. (More generally, if A is any $m \times n$ matrix over F, then $\mu_A : F^m \rightarrow F^n$ has matrix A relative to the standard bases ℓ_s and ℓ'_s of F^m and F^n.)

The matrix A specifies α without revealing any of its properties: the standard bases are useful in setting up the problem, but they are useless for finding its solution! The 'trick' in this case (and in all similar cases) consists of referring the linear mapping α to bases which are 'made to measure', that is, they are specially constructed to reveal the anatomy of α. The reader may verify that the rows r_1, r_2, r_3, r_4 of A are subject to the linear dependence relations

$$3r_1 - r_2 - r_3 = 0, \qquad 2r_1 - 2r_2 + r_3 - r_4 = 0$$

(see (8.19)(b) for a direct method of deriving such relations) and so $(3, -1, -1, 0)\alpha = 0$ and $(2, -2, 1, -1)\alpha = 0$. For technical reasons we number these vectors $v_3 = (3, -1, -1, 0)$ and $v_4 = (2, -2, 1, -1)$ and extend in any way to a basis ℓ of $\mathbb{Q}^4$ of the form v_1, v_2, v_3, v_4; let us choose $v_1 = (1, 0, 0, 0)$, $v_2 = (0, 1, 0, 0)$. As $v'_1 = (v_1)\alpha = (5, 4, 4)$ and $v'_2 = (v_2)\alpha = (7, 6, 5)$ are linearly independent vectors of $\mathbb{Q}^3$, we can extend to v'_1, v'_2, v'_3 forming the basis ℓ' of $\mathbb{Q}^3$ by

Fig. 7.3

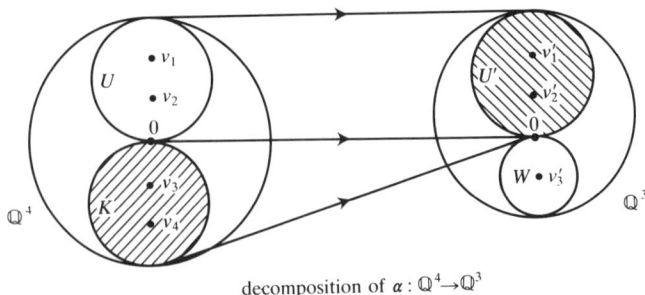

decomposition of $\alpha: \mathbb{Q}^4 \to \mathbb{Q}^3$

taking $v_3' = (1, 0, 0)$. The bases b and b' are linked by α as follows:

$$
\begin{aligned}
(v_1)\alpha &= v_1' + 0v_2' + 0v_3' \\
(v_2)\alpha &= 0v_1' + v_2' + 0v_3' \\
(v_3)\alpha &= 0v_1' + 0v_2' + 0v_3' \\
(v_4)\alpha &= 0v_1' + 0v_2' + 0v_3'
\end{aligned}
\quad \text{and so} \quad
B = \begin{pmatrix} 1 & 0 & 0 \\ 0 & 1 & 0 \\ 0 & 0 & 0 \\ 0 & 0 & 0 \end{pmatrix}
$$

is by (7.22) the matrix of α relative to b and b'. In terms of b and b', the effect of α is expressed by the equation

$$(a_1v_1 + a_2v_2 + a_3v_3 + a_4v_4)\alpha = a_1v_1' + a_2v_2' \quad \text{for} \quad a_i \in \mathbb{Q}.$$

Therefore the 2-dimensional subspace $K = \langle v_3, v_4 \rangle$ of $\mathbb{Q}^4$ is *collapsed* by α to the trivial subspace 0 of $\mathbb{Q}^3$, that is, every vector $a_3v_3 + a_4v_4$ of K is mapped by α to the zero vector of $\mathbb{Q}^3$ (K is the **kernel** (8.1) of α). In contrast, the 2-dimensional subspace $U = \langle v_1, v_2 \rangle$, which is a complement of K in $\mathbb{Q}^4$, is not collapsed at all, for it is mapped *isomorphically* onto the 2-dimensional subspace $U' = \langle v_1', v_2' \rangle$ of $\mathbb{Q}^3$, meaning that α restricted to U is an isomorphism $U \cong U'$; in fact $(a_1v_1 + a_2v_2)\alpha = a_1v_1' + a_2v_2'$, corresponding to the 2×2 identity matrix in the top-left of B (U' is the **image** (8.1) of α). The picture (Fig. 7.3) is completed by a complement $W = \langle v_3' \rangle$ of U' in $\mathbb{Q}^3$, for the decompositions $\mathbb{Q}^4 = U \oplus K$ and $\mathbb{Q}^3 = U' \oplus W$ effectively decompose α (into an isomorphism part and a trivial part).

   The above example gives the reader a bird's-eye view of much of the theory in the next chapter—for instance, using kernels and images we shall see (8.3) that *every* linear mapping of finite-dimensional vector spaces can be decomposed in the above way.

Definition 7.24   Let b denote the basis $v_1, v_2, \ldots, v_m$ of the vector space V over the field F, and let $\kappa_b : V \cong F^m$ denote the isomorphism defined by $(v_i)\kappa_b = e_i$ $(1 \leq i \leq m)$. Then $(v)\kappa_b$ is called the **co-ordinate vector** of v relative to the basis b.

The isomorphism κ_{ℓ} (which exists by (7.19)) maps the i th vector in the given basis ℓ of V to the i th vector in the standard basis of F^m. Using the linearity of κ_{ℓ} we obtain

$$(a_1 v_1 + a_2 v_2 + \ldots + a_m v_m)\kappa_{\ell} = (a_1, a_2, \ldots, a_m) \quad \text{for} \quad a_i \in F$$

and so κ_{ℓ} effectively co-ordinatizes V.

For example, the vectors $v_1 = (1, 0, -1)$ and $v_2 = (0, 1, -1)$ form a basis ℓ of $V = \{(x, y, z) \in \mathbb{R}^3 : x + y + z = 0\}$. Each vector in V is uniquely expressible in the form $x v_1 + y v_2$ and the ordered pair (x, y) is its co-ordinate vector relative to ℓ; for instance $(7, -5, -2)\kappa_{\ell} = (7, -5)$. The basis ℓ has been used to set up a system of co-ordinates in the plane V, and $\kappa_{\ell} : V \cong \mathbb{R}^2$ tells us that V is abstractly identical to the cartesian plane $\mathbb{R}^2$.

We show next that the correspondence between a linear mapping and its matrix relative to given bases can be concisely expressed using the co-ordinatizing isomorphisms κ_{ℓ}.

Proposition 7.25 Let the linear mapping $\alpha : V \to V'$ have matrix A relative to the bases ℓ and ℓ' of the vector spaces V and V' over the field F. Then

$$\kappa_{\ell}\mu_A = \alpha \kappa_{\ell'}.$$

Proof Suppose that the basis ℓ consists of $v_1, v_2, \ldots, v_m$ and the basis ℓ' consists of $v'_1, v'_2, \ldots, v'_n$. Consider the diagram

$$
\begin{array}{ccc}
V & \xrightarrow{\alpha} & V' \\
\kappa_{\ell} \downarrow & & \downarrow \kappa_{\ell'} \\
F^m & \xrightarrow{\mu_A} & F^n
\end{array}
$$

of linear mappings. We must show that following the arrows from V to F^n by the two possible routes (via F^m or via V') produces the same composite linear mapping; such diagrams are called **commutative**.

Now $(v_i)\kappa_{\ell}\mu_A = (c_i)\mu_A = e_i A = (\text{row } i \text{ of } A)$ by (7.24). Writing $A = [a_{ij}]$, from (7.22) we obtain

$$(v_i)\alpha = a_{i1}v'_1 + a_{i2}v'_2 + \ldots + a_{in}v'_n$$

which tells us that $(v_i)\alpha$ has co-ordinate vector $(a_{i1}, a_{i2}, \ldots, a_{in})$ relative to ℓ', that is, $(v_i)\alpha\kappa_{\ell'} = (\text{row } i \text{ of } A)$. Therefore $(v_i)\kappa_{\ell}\mu_A = (v_i)\alpha\kappa_{\ell'}$ for $1 \leq i \leq m$; so the linear mappings $\kappa_{\ell}\mu_A$ and $\alpha\kappa_{\ell'}$ agree on the basis ℓ of V, and hence they are equal by (7.16), that is, $\kappa_{\ell}\mu_A = \alpha\kappa_{\ell'}$. □

The equation $\kappa_{\ell}\mu_A = \alpha\kappa_{\ell'}$ says that μ_A is the co-ordinate form of α: for if v has co-ordinate vector x relative to ℓ, then

$$(v)\alpha\kappa_{\ell'} = (v)\kappa_{\ell}\mu_A = (x)\mu_A = xA$$

showing that $(v)\alpha$ has co-ordinate vector xA relative to ℓ'. In other words, every linear mapping α (of non-trivial finite-dimensional vector spaces) is no more than a thinly-disguised μ_A; as $\alpha = \kappa_{\ell}\mu_A\kappa_{\ell'}^{-1}$, the disguise amounts to premultiplication by κ_{ℓ} and postmultiplication by $\kappa_{\ell'}^{-1}$.

Now we establish the properties of the correspondence between α and A.

Corollary 7.26

Let V and V' be vector spaces of dimensions m and n over F with bases ℓ and ℓ'. The correspondence

$$\text{Hom}(V, V') \cong {}^mF^n \quad \text{where} \quad \kappa_{\ell}\mu_A = \alpha\kappa_{\ell'}$$
$$\alpha \mapsto A$$

is an isomorphism.

Proof

From (7.25), the above correspondence is between α and its matrix A relative to ℓ and ℓ'. This correspondence is bijective because $A \mapsto \alpha = \kappa_{\ell}\mu_A\kappa_{\ell'}^{-1}$ is its inverse (α is *the* linear mapping of V to V' having matrix A relative to ℓ and ℓ'). Addition is respected, for suppose $\alpha \mapsto A$ and $\beta \mapsto B$, which means $\kappa_{\ell}\mu_A = \alpha\kappa_{\ell'}$ and $\kappa_{\ell}\mu_B = \beta\kappa_{\ell'}$. Therefore

$$\kappa_{\ell}\mu_{A+B} = \kappa_{\ell}(\mu_A + \mu_B) = \kappa_{\ell}\mu_A + \kappa_{\ell}\mu_B$$
$$= \alpha\kappa_{\ell'} + \beta\kappa_{\ell'} = (\alpha + \beta)\kappa_{\ell'}$$

as linear mappings obey the distributive law (we leave the reader to verify this abstract version of (7.7)(a)) and so $\alpha + \beta \mapsto A + B$. Also

$$\kappa_{\ell}\mu_{aA} = \kappa_{\ell}(a\mu_A) = a(\kappa_{\ell}\mu_A) = a(\alpha\kappa_{\ell'}) = (a\alpha)\kappa_{\ell'}$$

showing $a\alpha \mapsto aA$ for $a \in F$. So the correspondence $\alpha \mapsto A$ is an isomorphism by (7.18). $\qquad\square$

As (7.23) suggests, it is the flexibility in the choice of bases ℓ and ℓ' which makes the above correspondence superior to its pure co-ordinate form $\text{Hom}(F^m, F^n) \cong {}^mF^n$; (7.26) highlights the impartiality of linear mappings towards bases—one basis is as good as another as far as linear mappings are concerned—whereas matrices are prejudiced, for they prefer the standard bases above all others.

Our next corollary is fundamental: it connects composition of linear mappings with matrix multiplication in a very useful way.

Corollary
7.27

Let V, V', V'' be vector spaces over F with bases b, b', b'' and let $\alpha : V \to V'$ and $\beta : V' \to V''$ be linear mappings such that α has matrix A relative to b and b', and β has matrix B relative to b' and b''. Then $\alpha\beta : V \to V''$ has matrix AB relative to b and b''.

Proof

Notice that the basis b' of V' serves as second basis in the construction of A and as first basis in the construction of B. From (7.25) we know $\kappa_b \mu_A = \alpha \kappa_{b'}$, $\kappa_{b'} \mu_B = \beta \kappa_{b''}$, and so two commutative diagrams can be placed side by side to give:

$$
\begin{array}{ccccc}
V & \xrightarrow{\alpha} & V' & \xrightarrow{\beta} & V'' \\
{\scriptstyle \kappa_b}\downarrow & & {\scriptstyle \kappa_{b'}}\downarrow & & {\scriptstyle \kappa_{b''}}\downarrow \\
F^l & \xrightarrow{\mu_A} & F^m & \xrightarrow{\mu_B} & F^n
\end{array}
$$

where $l = \dim V$, $m = \dim V'$, $n = \dim V''$, the vertical arrows denoting isomorphisms. As $\mu_{AB} = \mu_A \mu_B$ by (7.13), we obtain

$$\kappa_b \mu_{AB} = \kappa_b \mu_A \mu_B = \alpha \kappa_{b'} \mu_B = \alpha\beta \kappa_{b''}$$

which shows that the 'composite' diagram

$$
\begin{array}{ccc}
V & \xrightarrow{\alpha\beta} & V'' \\
{\scriptstyle \kappa_b}\downarrow & & \downarrow{\scriptstyle \kappa_{b''}} \\
F^l & \xrightarrow{\mu_{AB}} & F^n
\end{array}
$$

is commutative also; so $\alpha\beta$ has matrix AB relative to b and b''. □

To illustrate (7.27) we take up again the linear mapping $\alpha = \mu_A$ of (7.23).

Example
7.28

As before, let $A = \begin{pmatrix} 5 & 4 & 4 \\ 7 & 6 & 5 \\ 8 & 6 & 7 \\ 4 & 2 & 5 \end{pmatrix}$ and $B = \begin{pmatrix} 1 & 0 & 0 \\ 0 & 1 & 0 \\ 0 & 0 & 0 \\ 0 & 0 & 0 \end{pmatrix}$. We form the

matrices

$$P = \begin{pmatrix} v_1 \\ v_2 \\ v_3 \\ v_4 \end{pmatrix} = \begin{pmatrix} 1 & 0 & 0 & 0 \\ 0 & 1 & 0 & 0 \\ 3 & -1 & -1 & 0 \\ 2 & -2 & 1 & -1 \end{pmatrix} \quad \text{and} \quad Q = \begin{pmatrix} v_1' \\ v_2' \\ v_3' \end{pmatrix} = \begin{pmatrix} 5 & 4 & 4 \\ 7 & 6 & 5 \\ 1 & 0 & 0 \end{pmatrix}.$$

The rows of P are the vectors v_1, v_2, v_3, v_4 of the basis b of $\mathbb{Q}^4$, the rows of Q being the vectors v_1', v_2', v_3' of the basis b' of $\mathbb{Q}^3$. Let the reader be awake, for our next move is simple but subtle! By (7.22) the equations

$$(v_1)\iota = v_1 = e_1$$
$$(v_2)\iota = v_2 = \qquad e_2$$
$$(v_3)\iota = v_3 = 3e_1 - \ e_2 - e_3$$
$$(v_4)\iota = v_4 = 2e_1 - 2e_2 + e_3 - e_4$$

show that P is the matrix of the identity mapping ι of $\mathbb{Q}^4$ relative to b and the standard basis b_s of $\mathbb{Q}^4$. From (7.23) we know that A is the matrix of α relative to the standard bases b_s (of $\mathbb{Q}^4$) and b_s' (of $\mathbb{Q}^3$). Therefore by (7.27) the composite mapping $\iota\alpha$ $(=\alpha)$ has matrix PA relative to b and b_s'; in other words:

> Changing the first basis causes the
> matrix A of α to change to PA.

The matrix P is invertible, because its inverse P^{-1} is the matrix of $\iota^{-1} = \iota$ relative to b_s and b (that is, b and b_s in the opposite order); in fact, rearranging the above equations to express each e_i in terms of the vectors v_j gives

$$e_1 = v_1$$
$$e_2 = \qquad v_2 \qquad \text{and so} \quad P^{-1} = \begin{pmatrix} 1 & 0 & 0 & 0 \\ 0 & 1 & 0 & 0 \\ 3 & -1 & -1 & 0 \\ 5 & -3 & -1 & -1 \end{pmatrix}.$$
$$e_3 = 3v_1 - \ v_2 - v_3$$
$$e_4 = 5v_1 - 3v_2 - v_3 - v_4$$

Our remarks about P apply equally to Q, which may be regarded as the matrix of the identity mapping ι of $\mathbb{Q}^3$ relative to b' and b_s'. From (7.23) we know that B is the matrix of α relative to b and b'. Therefore BQ is the matrix of the composite mapping $\alpha\iota$ $(=\alpha)$ relative to b and b_s' by (7.27), and so:

> Changing the second basis causes the
> matrix B of α to change to BQ.

The matrix Q is invertible: in fact

$$Q^{-1} = \tfrac{1}{4}\begin{pmatrix} 0 & 0 & 4 \\ -5 & 4 & -3 \\ 6 & -4 & -2 \end{pmatrix}$$

as the reader may verify by multiplying Q and Q^{-1} together. One final point: we have calculated the matrix of α relative to b and b_s' in two ways, obtaining PA and BQ. Therefore $PA = BQ$ and so

$PAQ^{-1} = B$. Summing up:

> Changing both bases causes the matrix A of α
> to change to PAQ^{-1}, where P and Q are
> invertible matrices.

These ideas are developed more fully in (8.32). We now discuss Hom(V, V), the set of linear mappings of V to itself. It is straightforward to verify that Hom(V, V), together with the operations of addition (7.20) and composition (1.18) of linear mappings, is a ring. Notice that the identity mapping ι of V is the 1-element of Hom(V, V).

Let V be an n-dimensional vector space over the field F where $n > 0$. We prepare to show that each basis ℓ of V defines a ring isomorphism of Hom(V, V) to the ring $M_n(F)$ of $n \times n$ matrices over F. Let $\alpha : V \to V$ be a linear mapping. The matrix of α relative to ℓ and ℓ', where ℓ and ℓ' are different bases of V, may of course be formed; however, it suits our present purpose to take $\ell = \ell'$, that is, we use the *same* bases in the two copies of V to form a matrix corresponding to α.

Definition 7.29

Let $v_1, v_2, \ldots, v_n$ form the basis ℓ of the vector space V over the field F and let $\alpha : V \to V$ be a linear mapping. The $n \times n$ matrix $A = [a_{ij}]$, where $(v_i)\alpha = \sum_{j=1}^{n} a_{ij}v_j$ $(1 \le i \le n)$, is called **the matrix of α relative to ℓ.**

To form the matrix A of α relative to ℓ, the image by α of each of $v_1, v_2, \ldots, v_n$ must be expressed as a linear combination of these same vectors; the scalars in these linear combinations form the rows of A. The usefulness of (7.29) lies in its flexibility—the basis ℓ can be chosen to suit the linear mapping α in hand.

Corollary 7.30

Let ℓ be a basis of the n-dimensional vector space V over the field F. The correspondence

$$\text{Hom}(V, V) \cong M_n(F) \quad \text{where} \quad \kappa_\ell \mu_A = \alpha \kappa_\ell$$
$$\alpha \mapsto A$$

is a ring isomorphism.

Proof

From (7.25), the equation $\kappa_\ell \mu_A = \alpha \kappa_\ell$ expresses the fact that A is the matrix of α relative to ℓ. We are dealing with a special case of (7.26), namely where $V = V'$ and $\ell = \ell'$; therefore the above correspondence is bijective and respects addition. Suppose $\alpha \mapsto A$

and $\beta \mapsto B$. Applying (7.27) with $V = V' = V''$ and $\mathscr{b} = \mathscr{b}' = \mathscr{b}''$ gives $\alpha\beta \mapsto AB$, showing that the above correspondence respects multiplication. As $\iota \mapsto I$, the 1-elements of the rings $\mathrm{Hom}(V, V)$ and $\mathrm{M}_n(F)$ correspond and so the correspondence is a ring isomorphism by (5.3). $\square$

As an illustration of (7.30), consider $\alpha = \mu_A : \mathbb{Q}^2 \to \mathbb{Q}^2$ where $A = \begin{pmatrix} 4 & 9 \\ -1 & -2 \end{pmatrix}$, and let us suppose that we are required to carry out calculations involving A, for instance, we may wish to calculate $A^{63} + 5A^{20}$. A direct approach would be tedious, leaving one exhausted and none the wiser! One way round the problem is to use a convenient basis (how such bases are found will be discussed in chapter 10); in this case we use $\mathscr{b}$ consisting of $v_1 = (0, -1)$, $v_2 = (1, 3)$. As

$(v_1)\alpha = (0, -1)A = (1, 2) = v_1 + v_2$ we obtain $B = \begin{pmatrix} 1 & 1 \\ 0 & 1 \end{pmatrix}$

$(v_2)\alpha = (1, 3)A \quad = (1, 3) = \qquad v_2$

as the matrix of α relative to $\mathscr{b}$. Now α has matrix A relative to the standard basis $\mathscr{b}_s$ of $\mathbb{Q}^2$ and

$$P = \begin{pmatrix} v_1 \\ v_2 \end{pmatrix} = \begin{pmatrix} 0 & -1 \\ 1 & 3 \end{pmatrix}$$

is the matrix of the identity mapping ι relative to $\mathscr{b}$ and $\mathscr{b}_s$. Therefore $PA = BP$ is the matrix of $\iota\alpha = \alpha = \alpha\iota$ relative to $\mathscr{b}$ and $\mathscr{b}_s$ and so $PAP^{-1} = B$; in other words:

Changing the basis causes the matrix
A of α to change to PAP^{-1}.

The ring isomorphism (7.30) depends only on $\mathscr{b}$, and in this case has the explicit form

$\mathrm{Hom}(\mathbb{Q}^2, \mathbb{Q}^2) = \mathrm{M}_2(\mathbb{Q})$

$\mu_C \mapsto PCP^{-1}$

which is valid for all matrices C in $\mathrm{M}_2(\mathbb{Q})$. As this correspondence respects the vector space and ring operations and $\mu_A \mapsto B$, we obtain $\mu_{A^{63}+5A^{20}} \mapsto B^{63} + 5B^{20}$, that is,

$$P(A^{63} + 5A^{20})P^{-1} = B^{63} + 5B^{20}.$$

So, when faced with a calculation involving A, carry out the same (but easier) calculation with B instead. As

$$B^n = \begin{pmatrix} 1 & n \\ 0 & 1 \end{pmatrix} \quad \text{and} \quad P^{-1} = \begin{pmatrix} 3 & 1 \\ -1 & 0 \end{pmatrix}$$

we obtain

$$B^{63} + 5B^{20} = \begin{pmatrix} 1 & 63 \\ 0 & 1 \end{pmatrix} + 5\begin{pmatrix} 1 & 20 \\ 0 & 1 \end{pmatrix} = \begin{pmatrix} 6 & 163 \\ 0 & 6 \end{pmatrix},$$

and hence

$$A^{63} + 5A^{20} = P^{-1}(B^{63} + 5B^{20})P = \begin{pmatrix} 3 & 1 \\ -1 & 0 \end{pmatrix}\begin{pmatrix} 6 & 163 \\ 0 & 6 \end{pmatrix}\begin{pmatrix} 0 & -1 \\ 1 & 3 \end{pmatrix}$$

$$= \begin{pmatrix} 495 & 1467 \\ -163 & -483 \end{pmatrix}.$$

Exercises 7.3

1. Find the matrix of $\mu_A : \mathbb{Q}^2 \to \mathbb{Q}^3$ where $A = \begin{pmatrix} 1 & 2 & 3 \\ 3 & 6 & 9 \end{pmatrix}$ relative to the following pairs of bases:

   (a) $(1,0), (3,-1)$; $(1,0,0), (0,1,0), (0,0,1)$.
   (b) $(1,0), (0,1)$; $(1,2,3), (1,0,0), (0,1,0)$.
   (c) $(1,0), (3,-1)$; $(1,2,3), (1,0,0), (0,1,0)$.
   (d) $(3,-1), (1,0)$; $(1,0,0), (0,1,0), (1,2,3)$.

Find bases of $\mathbb{Q}^2$ and $\mathbb{Q}^3$ relative to which μ_A has matrix

$$\begin{pmatrix} 1 & 1 & 1 \\ 1 & 1 & 1 \end{pmatrix}.$$

2. (a) Use the method of (7.23) to find an invertible 3×3 matrix P and an invertible 2×2 matrix Q such that

$$P\begin{pmatrix} 1 & 2 \\ 3 & 4 \\ 5 & 6 \end{pmatrix} = \begin{pmatrix} 1 & 0 \\ 0 & 1 \\ 0 & 0 \end{pmatrix}Q.$$

   (b) Find invertible 3×3 matrices P and Q such that

$$P\begin{pmatrix} 1 & -1 & -4 \\ -1 & 1 & 4 \\ 3 & -3 & 2 \end{pmatrix} = \begin{pmatrix} 1 & 0 & 0 \\ 0 & 1 & 0 \\ 0 & 0 & 0 \end{pmatrix}Q.$$

3. Find the matrix of $\mu_A : \mathbb{Q}^2 \to \mathbb{Q}^2$ where $A = \begin{pmatrix} 6 & -5 \\ 5 & -4 \end{pmatrix}$ relative to the basis $(1,0)$ and $(1,-1)$ of $\mathbb{Q}^2$. Hence determine A^{2000} and find a 2×2 matrix B with integer entries such that $B^5 = A$.

4. Let V denote the vector space of polynomials of degree at most 3 over $\mathbb{Q}$ and let $\Delta : V \to V$ denote the **difference operator**, that is,

$$(f(x))\Delta = f(x+1) - f(x) \quad \text{for all polynomials } f \text{ in } V$$

(for instance $(x^2)\Delta = (x+1)^2 - x^2 = 2x + 1$). Show that Δ is linear and find its matrix relative to the basis

(i) x^3, x^2, x, x^0 of V,
(ii) $x(x-1)(x-2), x(x-1), x, x^0$ of V.

Find the matrices of $\Delta^2, \Delta^3, \Delta^4$ relative to these bases.
Let $\delta : V \to V$ denote formal differentiation (V as above), that is,

$$(a_0 + a_1x + a_2x^2 + a_3x^3)\delta = a_1 + 2a_2x + 3a_3x^2 \quad (a_i \in \mathbb{Q}).$$

Show that δ is linear and find its matrix relative to each of the above bases. Hence find an **automorphism** α of V (that is, $\alpha : V \cong V$) such that $\alpha\Delta = \delta\alpha$.

5. (a) Let $\alpha, \beta : V \to V'$ be linear mappings. Show that $\alpha + \beta$ and $a\alpha$ are linear, for each scalar a. If $\gamma : V' \to V''$ is linear, prove the distributive law: $(\alpha + \beta)\gamma = \alpha\gamma + \beta\gamma$.
(b) Let U be a subspace of V and let $\alpha : V \to V'$ be a linear mapping. Show that $(U)\alpha = \{(u)\alpha : u \in U\}$ is a subspace of V'.
(c) Let U be a subspace of V. Show that

$$S = \{\alpha \in \operatorname{Hom}(V, V) : (U)\alpha \subseteq U\}$$

is both a subspace and a subring of $\operatorname{Hom}(V, V)$. Taking $V = \mathbb{Q}^2$ and $U = \langle(1, 0)\rangle$, find the condition on $A = \begin{pmatrix} a & b \\ c & d \end{pmatrix}$ for μ_A to be in S.

6. (a) The linear mapping $\alpha : V \to V$ has matrix $\begin{pmatrix} a & b \\ c & d \end{pmatrix}$ relative to the basis v_1 and v_2 of the 2-dimensional vector space V. Find the matrix of α relative to the following bases of V:

(i) v_2, v_1 (ii) $-v_1, -v_2$ (iii) $v_1, -v_2$ (iv) $v_1 + v_2, v_2$.

(b) Let P be an invertible $n \times n$ matrix over F. Show that α, defined by $(A)\alpha = PAP^{-1}$ for all A in $M_n(F)$, is an automorphism (5.5) of the ring $M_n(F)$.

(c) Show that the correspondence $\alpha : \begin{pmatrix} a & b \\ c & d \end{pmatrix} \to \begin{pmatrix} d & c \\ b & a \end{pmatrix}$ is an automorphism of $M_2(F)$. Find an automorphism of $M_2(\mathbb{Q})$ which maps $\begin{pmatrix} 1 & 0 \\ 0 & 2 \end{pmatrix}$ to $\begin{pmatrix} 1 & 1 \\ 0 & 2 \end{pmatrix}$.

8 Rank and row-equivalence

We continue our discussion of linear mappings and matrices to discover that much of the theory is expressible in terms of a single non-negative integer, namely the **rank** of the linear mapping or matrix under consideration.

The rank of the linear mapping $\alpha : V \to V'$ is the dimension of its **image** im $\alpha = \{(v)\alpha : v \in V\}$, the subspace of V' consisting of the images by α of the vectors in V. For instance $\alpha : \mathbb{R}^3 \to \mathbb{R}^2$, defined by $(x_1, x_2, x_3)\alpha = (x_1, 0)$, maps $\mathbb{R}^3$ onto the 1-dimensional x_1-axis in $\mathbb{R}^2$; so im $\alpha = \langle (1, 0) \rangle$ and $1 = \text{rank } \alpha$ in this case. In general rank α measures the non-degeneracy of the linear mapping α; if $\alpha : V \to V'$ is regarded as conveying information from V to V', then rank α is a measure of the *amount* of information conveyed by α.

The rank of a matrix A is the dimension of its **row space**, the space spanned by the rows of A. We show that **column space** A—the space spanned by the columns of A—has the same dimension as row space A and so the rank of a matrix is an unbiased concept. Of particular importance are the **invertible** matrices, for they relate any two bases of a given vector space to each other; it turns out that an $n \times n$ matrix is invertible if and only if it is of rank n.

In the second section of this chapter we develop a practical method of determining rank, finding linear dependence relations, inverting matrices, and solving systems of linear equations—it is nothing less than a panacea for the ills of matrix theory! The method springs from the study of row-equivalent matrices, that is, matrices having the same row space (and the same number of rows).

Rank

We begin by introducing two important subspaces associated with a linear mapping.

Definition 8.1   Let $\alpha : V \to V'$ be a linear mapping.

The subspace $\{v \in V : (v)\alpha = 0\}$ of V is called the **kernel** of α and denoted by ker α.

Fig. 8.1

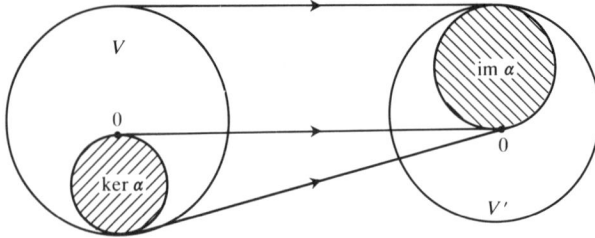

kernel and image of $\alpha : V \to V'$

The subspace $\{(v)\alpha : v \in V\}$ of V' is called the **image** of α and denoted by im α (or $(V)\alpha$).

We show in (8.2) that ker α and im α are indeed subspaces. Regarding α as conveying information from V to V', then im α is the information received by V', whereas ker α is the information lost in transit; more precisely, ker α consists of those vectors in V which are mapped by α to the zero vector of V' (Fig. 8.1). To illustrate these concepts, consider the formal differentiation mapping $\delta : \mathbb{Q}[x] \to \mathbb{Q}[x]$, that is, $(\sum a_i x^i)\delta = \sum i a_i x^{i-1}$, the summations being over all non-negative integers i. The kernel of δ consists of those polynomials which produce zero on being differentiated, that is, the 1-dimensional subspace of constant polynomials; so $\langle x^0 \rangle = \ker \delta$. As every polynomial can be integrated (every polynomial in $\mathbb{Q}[x]$ is expressible in the form $(f)\delta$ for some $f \in \mathbb{Q}[x]$), we see $\mathbb{Q}[x] = \operatorname{im} \delta$; in other words, δ is surjective.

Proposition 8.2   Let $\alpha : V \to V'$ be a linear mapping. Then ker α is a subspace of V, and im α is a subspace of V'. Further, α is injective if and only if $0 = \ker \alpha$, and α is surjective if and only if $V' = \operatorname{im} \alpha$.

Proof   We use (6.6). For u and v in ker α and the scalar a, we have $(u+v)\alpha = (u)\alpha + (v)\alpha = 0 + 0 = 0$, $(au)\alpha = a((u)\alpha) = a0 = 0$; so $u + v$, au belong to ker α. The equation $(0)\alpha = 0$ says $0 \in \ker \alpha$ (and $0 \in \operatorname{im} \alpha$); so ker α is a subspace of V by (6.6).

Consider two typical vectors $(u)\alpha$ and $(v)\alpha$ in im α $(u, v \in V)$. Then $(u)\alpha + (v)\alpha = (u+v)\alpha$ and $a((u)\alpha) = (au)\alpha$, showing that $(u)\alpha + (v)\alpha$ and $a((u)\alpha)$ belong to im α since $u + v$ and au belong to V. So im α is a subspace of V' by (6.6).

Suppose α to be injective and let $v \in \ker \alpha$. Then $(v)\alpha = 0 = (0)\alpha$, and so $v = 0$. Therefore $0 = \ker \alpha$ (ker α consists of the zero vector alone). Conversely suppose $0 = \ker \alpha$ and let $(u)\alpha = (v)\alpha$ for

$u, v \in V$. Then $(u - v)\alpha = (u)\alpha - (v)\alpha = 0$ showing that $u - v$ belongs to ker α. Therefore $u - v = 0$, that is, $u = v$; so α is injective.

Finally, $V' = \text{im } \alpha$ means that each vector in V' is the image by α of some vector in V, in other words, α is surjective. $\qquad \square$

Therefore $\alpha : V \to V'$ being injective means ker α is as small as possible (the trivial subspace of V), while α surjective means im α is as large as possible (the whole space V').

As preparation for our next theorem, consider again the differentiation mapping $\delta : \mathbb{Q}[x] \to \mathbb{Q}[x]$ and let $U = \{f \in \mathbb{Q}[x] : f(0) = 0\}$, that is, U is the subspace of polynomials having zero constant term; then U is a complement (6.27) in $\mathbb{Q}[x]$ of the subspace ker δ of constant polynomials, and so $\mathbb{Q}[x] = U \oplus \ker \delta$ (6.28) for every polynomial g is uniquely expressible as the sum of a polynomial in U and a polynomial in ker δ, namely, $g = (g - g(0)) + g(0)$. Let us *restrict* δ to U, obtaining $\delta\restriction_U$ (read: δ restricted to U), which is the mapping of U to $\mathbb{Q}[x]$ defined by $(f)(\delta\restriction_U) = (f)\delta$ for all f in U. In other words $\delta\restriction_U$ is formal differentiation of polynomials *with zero constant term* and so $(x^2 + x)(\delta\restriction_U) = (x^2 + x)\delta = 2x + 1$, whereas $(x^2 + x + 1)(\delta\restriction_U)$ is not defined. As δ is linear, so also is $\delta\restriction_U$. However $\delta\restriction_U$ has the advantage of being bijective, for it has an inverse—each polynomial in $\mathbb{Q}[x]$ can be integrated uniquely to a polynomial with zero constant term. Therefore $\delta\restriction_U$ is an isomorphism of U to $\mathbb{Q}[x]$, that is,

$$\delta\restriction_U : U \cong \mathbb{Q}[x].$$

Theorem 8.3

Let $\alpha : V \to V'$ be a linear mapping and let U be a complement of ker α in V (so $V = U \oplus \ker \alpha$). Then the restriction of α to U is an isomorphism $\alpha\restriction_U : U \cong \text{im } \alpha$.

Proof

As suggested in Fig. 8.2, $\alpha\restriction_U$ denotes the mapping of U to im α

Fig. 8.2

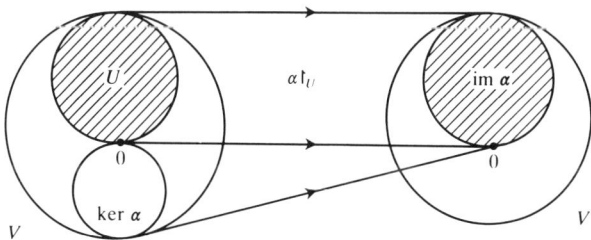

the isomorphism $\alpha\restriction_U$

defined by $(u)(\alpha\restriction_U) = (u)\alpha$ for all $u \in U$. As α is linear, so also is $\alpha\restriction_U$. To show that $\alpha\restriction_U$ is injective, let $u \in \ker(\alpha\restriction_U)$. Then $u \in U$ and $(u)\alpha = 0$ which combine to give $u \in U \cap \ker \alpha$. But $0 = U \cap \ker \alpha$ as U and $\ker \alpha$ are complementary subspaces. Therefore $u = 0$, showing $0 = \ker(\alpha\restriction_U)$. By (8.2), $\alpha\restriction_U$ is injective.

To show that $\alpha\restriction_U$ is surjective, consider $v' \in \operatorname{im} \alpha$. So $v' = (v)\alpha$ for some $v \in V$. As $V = U + \ker \alpha$, we see $v = u + w$ for $u \in U$ and $w \in \ker \alpha$. Therefore $v' = (v)\alpha = (u + w)\alpha = (u)\alpha + (w)\alpha = (u)\alpha + 0 = (u)\alpha = (u)(\alpha\restriction_U)$. So each vector of $\operatorname{im} \alpha$ is the image by $\alpha\restriction_U$ of some vector $u \in U$, that is, $\alpha\restriction_U$ is surjective. By (7.18), $\alpha\restriction_U$ is an isomorphism. $\square$

The above theorem is the vector space counterpart of (5.18) (the complement U of $\ker \alpha$ in V replaces the (conceptually harder) factor space $V/\ker \alpha$); it tells us how to conjure up an isomorphism which does the same job as the original linear mapping, but does it more efficiently. The reader should review (7.23) in the light of (8.3).

Definition 8.4

Let $\alpha : V \to V'$ be a linear mapping. If $\operatorname{im} \alpha$ is finite-dimensional, then $\dim(\operatorname{im} \alpha)$ is called the **rank** of α. If $\ker \alpha$ is finite-dimensional, then $\dim(\ker \alpha)$ is called the **nullity** of α.

The differentiation mapping $\delta : \mathbb{Q}[x] \to \mathbb{Q}[x]$ has infinite rank (that is, $\operatorname{im} \delta$ is not finite-dimensional); nullity $\alpha = 1$ as $\ker \delta = \langle x^0 \rangle$ is 1-dimensional.

Consider $\mu_A : \mathbb{Q}^3 \to \mathbb{Q}^2$ where $A = \begin{pmatrix} 1 & 1 \\ -1 & 0 \\ 1 & -1 \end{pmatrix}$. By (7.13),

$$(x_1, x_2, x_3)\mu_A = (x_1, x_2, x_3)A = (x_1 - x_2 + x_3, x_1 - x_3).$$

By (8.1), $\ker \mu_A = \{(x_1, x_2, x_3) \in \mathbb{Q}^3 : (x_1, x_2, x_3)\mu_A = (0, 0)\}$, and so $\ker \mu_A$ consists of the rational solutions of the simultaneous equations $x_1 - x_2 + x_3 = 0$ and $x_1 - x_3 = 0$ (note that the coefficients of the unknowns x_i correspond to the entries in the *columns* of A). So for (x_1, x_2, x_3) in $\ker \mu_A$ we have

$$(x_1, x_2, x_3) = (x_1, 2x_1, x_1) = x_1(1, 2, 1)$$

on eliminating x_2 and x_3; therefore $\ker \mu_A = \langle (1, 2, 1) \rangle$ and so nullity $\mu_A = 1$. Let r_i denote row i of A; so $r_1 = (1, 1)$, $r_2 = (-1, 0)$,

$r_3 = (1, -1)$. As im μ_A consists of all vectors in $\mathbb{Q}^2$ of the form

$$(x_1, x_2, x_3)\mu_A = (x_1, x_2, x_3)A = (x_1, x_2, x_3)\begin{pmatrix} r_1 \\ r_2 \\ r_3 \end{pmatrix} = x_1 r_1 + x_2 r_2 + x_3 r_3$$

we see im $\mu_A = \langle r_1, r_2, r_3 \rangle$. So im μ_A is the subspace of $\mathbb{Q}^2$ spanned by the rows of A. Here im $\mu_A = \mathbb{Q}^2$ and so rank $\mu_A = 2$. Notice one other fact: each non-zero vector in ker μ_A gives rise to a linear dependence relation between the rows of A; in this case

$$0 = (1, 2, 1)\mu_A = r_1 + 2r_2 + r_3.$$

The sceptical reader may like to carry out a spot check on (8.3) as follows: extend $v_3 = (1, 2, 1)$ to a basis v_1, v_2, v_3 of $\mathbb{Q}^3$ and then verify that $(v_1)\alpha$ and $(v_2)\alpha$ are linearly independent (v_1 and v_2 form a basis of a complement U of ker α in $\mathbb{Q}^3$ and $(v_1)\alpha$ and $(v_2)\alpha$ form a basis of im α by (8.3)).

The nullity of α is a measure of its degeneracy (the amount of information lost by α); in particular, $0 = $ nullity α if and only if α is injective. On the other hand, $0 = $ rank α if and only if $\alpha = 0$. We show next that there is a simple connection between rank and nullity.

Corollary 8.5

Let $\alpha : V \to V'$ be a linear mapping where V is finite-dimensional. Then im α and ker α are finite-dimensional and

rank $\alpha + $ nullity $\alpha = \dim V$.

Proof

By (8.2), ker α is a subspace of the finite-dimensional vector space V. So ker α is itself finite-dimensional by (6.26). By (6.30) ker α has a complement U in V and $\dim U + \dim(\ker \alpha) = \dim V$. Let U have basis $u_1, u_2, \ldots, u_r$; applying (7.16) to the isomorphism $\alpha\restriction_U$ of (8.3), we see that $(u_1)\alpha, (u_2)\alpha, \ldots, (u_r)\alpha$ is a basis of im α, and so im α is finite-dimensional. In fact $\dim U = \dim(\text{im } \alpha) = $ rank α, and as $\dim(\ker \alpha) = $ nullity α we obtain rank $\alpha + $ nullity $\alpha = \dim V$. □

As nullity α is a non-negative integer, (8.5) tells us that the dimension of $(V)\alpha = $ im α cannot be greater than $\dim V$, that is, rank $\alpha \leqslant \dim V$. If V' is finite-dimensional, then rank $\alpha \leqslant \dim V'$, on applying (6.26) to the subspace im α of V'. So for $\alpha : V \to V'$ we obtain

$$0 \leqslant \text{rank } \alpha \leqslant \min\{\dim V, \dim V'\}.$$

Table 8.1

$\alpha : F^2 \to F^2$		$\alpha : F^2 \to F^3$		$\alpha : F^3 \to F^2$		$\alpha : F^3 \to F^3$	
nullity	rank	nullity	rank	nullity	rank	nullity	rank
0	2	0	2	1	2	0	3
1	1	1	1	2	1	1	2
2	0	2	0	3	0	2	1
						3	0

The classification by rank of linear mappings of F^m to F^n (m, $n = 2$ or 3) is set out in Table 8.1. The reader should make up a simple linear mapping of each type (e.g. $\alpha : F^2 \to F^3$, defined by $(x_1, x_2)\alpha = (x_1, 0, x_1)$, has $1 = \text{rank } \alpha = \text{nullity } \alpha$).

We deal next with the rank of a composite linear mapping.

Corollary 8.6

Let $\alpha : V \to V'$ and $\beta : V' \to V''$ be linear mappings where V' has finite dimension n. Then im α, im β, im $\alpha\beta$ are finite-dimensional and

$$\text{rank } \alpha + \text{rank } \beta - n \leq \text{rank } \alpha\beta \leq \min\{\text{rank } \alpha, \text{rank } \beta\}.$$

Proof

By (8.5), im β is finite-dimensional; as im $\alpha \subseteq V'$ and im $\alpha\beta \subseteq$ im β, by (6.26) im α and im $\alpha\beta$ are finite-dimensional and rank $\alpha\beta \leq \text{rank } \beta$.

Write $W = \text{im } \alpha$ and apply (8.5) to the restriction of β to W, that is, to $\beta\restriction_W : W \to V''$. As im $\alpha\beta$ and $W \cap \text{ker } \beta$ are the image and kernel of $\beta\restriction_W$, we obtain

$$\text{rank } \alpha\beta + \dim(W \cap \text{ker } \beta) = \text{rank } \alpha$$

and so rank $\alpha\beta \leq \text{rank } \alpha$; therefore rank $\alpha\beta \leq \min\{\text{rank } \alpha, \text{rank } \beta\}$. Using the above equation and applying (8.5) to β gives

$$\text{rank } \alpha - \text{rank } \alpha\beta = \dim(W \cap \text{ker } \beta) \leq \dim(\text{ker } \beta)$$
$$= \text{nullity } \beta = n - \text{rank } \beta$$

which rearranges to rank $\alpha + \text{rank } \beta - n \leq \text{rank } \alpha\beta$.   □

It is time to come down to earth and discuss matrix rank, which provides concrete interpretations of (8.5) and (8.6).

Definition 8.7

Let A be an $m \times n$ matrix over the field F. Let $r_1, r_2, \ldots, r_m$ denote the rows of A and let $c_1, c_2, \ldots, c_n$ denote the columns of A. The subspace $\langle r_1, r_2, \ldots, r_m \rangle$ of F^n is called the **row space** of A. The subspace $\langle c_1, c_2, \ldots, c_n \rangle$ of ^mF is called the **column space** of A.

So the row space of a matrix is the space spanned by its rows; the column space of a matrix is the space spanned by its columns. (It is convenient to regard ^mF as a *right* vector space, that is, scalars are written on the right of column vectors.)

Consider $A = \begin{pmatrix} 3 & 2 & 1 \\ 6 & 1 & 5 \\ 5 & 4 & 1 \end{pmatrix}$ over $\mathbb{Q}$.

Then $c_1 = \begin{pmatrix} 3 \\ 6 \\ 5 \end{pmatrix}$, $c_2 = \begin{pmatrix} 2 \\ 1 \\ 4 \end{pmatrix}$, $c_3 = \begin{pmatrix} 1 \\ 5 \\ 1 \end{pmatrix}$.

The column space of A is $\langle c_1, c_2, c_3 \rangle = \langle c_1, c_2 \rangle$ as $c_3 = c_1 - c_2$. It is a surprising fact (8.8) that the row space and the column space of A have the *same* dimension (2 in this case); so $r_1 = (3, 2, 1)$, $r_2 = (6, 1, 5)$, $r_3 = (5, 4, 1)$ must be linearly dependent (actually $19r_1 = 2r_2 + 9r_3$). The reader may construct any 3×3 matrix, deliberately arranging for its columns to be linearly dependent; then— as if by magic—the rows will be found to be linearly dependent also!

Proposition 8.8 Let A be a matrix over a field. Then the row space of A and the column space of A have equal dimension.

Proof Let $A = [a_{ij}]$ be an $m \times n$ matrix over the field F and write $r_i = (\text{row } i \text{ of } A)$, $c_j = (\text{column } j \text{ of } A)$. Let $\langle r_1, r_2, \ldots, r_m \rangle$ have dimension s and $\langle c_1, c_2, \ldots, c_n \rangle$ have dimension t. By (6.20)(a), a selection $c_{j_1}, c_{j_2}, \ldots, c_{j_t}$ of the columns of A forms a basis of the column space of A; as each column of A is a linear combination of these selected columns, there are scalars b_{kj} $(1 \leq k \leq t)$ with

$$c_j = c_{j_1}b_{1j} + c_{j_2}b_{2j} + \ldots + c_{j_t}b_{tj} \qquad (1 \leq j \leq n).$$

Let $B = [b_{kj}]$ be the $t \times n$ matrix formed by these scalars and let $r'_k = (\text{row } k \text{ of } B)$. Comparing i th entries in the above column equation gives

$$a_{ij} = a_{ij_1}b_{1j} + a_{ij_2}b_{2j} + \ldots + a_{ij_t}b_{tj} \qquad (1 \leq i \leq m, \ 1 \leq j \leq n).$$

These mn scalar equations can be reconstituted into m row equations, namely

$$r_i = a_{ij_1}r'_1 + a_{ij_2}r'_2 + \ldots + a_{ij_t}r'_t \qquad (1 \leq i \leq m)$$

since the j th entries on each side agree. We have shown that each row of A is a linear combination of the rows of B. Hence row space A is contained in row space B. As B has only t rows, by (6.26) $s = \dim(\text{row space } A) \leqslant \dim(\text{row space } B) \leqslant t$, and so $s \leqslant t$.

Starting with a basis of the row space of A, we may repeat the above argument, interchanging the roles of row and column throughout, and conclude $t \leqslant s$. Therefore $s = t$. $\qquad\Box$

Definition 8.9

Let A be a matrix over the field F. The dimension of row space A is called the **rank** of A.

From (8.8) we see that the rank of A is also the dimension of the column space of A. Notice that $0 = \text{rank } A \Leftrightarrow A = 0$ (every entry in A is zero). Over $\mathbb{Q}$, the matrices

$$\begin{pmatrix} 1 & -1 & 2 \\ -2 & 2 & -4 \\ -1 & 1 & -2 \end{pmatrix} \text{ and } \begin{pmatrix} 1 & 0 & 0 \\ 0 & 0 & 0 \end{pmatrix} \text{ are of rank 1,}$$

whereas $\begin{pmatrix} 3 & 2 & 1 \\ 6 & 1 & 5 \\ 5 & 4 & 1 \end{pmatrix}$ and $\begin{pmatrix} 1 & 0 & 2 & 3 \\ 0 & 1 & 4 & 5 \\ 0 & 0 & 0 & 0 \end{pmatrix}$

both have rank 2. The 3×3 identity matrix I has rank 3, for its rows e_1, e_2, e_3 are linearly independent. As rank is unbiased (rows and columns have equal status as far as rank is concerned) we see $\text{rank } A = \text{rank } A^\mathsf{T}$.

Next, (8.5) and (8.6) are expressed in matrix terms.

Corollary 8.10

Let A be an $m \times n$ matrix over F. Then the row vectors x satisfying $xA = 0$ form an $(m - \text{rank } A)$-dimensional subspace of F^m.

Let B be an $n \times t$ matrix over F. Then

$$\text{rank } A + \text{rank } B - n \leqslant \text{rank } AB \leqslant \min\{\text{rank } A, \text{rank } B\}.$$

Proof

Consider $\mu_A : F^m \to F^n$ (see (7.13)). As usual write $r_i = (\text{row } i \text{ of } A)$. For $x = (x_1, x_2, \ldots, x_m)$ in F^m we have

$$(x)\mu_A = xA = x_1 r_1 + x_2 r_2 + \ldots + x_m r_m$$

showing that $\text{im } \mu_A = \langle r_1, r_2, \ldots, r_m \rangle = \text{row space } A$, and $\text{rank } \mu_A = \text{rank } A$ on comparing dimensions. Applying (8.5) to μ_A shows that $\ker \mu_A = \{x \in F^m : xA = 0\}$ is a subspace of F^m of dimension nullity $\mu_A = \dim F^m - \text{rank } \mu_A = m - \text{rank } A$.

From the preceding paragraph, $\text{rank } \mu_A = \text{rank } A$, and so $\text{rank } \mu_B = \text{rank } B$, $\text{rank } \mu_A \mu_B = \text{rank } \mu_{AB} = \text{rank } AB$. Applying (8.6) with α and β replaced by μ_A and μ_B now produces the inequalities of (8.10).   □

The row equation $xA = 0$ is made up of n simultaneous scalar equations

$$xc_j = x_1a_{1j} + x_2a_{2j} + \ldots + x_ma_{mj} = 0 \qquad (1 \leqslant j \leqslant n)$$

in the m unknowns $x_1, x_2, \ldots, x_m$ (there is one equation for each column c_j of A). A systematic method of solution will be given in (8.29); however, (8.10) provides an abstract description of the solution space: as rank $A = \dim \langle c_1, c_2, \ldots, c_n \rangle$, above there are rank A independent equations (those corresponding to a basis of column space A selected from $c_1, c_2, \ldots, c_n$) and the remaining $n - \text{rank } A$ equations are redundant (for every equation is a linear combination of any rank A independent equations). Leaving the details until later, any rank A independent equations can be used to eliminate rank A of the unknowns x_i; so the general solution contains the remaining $m - \text{rank } A$ unknowns. In other words (8.10) states:

The solutions x of $xA = 0$ form a space of
dimension: number of unknowns minus rank A.

For example

$$(x_1, x_2, x_3)\begin{pmatrix} 3 & 2 & 1 \\ 6 & 1 & 5 \\ 5 & 4 & 1 \end{pmatrix} = (0, 0, 0)$$

is made up of $3x_1 + 6x_2 + 5x_3 = 0$, $2x_1 + x_2 + 4x_3 = 0$, $x_1 + 5x_2 + x_3 = 0$. Each two of these equations are independent, but as $c_1 - c_2 - c_3 = 0$, the remaining equation is redundant. Ignoring the first equation (it's the sum of the others) we obtain $9x_2 = 2x_3$ on eliminating x_1. Therefore

$$(x_1, x_2, x_3) = (-5x_2 - x_3, x_2, x_3) = x_2(-\tfrac{19}{2}, 1, \tfrac{9}{2})$$

showing that $\langle (-19, 2, 9) \rangle$ is the solution space, that is, the solutions (x_1, x_2, x_3) are precisely the scalar multiples of $(-19, 2, 9)$. We have derived the linear dependence relation $19r_1 = 2r_2 + 9r_3$ between the rows of the above matrix and verified that the solutions form a space of dimension $1 = $ (number of unknowns (3) minus rank (2)).

As an illustration of the inequalities (8.10) bounding rank AB, let A be a 3×2 matrix and B a 2×3 matrix with rank $A = \text{rank } B = 2$. Then (8.10) tells us: $2 + 2 - 2 \leqslant \text{rank } AB \leqslant \min\{2, 2\}$, and so rank $AB = 2$. For instance

$$\begin{pmatrix} 17 & 22 & 27 \\ 22 & 29 & 36 \\ 27 & 36 & 45 \end{pmatrix} = \begin{pmatrix} 1 & 4 \\ 2 & 5 \\ 3 & 6 \end{pmatrix} \begin{pmatrix} 1 & 2 & 3 \\ 4 & 5 & 6 \end{pmatrix}$$

has rank 2. Generally however these inequalities are insufficient to determine rank AB in terms of rank A and rank B; for instance, if $1 = \text{rank } A = \text{rank } B$, then rank AB may be 0 or 1 by (8.10). The inequality

$$\text{rank } AB \leqslant \min\{\text{rank } A, \text{ rank } B\}$$

says that the rank of a product of matrices is not greater than the rank of any factor; this has special significance for invertible matrices, which we now discuss.

Definition 8.11   Let P be an $n \times n$ matrix over the field F. Then P is called **invertible** over F if there is an $n \times n$ matrix P^{-1} over F satisfying

$$P^{-1}P = I = PP^{-1}.$$

So P is invertible over F if and only if P has an inverse in the matrix ring $M_n(F)$ (7.9), in which case we may refer to *the* inverse P^{-1} (2.11) of P. As P^{-1} has inverse P, invertible matrices occur in pairs, for example:

$$\begin{pmatrix} 2 & 0 \\ 0 & 3 \end{pmatrix}, \begin{pmatrix} \frac{1}{2} & 0 \\ 0 & \frac{1}{3} \end{pmatrix}; \quad \begin{pmatrix} 1 & 3 \\ 2 & 4 \end{pmatrix}, \begin{pmatrix} -2 & \frac{3}{2} \\ 1 & -\frac{1}{2} \end{pmatrix}; \quad \begin{pmatrix} 1 & 1 \\ 0 & -1 \end{pmatrix}, \begin{pmatrix} 1 & 1 \\ 0 & -1 \end{pmatrix}.$$

Let P and Q be invertible $n \times n$ matrices over F. Then

$$Q^{-1}P^{-1}PQ = Q^{-1}IQ = Q^{-1}Q = I,$$
$$PQQ^{-1}P^{-1} = PIP^{-1} = PP^{-1} = I,$$

showing that PQ is invertible with inverse $Q^{-1}P^{-1}$. Therefore:

$$(PQ)^{-1} = Q^{-1}P^{-1}.$$

In words: the inverse of a product of invertible matrices is the product of their inverses in the *opposite* order.

We show next that invertible matrices P can be recognized as such by their rank, and that they correspond to invertible linear mappings μ_P.

Proposition 8.12   Let P be an $n \times n$ matrix over the field F. Then the following statements are logically equivalent:

(i) P is invertible over F,   (ii) P has rank n,

(iii) μ_P is invertible, and when true, $\mu_P^{-1} = \mu_{P^{-1}}$.

Proof

(i) $\Rightarrow$ (ii). Suppose that P is invertible over F. Then $P^{-1}P = I$ where P^{-1} is an $n \times n$ matrix over F. As $n = \operatorname{rank} I$ (the rows of I form the standard basis of F^n), from (8.10) we obtain:

$$n = \operatorname{rank} I = \operatorname{rank} P^{-1}P \leqslant \min\{\operatorname{rank} P^{-1}, \operatorname{rank} P\} \leqslant \operatorname{rank} P.$$

However, $n \geqslant \operatorname{rank} P$ as P has only n rows, and so $n = \operatorname{rank} P$.

(ii) $\Rightarrow$ (iii). Suppose $n = \operatorname{rank} P$. The n rows of P are linearly independent and so, by (6.24), form a basis of F^n. As $(e_i)\mu_P = e_i P = (\text{row } i \text{ of } P)$ for $1 \leqslant i \leqslant n$, the linear mapping $\mu_P : F^n \to F^n$ is bijective by (7.16). By (1.23) μ_P has an inverse, that is, μ_P is invertible.

(iii) $\Rightarrow$ (i). Suppose that μ_P is invertible. By (7.15)(d) $\mu_P^{-1} : F^n \to F^n$ is linear, and so there is an $n \times n$ matrix Q over F such that $\mu_Q = \mu_P^{-1}$ by (7.17). (The reader should have the feeling that Q must be P^{-1}; so let us show that (8.11) is satisfied with Q in the role of P^{-1}.) As μ_I is the identity mapping of F^n,

$$\mu_{QP} = \mu_Q\mu_P = \mu_P^{-1}\mu_P = \mu_I$$

by (7.13), and hence $QP = I$. In a similar way $\mu_P\mu_P^{-1} = \mu_I$ leads to $PQ = I$. Therefore $Q = P^{-1}$ showing that P is invertible over F and $\mu_{P^{-1}} = \mu_P^{-1}$. $\qquad\square$

From (8.12), the $n \times n$ matrix P is invertible over F if and only if the rows of P form a basis of F^n. (We shall see in Chapter 9 that P is invertible over F is and only if its determinant is non-zero.)

Suppose we know $PQ = I$, where P and Q are $n \times n$ matrices. Is it necessarily true that P and Q are inverses of each other? The answer is: yes, because $n = \operatorname{rank} P$ by (8.10) and so P has an inverse P^{-1} by (8.12). Therefore $Q = IQ = P^{-1}PQ = P^{-1}I = P^{-1}$, showing that Q is the inverse of P (and so $QP = I$).

For example, consider $P = \begin{pmatrix} 1 & 4 & 5 \\ 0 & 2 & 6 \\ 0 & 0 & 3 \end{pmatrix}$. As rank $P = 3$, P has an inverse by (8.12). Let r_i denote row i of P^{-1}. We dissect the equation PP^{-1} into rows:

$$PP^{-1} = \begin{pmatrix} 1 & 4 & 5 \\ 0 & 2 & 6 \\ 0 & 0 & 3 \end{pmatrix}\begin{pmatrix} r_1 \\ r_2 \\ r_3 \end{pmatrix} = \begin{pmatrix} r_1 + 4r_2 + 5r_3 \\ 2r_2 + 6r_3 \\ 3r_3 \end{pmatrix} = \begin{pmatrix} e_1 \\ e_2 \\ e_3 \end{pmatrix} = I$$

Comparing rows (starting at the bottom) gives $r_3 = (0, 0, \frac{1}{3})$, $r_2 = \frac{1}{2}e_2 - 3r_3 = (0, \frac{1}{2}, -1)$, $r_1 = e_1 - 4r_2 - 5r_3 = (1, -2, \frac{7}{3})$ and so

$$P^{-1} = \begin{pmatrix} 1 & -2 & \frac{7}{3} \\ 0 & \frac{1}{2} & -1 \\ 0 & 0 & \frac{1}{3} \end{pmatrix}.$$

The reader may verify that P^{-1} satisfies $P^{-1}P = I$ also.

Later in the chapter, a systematic method of inverting matrices is found. Here we establish the basis-dependent correspondence between abstract invertible linear mappings and invertible matrices.

Corollary 8.13

Let the linear mapping $\alpha : V \rightarrow V'$ have matrix P relative to the bases $\textit{b}$ and $\textit{b}'$ of V and V'. Then α is invertible if and only if P is invertible, in which case α^{-1} has matrix P^{-1} relative to $\textit{b}'$ and $\textit{b}$.

Proof

Let us return to (7.25), which tells us (on replacing A by P) that $\kappa_{\textit{b}}\mu_P = \alpha\kappa_{\textit{b}'}$. Suppose that α is invertible; then $\mu_P = \kappa_{\textit{b}}^{-1}\alpha\kappa_{\textit{b}'}$, being a composition of invertible mappings, is itself invertible; therefore P is invertible by (8.12). Multiplying $\kappa_{\textit{b}}\mu_P = \alpha\kappa_{\textit{b}'}$ on the left by α^{-1} and on the right by $\mu_P^{-1} = \mu_{P^{-1}}$ produces $\kappa_{\textit{b}'}\mu_{P^{-1}} = \alpha^{-1}\kappa_{\textit{b}}$, which means that α^{-1} has matrix P^{-1} relative to $\textit{b}'$ and $\textit{b}$ by (7.25).

Conversely, suppose that P is invertible. By (8.12) μ_P is invertible and so $\alpha = \kappa_{\textit{b}}\mu_P\kappa_{\textit{b}'}^{-1}$, being a composition of invertible mappings, is itself invertible. $\qquad\square$

We specialize (8.13) by taking $V = V'$ and $\alpha = \iota$ (the identity mapping of V) which is certainly invertible—in fact ι is self-inverse. Let $v_1, v_2, \ldots, v_n$ and $v_1', v_2', \ldots, v_n'$ form the bases $\textit{b}$ and $\textit{b}'$ of V where $n = \dim V$. Each vector v_i in $\textit{b}$ can be expressed as a linear combination of the vectors v_j' in $\textit{b}'$:

$$v_i = \sum_{j=1}^{n} p_{ij}v_j' \qquad (1 \le i \le n)$$

the $n \times n$ matrix $P = [p_{ij}]$ being the matrix of ι relative to $\textit{b}$ and $\textit{b}'$ by (7.22); P is called the **transition** matrix from $\textit{b}$ to $\textit{b}'$, and is said to **relate** $\textit{b}$ to $\textit{b}'$. By (8.13) P is invertible and so:

Bases of a given n-dimensional vector space over F are related by invertible $n \times n$ matrices over F.

Also, the matrix which relates $\textit{b}'$ to $\textit{b}$, that is, the matrix of ι relative to $\textit{b}'$ and $\textit{b}$, is P^{-1} by (8.13).

Consider the vector space V of polynomials f over $\mathbb{Q}$ such that $f(0) = 0$ and $\deg f \leqslant n$. V has basis $\mathscr{b}$ consisting of the powers $x, x^2, \ldots, x^n$, and basis $\mathscr{b}'$ consisting of the factorials $(x)_1, (x)_2, \ldots, (x)_n$ where $(x)_i = x(x-1)(x-2)\ldots(x-i+1)$. In the case $n = 4$:

$$(x)_1 = \quad x$$
$$(x)_2 = \quad -x + \quad x^2$$
$$(x)_3 = \quad 2x - \quad 3x^2 + \quad x^3$$
$$(x)_4 = -6x + 11x^2 - 6x^3 + x^4$$

and so $P = \begin{pmatrix} 1 & 0 & 0 & 0 \\ -1 & 1 & 0 & 0 \\ 2 & -3 & 1 & 0 \\ -6 & 11 & -6 & 1 \end{pmatrix}$

is the matrix relating factorials to powers. The inverse P^{-1} can be calculated row by row to give

$$P^{-1} = \begin{pmatrix} 1 & 0 & 0 & 0 \\ 1 & 1 & 0 & 0 \\ 1 & 3 & 1 & 0 \\ 1 & 7 & 6 & 1 \end{pmatrix}$$

which relates powers to factorials (the last row of P^{-1} shows that $x^4 = (x)_1 + 7(x)_2 + 6(x)_3 + (x)_4$). For arbitrary n, the (i, j)-entries in P and P^{-1} are denoted by $s(i, j)$ and $S(i, j)$ respectively and called **Stirling numbers** of the first and second kind; these numbers occur in combinatorial theory (there are exactly $j! \, S(i, j)$ surjections $\alpha : X \to Y$ where $|X| = i$ and $|Y| = j$).

Exercises 8.1

1. (a) For each of the following matrices A over $\mathbb{Q}$, find a basis of row space A, a basis of column space A, and state rank A.

$$\begin{pmatrix} 1 & 2 & 1 \\ 2 & 1 & -4 \\ -1 & 1 & 5 \end{pmatrix}, \quad \begin{pmatrix} 1 & 2 & 4 & 1 \\ 2 & 1 & -1 & 1 \\ 1 & 3 & 7 & 1 \end{pmatrix},$$

$$\begin{pmatrix} 1 \\ 2 \\ 3 \end{pmatrix}(4, \quad 5, \quad 6), \quad \begin{pmatrix} 1 & 2 & 1 \\ 2 & 1 & -4 \\ -1 & 1 & 4 \end{pmatrix}.$$

(b) For each of the above matrices A, determine bases of $\operatorname{im} \mu_A$ and $\ker \mu_A$, and verify that rank μ_A + nullity $\mu_A = 3$.

2. (a) Find the inverses of the following matrices over $\mathbb{Q}$.

$$\begin{pmatrix} 5 & 2 \\ 2 & 1 \end{pmatrix}, \quad \begin{pmatrix} -1 & 0 \\ 6 & 2 \end{pmatrix}, \quad \begin{pmatrix} 1 & 2 & 3 \\ 0 & -1 & 4 \\ 0 & 0 & 1 \end{pmatrix}, \quad \begin{pmatrix} 2 & 2 & 3 \\ 1 & -1 & 1 \\ 2 & 1 & 3 \end{pmatrix}.$$

(b) Find $(PQ)^{-1}$ where P and Q are (i) the first pair, (ii) the last pair, of matrices above.

3. (a) Let P be an invertible matrix over a field. Use (7.11) and (8.11) to show that P^{T} is invertible and $(P^{\mathrm{T}})^{-1} = (P^{-1})^{\mathrm{T}}$. Deduce that P is symmetric if and only if P^{-1} is symmetric. Find the inverse of

$$\begin{pmatrix} 1 & 2 & 3 \\ 2 & 2 & 3 \\ 3 & 3 & 3 \end{pmatrix} \text{ over } \mathbb{Q}.$$

(b) Let $P = \begin{pmatrix} a & b \\ c & d \end{pmatrix}$ be a matrix over the field F. Show that P is invertible over F if and only if $ad - bc \neq 0$, in which case

$$P^{-1} = (1/(ad - bc))\begin{pmatrix} d & -b \\ -c & a \end{pmatrix}.$$

4. Let A and B be $m \times n$ and $n \times t$ matrices over F.

(a) Show that each row of AB is a linear combination of the rows of B. Deduce that row space AB is contained in row space B. Establish a (similar) relation between column space AB and column space A.

(b) Adapt the proof of (8.6) to show that rank A + rank $B - n =$ rank AB if and only if ker $\mu_B \subseteq$ im μ_A.

5. Let C be an $m \times n$ matrix over F. Show that there are $m \times 1$ and $1 \times n$ matrices A and B over F with $AB = C$ if and only if rank $C \leqslant 1$. If $r =$ rank C, show that there are $m \times r$ and $r \times n$ matrices A and B over F with $AB = C$.

Find all matrices A and B as above, if $F = \mathbb{Z}_2$ and C is

(i) $\begin{pmatrix} 1 & 1 & 0 \\ 0 & 0 & 0 \\ 1 & 1 & 0 \end{pmatrix}$ (ii) $\begin{pmatrix} 1 & 0 & 1 \\ 1 & 1 & 0 \\ 0 & 1 & 1 \end{pmatrix}.$

6. (a) The $m \times m$, $n \times n$, $m \times n$ matrices P, Q, A over F are placed together as shown to form the $(m + n) \times (m + n)$ matrix $X = \left(\begin{array}{c|c} P & A \\ \hline 0 & Q \end{array}\right)$ where 0 denotes the $n \times m$ zero matrix. Show that X is invertible if and only if P and Q are invertible, in which case

$$X^{-1} = \left(\begin{array}{c|c} P^{-1} & -P^{-1}AQ^{-1} \\ \hline 0 & Q^{-1} \end{array}\right).$$

(b) Show that a matrix over F of the form

$$\begin{pmatrix} P & 0 & 0 \\ \hline A & Q & 0 \\ \hline B & C & R \end{pmatrix},$$

where P, Q, R are invertible, is itself invertible and express its inverse in a similar (partitioned) form.

(c) Let P, Q, R, S be $n \times n$ matrices over F with P and Q invertible. Show that the $2n \times 2n$ matrix

$$\left(\begin{array}{c|c} P & Q \\ \hline R & S \end{array}\right)$$

is invertible if and only if $RP^{-1} - SQ^{-1}$ is invertible, and find the partitioned form of its inverse in this case.

7. Let $\alpha : V \to V'$ and $\beta : V' \to V''$ be linear mappings of finite-dimensional spaces.

(a) Show that α is injective if and only if rank $\alpha = \dim V$, and α is surjective if and only if rank $\alpha = \dim V'$.

(b) Prove **Sylvester's law of nullity:**

nullity α + nullity $\beta \geqslant$ nullity $\alpha\beta$.

8. The linear mapping $\alpha : V \to V$ of the n-dimensional space V satisfies $\alpha^2 = 0$. Show that rank $\alpha \leqslant n/2$. Find a basis of V relative to which α has matrix

$$\left(\begin{array}{c|c} 0 & I \\ \hline 0 & 0 \end{array}\right)$$

where I is the $r \times r$ identity matrix and $r = $ rank α. (Consider a basis of a complement of ker α in V and its image by α.)

Verify that $A = \begin{pmatrix} 1 & 2 & 1 \\ -1 & -2 & -1 \\ 1 & 2 & 1 \end{pmatrix}$ over $\mathbb{Q}$ satisfies $A^2 = 0$. Find a basis b of $\mathbb{Q}^3$ of the form v_1, v_2, $v_1 A$ where $v_1 \notin$ ker μ_A and $v_2 \in$ ker μ_A. Verify that the invertible matrix P having the vectors of b as its rows satisfies

$$PAP^{-1} = \begin{pmatrix} 0 & 0 & 1 \\ 0 & 0 & 0 \\ 0 & 0 & 0 \end{pmatrix}.$$

9. (a) Let K be an ideal (5.14) of the ring $M_2(\mathbb{Q})$ (7.9). If $\begin{pmatrix} a & b \\ c & d \end{pmatrix}$ belongs to K, show that the following matrices also belong to K:

$$\begin{pmatrix} a & 0 \\ 0 & 0 \end{pmatrix}, \quad \begin{pmatrix} 0 & 0 \\ 0 & a \end{pmatrix}, \quad \begin{pmatrix} a & 0 \\ 0 & a \end{pmatrix}, \quad \begin{pmatrix} b & a \\ d & c \end{pmatrix}, \quad \begin{pmatrix} c & d \\ a & b \end{pmatrix}, \quad \begin{pmatrix} d & c \\ b & a \end{pmatrix}.$$

If $K \neq 0$, show that K contains an invertible matrix. Hence prove that either $K = 0$ or $K = M_2(\mathbb{Q})$.

(b) Let R denote the subring of $M_2(\mathbb{Q})$ consisting of all matrices of the form $\begin{pmatrix} a & b \\ 0 & c \end{pmatrix}$. Show that the matrices with $a = 0$ form an ideal of R. Find two further ideals of R which are non-zero and proper ($\neq R$). Prove that R has exactly five ideals.

Row-equivalence

Although four vectors of $\mathbb{Q}^3$ are necessarily linearly dependent, it may be difficult to spot a linear dependence relation between them. Similarly, if we wish to find the rank of a given matrix, the theoretical properties of rank are unlikely to help. Here we develop a procedure for solving such problems based on the simple idea of eliminating variables from linear equations by the subtraction of a suitable multiple of one equation from another; if the elimination is systematically carried out then ultimately the solution of the problem in hand will appear before our very eyes! In the formulation given here, equations correspond to rows of the coefficient matrix and so manipulation of equations corresponds to performing row operations on this matrix.

Definition 8.14   The $m \times n$ matrices A and B over the field F are called **row-equivalent** if there is an invertible $m \times m$ matrix P over F such that

$$PA = B.$$

For instance,

$$A = \begin{pmatrix} 1 & 2 & 3 \\ 4 & 5 & 6 \end{pmatrix} \quad \text{and} \quad B = \begin{pmatrix} 4 & 5 & 6 \\ 1 & 2 & 3 \end{pmatrix}$$

over $\mathbb{Q}$ are row-equivalent, as $PA = B$ where

$$P = \begin{pmatrix} 0 & 1 \\ 1 & 0 \end{pmatrix};$$

so **premultiplication** (multiplication on the left) by P interchanges the rows and hence A and B have the same row space, namely $\langle (1, 2, 3), (4, 5, 6) \rangle$. We begin our study of row-equivalence by establishing its close connection with row space.

Proposition 8.15   The $m \times n$ matrices A and B over the field F are row-equivalent if and only if row space A = row space B.

Proof

Suppose first that A and B are row-equivalent. By (8.14) there is an invertible $m \times m$ matrix $P = [p_{ij}]$ over F with $PA = B$. Comparing i th rows gives

$$p_{i1}r_1 + p_{i2}r_2 + \ldots + p_{im}r_m = (\text{row } i \text{ of } B)$$

where $r_j = (\text{row } j \text{ of } A)$. So each row of B is a linear combination of the rows of A, which means row space $B \subseteq$ row space A. As $P^{-1}B = A$, we may repeat the preceding argument with A and B interchanged to obtain row space $A \subseteq$ row space B. Hence row space $A =$ row space B.

Conversely, suppose A and B have the same row space, which we denote by U'. Then $U' = \text{im } \mu_A = \text{im } \mu_B$ (see (8.10) proof). Let U' have basis $v_1', v_2', \ldots, v_r'$ and let U be a complement of $\ker \mu_A$ in F^m. As μ_A restricted to U is an isomorphism $U \cong U'$ by (8.4), we see that U has basis $v_1, v_2, \ldots, v_r$ where $v_i A = v_i'$ ($1 \le i \le r$). Let $\ker \mu_A$ have basis $v_{r+1}, \ldots, v_m$; then $v_1, v_2, \ldots, v_r, v_{r+1}, \ldots, v_m$ is a basis of $F^m = U \oplus \ker \mu_A$ by (6.29). Let P be the $m \times m$ matrix over F with $v_i = (\text{row } i \text{ of } P)$ for $1 \le i \le m$; then P is invertible over F by (8.12), as rank $P = m$. Therefore

$$(\text{row } i \text{ of } PA) = e_i PA = v_i A = \begin{cases} v_i' & \text{for } i \le r, \\ 0 & \text{for } i > r. \end{cases}$$

In other words $PA = C$ where the first r rows of C are $v_1', v_2', \ldots, v_r'$ and the remaining rows of C are zero. As A and B have the same status, working with B in place of A, an invertible matrix Q over F can be found such that $QB = C$. So $PA = QB$ and hence $Q^{-1}PA = B$, showing that A and B are row-equivalent, for $Q^{-1}P$ is invertible over F. □

As row-equivalent matrices have the same row space, they have the same rank (the reader may prove that A and AQ have the same column space for Q invertible, and hence rank $A =$ rank AQ). Therefore:

Multiplication of a matrix by any invertible matrix leaves its rank unchanged.

It follows from (8.15) (or may be proved directly from (8.14)) that row-equivalence is an equivalence relation on the set $^m F^n$ of all $m \times n$ matrices over F, for two such matrices are row-equivalent if and only if they have the same row space. Further, the row-equivalence classes of $m \times n$ matrices A of rank r over F correspond to the r-dimensional subspaces U' of F^n (the row-equivalence class of A corresponds to $U' =$ row space A) where $r \le m$.

For example, let $F = \mathbb{Z}_2$ and $U' = \langle (1, 1, 0), (0, 1, 1) \rangle$. There are six 2×3 matrices over $\mathbb{Z}_2$ having U' as their row space, namely

$$\begin{pmatrix} 1 & 1 & 0 \\ 0 & 1 & 1 \end{pmatrix}, \quad \begin{pmatrix} 1 & 1 & 0 \\ 1 & 0 & 1 \end{pmatrix}, \quad \begin{pmatrix} 1 & 0 & 1 \\ 0 & 1 & 1 \end{pmatrix}$$

and the three matrices obtained from these by interchanging rows; these six matrices form a row-equivalence class. Similarly

$$\begin{pmatrix} 1 & 1 & 0 \\ 1 & 1 & 0 \end{pmatrix}, \quad \begin{pmatrix} 1 & 1 & 0 \\ 0 & 0 & 0 \end{pmatrix}, \quad \begin{pmatrix} 0 & 0 & 0 \\ 1 & 1 & 0 \end{pmatrix}$$

form a row-equivalence class, for they are the only 2×3 matrices over $\mathbb{Z}_2$ having $\langle (1, 1, 0) \rangle$ as their row space. In fact there are 15 row equivalence classes of 2×3 matrices over $\mathbb{Z}_2$, corresponding to the 15 subspaces U' of $\mathbb{Z}_2^3$ with dim $U' \leqslant 2$.

We write $A \equiv B$ if A and B are row-equivalent. Our purpose is to determine the *simplest* matrix which is row-equivalent to the given matrix A, and to do this in a practical way; the process amounts to simplifying A step by step until no more simplification (reduction) is possible. The final matrix E is called the **row-reduced echelon form** (8.21) of the original matrix A. Each step in the reduction process consists of applying a simple row operation as we now explain.

Definition 8.16 The following are called **elementary row operations**:

(i) interchanging two rows,
(ii) multiplying a row by a *non-zero* scalar,
(iii) adding to one row a scalar multiple of another row.

The row space of a matrix is unchanged by operations of this kind, that is, elementary row operations change each matrix into a row-equivalent matrix by (8.15). We now prepare to give a direct proof of this fact.

Definition 8.17 The matrices, which arise on applying a *single* elementary row operation to the identity matrix I, are called **elementary** matrices.

For instance, $\begin{pmatrix} 1 & 0 \\ b & 1 \end{pmatrix}$ is an elementary matrix as it results from applying the elementary row operations $r_2 + br_1$ (addition to row 2 of $b \times$ (row 1)) to $I = \begin{pmatrix} 1 & 0 \\ 0 & 1 \end{pmatrix}$; we say $\begin{pmatrix} 1 & 0 \\ b & 1 \end{pmatrix}$ *corresponds* to

$r_2 + br_1$. The elementary 2×2 matrices over F are:

$$\begin{pmatrix} 0 & 1 \\ 1 & 0 \end{pmatrix}, \quad \begin{pmatrix} a & 0 \\ 0 & 1 \end{pmatrix}, \quad \begin{pmatrix} 1 & 0 \\ 0 & a \end{pmatrix}, \quad \begin{pmatrix} 1 & b \\ 0 & 1 \end{pmatrix}, \quad \begin{pmatrix} 1 & 0 \\ b & 1 \end{pmatrix}$$

where $a, b \in F$ and $a \neq 0$. Notice that these matrices are invertible and their inverses are elementary matrices; for example

$$\begin{pmatrix} 1 & 0 \\ b & 1 \end{pmatrix}^{-1} = \begin{pmatrix} 1 & 0 \\ -b & 1 \end{pmatrix}$$

corresponds to $r_2 - br_1$.

We show next that premultiplying by an elementary matrix amounts to carrying out the corresponding elementary row operation: for instance

$$\begin{pmatrix} 1 & 0 & 2 \\ 0 & 1 & 0 \\ 0 & 0 & 1 \end{pmatrix} \begin{pmatrix} a & b \\ c & d \\ e & f \end{pmatrix} = \begin{pmatrix} a+2e & b+2f \\ c & d \\ e & f \end{pmatrix}$$

shows how premultiplication carries out the command: to row 1 add $2 \times$ (row 3).

Lemma 8.18

Let P be an elementary $m \times m$ matrix and A an $m \times n$ matrix over F. If the elementary row operation corresponding to P is applied to A, then PA is the resulting matrix. Further, P is invertible and P^{-1} is also elementary.

Proof

There are three types of elementary matrices corresponding to the three types of elementary row operations (8.16), and we deal with each type in turn.

(i) Let P be obtained from the $m \times m$ identity matrix I by interchanging row j and row k, where j and k are given integers with $1 \leq j, k \leq m$. Then the rows of P are easily described using the rows $e_1, e_2, \ldots, e_m$ of I; in fact, $e_j P = e_k$, $e_k P = e_j$, and $e_i P = e_i$ for $i \neq j, k$. Therefore $e_j PA = e_k A$ (row j of PA is row k of A), $e_k PA = e_j A$ (row k of PA is row j of A), and $e_i PA = e_i A$ (row i of PA is row i of A) for $i \neq j, k$. So PA is the result of interchanging row j and row k of A. Also $e_j P^2 = e_k P = e_j$, $e_k P^2 = e_j P = e_k$, and $e_i P^2 = e_i P = e_i$ for $i \neq j, k$, showing that $P^2 = I$. Therefore $P^{-1} = P$ in this case.

(ii) Let P be obtained from I by multiplying row j by the non-zero scalar a. Then $e_j P = ae_j$ and $e_i P = e_i$ for $i \neq j$. Hence $e_j PA = ae_j A$, $e_i PA = e_i A$ for $i \neq j$, showing that PA is obtained from

A by multiplying row j by a. In this case P^{-1} arises from I on multiplying row j by a^{-1}; so P is invertible and P^{-1} is elementary.

(iii) Let P be obtained from I by the addition to row j of $b \times (\text{row } k)$ where $j \neq k$. Then $e_j P = e_j + b e_k$ and $e_i P = e_i$ for $i \neq j$. Hence $e_j PA = e_j A + b e_k A$ and $e_i PA = e_i A$ for $i \neq j$, which tell us that PA is obtained from A by the addition to row j of $b \times (\text{row } k)$. In this case P^{-1} is obtained from I on subtracting $b \times (\text{row } k)$ from row j; so P is invertible and P^{-1} is elementary.   □

From (8.18) we see explicitly that elementary row operations change a matrix into a row-equivalent matrix; further, the composition of a finite number of such operations has the same property. To give the reader a sense of direction, let us pause and work some numerical examples.

Example 8.19(a)   Let $A = \begin{pmatrix} 1 & 2 & 4 \\ 2 & 5 & 9 \\ 1 & 6 & 8 \end{pmatrix}$ over $\mathbb{Q}$. We apply elementary row operations in order to 'simplify' A, our first objective being a matrix having $\begin{pmatrix} 1 \\ 0 \\ 0 \end{pmatrix}$ as its first column; this may be achieved as follows:

$$\begin{pmatrix} 1 & 2 & 4 \\ 2 & 5 & 9 \\ 1 & 6 & 8 \end{pmatrix} \underset{r_2 - 2r_1}{\equiv} \begin{pmatrix} 1 & 2 & 4 \\ 0 & 1 & 1 \\ 1 & 6 & 8 \end{pmatrix} \underset{r_3 - r_1}{\equiv} \begin{pmatrix} 1 & 2 & 4 \\ 0 & 1 & 1 \\ 0 & 4 & 4 \end{pmatrix}.$$

We now reduce further; without disturbing the first column, our aim is to make the second column as simple as possible:

$$\begin{pmatrix} 1 & 2 & 4 \\ 0 & 1 & 1 \\ 0 & 4 & 4 \end{pmatrix} \underset{r_1 - 2r_2}{\equiv} \begin{pmatrix} 1 & 0 & 2 \\ 0 & 1 & 1 \\ 0 & 4 & 4 \end{pmatrix} \underset{r_3 - 4r_2}{\equiv} \begin{pmatrix} 1 & 0 & 2 \\ 0 & 1 & 1 \\ 0 & 0 & 0 \end{pmatrix} = E \text{ say.}$$

No further reduction is possible (the third column cannot be simplified without making the first two columns more complicated), and E is the row-reduced echelon form of A, that is, E is the simplest matrix which is row-equivalent to A. As A and E have the same row space, row space $A = $ row space $E = \langle (1, 0, 2), (0, 1, 1) \rangle$ and rank $A = 2$.

Suppose we wish to find all rational solutions x_1, x_2, x_3 of the system of equations

$$x_1 + 2x_2 + 4x_3 = 0$$
$$2x_1 + 5x_2 + 9x_3 = 0, \quad \text{that is,} \quad Ax = 0 \quad \text{where} \quad x = \begin{pmatrix} x_1 \\ x_2 \\ x_3 \end{pmatrix}.$$
$$x_1 + 6x_2 + 8x_3 = 0$$

In elementary algebra one would subtract multiples of equations from other equations, obtaining a more manageable system; in other words, row operations would be used to simplify the coefficient matrix A of the above system. This is exactly what we have done! In this case the more manageable system is $Ex = 0$, that is,

$$\begin{array}{r} x_1 \quad\quad + 2x_3 = 0 \\ x_2 + \ x_3 = 0 \end{array} \quad \text{giving} \quad x = \begin{pmatrix} x_1 \\ x_2 \\ x_3 \end{pmatrix} = \begin{pmatrix} -2 \\ -1 \\ 1 \end{pmatrix} x_3.$$

The above equations express x_1 and x_2 in terms of the arbitrary rational number x_3, and so the solutions x of $Ax = 0$ (which are the same as the solutions of $Ex = 0$) form the 1-dimensional subspace $\left\langle \begin{pmatrix} -2 \\ -1 \\ 1 \end{pmatrix} \right\rangle$ of $^3\mathbb{Q}$. These solutions have another interpretation: writing c_j for column j of A, the column equation $Ax = 0$ can be expressed as

$$(c_1, c_2, c_3) \begin{pmatrix} x_1 \\ x_2 \\ x_3 \end{pmatrix} = 0,$$

that is, $c_1 x_1 + c_2 x_2 + c_3 x_3 = 0$; in other words, every non-zero solution of $Ax = 0$ is effectively a linear dependence relation between the columns of A (it is the transpose of the discussion following (8.10)). In this case

$$c_1(-2) + c_2(-1) + c_3 = 0$$

and what is more, this linear dependence relation is satisfied by the columns of *every* matrix which is row-equivalent to A (8.20).

Example 8 19(h)
Suppose we wish to find a linear dependence relation between the vectors $v_1 = (1, -1, 1)$, $v_2 = (4, -2, 1)$, $v_3 = (3, -1, 1)$, $v_4 = (5, 1, 5)$ of $\mathbb{Q}^3$. We have just found a method of deriving linear dependence relations between the columns of a matrix, so *transpose* each of the above rows and use the resulting columns to form a

matrix A which may be reduced as before:

$$A = \begin{pmatrix} 1 & 4 & 3 & 5 \\ -1 & -2 & -1 & 1 \\ 1 & 1 & 1 & 5 \end{pmatrix} \underset{\substack{r_2 + r_1 \\ r_3 - r_1}}{\equiv} \begin{pmatrix} 1 & 4 & 3 & 5 \\ 0 & 2 & 2 & 6 \\ 0 & -3 & -2 & 0 \end{pmatrix}$$

$$\underset{\frac{1}{2}r_2}{\equiv} \begin{pmatrix} 1 & 4 & 3 & 5 \\ 0 & 1 & 1 & 3 \\ 0 & -3 & -2 & 0 \end{pmatrix}$$

where the elementary row operations $r_2 + r_1$, $r_3 - r_1$, and $\frac{1}{2}r_2$ (multiplication of row 2 by $\frac{1}{2}$) are performed in order. After a little practice, the reader will be able to perform several elementary row operations at once. But be careful! It is easy to get carried away and unintentionally perform an *illegal* sequence of operations; as a guard against this and a practical policy:

Keep one row unchanged at each stage of the reduction.

Continuing with the reduction in hand:

$$\begin{pmatrix} 1 & 4 & 3 & 5 \\ 0 & 1 & 1 & 3 \\ 0 & -3 & -2 & 0 \end{pmatrix} \underset{\substack{r_1 - 4r_2 \\ r_3 + 3r_2}}{\equiv} \begin{pmatrix} 1 & 0 & -1 & -7 \\ 0 & 1 & 1 & 3 \\ 0 & 0 & 1 & 9 \end{pmatrix}$$

$$\underset{\substack{r_1 + r_3 \\ r_2 - r_3}}{\equiv} \begin{pmatrix} 1 & 0 & 0 & 2 \\ 0 & 1 & 0 & -6 \\ 0 & 0 & 1 & 9 \end{pmatrix} = E.$$

No further reduction is possible and, as before, E is the row-reduced echelon form of A. From E we read off the column relation

$$c_4 = 2c_1 - 6c_2 + 9c_3$$

which applies to *all* the above matrices as the reader may verify; this relation provides a check on the calculations—if it doesn't hold, there's a mistake somewhere! In particular the columns of the original matrix A are subject to the above relation, which, on transposing gives $v_4 = 2v_1 - 6v_2 + 9v_3$. Similarly, as the first three columns of E are linearly independent, the first three columns of A are likewise linearly independent (8.20); therefore

$$\begin{pmatrix} 1 \\ -1 \\ 1 \end{pmatrix}, \quad \begin{pmatrix} 4 \\ -2 \\ 1 \end{pmatrix}, \quad \begin{pmatrix} 3 \\ -1 \\ 1 \end{pmatrix}$$

is a basis of column space A.

Example 8.19(c)

As preparation for the general case, let us reduce the matrix A below:

$$A = \begin{pmatrix} 1 & 2 & 1 & 2 \\ 1 & 3 & 3 & 1 \\ 2 & 1 & -4 & 8 \end{pmatrix} \underset{\substack{r_2 - r_1 \\ r_3 - 2r_1}}{\equiv} \begin{pmatrix} 1 & 2 & 1 & 2 \\ 0 & 1 & 2 & -1 \\ 0 & -3 & -6 & 4 \end{pmatrix}$$

$$\underset{\substack{r_1 - 2r_2 \\ r_3 + 3r_2}}{\equiv} \begin{pmatrix} 1 & 0 & -3 & 4 \\ 0 & 1 & 2 & -1 \\ 0 & 0 & 0 & 1 \end{pmatrix}.$$

From the last matrix we see that c_1 and c_2 are linearly independent, but $c_3 = -3c_1 + 2c_2$. As column 3 is a linear combination of preceding columns, no further simplification of this column is possible; however we can simplify the next column, column 4, which is *not* a linear combination of preceding columns without disturbing columns 1–3:

$$\begin{pmatrix} 1 & 0 & -3 & 4 \\ 0 & 1 & 2 & -1 \\ 0 & 0 & 0 & 1 \end{pmatrix} \underset{\substack{r_1 - 4r_3 \\ r_2 + r_3}}{\equiv} \begin{pmatrix} 1 & 0 & -3 & 0 \\ 0 & 1 & 2 & 0 \\ 0 & 0 & 0 & 1 \end{pmatrix} = E.$$

In this case columns 1, 2, 4 of A form a basis of column space A, for they are exactly the columns of A which are not linear combinations of preceding columns of A. We are using the familiar technique (6.20)(a) of deriving a basis from an ordered set of vectors which span; the novelty is that by looking at E we can see immediately which columns of A are to be deleted (just column 3 in this case) and which are to be retained. One final remark before we return to the theory: the columns of E which are not linear combinations of preceding columns of E are precisely $e_1^T, e_2^T, \ldots, e_r^T$, that is, the first r columns of I, where $r = \text{rank } E$.

Lemma 8.20

Let A and B be row-equivalent $m \times n$ matrices over F and let $x \in {}^nF$. Then $Ax = 0$ if and only if $Bx = 0$.

Proof

By (8.14) there is an invertible $m \times m$ matrix P over F such that $PA = B$. Premultiplying $Ax = 0$ by P gives $Bx = 0$. Premultiplying $Bx = 0$ by P^{-1} gives $Ax = 0$. Therefore $Ax = 0$ if and only if $Bx = 0$. $\qquad \square$

Let c_j and c_j' denote the jth columns of A and B as in (8.20) and let x_j be the jth entry in the column vector x. Then $Ax = 0 \Leftrightarrow Bx = 0$

Fig. 8.3

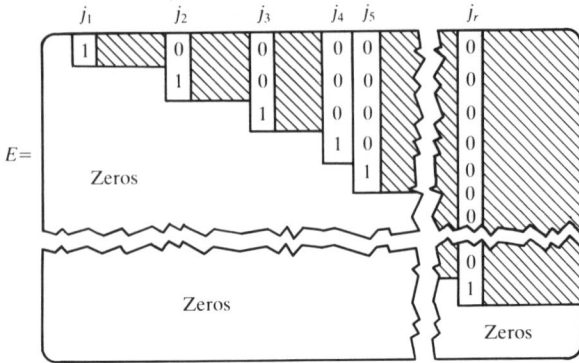

can be expressed:

$$c_1 x_1 + c_2 x_2 + \ldots + c_n x_n = 0 \quad \Leftrightarrow \quad c_1' x_1 + c_2' x_2 + \ldots + c_n' x_n = 0.$$

Therefore (8.20) has the following interpretation.

If A and B are row-equivalent, then the columns
of A are subject to the same linear dependence
relations as the columns of B.

In particular, the columns of A are linearly independent if and only
if the columns of the row-equivalent matrix B are linearly
independent.

What is meant by saying that the matrix E is in row-reduced
echelon form? The general pattern of E is shown in Fig. 8.3 where
the shaded areas denote blocks of arbitrary elements. Notice that
the first r rows of E are non-zero ($r = \operatorname{rank} E$) and the remaining
rows are zero; the first non-zero entry in row i is 1 (the (i, j_i)-entry)
and the remaining entries in column j_i are zero ($1 \leqslant i \leqslant r$).

In fact there is a concise description of E in terms of columns: we
may survey any matrix from left to right and delete those columns
which are linear combinations of preceding columns; the remaining
columns form a basis of the column space of the matrix (6.20)(a). In
the case of E (Fig. 8.3), this process singles out columns
$j_1, j_2, \ldots, j_r$ ($j_1 < j_2 < \ldots < j_r$), for these are precisely the columns
of E which are not linear combinations of preceding columns;
further these r special columns of E are the first r columns of the
identity matrix I.

Definition 8.21

Let E be an $m \times n$ matrix over a field. Let the i th column of E
which is *not* a linear combination of preceding columns be col-
umn j_i. If column j_i of E is e_i^T (column i of the $m \times m$ identity

matrix I) for $1 \le i \le r$, where $r = \text{rank } E$, then E is said to be in **row-reduced echelon form**.

The following matrix over $\mathbb{Q}$ is in row-reduced echelon form:

$$E = \begin{pmatrix} 0 & 1 & * & 0 & * & * & 0 & * \\ 0 & 0 & 0 & 1 & * & * & 0 & * \\ 0 & 0 & 0 & 0 & 0 & 0 & 1 & * \end{pmatrix}$$

where the entries $*$ are arbitrary rational numbers; here $j_1 = 2$ (by convention, the zero first column is a linear combination of preceding columns), $j_2 = 4$ and $j_3 = 7$, since it is the second, fourth, and seventh columns of E (and only these columns) which are not linear combinations of preceding columns.

Let E be a matrix in row-reduced echelon form as in (8.21). As the columns of E preceding column j_i are linear combinations of $e_1^T, e_2^T, \ldots, e_{i-1}^T$, we see that the (i, j)-entry in E is zero for $j < j_i$ $(1 \le i \le r)$; as $e_1^T, e_2^T, \ldots, e_r^T$ form a basis of column space E, the (i, j)-entry is zero $(r < i \le m, \; 1 \le j \le n)$. Therefore E is as described in Fig. 8.3.

We establish next the main property of row-equivalence by means of an algorithm, which works across a given matrix A from left to right in blocks of columns leaving in its wake a matrix E in row-reduced echelon form; in fact, this algorithm is no more than a straightforward generalization of the reduction procedure used in (8.19).

Theorem 8.22

Let A be an $m \times n$ matrix over the field F. Then A is row-equivalent to a unique matrix E in row-reduced echelon form. Further, A can be changed into E by at most mr elementary row operations where $r = \text{rank } A$.

Proof

Let $r = \text{rank } A$ and let column j_i be the i th column of A which is not a linear combination of preceding columns $(1 \le i \le r)$. We show by induction on i that a matrix E_i, having its first j_i columns in row-reduced echelon form, can be obtained from A by at most mi elementary row operations; by (8.18) E_i is row-equivalent to A. It is convenient (and not cheating) to start the induction at $i = 0$, taking $E_0 = A$ and $j_0 = 0$. Suppose that E_{i-1} (Fig. 8.4) has been obtained $(0 \le i - 1 < r)$. Let D be the $(m - i + 1) \times (n - j_{i-1})$ matrix in the lower-right corner of E_{i-1}. As $i - 1 < r = \text{rank } E_{i-1}$, we see $D \ne 0$. In fact column j_i of E_{i-1} is the first column containing a non-zero entry in D, because the

Fig. 8.4

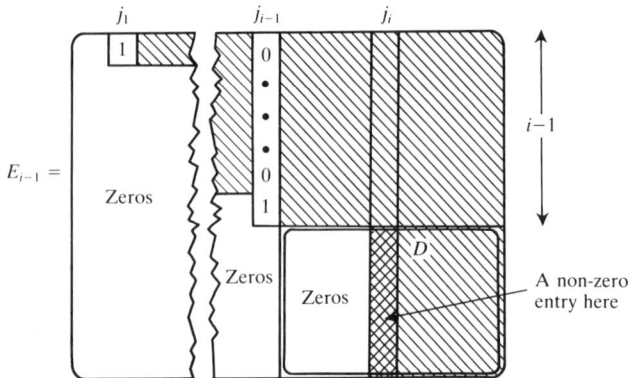

columns of E_{i-1} are subject to the same linear dependence relations as the columns of A by (8.20); so column j_i of E_{i-1} does not belong to $\langle e_1^T, e_2^T, \ldots, e_{i-1}^T \rangle$ (the subspace of column vectors in ${}^m F$ with their last $m-i+1$ entries zero), whereas the preceding columns of E do belong to this subspace. To change column j_i of E into e_i^T we carry out, in sequence, the following elementary row operations:

(i) Interchange, if necessary, two of the last $m-i+1$ rows of E_{i-1}, producing a matrix with non-zero (i, j_i)-entry.

(ii) Multiply row i by the inverse of the (i, j_i)-entry, producing a matrix having 1 as (i, j_i)-entry.

(iii) From row k subtract $((k, j_i)$-entry$) \times$ row i for all $k \neq i$, producing a matrix E_i having e_i^T as column j_i.

So at most m elementary row operations are required to change E_{i-1} into E_i (if (i) is necessary, then one of the $m-1$ operations (iii) is superfluous); therefore at most $m(i-1) + m = mi$ operations are needed to change A into E_i. As the first j_i columns of E_i are in row-reduced echelon form, we have completed the inductive step.

Ultimately (after at most mr elementary row operations have done their work) we arrive at E_r having its first j_r columns in row-reduced echelon form; as these columns span column space E_r, by (8.21) the *whole* matrix E_r is in row-reduced echelon form!

To prove uniqueness, let E be in row-reduced echelon form and row-equivalent to A. Then E and E_r are row-equivalent to each other, and so there is an invertible $m \times m$ matrix P over F with $PE = E_r$. Comparing j_i th columns gives $Pe_i^T = e_i^T$ $(1 \leq i \leq r)$. Hence $(P - I)e_i^T = 0$, showing that the first r columns of $P - I$ are zero; as the last $m - r$ rows of E are zero, the matrices $P - I$ and E fail to make contact when multiplied together, that is, $(P - I)E = 0$. Therefore $E_r = PE = E$. ☐

It is not necessary to stick rigidly to the algorithm of (8.22), as long as every operation in the reduction is built up from elementary row operations; indeed the intermediate matrices $E_1, E_2, \ldots, E_{r-1}$ are not, in general, unique.

Definition 8.23

The matrix E of (8.22) is called the **row-reduced echelon form** of A.

How can we tell whether or not two matrices are row-equivalent? The answer is simple: reduce them both and see!

Corollary 8.24

Let A and B be $m \times n$ matrices over F. Then A and B are row-equivalent if and only if they have the same row-reduced echelon form.

Proof

Let A and B have the same row-reduced echelon form E. Then $A \equiv E$ and $B \equiv E$ by (8.22), from which we deduce $A \equiv B$, that is, A is row-equivalent to B.

Conversely suppose $A \equiv B$ and let E denote the row-reduced echelon form of B. Then $B \equiv E$ and so $A \equiv E$, showing by (8.22) that E is the row-reduced echelon form of A. ☐

We have just passed our first milestone in matrix theory, since (8.22) and (8.24) provide us with a practical method of deciding row-equivalence.

Exercises 8.2

1. Find the row-reduced echelon forms of the following matrices over $\mathbb{Q}$, and hence decide which pairs of these matrices are row-equivalent.

$$\begin{pmatrix} 2 & 3 & 5 \\ 1 & 2 & 1 \\ 1 & 1 & 4 \end{pmatrix}, \quad \begin{pmatrix} 2 & 4 & 5 \\ 3 & 6 & 9 \\ 1 & 2 & 3 \end{pmatrix}, \quad \begin{pmatrix} 1 & -1 & 10 \\ 3 & -2 & 27 \\ 2 & -1 & 17 \end{pmatrix}, \quad \begin{pmatrix} 1 & 2 & 4 \\ 1 & 2 & 5 \\ 1 & 2 & 6 \end{pmatrix}.$$

2. Answer question 1 in the case of the following matrices over $\mathbb{Q}$:

$$\begin{pmatrix} 1 & 2 & 3 & 8 \\ 2 & 1 & 2 & 6 \\ 1 & 1 & 3 & 6 \end{pmatrix}, \quad \begin{pmatrix} 1 & 2 & 1 & 5 \\ 2 & 4 & -1 & 7 \\ 2 & 0 & 1 & 3 \end{pmatrix}, \quad \begin{pmatrix} 1 & 3 & 1 & 8 \\ 1 & 2 & 1 & 6 \\ -2 & 4 & 1 & 7 \end{pmatrix}.$$

3. For each of the matrices A in question 2, determine rank A, a basis of row space A, and a basis of column space A.

4. In each case, determine, as in (8.19)(b), a linear dependence relation between the given vectors over $\mathbb{Q}$.

(a) $(1, 1, 1)$, $(1, 2, 4)$, $(1, 2, 3)$, $(4, 5, 5)$;
(b) $(4, 5, 7)$, $(2, 3, 5)$, $(7, 6, 4)$, $(3, 6, 11)$;
(c) $(1, 1, 1, 3)$, $(2, 3, 2, 3)$, $(1, 2, 2, 4)$, $(1, 4, 3, 2)$.

5. Find, as in (8.19)(a), a basis of the solution space of the following systems over $\mathbb{Q}$.

(a) $x_1 + 3x_2 + 4x_3 = 0$
$\quad 4x_1 + 3x_2 + 7x_3 = 0$
$\quad 3x_1 + 2x_2 + 5x_3 = 0$

(b) $x_1 + x_2 + x_3 + 2x_4 = 0$
$\quad 2x_1 + 4x_2 + 3x_3 - 3x_4 = 0$
$\quad 2x_1 + 3x_2 + 3x_3 + x_4 = 0$.

6. (a) List the sixteen 3×3 matrices over $\mathbb{Z}_2$ which are in row-reduced echelon form (the zero matrix should be included). List the 2×3 matrices over $\mathbb{Z}_2$ which are in row-reduced echelon form. How many 4×3 matrices over $\mathbb{Z}_2$ in row-reduced echelon form are there?

(b) Let F be a field and suppose $m \geq n$. Find a bijective correspondence between $m \times n$ matrices over F in row-reduced echelon form and $n \times n$ matrices over F in row-reduced echelon form.

7. The $m \times n$ matrices A, B over F are such that $Ax = 0$ if and only if $Bx = 0$ where $x \in {}^nF$. Prove that A and B are row-equivalent.

8. (Co-ordinate version of (6.31).) Let U_1 and U_2 be subspaces of F^m with bases $v_1, v_2, \ldots, v_s$ and $v_{s+1}, v_{s+2}, \ldots, v_n$. Form the $m \times n$ matrix A over F having v_j^T as column j $(1 \leq j \leq n)$. Let $E = [e_{ij}]$ be the row-reduced echelon form of A and let $j_1, j_2, \ldots, j_r$ have their usual meaning (8.21). Prove that

(a) $j_i = i$ $(1 \leq i \leq s)$,
(b) $U_1 + U_2$ has basis $v_{j_1}, v_{j_2}, \ldots, v_{j_r}$,
(c) $v_j = \sum_{i=1}^r e_{ij} v_{j_i}$ and deduce that $u_j = v_j - \sum_{i=s+1}^r e_{ij} v_{j_i}$ belongs to $U_1 \cap U_2$ for $j \neq j_1, j_2, \ldots, j_r$.
(d) the $n - r$ vectors u_j $(j \neq j_1, j_2, \ldots, j_r)$ form a basis of $U_1 \cap U_2$.

9. Taking $F = \mathbb{Q}$, use the method of question 8 above to find bases of $U_1 + U_2$ and $U_1 \cap U_2$ in the following cases:

(a) $U_1 = \langle (1, 1, 2, 1), (1, 2, 2, 1), (1, 1, 1, 1) \rangle$,
$\quad U_2 = \langle (1, 2, 1, 1), (6, 7, 9, 6) \rangle$.

(b) $U_1 = \langle (1, 2, 1, 2), (2, 3, 1, 2), (2, 1, 1, 1) \rangle$,
$\quad U_2 = \langle (2, 1, 2, 1), (2, -2, 3, 1), (4, -2, 2, -3) \rangle$.

Row-reduction and inversion

We turn now to another aspect of row-equivalence. Let A be an $m \times n$ matrix over the field F and let E be the row-reduced echelon form of A (8.23). As A and E are row-equivalent, there is an invertible $m \times m$ matrix P with $PA = E$. How can such a matrix P be found? From (8.22) we know that it is possible to change A into E by means of a finite number of elementary row operations. Let $P_1, P_2, \ldots, P_t$ be the elementary matrices (8.17) corresponding to the elementary row operations used in the reduction of A to E. Then

$$P_t \ldots P_2 P_1 A = E$$

for by (8.18) the reduction process amounts to first premultiplying A by P_1 obtaining $P_1 A$, then $P_1 A$ is premultiplied by P_2 giving $P_2 P_1 A$, and so on. The product $P_t \ldots P_2 P_1$ can be conveniently calculated by applying the elementary row operations involved to the $m \times (n + m)$ matrix $[A : I]$ formed by adjoining to A the $m \times m$ identity matrix I; since, setting $P = P_t \ldots P_2 P_1$ we have

$$P_t \ldots P_2 P_1 [A : I] = P[A : I] = [PA : PI] = [E : P]$$

which expresses the (partial) reduction of $[A : I]$. The method therefore consists of applying the reduction algorithm (8.22) to $[A : I]$ until the first n columns are in row-reduced echelon form; then E, the row-reduced echelon form of A, forms the first part of this matrix, the second part being an invertible matrix P with $PA = E$.

Example 8.25

Consider again the 3×3 matrix A of (8.19)(a). We reduce as before, but starting with the 3×6 matrix $[A : I]$.

$$[A : I] = \begin{pmatrix} 1 & 2 & 4 & 1 & 0 & 0 \\ 2 & 5 & 9 & 0 & 1 & 0 \\ 1 & 6 & 8 & 0 & 0 & 1 \end{pmatrix} \underset{\substack{r_2 - 2r_1 \\ r_3 - r_1}}{\equiv} \begin{pmatrix} 1 & 2 & 4 & 1 & 0 & 0 \\ 0 & 1 & 1 & -2 & 1 & 0 \\ 0 & 4 & 4 & -1 & 0 & 1 \end{pmatrix}$$

$$\underset{\substack{r_1 - 2r_2 \\ r_3 - 4r_2}}{\equiv} \begin{pmatrix} 1 & 0 & 2 & 5 & -2 & 0 \\ 0 & 1 & 1 & -2 & 1 & 0 \\ 0 & 0 & 0 & 7 & -4 & 1 \end{pmatrix} = [E : P],$$

and so

$$P = \begin{pmatrix} 5 & -2 & 0 \\ -2 & 1 & 0 \\ 7 & -4 & 1 \end{pmatrix}.$$

The reader may verify $PA = E$. Notice that the row operations apply right across the above 3×6 matrices (don't forget to alter the second parts accordingly). There is no need to *completely* reduce the whole matrix $[A : I]$, but only the first three columns in this case—a complete reduction is not wrong—merely a waste of time! The last three columns of the above 3×6 matrices provide us with a tally of the row operations used up to that point; more precisely, they are the products (in reverse order) of the corresponding elementary matrices. Therefore

$$\begin{pmatrix} 1 & 0 & 0 \\ -2 & 1 & 0 \\ -1 & 0 & 1 \end{pmatrix} = \overset{r_3 - r_1}{\begin{pmatrix} 1 & 0 & 0 \\ 0 & 1 & 0 \\ -1 & 0 & 1 \end{pmatrix}} \overset{r_2 - 2r_1}{\begin{pmatrix} 1 & 0 & 0 \\ -2 & 1 & 0 \\ 0 & 0 & 1 \end{pmatrix}}$$

and

$$P = \overset{r_3 - 4r_2}{\begin{pmatrix} 1 & 0 & 0 \\ 0 & 1 & 0 \\ 0 & -4 & 1 \end{pmatrix}} \overset{r_1 - 2r_2}{\begin{pmatrix} 1 & -2 & 0 \\ 0 & 1 & 0 \\ 0 & 0 & 1 \end{pmatrix}} \overset{r_3 - r_1}{\begin{pmatrix} 1 & 0 & 0 \\ 0 & 1 & 0 \\ -1 & 0 & 1 \end{pmatrix}} \overset{r_2 - 2r_1}{\begin{pmatrix} 1 & 0 & 0 \\ -2 & 1 & 0 \\ 0 & 0 & 1 \end{pmatrix}}$$

where above each elementary matrix we have written the corresponding elementary row operation (remember (8.17): applying an elementary row operation to I produces the corresponding elementary matrix).

An elementary matrix is an invertible matrix which is *almost* equal to I, that is, all except (possibly) one of its entries agree with the entries in I. Nevertheless we shall see (8.27) that every invertible matrix over a field can be expressed as a product of elementary matrices. In fact, an important special case of (8.22) is that of an invertible $n \times n$ matrix A. By (8.12) rank $A = n$, which means that the n columns of A are linearly independent; so $j_1 = 1$, $j_2 = 2, \ldots, j_n = n$ in this case, that is, *every* column is a special column. Therefore column $j_i =$ column $i = e_i^T$ in E, the row-reduced echelon form of A ($1 \leqslant i \leqslant n$); in other words:

The identity matrix I is the row-reduced
echelon form of every invertible matrix A.

Further, the invertible matrix P, such that PA is in row-reduced echelon form, satisfies $PA = I$ and so $P = A^{-1}$. Premultiplying the $n \times 2n$ matrix $[A : I]$ by P gives

$$P[A : I] = [PA : PI] = [I : A^{-1}]$$

and as $[I : A^{-1}]$ is in row-reduced echelon form we see that:

> If A is invertible, then $[I : A^{-1}]$ is
> the row-reduced echelon form of $[A : I]$

and so the reduction algorithm (8.22) can be used to calculate inverses.

Example 8.26

Consider $A = \begin{pmatrix} 1 & 1 & 2 & 2 \\ 4 & 5 & 9 & 9 \\ 2 & 1 & 4 & 3 \\ 1 & 7 & 5 & 9 \end{pmatrix}$ over $\mathbb{Q}$. We form the 4×8 matrix

$[A : I]$ and reduce it (the reader should be getting used to the reduction procedure by now); as the first parts (columns 1–4) of the 4×8 matrices below reduce from A to I, showing that A is invertible over $\mathbb{Q}$, the second parts 'increase' from I to A^{-1}.

$$[A : I] = \begin{pmatrix} 1 & 1 & 2 & 2 & 1 & 0 & 0 & 0 \\ 4 & 5 & 9 & 9 & 0 & 1 & 0 & 0 \\ 2 & 1 & 4 & 3 & 0 & 0 & 1 & 0 \\ 1 & 7 & 5 & 9 & 0 & 0 & 0 & 1 \end{pmatrix}$$

$$\equiv \begin{pmatrix} 1 & 1 & 2 & 2 & 1 & 0 & 0 & 0 \\ 0 & 1 & 1 & 1 & -4 & 1 & 0 & 0 \\ 0 & -1 & 0 & -1 & -2 & 0 & 1 & 0 \\ 0 & 6 & 3 & 7 & -1 & 0 & 0 & 1 \end{pmatrix}$$

$$\equiv \begin{pmatrix} 1 & 0 & 1 & 1 & 5 & -1 & 0 & 0 \\ 0 & 1 & 1 & 1 & -4 & 1 & 0 & 0 \\ 0 & 0 & 1 & 0 & -6 & 1 & 1 & 0 \\ 0 & 0 & -3 & 1 & 23 & -6 & 0 & 1 \end{pmatrix}$$

$$\equiv \begin{pmatrix} 1 & 0 & 0 & 1 & 11 & -2 & -1 & 0 \\ 0 & 1 & 0 & 1 & 2 & 0 & -1 & 0 \\ 0 & 0 & 1 & 0 & -6 & 1 & 1 & 0 \\ 0 & 0 & 0 & 1 & 5 & -3 & 3 & 1 \end{pmatrix}$$

$$\equiv \begin{pmatrix} 1 & 0 & 0 & 0 & 6 & 1 & -4 & -1 \\ 0 & 1 & 0 & 0 & -3 & 3 & -4 & -1 \\ 0 & 0 & 1 & 0 & -6 & 1 & 1 & 0 \\ 0 & 0 & 0 & 1 & 5 & -3 & 3 & 1 \end{pmatrix} = [I : A^{-1}].$$

Therefore

$$A^{-1} = \begin{pmatrix} 6 & 1 & -4 & -1 \\ -3 & 3 & -4 & -1 \\ -6 & 1 & 1 & 0 \\ 5 & -3 & 3 & 1 \end{pmatrix}.$$

In Chapter 9 we shall derive a formula for the entries in A^{-1} using determinants; however, this formula is primarily of theoretical value, and of little use in calculating inverses of 4×4 or larger matrices. By contrast, computers can easily be programmed to carry out the reduction algorithm (8.22), and the above method of finding inverses is widely used. The reader should realize an important computational point: given an invertible matrix A over $\mathbb{Q}$ having integer entries, then the entries in A^{-1} will, in general, be fractional (their denominators are closely related to the determinant $|A|$ of A); we have avoided this difficulty so far by carefully selecting our examples! But fractions can be avoided legitimately, or at least postponed until the last step, as in the following reduction of $[A : I]$:

$$\begin{pmatrix} 1 & 5 & 2 & 1 & 0 & 0 \\ 1 & 7 & 4 & 0 & 1 & 0 \\ 1 & 1 & 8 & 0 & 0 & 1 \end{pmatrix} \equiv \begin{pmatrix} 1 & 5 & 2 & 1 & 0 & 0 \\ 0 & 2 & 2 & -1 & 1 & 0 \\ 0 & -4 & 6 & -1 & 0 & 1 \end{pmatrix}.$$

Instead of dividing row 2 by 2, multiply row 1 by 2 (we are aiming to create an integer multiple of I in the first part of the matrix while avoiding fractions):

$$\begin{pmatrix} 2 & 10 & 4 & 2 & 0 & 0 \\ 0 & 2 & 2 & -1 & 1 & 0 \\ 0 & -4 & 6 & -1 & 0 & 1 \end{pmatrix} \equiv \begin{pmatrix} 2 & 0 & -6 & 7 & -5 & 0 \\ 0 & 2 & 2 & -1 & 1 & 0 \\ 0 & 0 & 10 & -3 & 2 & 1 \end{pmatrix}.$$

Instead of dividing row 3 by 5, multiply rows 1 and 2 by 5 ($10I$ can then be created in the first part of the matrix):

$$\begin{pmatrix} 10 & 0 & -30 & 35 & -25 & 0 \\ 0 & 10 & 10 & -5 & 5 & 0 \\ 0 & 0 & 10 & -3 & 2 & 1 \end{pmatrix} \equiv \begin{pmatrix} 10 & 0 & 0 & 26 & -19 & 3 \\ 0 & 10 & 0 & -2 & 3 & -1 \\ 0 & 0 & 10 & -3 & 2 & 1 \end{pmatrix}.$$

Dividing the last matrix by 10 produces $[I : A^{-1}]$ where

$$A^{-1} = \tfrac{1}{10} \begin{pmatrix} 26 & -19 & 3 \\ -2 & 3 & -1 \\ -3 & 2 & 1 \end{pmatrix}.$$

Corollary 8.27   Let A be an invertible $n \times n$ matrix over a field. Then A is expressible as a product of at most n^2 elementary matrices.

Proof   By (8.22), A can be reduced to I by means of t elementary row operations, where $t \leqslant n^2$. By (8.18), $P_t \ldots P_2 P_1 A = I$ where $P_1, P_2, \ldots, P_t$ are the corresponding t elementary matrices. Therefore

$$A = (P_t \ldots P_2 P_1)^{-1} = P_1^{-1} P_2^{-1} \ldots P_t^{-1}$$

expresses A as a product of elementary matrices, for $P_1^{-1}, P_2^{-1}, \ldots, P_t^{-1}$ are elementary by (8.18).   □

For instance, consider $A = \begin{pmatrix} 3 & 4 \\ 5 & 6 \end{pmatrix}$ over $\mathbb{Q}$. First reduce A to I, noting the elementary row operations used:

$$\begin{pmatrix} 3 & 4 \\ 5 & 6 \end{pmatrix} \underset{\frac{1}{3}r_1}{\equiv} \begin{pmatrix} 1 & \frac{4}{3} \\ 5 & 6 \end{pmatrix} \underset{r_2 - 5r_1}{\equiv} \begin{pmatrix} 1 & \frac{4}{3} \\ 0 & -\frac{2}{3} \end{pmatrix} \underset{-\frac{3}{2}r_2}{\equiv} \begin{pmatrix} 1 & \frac{4}{3} \\ 0 & 1 \end{pmatrix} \underset{r_1 - \frac{4}{3}r_2}{\equiv} \begin{pmatrix} 1 & 0 \\ 0 & 1 \end{pmatrix}.$$

From the proof of (8.27), A is the product (in order) of the elementary matrices corresponding to the inverses of the above row operations, that is,

$$\begin{pmatrix} 3 & 4 \\ 5 & 6 \end{pmatrix} = \overset{3r_1}{\begin{pmatrix} 3 & 0 \\ 0 & 1 \end{pmatrix}} \overset{r_2 + 5r_1}{\begin{pmatrix} 1 & 0 \\ 5 & 1 \end{pmatrix}} \overset{-\frac{2}{3}r_2}{\begin{pmatrix} 1 & 0 \\ 0 & -\frac{2}{3} \end{pmatrix}} \overset{r_1 + \frac{4}{3}r_2}{\begin{pmatrix} 1 & \frac{4}{3} \\ 0 & 1 \end{pmatrix}}$$

since the inverse of $\frac{1}{3}r_1$ (multiply row 1 by $\frac{1}{3}$) is $3r_1$ (multiply row 1 by 3), and the inverse of $r_2 - 5r_1$ is $r_2 + 5r_1$, etc.

Linear systems

Next we study systems of linear equations; as suggested in (8.19)(a), the reduction algorithm (8.22) provides a practical way of finding the general solution of such systems. Consider the system of m simultaneous equations in the n unknowns $x_1, x_2, \ldots, x_n$:

$$a_{11}x_1 + a_{12}x_2 + \ldots + a_{1n}x_n = b_1$$
$$a_{21}x_1 + a_{22}x_2 + \ldots + a_{2n}x_n = b_2$$
$$\vdots \qquad \vdots \qquad \qquad \vdots \qquad \vdots$$
$$a_{m1}x_1 + a_{m2}x_2 + \ldots + a_{mn}x_n = b_m.$$

Assuming that a_{ij} and b_i are given elements of a field F, we set ourselves the task of describing the set of solutions $x_1, x_2, \ldots, x_n$ where $x_j \in F$. The above system can be expressed by the column equation

$$Ax = b$$

where $A = [a_{ij}]$ is the $m \times n$ **coefficient matrix**, b is the $m \times 1$ column of constants appearing on the right-hand side (so b_i is the i th entry in $b \in {}^mF$) and x is the $n \times 1$ column of unknowns (so x_j is the j th entry in $x \in {}^nF$). We must describe x in terms of A and b. The $m \times (n + 1)$ matrix $[A \vdots b]$, formed by adjoining to A the column b, is called the **augmented matrix** of the system $Ax = b$. The system $Ax = 0$ is called the **associated homogeneous system**.

Let $[E \vdots e]$ denote the row-reduced echelon form of the $m \times (n + 1)$ augmented matrix $[A \vdots b]$. By (8.22) there is an invertible $m \times m$ matrix P over F with $P[A \vdots b] = [E \vdots e]$; so $PA = E$ (which is the row-reduced echelon form of the coefficient matrix A) and $Pb = e$. Also

$$Ax = b \quad \Leftrightarrow \quad Ex = e$$

for premultiplying $Ax = b$ by P produces $Ex = e$, and premultiplying $Ex = e$ by P^{-1} gets us back to $Ax = b$; in other words, the systems $Ax = b$ and $Ex = e$ have the same set of solutions. However the reduced system $Ex = e$ is more manageable—indeed one can see by inspection if there are any solutions at all, and when the system is **consistent,** that is, when there are solutions, how the general solution can be found.

Example 8.28(a) Working over the real field $\mathbb{R}$, consider the system

$$x_1 + 4x_2 + 3x_3 = 5$$
$$-x_1 - 2x_2 - x_3 = 1$$
$$x_1 + x_2 + x_3 = 5$$

which has

$$\begin{pmatrix} 1 & 4 & 3 & 5 \\ -1 & -2 & -1 & 1 \\ 1 & 1 & 1 & 5 \end{pmatrix}$$

as augmented matrix. This matrix reduces (8.19)(b) to

$$\begin{pmatrix} 1 & 0 & 0 & 2 \\ 0 & 1 & 0 & -6 \\ 0 & 0 & 1 & 9 \end{pmatrix}$$

and so

$$x_1 \qquad = \quad 2$$
$$x_2 \quad = -6$$
$$x_3 = \quad 9$$

is the reduced system. The reduction process has solved the system for us! There is a geometric interpretation: the original equations represent planes in $\mathbb{R}^3$ having a unique point in common, namely $(2, -6, 9)$; the reduced system consists of the equations of the planes through $(2, -6, 9)$ which are parallel to the co-ordinate planes $x_1 = 0$, $x_2 = 0$, $x_3 = 0$.

Example 8.28(b) Consider the system over $\mathbb{R}$

$$x_1 + 2x_2 + \ x_3 = 2$$
$$x_1 + 3x_2 + 3x_3 = 1$$
$$2x_1 + \ x_2 - 4x_3 = 8$$

which has

$$\begin{pmatrix} 1 & 2 & 1 & 2 \\ 1 & 3 & 3 & 1 \\ 2 & 1 & -4 & 8 \end{pmatrix}$$

as augmented matrix. In (8.19)(c), we saw that this matrix reduces to

$$\begin{pmatrix} 1 & 0 & -3 & 4 \\ 0 & 1 & 2 & -1 \\ 0 & 0 & 0 & 1 \end{pmatrix}$$

and so

$$x_1 \qquad -3x_3 = \quad 4$$
$$x_2 + 2x_3 = -1$$
$$0 - \quad 1$$

is the reduced system. The (absurd!) equation $0 = 1$ tells us that the reduced system is **inconsistent** (it has no solutions), and hence the original system is also inconsistent. In geometric terms, the original equations represent planes in $\mathbb{R}^3$ having no common point. The associated homogeneous system consists of the equations of the

Fig. 8.5

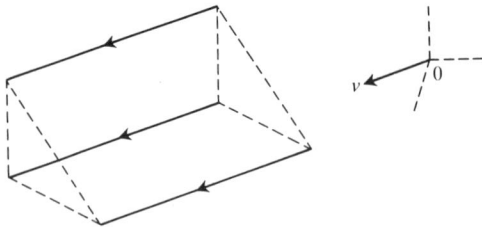

corresponding parallel planes through the origin, that is

$$x_1 + 2x_2 + x_3 = 0$$
$$x_1 + 3x_2 + 3x_3 = 0$$
$$2x_1 + x_2 - 4x_3 = 0$$

which reduces to

$$x_1 \quad - 3x_3 = 0$$
$$x_2 + 2x_3 = 0$$

and so has general solution $(x_1, x_2, x_3) = (3x_3, -2x_3, x_3)$ on eliminating x_1 and x_2; in other words, these planes through the origin
intersect in the line $\langle(3, -2, 1)\rangle$. This means that the original planes
form a 'prism' (Fig. 8.5) having its edges (the intersections of pairs
of planes) parallel to the vector $v = (3, -2, 1)$.

We now return to the case of a system $Ax = b$ of m equations in n
unknowns over the field F. As before $[E : e]$ denotes the row-
reduced form of the augmented matrix $[A : b]$, c_{j_i} is the i th column
of A which is not a linear combination of preceding columns, and
$r = \text{rank } A$. Our next proposition completes the task, for it provides
a complete description of the unknown entries x_j in the column
vector x.

Proposition 8.29 The system $Ax = b$ is consistent if and only if $\text{rank } A = \text{rank}[A : b]$. In this case, the reduced system $Ex = e$ expresses the r unknowns $x_{j_1}, x_{j_2}, \ldots, x_{j_r}$ in terms of the remaining $n - r$ unknowns x_j $(j \neq j_i)$ which are arbitrary elements of F.

Proof Writing $c_j = (\text{column } j \text{ of } A)$, the system $Ax = b$ is consistent if
and only if $c_1 x_1 + c_2 x_2 + \ldots + c_n x_n = b$ for some $x_j \in F$, that is, if
and only if b belongs to the column space of A. In other words
$Ax = b$ is consistent if and only if column space$[A : b]$ is no larger
than its subspace column space A. Therefore consistency is the
same as $\text{rank}[A : b] = \text{rank } A$, on comparing dimensions by (6.26).

Suppose $\mathrm{rank}[A \vdots b] = \mathrm{rank}\, A = r$. From $P[A \vdots b] = [E \vdots e]$ where P is invertible we deduce $\mathrm{rank}[E \vdots e] = \mathrm{rank}\, E = r$ by (8.15). As $Ax = b \Leftrightarrow Ex = e$, it is good enough (and much easier) to deal with the reduced system $Ex = e$. Let e_{ij} denote the (i, j)-entry in $[E \vdots e]$ for $1 \leqslant i \leqslant m$, $1 \leqslant j \leqslant n + 1$. Equation i of $Ex = e$ can be rearranged to

$$x_{j_i} = -\sum_j e_{ij} x_j + e_{in+1} \qquad (1 \leqslant i \leqslant r)$$

where the summation is over integers $j \neq j_1, j_2, \ldots, j_r$ and $1 \leqslant j \leqslant n$. So the first r equations of $Ex = e$ express the unknowns $x_{j_1}, x_{j_2}, \ldots, x_{j_r}$ in terms of the remaining $n - r$ unknowns x_j which (since they are not related among themselves) are arbitrary elements of F. The last $m - r$ equations of $Ex = e$ tell us (repeatedly!) that $0 = 0$, and so may be disregarded. $\qquad \square$

Consider the system over $\mathbb{R}$

$$x_1 + 3x_2 - x_3 - 3x_4 = -2$$
$$2x_1 + 6x_2 - x_3 - x_4 = 2$$
$$x_1 + 3x_2 + x_3 + 7x_4 = 10.$$

To solve the system, reduce the augmented matrix:

$$\begin{pmatrix} 1 & 3 & -1 & -3 & -2 \\ 2 & 6 & -1 & -1 & 2 \\ 1 & 3 & 1 & 7 & 10 \end{pmatrix} \equiv \begin{pmatrix} 1 & 3 & -1 & -3 & -2 \\ 0 & 0 & 1 & 5 & 6 \\ 0 & 0 & 2 & 10 & 12 \end{pmatrix} \equiv \begin{pmatrix} 1 & 3 & 0 & 2 & 4 \\ 0 & 0 & 1 & 5 & 6 \\ 0 & 0 & 0 & 0 & 0 \end{pmatrix}.$$

From the last matrix we see $j_1 = 1$, $j_2 = 3$, $r = 2$. The system is consistent and the reduced system equations $x_1 + 3x_2 + 2x_4 = 4$, $x_3 + 5x_4 = 6$ can be used to eliminate x_1 and x_3 giving the general solution:

$$x = \begin{pmatrix} x_1 \\ x_2 \\ x_3 \\ x_4 \end{pmatrix} = \underbrace{\begin{pmatrix} -3 \\ 1 \\ 0 \\ 0 \end{pmatrix} x_2 + \begin{pmatrix} -2 \\ 0 \\ -5 \\ 1 \end{pmatrix} x_4}_{y} + \underbrace{\begin{pmatrix} 4 \\ 0 \\ 6 \\ 0 \end{pmatrix}}_{z}.$$

Notice that $x = y + z$ where y is the general solution of the associated homogeneous system and z is a particular solution of the given system. The solutions x form a plane in $^4\mathbb{R}$, while the solutions y form the parallel plane (2-dimensional subspace) through the origin.

This section ends with a discussion of **Gaussian elimination**, which is an efficient method of solving systems $Ax = b$. For simplicity, we suppose that A is an invertible $n \times n$ matrix. The method consists of row-reducing $[A : b]$ to $[U : u]$, where U is **upper triangular** (the (i, j)-entries in U are zero for $i > j$); the original system $Ax = b$ becomes $Ux = u$, from which the entries in x can be determined successively in reverse order $x_n, x_{n-1}, \ldots, x_1$.

Example 8.30

Working over $\mathbb{Q}$, consider the system

$$
\begin{aligned}
2x_1 + x_2 + x_3 - x_4 &= 6 \\
4x_1 + x_2 + 4x_3 - x_4 &= 13 \\
-2x_1 - 2x_2 + 7x_3 + 4x_4 &= 5 \\
4x_1 + x_2 - 8x_3 - 7x_4 &= -11.
\end{aligned}
$$

It is not necessary to know in advance that the coefficient matrix is invertible, for this will become clear during the reduction process: A is invertible if and only if all the diagonal entries (the (i, i)-entries) in U are non-zero. We produce (column by column) an upper triangular matrix by applying row operations to the augmented matrix of the above system:

$$
\begin{pmatrix}
2 & 1 & 1 & -1 & 6 \\
4 & 1 & 4 & -1 & 13 \\
-2 & -2 & 7 & 4 & 5 \\
4 & 1 & -8 & -7 & -11
\end{pmatrix}
\equiv
\begin{pmatrix}
2 & 1 & 1 & -1 & 6 \\
0 & -1 & 2 & 1 & 1 \\
0 & -1 & 8 & 3 & 11 \\
0 & -1 & -10 & -5 & -23
\end{pmatrix}
$$

$$
\equiv
\begin{pmatrix}
2 & 1 & 1 & -1 & 6 \\
0 & -1 & 2 & 1 & 1 \\
0 & 0 & 6 & 2 & 10 \\
0 & 0 & -12 & -6 & -24
\end{pmatrix}
\equiv
\begin{pmatrix}
2 & 1 & 1 & -1 & 6 \\
0 & -1 & 2 & 1 & 1 \\
0 & 0 & 6 & 2 & 10 \\
0 & 0 & 0 & -2 & -4
\end{pmatrix}.
$$

The last matrix above is the augmented matrix of

$$
\begin{aligned}
2x_1 + x_2 + x_3 - x_4 &= 6 \\
- x_2 + 2x_3 + x_4 &= 1 \\
6x_3 + 2x_4 &= 10 \\
- 2x_4 &= -4
\end{aligned}
$$

from which we deduce $x_4 = 2$, $x_3 = 1$, $x_2 = 3$, $x_1 = 2$. Notice one further fact: let L denote the invertible matrix such that $LA = U$ (it will be clear in a moment why the notation L, rather than P, is used here). L may be found directly by applying the row operations used above to the 4×8 matrix $[A : I]$ obtaining $[LA : LI] = [U : L]$; in this case

$$L = \begin{pmatrix} 1 & 0 & 0 & 0 \\ -2 & 1 & 0 & 0 \\ 3 & -1 & 1 & 0 \\ 6 & -3 & 2 & 1 \end{pmatrix}$$

that is, L is **lower triangular** with diagonal entries equal 1, and so A factorizes in the form $A = L^{-1}U$.

Gaussian elimination may be compared with the method (8.29) of finding the solution of $Ax = b$, where A is an invertible $n \times n$ matrix; in terms of the number of scalar multiplications required, Gaussian elimination is usually preferable.

Dealing with Gaussian elimination, the first step in the reduction of $[A : b] = [a_{ij}]$ to $[U : u]$ involves, in general, $n^2 - 1$ multiplications (we use multiplication to mean a scalar multiplication or division): for creating the first column of an upper triangular matrix involves dividing a_{i1} by a_{11} (we achieve $a_{11} \neq 0$ by interchanging rows) for $1 < i \leq n$ giving $n - 1$ multiplications, then multiplying a_{i1}/a_{11} by a_{1j} for $1 < j \leq n + 1$ giving $(n - 1)n$ multiplications.

The procedure is repeated on the $(n - 1) \times n$ matrix obtained by deleting row 1 and column 1, and so the complete reduction involves

$$n^2 + (n - 1)^2 + \ldots + 1 - n = n(n - 1)(2n + 5)/6$$

multiplications in general. The **back substitution**, that is, the calculation of $x_n, x_{n-1}, \ldots, x_1$ from $Ux = u$, involves a mere

$$1 + 2 + \ldots + n = n(n + 1)/2$$

multiplications, giving a total of $n(n^2 + 3n - 1)/3$ multiplications in the solution of $Ax = b$ by Gaussian elimination.

On the other hand, (8.29) requires us to reduce $[A : b]$ to $[I : A^{-1}b]$ using (8.22). Let us suppose the first $i - 1$ columns of I have been created where $1 \leq i \leq n$. We assume (by interchanging rows) that the (i, i)-entry is non-zero. Changing the (i, i)-entry into 1 involves $n + 1 - i$ multiplications, and changing the remaining $n - 1$ entries in column i into zero involves $(n - 1)(n + 1 - i)$ multiplications. So creating the further column i of I involves $n(n + 1 - i)$ multiplications; hence a total of

$$n^2 + n(n - 1) + n(n - 2) + \ldots + n = n^2(n + 1)/2$$

multiplications are generally required in (8.29). As

$$n(n^2 + 3n - 1)/3 < n^2(n + 1)/2 \quad \Leftrightarrow \quad n > 2$$

we see that Gaussian elimination is preferable for $n \geq 3$.

Exercises 8.3

1. For each of the following matrices A over $\mathbb{Q}$, determine an invertible matrix P such that PA is in row-reduced echelon form.

$$\begin{pmatrix} 1 & 2 & 3 & 4 \\ 2 & 4 & 5 & 3 \\ 3 & 6 & 7 & 3 \end{pmatrix}, \quad \begin{pmatrix} 1 & 2 & 3 \\ 2 & 4 & 6 \\ 3 & 6 & 9 \end{pmatrix}, \quad \begin{pmatrix} 1 & 2 & 3 & 4 \\ 2 & 4 & 5 & 3 \\ 4 & 8 & 9 & 1 \end{pmatrix}.$$

2. Find the inverses of the following matrices over $\mathbb{Q}$ and check your answers.

$$\begin{pmatrix} 1 & 1 & 1 \\ 5 & 5 & 4 \\ 6 & 5 & 3 \end{pmatrix}, \quad \begin{pmatrix} 1 & 2 & 5 \\ 3 & 4 & 4 \\ 3 & 3 & 2 \end{pmatrix}, \quad \begin{pmatrix} 1 & 1 & 2 & 2 \\ 2 & 1 & 1 & 2 \\ 2 & 2 & 1 & 2 \\ 3 & 3 & 1 & 3 \end{pmatrix}, \quad \begin{pmatrix} 1 & -1 & 1 & 2 \\ 0 & 1 & 2 & -1 \\ 3 & 1 & 1 & 1 \\ 3 & 2 & 1 & 0 \end{pmatrix}.$$

3. Find the general solution of the following systems of equations over $\mathbb{R}$, and of the associated homogeneous systems. Interpret each case geometrically.

(a) $\begin{aligned} x_1 + x_2 + x_3 &= 1 \\ 2x_1 + 3x_2 + x_3 &= 4 \end{aligned}$

(b) $\begin{aligned} x_1 - 9x_2 - 7x_3 &= 10 \\ 3x_1 + x_2 - x_3 &= 2 \\ 2x_1 + 3x_2 + x_3 &= -1 \end{aligned}$

(c) $\begin{aligned} x_1 + x_2 + 4x_3 &= 3 \\ 5x_1 + 7x_2 + 10x_3 &= 4 \\ 2x_1 + 3x_2 + 3x_3 &= 1 \end{aligned}$

(d) $\begin{aligned} x_1 + 4x_2 + 5x_3 &= 2 \\ 2x_1 + 3x_2 + 6x_3 &= 1 \\ 2x_1 + 4x_2 + 7x_3 &= 3. \end{aligned}$

4. Determine the row-reduced form of the augmented matrix of the following system over $\mathbb{Q}$:

$$\begin{aligned} x_1 + x_2 + 5x_3 + x_4 + 7x_5 &= 3 \\ x_1 + 2x_2 + 8x_3 + x_4 + 12x_5 &= 3 \\ 2x_1 + x_2 + 7x_3 + x_4 + 11x_5 &= 2 \\ x_1 + 3x_2 + 11x_3 + 2x_4 + 15x_5 &= 7. \end{aligned}$$

Write down the general solution of this system and hence find the particular solutions satisfying $x_3^2 = x_5^2 = 1$.

5. Find the general solution of the system over $\mathbb{Q}$:

$$\begin{aligned} x_1 + x_2 + x_3 + 6x_4 &= 13 \\ 3x_1 + 4x_2 - x_3 + x_4 &= 8 \\ 5x_1 + 7x_2 - 2x_3 + x_4 &= 12. \end{aligned}$$

Hence find the minimum and maximum values of $x_1 + x_2 + x_3 + x_4$ given that the x_j are non-negative and satisfy the above equations.

6. Express $\begin{pmatrix} 1 & 2 \\ 3 & 4 \end{pmatrix}$ and $\begin{pmatrix} 0 & 2 \\ 3 & 4 \end{pmatrix}$ as products of elementary matrices over $\mathbb{Q}$.

7. List the six invertible 2×2 matrices over $\mathbb{Z}_2$ and verify that each is elementary. Express the matrix

$$\begin{pmatrix} 0 & 1 & 0 \\ 0 & 0 & 1 \\ 1 & 1 & 1 \end{pmatrix}$$

over $\mathbb{Z}_2$ as a product of four elementary matrices. Show that every invertible $n \times n$ matrix over $\mathbb{Z}_2$ is expressible as a product of at most $(n-1)^2$ elementary matrices over $\mathbb{Z}_2$.

8. Show that each elementary row operation (8.16) of type (i) can be expressed as a composition of elementary row operations of types (ii) and (iii).

9. Use Gaussian elimination to find the solution of the system over $\mathbb{Q}$:

$$\begin{aligned}
3x_1 + \ x_2 + 2x_3 + \ x_4 &= \ \ 12 \\
6x_1 + 3x_2 + 3x_3 + 5x_4 &= \ \ 12 \\
6x_1 + 3x_2 + \ x_3 + 4x_4 &= \ \ \ 5 \\
9x_1 + 5x_2 - 4x_3 + 9x_4 &= -28.
\end{aligned}$$

Find L and U (as in (8.30)) such that $L^{-1}U$ is the coefficient matrix of the above system.

10. (a) Show that the invertible $n \times n$ matrix A over F can be expressed in at most one way as $A = L^{-1}U$, where L and U are respectively lower and upper triangular $n \times n$ matrices, the (i, i)-entries in L being 1.

(b) Let A be an $n \times n$ matrix over F. Denote by A_r the $r \times r$ matrix which remains on deleting the last $n - r$ rows and last $n - r$ columns of A. Show that $A = L^{-1}U$, where L (and U) are invertible lower (and upper) triangular matrices over F, if and only if A_r is invertible for all r with $1 \leqslant r \leqslant n$.

Equivalence of matrices

We have deliberately concentrated on row operations in the previous sections to get the reader thoroughly used to them. However, it is now time to mention **elementary column operations**, that is, those operations obtained on replacing 'row' in (8.16) by 'column'. There is no need to go into great detail, as row operations and column operations are related to each other by matrix transposition; the transpose of an elementary matrix is again elementary (hence (8.17) is not truly biased towards rows), and since transposition reverses the order of a matrix product (see (7.11)(b)), we see that:

Postmultiplication by an elementary matrix performs the corresponding elementary *column* operation.

For example $c_2 + 3c_1$ denotes the elementary column operation: to column 2 add $3 \times$ (column 1). Applying this operation to the 2×2 identity matrix I produces the corresponding elementary matrix, namely $Q = \begin{pmatrix} 1 & 3 \\ 0 & 1 \end{pmatrix}$. Postmultiplying any $m \times 2$ matrix by Q carries out the operation $c_2 + 3c_1$:

$$\begin{pmatrix} a & b \\ c & d \\ e & f \end{pmatrix} \begin{pmatrix} 1 & 3 \\ 0 & 1 \end{pmatrix} = \begin{pmatrix} a & 3a + b \\ c & 3c + d \\ e & 3e + f \end{pmatrix}.$$

Applying matrix transposition to the theory of row-equivalence, one obtains the corresponding theory of column-equivalence: the $m \times n$ matrices A and B over F are called **column-equivalent** if there is an invertible $n \times n$ matrix Q over F with $A = BQ$; transposing (8.15), we see that A and B are column-equivalent if and only if their column spaces are equal. The reader who wants to be completely ambidextrous must learn to column-reduce matrices (this is no bad thing—for instance, it is more natural to use column reduction in (8.19)(b) and avoid matrix transposition).

Here we concern ourselves with a relatively crude type of equivalence. How simple can a matrix be made if *both* row and column operations are allowed? The answer is: very simple indeed! In fact we shall see that the rank of the matrix tells us all there is to know in this context (8.34).

Definition 8.31   The $m \times n$ matrices A and B over the field F are called **equivalent** if there are invertible matrices P and Q over F such that

$$PAQ^{-1} = B.$$

Taking $Q = I$ we see that row-equivalent matrices are equivalent; so row-equivalence is more refined (and its resolution more delicate) than equivalence as defined in (8.31). Similarly, taking $P = I$ in (8.31) we see that column-equivalent matrices are equivalent.

Our next proposition shows how matrix equivalence arises, namely by observing the change in the matrix of a given linear mapping resulting from a change of bases.

Proposition 8.32 Let the linear mapping $\alpha : V \to V'$ have matrix A relative to bases b and b' of V and V'. Then α has an equivalent matrix PAQ^{-1} relative to any other pair of bases b_1 and b'_1 of V and V'.

Proof Let ι_V (the identity mapping of V) have matrix P relative to the bases b_1 and b; let $\iota_{V'}$ (the identity mapping of V') have matrix Q relative to the bases b'_1 and b' of V'. As ι_V and $\iota_{V'}$ are invertible, P and Q are invertible also, by (8.13), and $\iota_{V'}$ $(= \iota_{V'}^{-1})$ has matrix Q^{-1} relative to b' and b'_1; therefore α $(= \iota_V \alpha \iota_{V'})$ has matrix PAQ^{-1} relative to b_1 and b'_1 by (7.27). □

So changing the bases of V and V' results in the matrix of α changing from A to PAQ^{-1} (see (7.28)). Incidentally, taking $b'_1 = b'$ (and hence $Q = I$), we see that changing the basis of V and keeping fixed the basis of V', results in the matrix of α changing from A to the row-equivalent matrix PA.

Notation Let I_r denote the $m \times n$ matrix having e_i (the i th vector in the standard basis of F^n) as row i $(1 \leq i \leq r)$, the remaining $m - r$ rows being zero.

The shape of I_r should be clear from the context; in the case of 2×3 matrices:

$$I_0 = \begin{pmatrix} 0 & 0 & 0 \\ 0 & 0 & 0 \end{pmatrix}, \qquad I_1 = \begin{pmatrix} 1 & 0 & 0 \\ 0 & 0 & 0 \end{pmatrix}, \qquad I_2 = \begin{pmatrix} 1 & 0 & 0 \\ 0 & 1 & 0 \end{pmatrix}.$$

Notice that I_r has rank r; in fact I_r is the *simplest* $m \times n$ matrix of rank r over F. At this point the reader should review (7.23) and (7.28), for our next proposition generalizes the procedure used there.

Proposition 8.33 Let V and V' be vector spaces of dimensions m and n over F and let $\alpha : V \to V'$ be a linear mapping of rank r. Then there are bases b and b' of V and V' relative to which α has the $m \times n$ matrix I_r.

Proof We use (8.3). Let U be a complement of $\ker \alpha$ in V. Then $\dim U = r$. Let U have basis $v_1, v_2, \ldots, v_r$ and let $\ker \alpha$ have basis $v_{r+1}, \ldots, v_m$. Then $v_1, v_2, \ldots, v_r, v_{r+1}, \ldots, v_m$ is a basis b

of V by (6.29). Write $v_i' = (v_i)\alpha$ $(1 \leqslant i \leqslant r)$; then $v_1', v_2', \ldots, v_r'$ form a basis of $\operatorname{im} \alpha$ by (8.3). Extend these vectors to $v_1', v_2', \ldots, v_r', v_{r+1}', \ldots, v_n'$ forming a basis ℓ' of V'. The images by α of the vectors in ℓ are related to the vectors in ℓ' in a very simple way:

$$(v_i)\alpha = v_i' \quad (1 \leqslant i \leqslant r), \qquad (v_i)\alpha = 0 \quad (r < i \leqslant m),$$

as $v_{r+1}, \ldots, v_m$ belong to $\ker \alpha$. The reader should write out these m equations in an array to see that α has matrix I_r relative to ℓ and ℓ' by (7.22). $\qquad \square$

We now put (8.32) and (8.33) together and extract the essence of matrix equivalence.

Theorem 8.34

Let A be an $m \times n$ matrix of rank r over the field F. Then A is equivalent to I_r. Two $m \times n$ matrices over F are equivalent if and only if they have the same rank.

Proof

We apply (8.32) and (8.33) to the linear mapping $\mu_A : F^m \to F^n$ of (7.13). As μ_A has matrix A relative to the standard bases of F^m and F^n, we see that μ_A has the equivalent matrix PAQ^{-1} relative to bases ℓ and ℓ' of F^m and F^n. Choosing ℓ and ℓ' in the 'clever' way (8.33) gives $PAQ^{-1} = I_r$.

Multiplication by invertible matrices does not change rank; therefore $\operatorname{rank}(PAQ^{-1}) = \operatorname{rank}(AQ^{-1}) = \operatorname{rank} A$, showing that equivalent matrices have equal rank. Conversely let A and B be $m \times n$ matrices over F with $\operatorname{rank} A = \operatorname{rank} B = r$ say. Then there are invertible matrices P, P_1, Q, Q_1 over F such that $PAQ^{-1} = I_r = P_1 B Q_1^{-1}$, since A and B are both equivalent to I_r. Therefore $P_2 A Q_2^{-1} = B$ where $P_2 = P_1^{-1}P$, $Q_2 = Q_1^{-1}Q$. As P_2 and Q_2 are invertible over F, we see that A and B are equivalent. $\qquad \square$

Equivalence (8.31) of matrices is, as the terminology suggests, an equivalence relation on the set $^mF^n$ of all $m \times n$ matrices over F. From (8.34) we see that the concept of rank is enough to determine equivalence; as the rank r of $m \times n$ matrices can take all integer values in the range $0 \leqslant r \leqslant \min\{m, n\}$, there are $1 + \min\{m, n\}$ equivalence classes of matrices in $^mF^n$.

Given an $m \times n$ matrix A over a field F, how can invertible matrices P and Q over F be found such that $PAQ^{-1} = I_r$? The technique of the previous section can be adapted to give the method. Form the $m \times (n + m)$ matrix $[A : I]$ and apply row operations to produce a matrix $[E' : P]$ such that the last $m - r$ rows of E' are zero (for instance E' could be E, the row-reduced echelon form of A, although it is not necessary to row reduce as far as E). Now $P[A : I] = [E' : P]$ and so $PA = E'$. To find Q notice that $PAQ^{-1} = E'Q^{-1} = I$, gives $E' = I,Q$ and so

the first r rows of E' agree with the first r rows of Q.

To form the complete invertible matrix Q, extend the first r rows of E' (which are linearly independent vectors) to a basis of F^n, and take Q to be the matrix (invertible by (8.12)) having the vectors of this basis as its rows. Then $PAQ^{-1} = I,$. One final comment: if $E' = E$ (the row-reduced echelon form of A), then the last $n - r$ rows of Q can be taken to be the standard basis vectors e_j of F^n where $j \neq j_1, j_2, \ldots, j_r$, because these $n - r$ vectors form a basis of a complement of row space E in F^n.

For example, let A be the 3×3 matrix of (8.25). Then

$$PA = \begin{pmatrix} 1 & 0 & 2 \\ 0 & 1 & 1 \\ 0 & 0 & 0 \end{pmatrix} \text{ and so } Q = \begin{pmatrix} 1 & 0 & 2 \\ 0 & 1 & 1 \\ 0 & 0 & 1 \end{pmatrix} \text{ satisfies } PAQ^{-1} = I_2.$$

As another illustration, consider the 3×4 matrix A of (8.28)(b). Since

$$PA = \begin{pmatrix} 1 & 0 & -3 & 4 \\ 0 & 1 & 2 & -1 \\ 0 & 0 & 0 & 1 \end{pmatrix},$$

we see $Q = \begin{pmatrix} 1 & 0 & -3 & 4 \\ 0 & 1 & 2 & -1 \\ 0 & 0 & 0 & 1 \\ 0 & 0 & 1 & 0 \end{pmatrix}$ satisfies $PAQ^{-1} = I_3$.

Exercises 8.4

1. For each of the following matrices A, determine invertible matrices P and Q over the rational field such that $PAQ^{-1} = I,$.

$$\begin{pmatrix} 1 & 2 & 3 \\ 2 & 4 & 6 \end{pmatrix}, \quad \begin{pmatrix} 1 & -2 & 3 \\ -2 & 4 & 4 \\ -1 & 2 & 5 \end{pmatrix}, \quad \begin{pmatrix} 1 & 2 & 1 & 7 \\ 2 & 4 & -1 & 2 \\ 3 & 6 & -2 & 1 \end{pmatrix}.$$

2. By column reducing the matrix with rows $r_1 = (1, 1, 1)$, $r_2 = (1, 2, 1)$, $r_3 = (2, 1, 4)$, $r_4 = (3, 2, 1)$, find a linear dependence relation between these vectors of $\mathbb{Q}^3$.

3. Let A be an $n \times n$ matrix over a field. Using $PAQ^{-1} = I,$, show that A is row-equivalent to a symmetric matrix.

Find an invertible matrix R over the rational field such that RA is symmetric if

$$A = \begin{pmatrix} 1 & 2 & -1 \\ 3 & 1 & 2 \\ -3 & 4 & -7 \end{pmatrix}.$$

4. Let E be an $m \times n$ matrix over F and suppose that E is in row-reduced echelon form (8.21). Let U be the subspace spanned by the $n - r$ vectors e_j of the standard basis of F^n where $r = \text{rank } E$ and $j \neq j_1, j_2, \ldots, j_r$. Prove that $U \cap (\text{row space } E) = 0$ and deduce that U is a complement of row space E in F^n. Find a complement of column space E in $^m F$.

5. Let A be an $m \times n$ matrix over F and let P and Q be invertible matrices such that $PAQ^{-1} = I_r$. Show that the last $m - r$ rows of P form a basis of $\ker \mu_A$. Show that the last $n - r$ columns of Q^{-1} form a basis of the solution space of the system $Ax = 0$.

6. (a) Show that there are five classes of column-equivalent 2×3 matrices over $\mathbb{Z}_2$. How many row-equivalence classes of such matrices are there?

(b) The 2×3 matrices A and B over $\mathbb{Z}_2$ are row-equivalent and column-equivalent. Are A and B necessarily equal?

(c) The $m \times n$ matrices A and B of rank 1 over F are row-equivalent and column-equivalent. Prove that A is a non-zero scalar multiple of B.

7. (a) Let A and B be $m \times n$ matrices over F. Show that A and B are column-equivalent if and only if $\ker \mu_A = \ker \mu_B$.

(b) Let A be an $m \times n$ matrix of rank r over a finite field F. Show that the number of matrices which are column-equivalent to A is $(q^n - 1)(q^n - q) \ldots (q^n - q^{r-1})$ where $|F| = q$. How many matrices are row-equivalent to A? Derive a formula for

(i) the number of $m \times n$ matrices of rank r over F,
(ii) the number of matrices which are row-equivalent and column-equivalent to A. (Hint: consider the case $A = I_r$.)

9 Groups and determinants

Here we digress from linear algebra in order to discuss the elementary theory of groups. A **group** is a system consisting of a set together with a *single* associative binary operation (usually thought of as multiplication) having an identity element and such that each element has an inverse within the system. For instance, the non-zero elements of the field $\mathbb{Z}_5$ form a group with multiplication table:

$\times$	$\bar{1}$	$\bar{2}$	$\bar{3}$	$\bar{4}$
$\bar{1}$	$\bar{1}$	$\bar{2}$	$\bar{3}$	$\bar{4}$
$\bar{2}$	$\bar{2}$	$\bar{4}$	$\bar{1}$	$\bar{3}$
$\bar{3}$	$\bar{3}$	$\bar{1}$	$\bar{4}$	$\bar{2}$
$\bar{4}$	$\bar{4}$	$\bar{3}$	$\bar{2}$	$\bar{1}$

In this case $\mathbb{Z}_5^* = \{\bar{1}, \bar{2}, \bar{3}, \bar{4}\}$ is the set of group elements, $\bar{1}$ is the identity element, $\bar{1}$ and $\bar{4}$ are self-inverse, while $\bar{2}$ and $\bar{3}$ are inverses of each other. Notice that the product of each pair of group elements is itself a group element, and the inverse of each group element is again a group element.

The concept of a group has an immediate unifying influence over the systems already familiar to us, and, with the applications of Lagrange's theorem (9.19), this is enough to justify its introduction. However, group theory is grounded in the study of **permutation groups** (9.11) and it is the usefulness of such concrete groups which motivates much of the abstract theory.

Later in the chapter we return to matrix theory and, using parity of permutations, introduce determinants. Suppose A is a square matrix over a field F; we shall see that A is invertible over F if and only if a certain element $|A|$ (the **determinant** of A) of F is non-zero. The task of inverting large matrices is not made any easier by this surprising property; nevertheless it will give us a clearer picture of the inversion process and help our work on diagonalization in Chapter 10.

Groups

Let G be a set with binary operation $\mu : G \times G \to G$. Generally we interpret μ as 'multiplication' on G, writing xy in place of $(x, y)\mu$. With this convention G is closed under multiplication, as $xy \in G$ for all $x, y \in G$. The system consisting of G and the operation of multiplication on G is denoted by $(G, \times)$, or simply by G (taking the binary operation for granted).

Definition
9.1

The system $(G, \times)$ is called a **group** if the following laws hold:
 1. **Associative law:** $(xy)z = x(yz)$ for all $x, y, z \in G$.
 2. **Existence of identity:** there is an element e in G satisfying $ex = x$ for all $x \in G$.
 3. **Existence of inverses:** for each x in G there is x^{-1} in G satisfying $x^{-1}x = e$.

Notice that the identity element e and x^{-1} appear on the *left* of x in their defining equations above. We begin by redressing the balance: although (9.1) is biased to the left, groups themselves are unbiased!

Proposition
9.2

Let $(G, \times)$ be a group. The identity element e of G is unique and satisfies $xe = x$ for all $x \in G$. The inverse x^{-1} of each element x in G is unique and satisfies $xx^{-1} = e$.

Proof

By law 3 of (9.1) with x replaced by $y = x^{-1}$, there is y^{-1} in G with $y^{-1}y = e$. As $yx = e$ we obtain

$$xx^{-1} = e(xx^{-1}) = (y^{-1}y)(xx^{-1}) = y^{-1}(y(xx^{-1}))$$
$$= y^{-1}((yx)x^{-1}) = y^{-1}(ex^{-1}) = y^{-1}x^{-1} = y^{-1}y = e$$

using the group laws (9.1). Therefore x^{-1} is a two-sided inverse of x, that is, $x^{-1}x = e = xx^{-1}$; hence x^{-1} is unique (see (2.11) ff.). As $xe = x(x^{-1}x) = (xx^{-1})x = ex = x$, we see that e is a two-sided identity, that is, $ex = x = xe$ for all $x \in G$; hence e is unique (as in (2.8)(f)). $\qquad \square$

We now discuss three types of group: symmetric groups, groups of units, and additive groups; these, together with their subgroups (9.9), are the main sources of groups.

Symmetric groups

Let X be a non-empty set and let G denote the set of all bijections $\alpha : X \to X$. As the composition of compatible bijections is itself a bijection, we see that $\alpha\beta \in G$ for all $\alpha, \beta \in G$. In fact G, together with the binary operation of composition, is a group, for the laws (9.1) may be verified:

1. $(\alpha\beta)\gamma = \alpha(\beta\gamma)$ for all $\alpha, \beta, \gamma \in G$, since composition is associative by (1.19).

2. The identity mapping ι of X is the identity element of G, as ι belongs to G and satisfies $\iota\alpha = \alpha$ for all $\alpha \in G$.

3. For each α in G, the inverse bijection α^{-1} belongs to G and satisfies $\alpha^{-1}\alpha = \iota$.

Definition 9.3

The elements of the above group G are called **permutations** of X. The group G itself is called the **symmetric group** on X and denoted by $S(X)$. If $X = \{1, 2, \ldots, n\}$, then G is denoted by S_n and called the **symmetric group** of degree n.

We shall be particularly interested in S_n, the group of all permutations α of the first n natural numbers.

Notation

For α in S_n, write

$$\alpha = \begin{pmatrix} 1 & 2 & \ldots & 3 \\ (1)\alpha & (2)\alpha & \ldots & (3)\alpha \end{pmatrix}.$$

In this notation, the symbols in the bottom row are merely a rearrangement of the integers $1, 2, \ldots, n$; the image by α of each integer in the top row is the integer directly below it in the bottom row.

Example 9.4

Consider the symmetric group S_3 of all permutations of $\{1, 2, 3\}$.

$$\alpha = \begin{pmatrix} 1 & 2 & 3 \\ 2 & 3 & 1 \end{pmatrix}$$

stands for the element α of S_3 with $(1)\alpha = 2$, $(2)\alpha = 3$, $(3)\alpha = 1$. Remembering that the group operation is composition, we see

$$\alpha^2 = \begin{pmatrix} 1 & 2 & 3 \\ 3 & 1 & 2 \end{pmatrix}$$

for $(1)\alpha^2 = 3$, $(2)\alpha^2 = 1$, $(3)\alpha^2 = 2$. Similarly

$$\alpha^3 = \begin{pmatrix} 1 & 2 & 3 \\ 1 & 2 & 3 \end{pmatrix}$$

showing that α^3 is the identity mapping of $\{1, 2, 3\}$, and so we write $\alpha^3 = e$. Let β denote the permutation of $\{1, 2, 3\}$ which

Table 9.1

×	e	α	α^2	β	$\alpha\beta$	$\alpha^2\beta$
e	e	α	α^2	β	$\alpha\beta$	$\alpha^2\beta$
α	α	α^2	e	$\alpha\beta$	$\alpha^2\beta$	β
α^2	α^2	e	α	$\alpha^2\beta$	β	$\alpha\beta$
β	β	$\alpha^2\beta$	$\alpha\beta$	e	α^2	α
$\alpha\beta$	$\alpha\beta$	β	$\alpha^2\beta$	α	e	α^2
$\alpha^2\beta$	$\alpha^2\beta$	$\alpha\beta$	β	α^2	α	e

interchanges 1 and 2 while fixing 3, that is,

$$\beta = \begin{pmatrix} 1 & 2 & 3 \\ 2 & 1 & 3 \end{pmatrix}.$$

Then $\beta^2 = e$, and composing α with β gives

$$\alpha\beta = \begin{pmatrix} 1 & 2 & 3 \\ 1 & 3 & 2 \end{pmatrix}.$$

Notice that

$$\alpha^2\beta = \begin{pmatrix} 1 & 2 & 3 \\ 3 & 2 & 1 \end{pmatrix} = \beta\alpha.$$

We have expressed the six permutations of $\{1, 2, 3\}$ in terms of α and β. The multiplication table (Table 9.1) of S_3 can now be written out. Notice that each row (except the rows beginning with e and β) is the row above premultiplied by α, and so the table may be filled in row by row. However, the row beginning with β requires some calculation using the relations $\alpha^3 = \beta^2 = e$, $\beta\alpha = \alpha^2\beta$; for instance

$$\beta\alpha^2 = \beta\alpha\alpha = \alpha^2\beta\alpha = \alpha^2\alpha^2\beta = \alpha^3\alpha\beta = e\alpha\beta = \alpha\beta.$$

The above 6×6 table forms a **Latin square**, that is, each group element occurs exactly once in each row and column of the table (for in an equation $xy = z$ between group elements, x and z determine y ($= x^{-1}z$); similarly y and z determine x ($= zy^{-1}$)).

As with matrices, the unconscious use of the commutative law should be avoided, because generally groups are non-commutative; the above example S_3 is non-commutative as $\alpha\beta \neq \beta\alpha$. Those groups which are commutative have a special name (after the mathematician Abel).

Definition 9.5 A group G is called **abelian** (or **commutative**) if $xy = yx$ for all $x, y \in G$.

So S_3 is non-abelian, but the symmetric group S_2 (it has only two elements) is abelian; $\mathbb{Z}_5^*$, mentioned in the introduction, is abelian.

Returning to the symmetric group S_n, we use the symbol

$$\begin{pmatrix} 1 & 2 & \cdots & n \\ (1)\alpha & (2)\alpha & \cdots & (n)\alpha \end{pmatrix}$$

with its columns in *any* order to specify α. For instance

$$\begin{pmatrix} 1 & 2 & 3 \\ 3 & 1 & 2 \end{pmatrix} = \begin{pmatrix} 3 & 2 & 1 \\ 2 & 1 & 3 \end{pmatrix}$$

as both symbols denote the permutation $1 \to 3$, $2 \to 1$, $3 \to 2$. When working out a composite permutation, the course of each integer must be traced (from left to right); this can be done by matching the top row of the second symbol with the bottom row of the first symbol. For example,

$$\begin{pmatrix} 1 & 2 & 3 & 4 \\ 2 & 3 & 4 & 1 \end{pmatrix} \begin{pmatrix} 1 & 2 & 3 & 4 \\ 2 & 1 & 4 & 3 \end{pmatrix}$$

$$= \begin{pmatrix} 1 & 2 & 3 & 4 \\ 2 & 3 & 4 & 1 \end{pmatrix} \begin{pmatrix} 2 & 3 & 4 & 1 \\ 1 & 4 & 3 & 2 \end{pmatrix} = \begin{pmatrix} 1 & 2 & 3 & 4 \\ 1 & 4 & 3 & 2 \end{pmatrix}$$

on rearranging the columms of the second symbol.

Notice that

$$\alpha^{-1} = \begin{pmatrix} (1)\alpha & (2)\alpha & \cdots & (n)\alpha \\ 1 & 2 & \cdots & n \end{pmatrix},$$

that is, interchanging the rows of the symbol (but maintaining vertical alignments) produces the inverse permutation:

$$\text{if } \alpha = \begin{pmatrix} 1 & 2 & 3 & 4 \\ 2 & 3 & 4 & 1 \end{pmatrix}, \text{ then } \alpha^{-1} = \begin{pmatrix} 2 & 3 & 4 & 1 \\ 1 & 2 & 3 & 4 \end{pmatrix}.$$

Finally we count the number of permutations α in S_n: there are n choices for $(1)\alpha$, so $n-1$ choices remain for $(2)\alpha$, leaving $n-2$ choices for $(3)\alpha$, and so on. So in all, there are $n(n-1)(n-2)\ldots(2)(1)$ permutations in S_n, that is,

$$|S_n| = n!.$$

Groups of units

Let R be a ring with 1-element e, and let G denote the set of units (2.11) of R; so G consists of those elements of R which have

Table 9.2. The group
of units of $\mathbb{Z}_8$

×	$\bar{1}$	$\bar{3}$	$\bar{5}$	$\bar{7}$
$\bar{1}$	$\bar{1}$	$\bar{3}$	$\bar{5}$	$\bar{7}$
$\bar{3}$	$\bar{3}$	$\bar{1}$	$\bar{7}$	$\bar{5}$
$\bar{5}$	$\bar{5}$	$\bar{7}$	$\bar{1}$	$\bar{3}$
$\bar{7}$	$\bar{7}$	$\bar{5}$	$\bar{3}$	$\bar{1}$

(two-sided) inverses in R. For x and y in G we have

$$(y^{-1}x^{-1})(xy) = e = (xy)(y^{-1}x^{-1})$$

showing that xy has inverse $(xy)^{-1} = y^{-1}x^{-1} \in R$; therefore $xy \in G$. In fact G is a group, the group operation being the restriction to G of the multiplication on R, for the laws (9.1) may be verified:

1. $(xy)z = x(yz)$ for all x, y, z in G, as multiplication on R is associative.

2. The identity element of G is the 1-element e of R.

3. For each x in G, the equations $x^{-1}x = e = xx^{-1}$ tell us that x^{-1} belongs to G.

We therefore refer to the **group of units** of the given ring. Taking $R = \mathbb{Z}$, we see $G = \{1, -1\}$. Taking $R = \mathbb{Z}_8$ gives $G = \{\bar{1}, \bar{3}, \bar{5}, \bar{7}\}$ and from the multiplication table (Table 9.2) of G, we see that each element of this group is self-inverse. More generally, the group of units of $\mathbb{Z}_n$ has $\phi(n)$ elements, where ϕ is Euler's function (see (5.26)).

As every non-zero element of a field has an inverse, we see that F^* is the group of units of the field F; the group F^* is abelian and called the **multiplicative group** of the field F. In particular $\mathbb{Z}_p^*$ (p prime) is a group with $p - 1$ elements, namely $\bar{1}, \bar{2}, \ldots, \overline{p-1}$.

The group of invertible $n \times n$ matrices over the field F (that is, the group of units of the ring $M_n(F)$ of all $n \times n$ matrices over F—see (7.9)), is called the **general linear group** $GL_n(F)$; this is the group of transition matrices (8.13) which relate bases of an n-dimensional vector space over F to each other. In particular—see (8.12)—each matrix P in $GL_n(F)$ corresponds to the basis ℓ of F^n consisting of the rows of P. Following (6.25) we found a formula for the number of bases of F^n where F is a finite field; therefore

$$|GL_n(F)| = (q^n - 1)(q^n - q) \ldots (q^n - q^{n-1}) \quad \text{where} \quad |F| = q$$

since the number of invertible $n \times n$ matrices over F is equal to the number of bases of F^n. For example, $GL_2(\mathbb{Z}_2)$ consists of the six invertible 2×2 matrices over $\mathbb{Z}_2$:

$$\begin{pmatrix} 1 & 0 \\ 0 & 1 \end{pmatrix}, \begin{pmatrix} 0 & 1 \\ 1 & 1 \end{pmatrix}, \begin{pmatrix} 1 & 1 \\ 1 & 0 \end{pmatrix}, \begin{pmatrix} 0 & 1 \\ 1 & 0 \end{pmatrix}, \begin{pmatrix} 1 & 0 \\ 1 & 1 \end{pmatrix}, \begin{pmatrix} 1 & 1 \\ 0 & 1 \end{pmatrix},$$

corresponding to the six bases of $\mathbb{Z}_2^2$:

$$e_1, e_2; \quad e_2, e_1 + e_2; \quad e_1 + e_2, e_1;$$

$$e_2, e_1; \quad e_1, e_1 + e_2; \quad e_1 + e_2, e_2.$$

Additive groups

Some abelian groups arise in additive (rather than multiplicative) notation, and we now restate (9.1) and (9.5) to help recognize such groups. As before, let G be a set and $\mu : G \times G \to G$ a binary operation on G. In place of $(x, y)\mu$ we write $x + y$, that is, μ is interpreted as 'addition' on G. As $x + y \in G$ for all $x, y \in G$, we see G is closed under addition. The system $(G, +)$, or simply G, is an abelian group if the following laws hold:

1. $(x + y) + z = x + (y + z)$ for all $x, y, z \in G$.
2. There is an element 0 in G satisfying $0 + x = x$ for all $x \in G$.
3. For each x in G there is $-x$ in G such that $-x + x = 0$.
4. $x + y = y + x$ for all $x, y \in G$.

The reader will realize that these laws are nothing new (in fact they are making their fourth debut!): they are the ring laws 1–4 of (2.2), the vector space laws 1–4 of (6.1), and the abelian group laws— (9.1) with (9.5)—in additive notation. Let $(R, +, \times)$ be a ring; ignoring the ring multiplication, we obtain an abelian group $(R, +)$ called the **additive group** of R. In particular $\mathbb{Z}, \mathbb{Q}, \mathbb{R}, \mathbb{C}$ are abelian groups, the group operation being addition. Let V be a vector space; in a similar way (ignoring scalar multiplication of vectors), we obtain the additive group $(V, +)$ of V. In particular $^m F^n$ (the set of all $m \times n$ matrices over F) is an abelian group, the group operation being matrix addition.

Returning to multiplicative notation, we introduce structure-preserving mappings of groups (cf. (5.1), (7.12)).

Definition 9.6   Let G and G' be groups. A mapping $\alpha : G \to G'$ is called a **(group) homomorphism** if $(xy)\alpha = ((x)\alpha)((y)\alpha)$ for all $x, y \in G$.

The modulus (2.24) of complex numbers provides us with an accessible example of a group homomorphism, namely $\alpha : \mathbb{C}^* \to \mathbb{R}^*$

defined by $(z)\alpha = |z|$; using the multiplicative property of the modulus, α satisfies (9.6).

Each ring homomorphism (5.1) gives rise (on ignoring part of the ring structure) to a group homomorphism of additive groups of rings and a group homomorphism of groups of units of rings. For instance, the natural ring homomorphism $\eta : \mathbb{Z} \to \mathbb{Z}_8$ gives (on ignoring multiplication) a group homomorphism of the additive group of $\mathbb{Z}$ to the additive group of $\mathbb{Z}_8$; ignoring non-units, we obtain the group homomorphism $\{1, -1\} \to \{\bar{1}, \bar{3}, \bar{5}, \bar{7}\}$ where $1 \to (1)\eta = \bar{1}$ and $-1 \to (-1)\eta = -\bar{1} = \bar{7}$. Similarly, each linear mapping (7.12) gives rise to a group homomorphism of additive groups of vector spaces.

The reader may wonder why the condition $(e)\alpha = e'$ (α maps the identity element of G to the identity element of G') is not required of a group homomorphism α; the reason is that α has this property automatically. For if z in G' satisfies $z^2 = z$, multiplying by z^{-1} gives $z = e'$; taking $z = (e)\alpha$, we see $z^2 = ((e)\alpha)((e)\alpha) = (e^2)\alpha = (e)\alpha = z$ on applying (9.6) with $x = y = e$, and so $(e)\alpha = e'$.

It now follows that the group homomorphism α of (9.6) satisfies $(x^{-1})\alpha = ((x)\alpha)^{-1}$ for all $x \in G$. The composition of compatible group homomorphisms is itself a group homomorphism; the inverse of a bijective group homomorphism is itself a group homomorphism (cf. (5.2), (7.15)(d)).

Definition 9.7

A bijective group homomorphism $\alpha : G \to G'$ is called a (**group**) **isomorphism** and denoted by $\alpha : G \cong G'$. Two groups are called **isomorphic** if there is a group isomorphism between them.

The reader should be driven to compare (9.7) with (5.3) and (7.18): in each case, isomorphisms are structure-preserving mappings having structure-preserving inverses.

Isomorphic groups (that is, abstractly identical groups) arise in widely different contexts. For example let G denote the additive group of the vector space V over $\mathbb{Z}_2$ with basis u, v. Let G' denote the group of units of $\mathbb{Z}_{12}$. The group tables (Table 9.3) of G and G' both have the same pattern showing that these groups are isomorphic; in fact $\alpha : G \cong G'$, defined by $(0)\alpha = \bar{1}$, $(u)\alpha = \bar{5}$, $(v)\alpha = \bar{7}$, $(u + v)\alpha = \overline{11}$, is an isomorphism, since α matches the elements of G with those of G' so that addition on G matches multiplication on G'. Table 9.2 has the same pattern also, and so the group of units of $\mathbb{Z}_8$ is isomorphic to the group of units of $\mathbb{Z}_{12}$. On the other hand, $\mathbb{Z}_5^*$ (having elements which are not self-inverse) is not isomorphic to the group of units of $\mathbb{Z}_8$ (nor to G or G').

Table 9.3

+	0	u	v	$u+v$
0	0	u	v	$u+v$
u	u	0	$u+v$	v
v	v	$u+v$	0	u
$u+v$	$u+v$	v	u	0

$\times$	$\bar{1}$	$\bar{5}$	$\bar{7}$	$\overline{11}$
$\bar{1}$	$\bar{1}$	$\bar{5}$	$\bar{7}$	$\overline{11}$
$\bar{5}$	$\bar{5}$	$\bar{1}$	$\overline{11}$	$\bar{7}$
$\bar{7}$	$\bar{7}$	$\overline{11}$	$\bar{1}$	$\bar{5}$
$\overline{11}$	$\overline{11}$	$\bar{7}$	$\bar{5}$	$\bar{1}$

Definition 9.8 A group with only a finite number of elements is called a **finite group**. The finite group G is said to have **order** $|G|$.

The order of a finite group is the number of its elements; therefore S_3 has order six, and, more generally S_n has order $n!$. Group theory generally is an extremely deep and difficult subject— we shall merely touch on a few aspects—and many questions about groups remain unresolved. For example the number of isomorphism classes of groups of order n is unknown for quite modest natural numbers n.

We now have a nodding acquaintance with a number of groups; however, many more are about to make their entrance! For just as vector spaces arise as subspaces of standard spaces, so many groups arise as subgroups of the groups we have already met; but there is a significant difference in that apparently harmless groups often have very intricate subgroups.

Let H be a subset of the group G and suppose that H is closed under multiplication: $xy \in H$ for all $x, y \in H$. Then the restriction to $H \times H$ of the group operation on G is a binary operation on H, called the **inherited operation** on H (see (5.6), (6.5)).

Definition 9.9 Let H be a subset of G and suppose that H is closed under multiplication. If H, with the inherited operation from G, is a group, then H is called a **subgroup** of G.

A subgroup is therefore a part of a group which is itself a group. The reader may compare our next proposition with (5.7) and (6.6).

Proposition 9.10 (Criterion for a subgroup.) Let H be a subset of the group G. Then H is a subgroup of G if and only if

(i) $xy \in H$ for all $x, y \in H$ (H is closed under the operation on G),

(ii) $e \in H$ (H contains the identity element of G),

(iii) $x^{-1} \in H$ for all $x \in H$ (H is closed under inversion).

Proof   Suppose that H is a subgroup of G. Then H satisfies (i) by (9.9), and being a group with operation inherited from G, by (9.1) law 2 there is e_1 in H satisfying $e_1^2 = e_1$; as e is the only element in G such that $x^2 = x$, we see $e = e_1 \in H$. So H satisfies (ii). By (9.1) law 3, for each x in H there is y, also in H, with $yx = e$; post-multiplying by x^{-1} gives $y = x^{-1}$ by (9.2). Therefore H satisfies (iii) as $x^{-1} = y \in H$.

Conversely, suppose H satisfies (i), (ii), (iii). We verify that the laws (9.1) hold in H using the operation inherited from G: law 1 holds in H as it holds in the parent group G; laws 2 and 3 hold in H as H satisfies conditions (ii) and (iii). Therefore H is a subgroup of G by (9.9).   □

For example, take $G = \mathbb{C}^*$ and $H = \{z \in \mathbb{C} : |z| = 1\}$. So the group of non-zero complex numbers is the parent group, and H is the subset of complex numbers with modulus 1. Is H a subgroup of G in this case? In other words, does H satisfy the conditions of (9.10)? The product of complex numbers of modulus 1 is itself of modulus 1, the complex number 1 (the identity element of $\mathbb{C}^*$) has modulus 1, and the inverse of a complex number of modulus 1 is itself of modulus 1. Therefore H does satisfy the conditions of (9.10): the complex numbers of modulus 1 form a subgroup of $\mathbb{C}^*$.

Definition 9.11   Let X be a non-empty set. Any subgroup of the symmetric group $S(X)$ is called a **permutation group** on X.

For example, let H denote the set of bijections $\alpha : \mathbb{R} \to \mathbb{R}$ which are **order-preserving** (that is, $x < y \Leftrightarrow (x)\alpha < (y)\alpha$); then H satisfies the conditions of (9.10) and so is a subgroup of $S(\mathbb{R})$. The automorphisms (5.5) of the ring R form a group, for they are the permutations of R which preserve the ring structure; indeed every algebraic structure gives rise to a group (its automorphism group) in this way.

Example 9.12   Consider a square object with its corners numbered as shown;

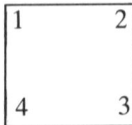

so 1 and 3 label opposite corners as do 2 and 4.

Let H denote the permutations of $\{1, 2, 3, 4\}$ which arise on picking up the object and replacing it in any way to cover the original area. For instance

$$\alpha = \begin{pmatrix} 1 & 2 & 3 & 4 \\ 2 & 3 & 4 & 1 \end{pmatrix}$$

belongs to H, for α arises on rotating the square clockwise through $\pi/2$. Similarly α^2, α^3, $\alpha^4 = e$, arise from rotations through π, $3\pi/2$, 2π (which means replacing the square exactly as we found it). The square can also be 'flipped' vertically, horizontally and diagonally, producing

$$\beta = \begin{pmatrix} 1 & 2 & 3 & 4 \\ 2 & 1 & 4 & 3 \end{pmatrix}, \ \alpha\beta, \ \alpha^2\beta, \ \alpha^3\beta.$$

In fact

$$H = \{e, \ \alpha, \ \alpha^2, \ \alpha^3, \ \beta, \ \alpha\beta, \ \alpha^2\beta, \ \alpha^3\beta\}$$

is a subgroup of S_4 (the group of all 24 permutations of $\{1, 2, 3, 4\}$). As H arises from the transformations of $\mathbb{R}^3$ which physically preserve the square, H is called (a **representation** of) the **symmetry group** of the square. In a similar way, every geometric object has its group of symmetries—the more symmetric the object, the larger its symmetry group.

Definition 9.13

The group G is called **cyclic** if it contains an element x, called a **generator** of G, such that each element of G is an integer power of x.

The subgroup $H = \{\ldots, \frac{1}{4}, \frac{1}{2}, 1, 2, 4, \ldots\} = \{2^m : m \in \mathbb{Z}\}$ of $\mathbb{Q}^*$ is cyclic, for each element of H is an integer power of 2; notice that each element of H is an integer power of $\frac{1}{2}$, and so 2 and $\frac{1}{2}$ are generators (in fact the only generators) of H.

The subgroup $H = \{i, i^2, i^3, i^4 = 1\}$ of $\mathbb{C}^*$ is cyclic, being generated by the complex number i. More generally, let n be a given positive integer and let $H = \{z \in \mathbb{C} : z^n = 1\}$, that is, H consists of the complex n th roots of 1. Then H is a subgroup of $\mathbb{C}^*$; as in (2.31), we see that H consists of the n complex numbers $z_m = \cos(2m\pi/n) + i \sin(2m\pi/n)$ for $1 \leq m \leq n$, and so H is cyclic being generated by z_1.

Let x belong to the group G. Then x generates the cyclic subgroup $H = \{x^m : m \in \mathbb{Z}\}$ of G (by convention $x^0 = e$ and so H is a subgroup of G by (9.10)). In fact the abstract nature of every cyclic group is completely specified by a suitable non-negative integer, as we now explain.

Definition 9.14

Let x belong to the group G. If $x^n \neq e$ for all positive integers n, then x is said to have **infinite order**. Otherwise x is said to have **finite order** n, where n is the *smallest* positive integer with $x^n = e$.

The complex number i has order 4 as $i^4 = 1$ (and i, i^2, i^3 are unequal 1). We show next that integer powers of a group element x behave like integers if x has infinite order, and like residue classes (mod n) of integers if x has finite order n.

Lemma 9.15

Let x belong to the group G and let l and m be integers.

(a) If x has infinite order, then $x^l = x^m \Leftrightarrow l = m$.
(b) If x has finite order n, then $x^l = x^m \Leftrightarrow l \equiv m \pmod{n}$.

Proof

(a) As l and m have equal status, we take $l \geq m$. Now $x^l = x^m \Leftrightarrow x^{l-m} = e$. From $x^{l-m} = e$ we deduce that $l - m$ *cannot* be positive, because x has infinite order (see (9.14)); as $l - m \geq 0$, there is only one way out: $l - m = 0$. Therefore $x^{l-m} = e \Leftrightarrow l - m = 0 \Leftrightarrow l = m$.

(b) We use (3.6) to divide $l - m$ by n; there are integers q and r with $l - m = nq + r$, $0 \leq r < n$. As $x^n = e$, we obtain $x^{l-m} = (x^n)^q x^r = e^q x^r = x^r$ (the laws of indices (2.12)(b) hold in this context). Now $x^l = x^m \Leftrightarrow x^{l-m} = e \Leftrightarrow x^r = e$. We have arrived at the crux: what does the equation $x^r = e$ tell us? As $r < n$ and n is the smallest positive integer with $x^n = e$ by (9.14), we deduce as above that r *cannot* be positive! As $r \geq 0$ the conclusion is inescapable: $r = 0$. Therefore $x^r = e \Leftrightarrow r = 0 \Leftrightarrow l \equiv m \pmod{n}$ by (3.16). $\qquad \square$

Taking $l = 0$ in (9.15)(b) we obtain:

$x^m = e \Leftrightarrow$ the order of x is a divisor of m.

This result is useful in determining the orders of group elements.

Suppose we wish to simplify 2^{1984} working modulo 23. One way (which is not recommended!) is to calculate 2^{1984} explicitly and find the remainder on division by 23. Alternatively, we can find the order of the element $\bar{2}$ in the group $\mathbb{Z}_{23}^*$, for this order tells us all there is to know about powers of 2 (mod 23). Now $2^6 \equiv 64 \equiv -5 \pmod{23}$ which on squaring gives $2^{12} \equiv 25 \equiv 2$ (mod 23). Therefore $(\bar{2})^{12} = \bar{2}$ in $\mathbb{Z}_{23}^*$ and so $(\bar{2})^{11} = \bar{1}$; hence $\bar{2}$ has order 11 (for the order of $\bar{2}$ is a positive divisor of 11). By (9.15)(b), the first eleven powers of $\bar{2}$ are distinct, further powers of $\bar{2}$ being repetitions of these. Now $1984 = 11 \times 180 + 4$, that is,

$1984 \equiv 4 \pmod{11}$ and so $(\bar{2})^{1984} = (\bar{2})^4$ which means $2^{1984} \equiv 16 \pmod{23}$.

More generally, let x in G have finite order n. By (9.15)(b) the cyclic subgroup of G generated by x consists of the n elements $x, x^2, \ldots, x^n = e$, corresponding to the n residue classes $\pmod n$ of integers; therefore:

> A group element of order n generates a subgroup of order n.

This establishes the connection between (9.8) and (9.14). From (9.15)(a) we see that a group element of infinite order generates an infinite subgroup (that is, a subgroup with an infinite number of elements).

Cyclic groups, being abelian, occur in additive notation. By (9.13) the group $(G, +)$ is cyclic if it contains an element x such that each element of G is an integer *multiple* of x. The additive group $(\mathbb{Z}, +)$ of integers is infinite cyclic (1 is a generator), and for each positive integer n the additive group $(\mathbb{Z}_n, +)$ is a finite cyclic group of order n (being generated by $\bar{1}$). These groups may be regarded as the **standard cyclic groups**, since, as we show next, each cyclic group is isomorphic to exactly one of them.

Corollary 9.16   Let $(G, \times)$ be a cyclic group.

    (a) If G is infinite, then G is isomorphic to $(\mathbb{Z}, +)$.

    (b) If G has finite order n, then G is isomorphic to $(\mathbb{Z}_n, +)$.

Proof   (a) Let x generate G. The mapping $\alpha : \mathbb{Z} \to G$, defined by $(m)\alpha = x^m$ for all $m \in \mathbb{Z}$, is a homomorphism (9.6), that is,

$$(l + m)\alpha = x^{l+m} = (x^l)(x^m) = ((l)\alpha)((m)\alpha) \quad \text{for } l, m \in \mathbb{Z}.$$

As x has infinite order, α is injective by (9.15)(a); α is surjective as x generates G. Therefore $\alpha : (\mathbb{Z}, +) \cong (G, \times)$ is an isomorphism (9.7).

    (b) Let x generate G; then x has order $n = |G|$. By (9.15)(b), an injective mapping $\alpha : \mathbb{Z}_n \to G$ is unambiguously defined by $(\bar{m})\alpha = x^m$ for $m \in \mathbb{Z}$. As above, α is surjective and a homomorphism; therefore $\alpha : (\mathbb{Z}_n, +) \cong (G, \times)$ is an isomorphism. $\square$

Summarizing, two cyclic groups are isomorphic if and only if they have the same order.

Notation   Let C_0 denote the isomorphism class of cyclic groups of infinite order. Let C_n denote the isomorphism class of cyclic groups having finite order n.

It is customary to adopt the above scheme when dealing with cyclic groups in additive notation. For example, the additive group of $\mathbb{Z}_6$ is a C_6 (meaning $(\mathbb{Z}_6, +)$ belongs to C_6); the cyclic subgroup of $(\mathbb{Z}_6, +)$ generated by $\bar{2}$ consists of $\bar{0}$, $\bar{2}$, $\bar{4}$, and so is a C_3. Trivial groups (groups with only one element) belong to C_1.

It may seem strange to denote infinite cyclic groups by C_0, although in fact we are adopting a comprehensive approach to all cyclic groups: working with integers modulo n produces C_n, and working modulo 0 produces C_0 ($l \equiv m \pmod{0} \Leftrightarrow 0 \,|\, l - m \Leftrightarrow l = m$; so congruence modulo 0 is simply 'equality'). The reader should compare C_0 with the concept characteristic 0 (5.23), for the same idea is present in both.

Exercises 9.1

1. For each of the rings $\mathbb{Z}_6$, $\mathbb{Z}_7$, $\mathbb{Z}_9$, $\mathbb{Z}_{10}$, write out the multiplication table of its group of units. Show that each of these groups is cyclic, and decide which pairs of groups are isomorphic.

2. List the powers of $\bar{2}$ in $\mathbb{Z}_{17}^*$ and find the order of $\bar{2}$. List the powers of $\bar{3}$ in $\mathbb{Z}_{17}^*$. Is $\mathbb{Z}_{17}^*$ cyclic?

3. Find r such that $0 < r < 31$ and $2^{1494} \equiv r \pmod{31}$.

4. Find the order of $\bar{3}$ in the group of units of $\mathbb{Z}_{244}$ and hence find r with $0 < r < 244$ such that $3^{8887} \equiv r \pmod{244}$.

5. Use (9.10) to decide which of the following are subgroups of the indicated parent group.

   (a) Parent: $(\mathbb{Q}^*, \times)$. $\{m/n \in \mathbb{Q}^* : m/n \text{ positive}\}$, $\{m/n \in \mathbb{Q}^* : m \text{ and } n \text{ odd}\}$, $\{m/n \in \mathbb{Q}^* : m \text{ and } n \text{ powers of } 2\}$.
   (b) Parent: $(\mathbb{C}^*, \times)$. $\{z \in \mathbb{C}^* : |z| \in \mathbb{Q}\}$, $\{z \in \mathbb{C}^* : z^n = 1 \text{ for some } n \in \mathbb{N}\}$, $\{x + iy \in \mathbb{C}^* : x, y \in \mathbb{Q}\}$.
   (c) Parent: the group G of invertible 2×2 matrices P over $\mathbb{R}$. $\{P \in G : P \text{ diagonal}\}$, $\{P \in G : P \text{ symmetric } (P^T = P)\}$, $\{P \in G : P \text{ orthogonal } (P^T = P^{-1})\}$.

6. (a) Write out the 8×8 multiplication table of the symmetry group H of the square (9.12) in terms of α and β. Find the order of each element of H and list the seven cyclic subgroups of H.

   (b) Write down the permutations in the symmetry group of (i) the rectangle, (ii) the rhombus, (iii) the parallelogram, with vertices numbered:

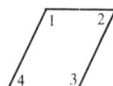

Hence find two non-cyclic subgroups of order 4 in H.

(c) By renumbering the vertices of the square, find three subgroups (H is one of them) of order 8 in S_4.

7. (a) Use (3.4)(b) to show that ± 1 are the only generators of $(\mathbb{Z}, +)$.

   (b) Use (3.9) to show that $\bar{r}$ generates $(\mathbb{Z}_n, +)$ if and only if the g.c.d. $(n, r) = 1$. Hence show that each finite cyclic group of order n has $\phi(n)$ generators (5.26).

   (c) List the generators of $(\mathbb{Z}_{28}, +)$. Show that $\bar{2}$ generates $\mathbb{Z}_{29}^*$ and hence find all the generators of $\mathbb{Z}_{29}^*$.

   (d) The group element x has finite order n. Show that x^m has order $n/(m, n)$. Deduce that the group element y has order $4n$ if and only if y^2 has order $2n$.

8. Let G denote the group of invertible 2×2 matrices over $\mathbb{Z}_2$. Express the six elements of G in the form $A^i B^j$ $(0 \leqslant i \leqslant 2, 0 \leqslant j \leqslant 1)$ where

$$A = \begin{pmatrix} 0 & 1 \\ 1 & 1 \end{pmatrix}, \qquad B = \begin{pmatrix} 0 & 1 \\ 1 & 0 \end{pmatrix}.$$

Verify that $A^3 = B^2 = I$ and $BA = A^2 B$. By comparing G with S_3 in (9.4), show that G and S_3 are isomorphic.

9. Show that the matrices of the form $\begin{pmatrix} a & b \\ 0 & 1 \end{pmatrix}$, where $a, b \in \mathbb{Z}_3$ and $a \neq 0$, form a subgroup H of the group G of all invertible 2×2 matrices over $\mathbb{Z}_3$. Find $|G|$ and $|H|$. Show that H contains an element of order 3 and an element of order 2. Is H abelian? Is H isomorphic to S_3?

10. Let H_1 and H_2 be subgroups of the group G. Using (9.10) show that $H_1 \cap H_2$ is a subgroup of G. Show that $H_1 \cup H_2$ is a subgroup of G if and only if either $H_1 \subseteq H_2$ or $H_2 \subseteq H_1$.

11. Let G be a finite group with $|G|$ even. Show that G contains at least one element of order 2.

12. (a) Let G be a group such that $x^2 = e$ for all $x \in G$. Show that G is abelian. Hence show that G is isomorphic to the additive group of a vector space over $\mathbb{Z}_2$ (Hint: let G be the set of vectors). If G is a finite group, deduce $|G| = 2^n$.

   (b) Let G be a group and let $\alpha : G \to G$ be defined by $(x)\alpha = x^{-1}$ for all $x \in G$. Show that α is an **automorphism** of G (an isomorphism $G \cong G$) if and only if G is abelian.

   (c) The identity mapping is the only automorphism of the non-trivial finite abelian group G. Show that G has order 2.

13. Let H be a subgroup of $(\mathbb{Z}, +)$. Express the subgroup conditions (9.10) in additive notation, and hence show that there is a unique non-negative integer n such that H is the cyclic subgroup generated by n.

Subgroups and cosets

Here we concentrate on a few particular aspects of groups, beginning with coset decomposition: just as each natural number n gives rise to a partition of $\mathbb{Z}$ (the partition into residue classes $(\bmod\, n)$), so each subgroup leads to a partitioning of its parent group.

Proposition 9.17 Let H be a subgroup of the group G. For x and y in G write $x \equiv y$ if $xy^{-1} \in H$. Then $\equiv$ is an equivalence relation on G and $Hx = \{hx : h \in H\}$ is the equivalence class of x.

Proof We regard the elements x and y of G as being equivalent if their quotient xy^{-1} belongs to the subgroup H. As $xx^{-1} = e \in H$, the reflexive law holds: $x \equiv x$ for all $x \in G$. Suppose $x \equiv y$, and so $xy^{-1} \in H$; as H contains the inverse of each of its elements, $yx^{-1} = (xy^{-1})^{-1} \in H$ showing $y \equiv x$. Therefore the symmetric law holds: $x \equiv y \Rightarrow y \equiv x$. Now suppose $x \equiv y$ and $y \equiv z$ $(x, y, z \in G)$. Both xy^{-1} and yz^{-1} belong to H, and as H is closed under multiplication we see $xz^{-1} = (xy^{-1})(yz^{-1}) \in H$, showing $x \equiv z$. Thus the transitive law holds, and so $\equiv$ is an equivalence relation on G by (1.25).

By (1.26), the equivalence class $\bar{x}$ consists of the elements y in G which are equivalent to x; in this case

$$\bar{x} = \{y \in G : y \equiv x\} = \{y \in G : yx^{-1} = h \text{ where } h \in H\}$$
$$= \{y \in G : y = hx \text{ where } h \in H\} = \{hx : h \in H\} = Hx. \qquad \square$$

The set Hx is the multiplicative analogue of the coset $K + x$ (5.16).

Definition 9.18 Let H be a subgroup of the group G. For each x in G the subset $Hx = \{hx : h \in H\}$ is called the **left coset** of H in G with **representative** x.

Some (50%?) mathematicians disagree with the above terminology (for them, Hx is a *right* coset). We favour (9.18) as Hx is closed under left-multiplication (premultiplication) by elements of H. Combining (1.30) and (9.17) the left cosets Hx, for x in G, form a partition of G.

For example, take $G = \mathbb{C}^*$, $H = \{z \in \mathbb{C} : |z| = 1\}$. The coset $H(1 + i)$ consists of all complex numbers $w = z(1 + i)$ where $|z| = 1$; now $|w| = |z|\,|1 + i| = \sqrt{2}$ and every complex number of modulus $\sqrt{2}$

Fig. 9.1

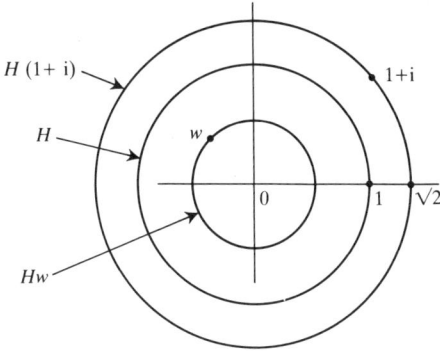

arises in this way. Therefore $H(1+\mathrm{i})$ is the circle of radius $\sqrt{2}$, centre the origin in the Argand diagram (Fig. 9.1).

More generally Hw is the circle consisting of complex numbers of modulus $|w|$, for each $w \in \mathbb{C}^*$. In this case $z \equiv w$ means $|z| = |w|$, and the partitioning of $\mathbb{C}^*$ is by the family of concentric circles, centre the origin. Notice that the subgroup H is itself a coset (the coset with representative 1).

The symmetric group S_3 has subgroup $H = \{e, \alpha, \alpha^2\}$; there are two left cosets of H in S_3:

$$\{e, \alpha, \alpha^2\} = He = H\alpha = H\alpha^2 \ (= H),$$
$$\{\beta, \alpha\beta, \alpha^2\beta\} = H\beta = H\alpha\beta = H\alpha^2\beta.$$

The same coset can be denoted in several ways, one for each of its representatives.

Expressing (9.17) in additive notation, let $(H, +)$ be a subgroup of the group $(G, +)$. Then an equivalence relation $\equiv$ is defined on G by: $x \equiv y$ if $x - y \in H$ (two elements of G are equivalent if their difference belongs to H), and the equivalence class of x is $H + x = \{h + x : h \in H\}$; in this context the notation $x \equiv y \pmod{H}$ is used (see (5.16)). For instance, let G be the additive group $\mathbb{Z}$ of integers and let H be the cyclic subgroup generated by the given natural number n (so H consists of all integer multiples of n); as $x - y \in H \Leftrightarrow n \mid (x - y)$ we see from (3.16) that $x \equiv y \pmod{H}$ means $x \equiv y \pmod{n}$, that is, the equivalence relation of (9.17) is, in this case, congruence modulo n. What is more, $H + x = \bar{x}$ (the coset $H + x$ is the residue class $\pmod{n}$ of x). In short, cosets are nothing to be frightened of, for we are already familiar with them under a different name.

In fact the reader will (almost) certainly be familiar with cosets in a geometric setting. Let U be a subspace of the vector space V; comparing additive groups, we see that $(U, +)$ is a subgroup of

Fig. 9.2

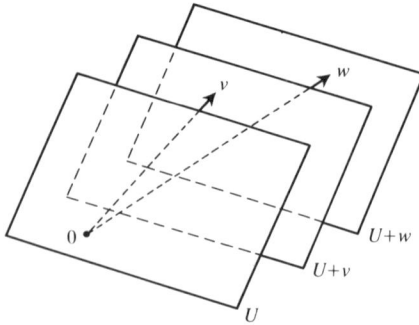

$(V, +)$. By (9.17) an equivalence relation $\equiv$ can be introduced on V as follows:

$$v \equiv v' \pmod{U} \quad \Leftrightarrow \quad v - v' \in U \qquad (v, v' \in V)$$

and $U + v = \{u + v : u \in U\}$ is the equivalence class of v. To picture the cosets $U + v$, let U be a 2-dimensional subspace of $V = \mathbb{R}^3$; then U is a plane through the origin and $U + v$ is the *parallel* plane through the point with position vector v (Fig. 9.2), that is, $U + v$ is the **translate** of U by the vector v. So the cosets of U in $\mathbb{R}^3$ are the family of planes parallel to U; these planes partition $\mathbb{R}^3$ into parallel 'slices', each point of $\mathbb{R}^3$ belonging to a unique slice parallel to U.

Returning to multiplicative notation, we now apply (9.17) to finite groups.

Corollary 9.19

(Lagrange's theorem.) Let H be a subgroup of the finite group G. Then $|H|$ is a divisor of $|G|$.

Proof

By (9.17), the left cosets of H form a partition of G (Fig. 9.3). As G is finite, there are altogether a finite number, k say, of distinct cosets of this kind, which we denote by $Hx_1, Hx_2, \ldots, Hx_k$. Writing $H = \{h_1, h_2, \ldots, h_l\}$ where $|H| = l$, we see $Hx = \{h_1x, h_2x, \ldots, h_lx\}$, and so $|Hx| = |H|$ for all $x \in G$; therefore each of the k left cosets of H in G contains exactly l elements. Counting up the elements of G gives

$$|G| = kl = k\,|H|$$

as each element of G belongs to exactly one of these cosets. $\qquad \square$

Fig. 9.3

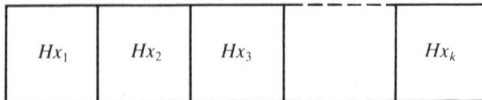

| Hx_1 | Hx_2 | Hx_3 | | Hx_k |

Partition of G into left cosets of H

The number of distinct left cosets of H in G is called the **index** of H in G; so if G is finite, then the index of H in G is $k = |G|/|H|$. For example, the subgroup H of (9.12), arising from the symmetries of the square, has index 3 $(= 24/8 = |S_4|/|H|)$ in S_4. However, it is possible for an infinite subgroup to have finite index in its parent: the subgroup of even integers has index 2 in $(\mathbb{Z}, +)$.

Lagrange's theorem says that the order of each subgroup of a finite group G is a divisor of $|G|$. Notice that the theorem does *not* say that each positive divisor of $|G|$ is necessarily the order of some subgroup of G. Indeed there is a group of order 12 having no subgroup of order 6 (see exercises 9.2, question 8). It is usually a difficult, if not impossible, task to find all the subgroups of a given group; nevertheless, we are now able to deal with one easy case.

Corollary 9.20

Let G be a group of prime order. Then G has only two subgroups, namely G and $\{e\}$. Further, G is cyclic, being generated by any non-identity element.

Proof

Write $p = |G|$ and let H be a subgroup of G. By Lagrange's theorem, $|H|$ is a divisor of p, and so either $|H| = 1$ or $|H| = p$. Every subgroup contains the identity e, and so $|H| = 1$ means $H = \{e\}$. The equation $|H| = p$ tells us that H contains all p elements of G, that is, $H = G$.

Consider x in G with $x \neq e$. Let H be the cyclic subgroup of G generated by x. As $x \in H$, we see $H \neq \{e\}$, and so $H = G$ by the above paragraph. Therefore G is cyclic with generator x. $\square$

Together with (9.16)(b), (9.20) tells us that every group of prime order p is isomorphic to $(\mathbb{Z}_p, +)$, and so these groups form the isomorphism class C_p. Although there is no difficulty in classifying groups of order p^2 (nevertheless we shall not do this), and groups of order pq or p^3 (p, q distinct primes) can be treated once more theory has been covered, this approach to finite groups quickly becomes intractable. On the other hand, the classification (into isomorphism classes) of finite *abelian* groups can be achieved with the help of more linear algebra; in particular, every finite abelian group is expressible as a direct sum of cyclic groups.

Corollary 9.21

Let G be a finite group. Then $x^{|G|} = e$ for all $x \in G$.

Proof

Let H be the cyclic subgroup of G generated by the given element x of G. Then x has order $|H|$ by (9.15)(b) and so $x^{|H|} = e$. By Lagrange's theorem (9.19), $|G| = |H| k$ for some integer k. Therefore $x^{|G|} = x^{|H|k} = e^k = e$. $\square$

We now mention some applications of (9.21). First take G to be the group of units of $\mathbb{Z}_n$; in this case $|G| = \phi(n)$, $e = \bar{1}$, $x = \bar{m}$ where g.c.d. $(m, n) = 1$. By (9.21) we obtain $(\bar{m})^{\phi(n)} = \bar{1}$, that is, since the bars denote residue classes modulo n,

Euler's theorem: $m^{\phi(n)} \equiv 1 \pmod{n}$ where $(m, n) = 1$.

For instance $\phi(10) = 4$ and so $1^4 \equiv 3^4 \equiv 7^4 \equiv 9^4 \equiv 1 \pmod{10}$. Setting $n = p$ (prime) in Euler's theorem gives, since $\phi(p) = p - 1$,

Fermat's theorem: $m^{p-1} \equiv 1 \pmod{p}$ where $(m, p) = 1$.

For instance $\phi(7) = 6$ and so $1^6 \equiv 2^6 \equiv 3^6 \equiv 4^6 \equiv 5^6 \equiv 6^6 \equiv 1 \pmod{7}$. If p is an odd prime, then Fermat's theorem with $m = 2$ tells us that p is a divisor of $2^{p-1} - 1$; turning the logic round and reverting to n in place of p, we obtain:

The integer $n > 2$ is not prime if n is
not a divisor of $2^{n-1} - 1$.

This gives a direct method of showing that certain natural numbers are not prime (but be careful: do not jump to the conclusion that n is prime simply because $n \mid 2^{n-1} - 1$).
Consider the polynomial $f = x^p - x$ over the field $\mathbb{Z}_p$ (p prime). The elements 0 and 1 (omitting bars) of $\mathbb{Z}_p$ are zeros of f. Does f have any further zeros in $\mathbb{Z}_p$? The answer is surprising: *every* element of $\mathbb{Z}_p$ is a zero of f. For $\mathbb{Z}_p^*$ is a group of order $p - 1$ and so $c^{p-1} = 1$ by (9.21) where $c \in \mathbb{Z}_p^*$; hence $c^p - c = 0$ for all c in $\mathbb{Z}_p$. As $p = \deg f$, we have found all the zeros of f by (4.29) and so

$$x^p - x = x(x - 1)(x - 2) \ldots (x - p + 1) \quad \text{over } \mathbb{Z}_p.$$

(The reader may verify this equation directly for the first few primes p by multiplying out the right-hand side.)
The above idea applies to any finite field, and provides an unbiased view of such fields. For instance, consider the field $E = \mathbb{F}_8$ with exactly eight elements (4.22)(b). Applying (9.21) to E^* we see that the elements c of E satisfy $c^8 = c$; therefore $f = x^8 - x$ regarded as a polynomial over the extension field E, splits into distinct linear factors

$$f = \prod_{c \in E} (x - c).$$

The splitting of f over E can be used to find the factorization of f into irreducible polynomials over $\mathbb{Z}_2$ (we did an analogous thing following (4.27); here $\mathbb{Z}_2$ and E take the place of $\mathbb{R}$ and $\mathbb{C}$). In (4.22)(b) E was constructed in a biased way using the irreducible polynomial $p = x^3 + x + 1$ over $\mathbb{Z}_2$; this led to the element j of E satisfying $j^3 + j + 1 = 0$, E^* being cyclic with generator j. Now

$$(x - j)(x - j^2)(x - j^4) = x^3 + x + 1,$$
$$(x - j^3)(x - j^5)(x - j^6) = x^3 + x^2 + 1,$$

and hence the factorization (4.19) of f over $\mathbb{Z}_2$ is

$$x^8 - x = x(x + 1)(x^3 + x + 1)(x^3 + x^2 + 1).$$

The polynomial $x^3 + x^2 + 1$ (which is irreducible over $\mathbb{Z}_2$ and could have been used in place of p to construct E) has turned up uninvited! So it does not matter which irreducible cubic over $\mathbb{Z}_2$ is used to construct E. What is more, E can be thought of (without prejudice) as a **splitting field** of f over $\mathbb{Z}_2$, that is, E is an extension field of $\mathbb{Z}_2$ such that f factorizes completely over E, and no subfield of E has this property; this important concept enlightens the general theory of fields.

We now return to the symmetric group S_n.

Definition 9.22 Let α be a permutation of $\{1, 2, \ldots, n\}$ and let i and j be integers with $1 \leq i < j \leq n$. Then the subset $\{(i)\alpha, (j)\alpha\}$ is called an **inversion** in α if $(i)\alpha > (j)\alpha$.

Using the notation $\alpha = \begin{pmatrix} 1 & 2 & \cdots & n \\ (1)\alpha & (2)\alpha & \cdots & (n)\alpha \end{pmatrix}$ with the top row in natural order, inversions in α are pairs of integers in the bottom row which are the 'wrong way round'. For instance $\alpha = \begin{pmatrix} 1 & 2 & 3 & 4 \\ 3 & 1 & 4 & 2 \end{pmatrix}$ contains three inversions, namely $\{3, 1\}$, $\{3, 2\}$, $\{4, 2\}$. A simple, but effective, method of finding the inversions in α (and not leaving any out) consists of joining up equal integers in the symbol for α while avoiding multiple and unnecessary intersections; each intersection then corresponds to an inversion. For example

are permutations containing 3 and 6 inversions respectively.

Definition 9.23 The permutation α in S_n is called **even/odd** according as α contains an even/odd number $N(\alpha)$ of inversions.

In the case of the above permutations, $N(\alpha) = 3$ and so α is an odd permutation in S_4, while $N(\beta) = 6$ showing that β is an even permutation in S_5. As (9.23) suggests, we are interested in the *parity* of $N(\alpha)$ and hence in the sign of $(-1)^{N(\alpha)}$. Our next job is to derive a formula for $(-1)^{N(\alpha)}$ which will enable us to work out the parity of composite permutations. To get the idea, notice that $\{(i)\alpha, (j)\alpha\}$, where $1 \leq i < j \leq n$, is an inversion in $\alpha \in S_n$ if and only if the rational number $(j-i)/((j)\alpha - (i)\alpha)$ is negative. Therefore the product $\prod((j-i)/((j)\alpha - (i)\alpha))$ of all these numbers (there are $n(n-1)/2$ of them) is positive/negative according as α is even/odd; for

$$\alpha = \begin{pmatrix} 1 & 2 & 3 & 4 \\ 3 & 1 & 4 & 2 \end{pmatrix},$$

this product is

$$\frac{(2-1)(3-1)(4-1)(3-2)(4-2)(4-3)}{(1-3)(4-3)(2-3)(4-1)(2-1)(2-4)}.$$

Without simplifying (that would obscure the issue), we see that each numerator $(j-i)$ occurs exactly once (apart from sign) as a denominator—either as $(j-i)$ or $(i-j)$. Taken overall, numerators and denominators cancel out, showing that the product has value ± 1. So in fact the product is ± 1 according as α is even/odd, that is,

$$(-1)^{N(\alpha)} = \prod_{i<j} \frac{j-i}{(j)\alpha - (i)\alpha}.$$

We are now ready to show that parity of permutations behaves as we might expect: the parity of a composite permutation $\alpha\beta$ is even/odd according as the parities of α and β are the same/different.

Lemma 9.24

Let α and β be permutations in the symmetric group S_n. Then $N(\alpha\beta) \equiv N(\alpha) + N(\beta) \pmod{2}$.

Proof

Let us shuffle the columns of the symbol

$$\begin{pmatrix} 1 & 2 & \dots & n \\ (1)\alpha & (2)\alpha & \dots & (n)\alpha \end{pmatrix}$$

to obtain

$$\alpha = \begin{pmatrix} k_1 & k_2 & \dots & k_n \\ l_1 & l_2 & \dots & l_n \end{pmatrix}.$$

Shuffling the factors of the above product correspondingly does not alter its value and leads to the formula: $(-1)^{N(\alpha)} = \prod((k_j - k_i)/(l_j - l_i))$ where, as before, the product is over all integers i and j with $1 \leq i < j \leq n$. Now β is specified by a symbol which 'begins' exactly as α's symbol 'ends', that is,

$$\beta = \begin{pmatrix} l_1 & l_2 & \cdots & l_n \\ m_1 & m_2 & \cdots & m_n \end{pmatrix},$$

and so $(-1)^{N(\beta)} = \prod((l_j - l_i)/(k_j - k_i))$. Multiplying these formulae together and cancelling the factors $l_j - l_i$ gives

$$(-1)^{N(\alpha)}(-1)^{N(\beta)} = \prod\left(\frac{k_j - k_i}{l_j - l_i}\right)\prod\left(\frac{l_j - l_i}{m_j - m_i}\right)$$

$$= \prod\frac{k_j - k_i}{m_j - m_i} = (-1)^{N(\alpha\beta)},$$

since $\alpha\beta = \begin{pmatrix} k_1 & k_2 & \cdots & k_n \\ m_1 & m_2 & \cdots & m_n \end{pmatrix}$. We have shown $(-1)^{N(\alpha)+N(\beta)} = (-1)^{N(\alpha\beta)}$ which by (9.15)(b) can be expressed $N(\alpha) + N(\beta) \equiv N(\alpha\beta) \pmod 2$ since -1 has order 2. $\quad\square$

Notice that (9.24) tells us that the mapping $\alpha \to N(\alpha)$ is a group homomorphism $S_n \to (\mathbb{Z}_2, +)$. If α and β are both odd permutations in S_n, then $N(\alpha)$ and $N(\beta)$ are odd integers and so $N(\alpha) + N(\beta)$ is even. Therefore $N(\alpha\beta)$ is even by (9.24), which means $\alpha\beta$ is an even permutation. For instance,

$$\alpha = \begin{pmatrix} 1 & 2 & 3 & 4 \\ 3 & 1 & 4 & 2 \end{pmatrix}$$

is odd but

$$\alpha^2 = \begin{pmatrix} 1 & 2 & 3 & 4 \\ 4 & 3 & 2 & 1 \end{pmatrix}$$

is even as $N(\alpha^2) = 6$.

We introduce next the simplest type of odd permutation.

Definition 9.25 Let k and l be distinct integers in $\{1, 2, \ldots, n\}$ and let τ be the permutation in S_n defined by $(k)\tau = l$, $(l)\tau = k$, $(i)\tau = i$ for $i \neq k, l$. Then τ is called a **transposition** and denoted by $(k\ l)$.

In other words, the transposition $(k\ l)$ interchanges k and l

while fixing all other integers i in $\{1, 2, \ldots, n\}$. Instead of

$$\tau = \begin{pmatrix} 1 & 2 & 3 & 4 & 5 \\ 1 & 4 & 3 & 2 & 5 \end{pmatrix}$$

we write $\tau = (2 \quad 4)$, suppressing all integers fixed (mapped to themselves) by τ. The notation $(k \quad l)$ should be used only for $k \neq l$, in which case $(k \quad l) = (l \quad k)$; the symmetric group S_n contains $n(n-1)/2$ transpositions.

Let us calculate the number $N(\tau)$ of inversions in the transposition $\tau = (k \quad l)$. Assuming $k < l$, we consider

$$\tau = (k \quad l) = \begin{pmatrix} \ldots & i_1 & \ldots & k & \ldots & i_2 & \ldots & l & \ldots & i_3 & \ldots \\ \ldots & i_1 & \ldots & l & \ldots & i_2 & \ldots & k & \ldots & i_3 & \ldots \end{pmatrix}$$

where the top row of the above symbol is in natural order. Each inversion in τ involves at least one of k and l as the ordering of the other integers is not disturbed by τ. In fact τ contains $l - k - 1$ inversions $\{i_2, k\}$ and $l - k - 1$ inversions $\{l, i_2\}$ arising from the integers i_2 with $k < i_2 < l$; there is just one more inversion in τ, namely $\{l, k\}$, as the integers $i_1 (<k)$ and $i_3 (>l)$ are not involved. Counting up we see that $\tau = (k \quad l)$ contains $N(\tau) = 2(l - k - 1) + 1$ inversions and so by (9.23):

> Transpositions are odd permutations.

Just as each invertible matrix over a field can be expressed as a product of elementary matrices, so each permutation α in S_n can be expressed as a product of transpositions; although there are many ways of doing this, the number of transpositions appearing in the product is always even/odd according as α is even/odd.

Lemma 9.26

Let α be a permutation in S_n. Then $\alpha = \tau_1 \tau_2 \ldots \tau_r$ where each τ is a transposition in S_n. Whenever α is expressed as the composition of r transpositions, then r has the same parity as α.

Proof

Let m be the number of integers i $(1 \leq i \leq n)$ which are fixed by α, that is, $(i)\alpha = i$. We argue by induction on $n - m$. If $n - m = 0$, then α is the identity permutation e, which by convention is regarded as the empty product of transpositions. Now let $n - m > 0$ and suppose all permutations in S_n fixing more than m integers i are expressible as a product of transpositions in S_n. As $n - m > 0$, there are distinct integers k and l with $(k)\alpha = l$ (k is not fixed by α, and, as α is injective, nor is l). Let $\tau = (k \quad l)$ and consider the composite permutation $\alpha\tau$; we show that $\alpha\tau$ fixes *more* integers than α. In fact $(i)\alpha = i$ implies $i \neq k, l$ and so

$(i)\alpha\tau = (i)\tau = i$, showing that $\alpha\tau$ fixes the m integers which α fixes; but also $(k)\alpha\tau = (l)\tau = k$, showing that $\alpha\tau$ fixes k also. By inductive hypothesis $\alpha\tau = \tau_1\tau_2 \ldots \tau_{r-1}$ where each τ_j is a transposition in S_n. Therefore $\alpha = \tau_1\tau_2 \ldots \tau_{r-1}\tau_r$ where $\tau_r = \tau = \tau^{-1}$, which completes the induction. Now $N(\tau_j) \equiv 1 \pmod 2$ for $1 \leqslant j \leqslant r$ as transpositions are odd, and so by (9.24)

$$N(\alpha) = N(\tau_1\tau_2 \ldots \tau_r) \equiv N(\tau_1) + N(\tau_2) + \ldots + N(\tau_r) \equiv r \pmod 2$$

showing that r and α have the same parity. □

To find transpositions τ_j in S_8 such that

$$\alpha = \begin{pmatrix} 1 & 2 & 3 & 4 & 5 & 6 & 7 & 8 \\ 5 & 8 & 7 & 2 & 3 & 6 & 1 & 4 \end{pmatrix} = \tau_1\tau_2 \ldots \tau_r$$

start by choosing any integer not fixed by α (we choose 1). As $(1)\alpha = 5$, let $\tau_1 = (1 \ \ 5)$ (τ_1 maps 1 'home' to 5, but maps 5 to 1). As $(5)\alpha = 3$, let $\tau_2 = (1 \ \ 3)$ (then $\tau_1\tau_2$ maps 1 and 5 'home' to 5 and 3 respectively, but maps 3 to 1). As $(3)\alpha = 7$, let $\tau_3 = (1 \ \ 7)$ (then $\tau_1\tau_2\tau_3$ maps 1, 5, and 3 'home' to 5, 3, and 7 respectively and (as it happens) maps 7 'home' to 1). We have completed the first stage in the 'factorization' of α. Notice that

$$\tau_1\tau_2\tau_3 = (1 \ \ (1)\alpha)(1 \ \ (1)\alpha^2)(1 \ \ (1)\alpha^3) \quad \text{and} \quad (1)\alpha^4 = 1.$$

Next continue by choosing any integer not mentioned already (not 1, 3, 5, 7) and not fixed by α (we choose 2); the above process is then repeated. As $(2)\alpha = 8$, let $\tau_4 = (2 \ \ 8)$; as $(8)\alpha = 4$, let $\tau_5 = (8 \ \ 4)$. Then $\tau_4\tau_5$ maps 2 and 8 'home' to 8 and 4 and (incidentally) 4 'home' to 2. The second stage in the 'factorization' of α is complete, and

$$\tau_4\tau_5 = (2 \ \ (2)\alpha)(2 \ \ (2)\alpha^2) \quad \text{with} \quad (2)\alpha^3 = 2.$$

All integers not fixed by α have now been accounted for, and so the process ends with $\alpha = \tau_1\tau_2\tau_3\tau_4\tau_5$. This process can be used to express any given permutation in S_n as a product of $r\,(<n)$ transpositions.

We now use parity of permutations to introduce an important subgroup of S_n.

Proposition 9.27 The even permutations in S_n form a subgroup A_n. For $n \geqslant 2$, half the permutations in S_n are even and half are odd; so $|A_n| = n!/2$.

Proof Let A_n denote the subset of even permutations in S_n. We apply (9.10), and so consider $\alpha, \beta \in A_n$. As $N(\alpha)$ and $N(\beta)$ are even, then $N(\alpha\beta)$ is even by (9.24); therefore $\alpha\beta \in A_n$. As $N(e) = 0$ (the identity permutation contains no inversions) we see $e \in A_n$. By (9.24) $N(\alpha^{-1}) + N(\alpha) \equiv 0 \pmod 2$ as $N(\alpha^{-1}\alpha) = N(e) = 0$, showing that α and α^{-1} have the same parity; so $\alpha \in A_n$ implies $\alpha^{-1} \in A_n$. By (9.10) A_n is a subgroup of S_n.

Suppose $n \geqslant 2$ and consider $\tau = (1 \quad 2)$ in S_n. As τ is odd, the cosets A_n and $A_n\tau$ are disjoint. In fact $A_n\tau$ is the subset of odd permutations β in S_n (for β odd $\Rightarrow \beta\tau (= \alpha$ say) even $\Rightarrow \beta = \alpha\tau \in A_n\tau$). So A_n and $A_n\tau$ partition S_n, and hence $|S_n| = |A_n| + |A_n\tau|$. Therefore $|A_n| = |A_n\tau| = n!/2$ as A_n and $A_n\tau$ contain the same number of permutations. □

Definition 9.28

The subgroup A_n of even permutations is called the **alternating group** of degree n.

The groups A_1 and A_2 are trivial, and A_3 is cyclic of order 3 with generator $(1 \quad 2)(1 \quad 3)$. The group A_4 has order 12, each of its elements being expressible as a product of two transpositions.

Exercises 9.2

1. For each of the following subgroups H of $(\mathbb{C}, +)$ describe (by a sketch in the Argand diagram) the coset $H + i$, a general coset $H + z$, and the partitioning (9.17) of $\mathbb{C}$ by the cosets of H.

(i) $H = \{z \in \mathbb{C} : z \text{ is real}\}$,
(ii) $H = \{z \in \mathbb{C} : z \text{ is imaginary}\}$,
(iii) $H = \{x + ix : x \text{ is real}\}$,
(iv) $H = \{x + iy : x + y \in \mathbb{Z}\}$.

2. For each of the following subgroups H of $(\mathbb{C}^*, \times)$ describe (as above) the coset $H(1 + i)$ and the partitioning of $\mathbb{C}^*$ by the cosets Hz.

(i) $H = \{z \in \mathbb{C}^* : z \text{ is real}\}$,
(ii) $H = \{z \in \mathbb{C}^* : z \text{ is real and positive}\}$,
(iii) $H = \{\pm 1\}$,
(iv) $H = \{x + iy \in \mathbb{C}^* : \text{either } x = 0 \text{ or } y = 0\}$.

3. (a) Verify Euler's theorem (see (9.21)) for $n = 14$.
(b) Verify Fermat's theorem for $p = 11$.
(c) Find the order of $\bar{2}$ in the group of units of $\mathbb{Z}_n$ where $n = 1105$ (Hint: factorize n). Deduce that $n \mid (2^{n-1} - 1)$.

4. (a) List the six monic irreducible polynomials of degrees 1 and 2 over $\mathbb{Z}_3$ and verify that their product is $x^9 - x$.
(b) Working over $\mathbb{Z}_2$, factorize $x^{16} - x$ and $x^{12} + x^8 + x^4 + 1$ into irreducible polynomials.

5. (a) Find the number of inversions in each of the permutations

$$\begin{pmatrix} 1 & 2 & 3 & 4 & 5 \\ 4 & 2 & 1 & 5 & 3 \end{pmatrix}, \quad \begin{pmatrix} 1 & 2 & 3 & 4 & 5 \\ 5 & 4 & 3 & 2 & 1 \end{pmatrix}.$$

Determine their parities and express both as products of transpositions.

(b) Express α as a product of transpositions, where

$$\alpha = \begin{pmatrix} 1 & 2 & 3 & 4 & 5 & 6 & 7 & 8 & 9 \\ 7 & 1 & 8 & 9 & 4 & 6 & 3 & 2 & 5 \end{pmatrix}.$$

(c) Determine the permutations α in S_4 such that $\alpha\beta = \beta\alpha$, where

$$\beta = \begin{pmatrix} 1 & 2 & 3 & 4 \\ 2 & 3 & 4 & 1 \end{pmatrix}.$$

6. (a) Let H be a subgroup of the group G. Show that the rule: $x \equiv y \Leftrightarrow x^{-1}y \in H$ defines an equivalence relation on G and that the equivalence class of x is $xH = \{xh : h \in H\}$ (the **right coset** of H in G with representative x).

Show that x and y belong to the same left coset of H in G if and only if x^{-1} and y^{-1} belong to the same right coset of H in G. Hence find a bijection between the set of left cosets and the set of right cosets of H in G. (The index of H in G is therefore an unbiased concept.)

(b) Let $H = \{\beta \in S_n : (n)\beta = n\}$. Show that H is a subgroup of the symmetric group S_n and that α and α' belong to the same left coset of H in S_n if and only if $(n)\alpha = (n)\alpha'$. Find the (similar) condition for α and α' to be in the same right coset of H in S_n.

Taking $n = 3$, partition S_3 into (i) left cosets of H, (ii) right cosets of H.

7. The vertices of a rectangle are assigned the colours red, yellow, green, blue, using each colour once. In how many *distinguishable* ways can this be done? (Two colourings are distinguishable if one cannot be moved into coincidence with the other.)

Answer the same question with 'rectangle' replaced by

(i) square, (ii) rhombus, (iii) parallelogram,
(iv) regular tetrahedron (pyramid with equilateral triangular faces).

Explain why in each case the answer is the index of the symmetry group of the figure in S_4.

8. List the 12 permutations in the alternating group A_4 together with their orders. Show that A_4 has a unique subgroup K of order 4.

Suppose A_4 has a subgroup H of order 6. Show that H cannot contain two permutations of order 2, and deduce that $H = \{e, \gamma, \alpha, \alpha^2, \beta, \beta^2\}$ where γ has order 2, and α and β have order 3. Show that this also is impossible by considering $\alpha\beta, \alpha\beta^2 \in H$.

Determinants

The reader may have met determinants of 2×2 matrices in connection with the solution of simultaneous equations. We now review this point by showing how a determinant arises from the linear substitution

$$X = ax + by$$
$$Y = cx + dy$$

expressing the variables X and Y in terms of x and y. Let us try to form the inverse substitution, that is, express x and y in terms of X and Y. Eliminating y (by multiplying the above equations by d and $-b$ respectively and adding), and similarly eliminating x, produces the equations

$$(ad - bc)x = \quad dX - bY$$
$$(ad - bc)y = -cX + aY$$

in which x and y have the same coefficient. In fact:

The scalar $ad - bc$ is the **determinant** of the matrix $A = \begin{pmatrix} a & b \\ c & d \end{pmatrix}$ and denoted by $|A|$.

If $|A| \neq 0$, then the original linear substitution is invertible, for

$$x = (1/|A|)(\ dX - bY)$$
$$y = (1/|A|)(-cX + aY)$$

is the inverse substitution; in this case A is invertible also and

$$A^{-1} = \begin{pmatrix} a & b \\ c & d \end{pmatrix}^{-1} = (1/|A|)\begin{pmatrix} d & -b \\ -c & a \end{pmatrix}$$

while if $|A| = 0$, neither the original linear substitution nor its matrix A is invertible. In short, the matrix A is invertible if and only if the scalar $|A|$ is invertible.

The reader should bear in mind throughout that, although out of habit we continue to work with matrices over a field, the theory of determinants applies equally well to matrices having entries in a given commutative ring.

Definition 9.29   Let $A = [a_{ij}]$ be an $n \times n$ matrix over the field F. The **determinant** of A is the scalar given by

$$|A| = \sum_{\alpha \in S_n} (-1)^{N(\alpha)} a_{1\,(1)\alpha} a_{2\,(2)\alpha} \cdots a_{n\,(n)\alpha}.$$

The determinant of an $n \times n$ matrix is therefore the sum of $n!$ terms corresponding to the $n!$ permutations α in S_n. The term corresponding to α is the product of n entries in A (the $(i, (i)\alpha)$-entries for $i = 1, 2, \ldots, n$) with sign prefix $(-1)^{N(\alpha)} = \pm 1$ according as α is even or odd; so the row suffices of each term $a_{1*}a_{2*} \cdots a_{n*}$ in $|A|$ appear in natural order, the column suffixes $a_{*(1)\alpha}a_{*(2)\alpha} \cdots a_{*(n)\alpha}$ being permuted by α. As F is closed under addition and multiplication, the determinant of a square matrix over F is itself an element of F. (Matrices which are not square do not have determinants!) Taking $n = 2$, from (9.29) we obtain (as before)

$$|A| = \begin{vmatrix} a_{11} & a_{12} \\ a_{21} & a_{22} \end{vmatrix} = a_{11}a_{22} - a_{12}a_{21}$$

since S_2 consists of the identity permutation (which is even and corresponds to $a_{11}a_{22}$) and the transposition $(1\ 2)$ (which is odd and corresponds to $-a_{12}a_{21}$). Taking $n = 3$ we obtain

$$\begin{vmatrix} a_{11} & a_{12} & a_{13} \\ a_{21} & a_{22} & a_{23} \\ a_{31} & a_{32} & a_{33} \end{vmatrix} = \begin{cases} a_{11}a_{22}a_{33} + a_{12}a_{23}a_{31} + a_{13}a_{21}a_{32} \\ -a_{11}a_{23}a_{32} - a_{12}a_{21}a_{33} - a_{13}a_{22}a_{31}, \end{cases}$$

that is, the determinant is the sum of six terms corresponding to the six permutations in S_3; for instance $a_{13}a_{21}a_{32}$ corresponds to the even permutation $\begin{pmatrix} 1 & 2 & 3 \\ 3 & 1 & 2 \end{pmatrix}$ and $-a_{13}a_{22}a_{31}$ corresponds to the odd permutation $\begin{pmatrix} 1 & 2 & 3 \\ 3 & 2 & 1 \end{pmatrix}$. Notice that the above equation can be arranged in the form:

$$\begin{vmatrix} a_{11} & a_{12} & a_{13} \\ a_{21} & a_{22} & a_{23} \\ a_{31} & a_{32} & a_{33} \end{vmatrix} = a_{11} \begin{vmatrix} a_{22} & a_{23} \\ a_{32} & a_{33} \end{vmatrix} - a_{12} \begin{vmatrix} a_{21} & a_{23} \\ a_{31} & a_{33} \end{vmatrix} + a_{13} \begin{vmatrix} a_{21} & a_{22} \\ a_{31} & a_{32} \end{vmatrix}$$

which is known as **expansion along row** 1 (see (9.33)) and provides a practical way of evaluating third-order determinants (determinants of 3×3 matrices). For instance

$$\begin{vmatrix} 9 & 2 & 6 \\ 3 & 1 & 8 \\ 4 & 7 & 5 \end{vmatrix} = 9 \begin{vmatrix} 1 & 8 \\ 7 & 5 \end{vmatrix} - 2 \begin{vmatrix} 3 & 8 \\ 4 & 5 \end{vmatrix} + 6 \begin{vmatrix} 3 & 1 \\ 4 & 7 \end{vmatrix}$$

$$= 9(-51) - 2(-17) + 6 \times 17 = -323.$$

The evaluation of determinants, working directly from (9.29), is generally a tedious business. So, without further delay, we establish the basic properties of determinants which are used in their manipulation.

Property I   The determinant of a square matrix over a field depends linearly on any one row, the remaining rows being held constant.

Proof   We first explain the statement. Let k and n be integers $(1 \leqslant k \leqslant n)$ and consider $n \times n$ matrices over the field F having $x = (x_1, x_2, \ldots, x_n) \in F^n$ as a variable row k and $(a_{i1}, a_{i2}, \ldots, a_{in}) \in F$ as a constant row i for all $i \neq k$. Denoting the determinant of such a matrix by $(x)\delta$, property I says that the mapping $\delta : F^n \to F$ is linear.

To prove property I (now we know what it means), let $y = (y_1, y_2, \ldots, y_n) \in F^n$; then using (9.29)

$$(x + y)\delta = \sum_{\alpha \in S_n} (-1)^{N(\alpha)} a_{1\,(1)\alpha} \ldots (x_{(k)\alpha} + y_{(k)\alpha}) \ldots a_{n\,(n)\alpha}$$

where $x_{(k)\alpha} + y_{(k)\alpha}$ appears in place of $a_{k\,(k)\alpha}$. As the distributive law holds in F, we obtain

$$(x + y)\delta =$$

$$\left\{ \begin{aligned} &\sum_{\alpha \in S_n} (-1)^{N(\alpha)} a_{1\,(1)\alpha} \ldots x_{(k)\alpha} \ldots a_{n\,(n)\alpha} \\ &+ \sum_{\alpha \in S_n} (-1)^{N(\alpha)} a_{1\,(1)\alpha} \ldots y_{(k)\alpha} \ldots a_{n\,(n)\alpha} \end{aligned} \right\} = (x)\delta + (y)\delta.$$

Similarly, as F has commutative multiplication, we obtain $(ax)\delta = a((x)\delta)$ for $a \in F$. Therefore δ is a linear mapping by (7.12).   ⊏

Property II   Let A be a square matrix over a field F. If A has two identical rows, then $|A| = 0$.

Proof   Let $A = [a_{ij}]$ be an $n \times n$ matrix over F such that row k is equal to row l, that is, $a_{kj} = a_{lj}\ (1 \leqslant j \leqslant n)$, with $k < l$. We use the transposition $\tau = (k\ l)$. Now $|A|$ is the sum of $n!$ terms and it is helpful to pair the terms corresponding to the permutations α and $\tau\alpha$ in S_n. (The partner of $\tau\alpha$ is $\tau\tau\alpha = \alpha$, and so the pairing is genuine; in fact $\{\alpha, \tau\alpha\}$ is a left coset of the cyclic subgroup generated by τ.) So $|A|$ is the sum of $n!/2$ pairs of terms. Consider the sum of a typical pair. Recalling that α and $\tau\alpha$ agree except that $(k)\tau\alpha = (l)\alpha$ and $(l)\tau\alpha = (k)\alpha$, we may write this sum:

$$(-1)^{N(\alpha)} a_{1\,(1)\alpha} \cdots a_{k\,(k)\alpha} \cdots a_{l\,(l)\alpha} \cdots a_{n\,(n)\alpha}$$

$$+ (-1)^{N(\tau\alpha)} a_{1\,(1)\alpha} \cdots a_{k\,(l)\alpha} \cdots a_{l\,(k)\alpha} \cdots a_{n\,(n)\alpha}.$$

Notice that $a_{k\,(k)\alpha} = a_{l\,(k)\alpha}$ and $a_{k\,(l)\alpha} = a_{l\,(l)\alpha}$, as rows k and l of A are identical; otherwise, corresponding factors in the above two terms are equal. As the transposition τ is odd, the permutations α and $\tau\alpha$ have opposite parities by (9.24), and so paired terms are prefixed by opposite signs but are otherwise equal; in other words, paired terms are negatives of each other and their sum is therefore zero. Hence $|A| = 0$, since $|A|$ is the sum of $n!/2$ zeros. □

As an illustration, consider $\delta : \mathbb{R}^3 \to \mathbb{R}$ defined by

$$(x)\delta = \begin{vmatrix} x_1 & x_2 & x_3 \\ 1 & -4 & 1 \\ 3 & -5 & -2 \end{vmatrix} \quad \text{where} \quad x = (x_1, x_2, x_3) \in \mathbb{R}^3.$$

Property I tells us that δ is linear; in particular $(10, 70, 80)\delta = 10((1, 7, 8)\delta)$, that is,

$$\begin{vmatrix} 10 & 70 & 80 \\ 1 & -4 & 1 \\ 3 & -5 & -2 \end{vmatrix} = 10 \begin{vmatrix} 1 & 7 & 8 \\ 1 & -4 & 1 \\ 3 & -5 & -2 \end{vmatrix}.$$

(Note the contrast with matrices: $10A$ means *all* entries in A are multiplied by 10.) Now $x = x_1 e_1 + x_2 e_2 + x_3 e_3$ using the standard basis of $\mathbb{R}^3$, and applying δ to this equation gives by property I:

$$\begin{vmatrix} x_1 & x_2 & x_3 \\ 1 & -4 & 1 \\ 3 & -5 & -2 \end{vmatrix} = x_1 \begin{vmatrix} 1 & 0 & 0 \\ 1 & -4 & 1 \\ 3 & -5 & -2 \end{vmatrix} + x_2 \begin{vmatrix} 0 & 1 & 0 \\ 1 & -4 & 1 \\ 3 & -5 & -2 \end{vmatrix} + x_3 \begin{vmatrix} 0 & 0 & 1 \\ 1 & -4 & 1 \\ 3 & -5 & -2 \end{vmatrix}$$

$$= 13x_1 + 5x_2 + 7x_3.$$

Property II tells us $(1, -4, 1)\delta = (3, -5, -2)\delta = 0$, and so $\langle(1, -4, 1),\ (3, -5, -2)\rangle = \ker \delta$ since rank $\delta = 1$ and nullity $\delta = 2$. Geometrically, $\ker \delta$ is the plane in $\mathbb{R}^3$ passing through the origin and the points $(1, -4, 1)$ and $(3, -5, -2)$ and has equation $13x_1 + 5x_2 + 7x_3 = 0$.

Property III $|I| = 1$, the identity matrix has determinant 1.

Proof   Take $A = I$ in (9.29); then $a_{ii} = 0$ $(1 \leq i \leq n)$ and $a_{ij} = 0$ for $i \neq j$. If $\alpha \neq e$, there is i such that $(i)\alpha \neq i$; hence the term in $|A|$ corresponding to α is zero, for it contains the factor $a_{i\,(i)\alpha} = 0$. Only the term corresponding to e remains, and so $|A| = (-1)^{N(e)} a_{11} a_{22} \ldots a_{nn} = 1$, as e is even and $a_{ii} = 1$.   □

Each $n \times n$ matrix A over F can be thought of as an n-tuple $r_1, r_2, \ldots, r_n$ of vectors in F^n, where $r_i = (\text{row } i \text{ of } A)$. As $|A|$ belongs to F, (9.29) defines a scalar-valued function (called the **determinant function**) of n-tuples of vectors in F^n. It is helpful to think of determinants in this way: their properties are easier to express and easier to understand. Property I is expressed by saying that the determinant function is a **multilinear form**, property II adds that the determinant is an **alternating form** which, by property III, takes the value 1 on the standard basis $e_1, e_2, \ldots, e_n$ of F^n. (In fact it is straightforward to show that properties I, II, III characterize the determinant function, that is, (9.29) defines the unique scalar-valued function of n-tuples of vectors in F^n having these properties.)

Of course, each $n \times n$ matrix A over F can be thought of as an n-tuple $c_1, c_2, \ldots, c_n$ of vectors in ^nF, where $c_j = (\text{column } j \text{ of } A)$. The next property tells us that rows and columns have equal status as far as determinants are concerned, for determinants are unchanged by matrix transposition.

Property IV   Let A be a square matrix over a field. Then $|A| = |A^T|$.

Proof   Let $A = [a_{ij}]$ be an $n \times n$ matrix over F. By (7.10) the (i, j)-entry in A^T is a_{ji}, and so by (9.29)

$$|A^T| = \sum_{\alpha \in S_n} (-1)^{N(\alpha)} a_{(1)\alpha 1} a_{(2)\alpha 2} \cdots a_{(n)\alpha n}$$

a typical term in $|A^T|$ having the column suffices in natural order and the row suffices permuted by α. As scalar multiplication is commutative we may rearrange the factors in each term so that the row suffices appear in natural order:

$$a_{(1)\alpha 1} a_{(2)\alpha 2} \cdots a_{(n)\alpha n} = a_{1\,(1)\alpha^{-1}} a_{2\,(2)\alpha^{-1}} \cdots a_{n\,(n)\alpha^{-1}}$$

since the first and second suffices are linked by the permutation

$$\begin{pmatrix} (1)\alpha & (2)\alpha & \cdots & (n)\alpha \\ 1 & 2 & \cdots & n \end{pmatrix} = \alpha^{-1}.$$

The mapping $S_n \to S_n$ given by $\alpha \to \alpha^{-1}$ is bijective (it is self-inverse), and therefore as α ranges over S_n so also does α^{-1}; hence

$$|A^T| = \sum_{\alpha^{-1} \in S_n} (-1)^{N(\alpha^{-1})} a_{1(1)\alpha^{-1}} a_{2(2)\alpha^{-1}} \cdots a_{n(n)\alpha^{-1}}$$

since α and α^{-1} have the same parity, that is, $(-1)^{N(\alpha)} = (-1)^{N(\alpha^{-1})}$. Replacing α^{-1} by α throughout (one summation dummy is as good as another) gives

$$|A^T| = \sum_{\alpha \in S_n} (-1)^{N(\alpha)} a_{1(1)\alpha} a_{2(2)\alpha} \cdots a_{n(n)\alpha} = |A|$$

by (9.29). $\qquad \square$

In spite of (9.29) being biased towards rows (the row suffices are kept in natural order while the column suffices are jumbled up), property IV assures us that determinants are nevertheless unbiased. Indeed any property of determinants, expressed in terms of rows, remains valid with 'row' replaced by 'column' throughout; for instance, if A has two identical columns, then A^T has two identical rows and hence $|A| = |A^T| = 0$ by properties II and IV.

Next we deduce two further properties of determinants which are useful in their evaluation.

Corollary 9.30   Let A be a square matrix over a field.

(a) Interchanging two rows of A changes $|A|$ to $-|A|$.
(b) Adding a scalar multiple of one row to another row of A leaves $|A|$ unchanged.

Proof   (a) Let A be an $n \times n$ matrix over F and write $r_i = (\text{row } i \text{ of } A)$. Suppose r_k and r_l are to be interchanged $(k \neq l)$. Let $(x, y)\delta$ be the determinant of the matrix obtained from A by substituting x and y in F^n in place of r_k and r_l. (δ being a scalar-valued mapping of ordered pairs of vectors is, by properties 1 and II, an example of an **alternating bilinear form**.) Expanding $(r_k + r_l, r_k + r_l)\delta$, which is the determinant of a matrix with two equal rows:

$$0 = (r_k + r_l, r_k + r_l)\delta = (r_k + r_l, r_k)\delta + (r_k + r_l, r_l)\delta$$
$$= (r_k, r_k)\delta + (r_l, r_k)\delta + (r_k, r_l)\delta + (r_l, r_l)\delta$$
$$= (r_l, r_k)\delta + (r_k, r_l)\delta = (r_l, r_k)\delta + |A|$$

using properties I and II. Therefore the matrix obtained from A by interchanging rows k and l has determinant $(r_l, r_k)\delta = -|A|$.

(b) Consider the effect of adding a times (row k) to row l ($a \in F$). Using δ as above, the determinant of the resulting matrix is

$$(r_k, ar_k + r_l)\delta = a((r_k, r_k)\delta) + (r_k, r_l)\delta$$
$$= (r_k, r_l)\delta = |A|. \qquad \square$$

In Chapter 10 we shall meet polynomials in determinant form, such as

$$f = \begin{vmatrix} 1-x & 1 & -1 \\ 2 & 3-x & 2 \\ 3 & -1 & 5-x \end{vmatrix}$$

requiring to be factorized. This can be done by expanding (along row 1 say) and then factorizing the resulting cubic, or preferably by extracting factors in the first place. In this case, *start* by adding row 3 to row 1 (which by (9.30)(b) does not alter f), extract the factor $4 - x$, subtract column 1 from column 3 (this does not alter the determinant) and *finish* by expanding:

$$f = \begin{vmatrix} 4-x & 0 & 4-x \\ 2 & 3-x & 2 \\ 3 & -1 & 5-x \end{vmatrix} = (4-x) \begin{vmatrix} 1 & 0 & 1 \\ 2 & 3-x & 2 \\ 3 & -1 & 5-x \end{vmatrix}$$

$$= (4-x) \begin{vmatrix} 1 & 0 & 0 \\ 2 & 3-x & 0 \\ 3 & -1 & 2-x \end{vmatrix} = (4-x) \begin{vmatrix} 3-x & 0 \\ -1 & 2-x \end{vmatrix}$$

$$= (2-x)(3-x)(4-x).$$

By collecting up its terms in various ways, we now derive the expansion of a determinant along any row or any column.

Definition 9.31 Let $A = [a_{ij}]$ be an $n \times n$ matrix over a field with $n > 1$. The determinant of the $(n-1) \times (n-1)$ matrix obtained from A by deleting row i and column j, multiplied by $(-1)^{i+j}$, is called the **cofactor** of a_{ij} in $|A|$ and denoted by A_{ij}.

We shall see shortly that A_{ij} is the coefficient of a_{ij} in the formula (9.29) for $|A|$ (and so there is a reason for the sign $(-1)^{i+j}$). In spite of the notation, keep in mind that A_{ij} is a scalar; in the case of a 3×3 matrix $A = [a_{ij}]$, deleting row 2 and column 3 gives

$$A_{23} = (-1)^{2+3} \begin{vmatrix} a_{11} & a_{12} \\ a_{31} & a_{32} \end{vmatrix} = -a_{11}a_{32} + a_{12}a_{31}$$

and $|A| = a_{11}A_{11} + a_{12}A_{12} + a_{13}A_{13}$ is the expansion along row 1.

We treat a special case next.

Lemma 9.32

Let $A = [a_{ij}]$ be an $n \times n$ matrix over a field with $n > 1$, such that $a_{in} = 0$ for $1 \le i < n$, and $a_{nn} = 1$. Then $|A| = A_{nn}$.

Proof

By hypothesis the last column of A is e_n^T and so

$$A = \begin{pmatrix} B & \vdots \\ \cdots\cdots\cdots\cdots \\ *\ldots* & 1 \end{pmatrix}$$

where B is the $(n-1) \times (n-1)$ matrix obtained from A by deleting the last row and last column, arbitrary entries are denoted by $*$ and zero entries are left blank. As $|B| = A_{nn}$ by (9.31), we must show $|A| = |B|$.

Let $H = \{\alpha \in S_n : (n)\alpha = n\}$, that is, H is the subgroup of permutations in S_n which fix the integer n. The elements of H are effectively permutations of $\{1, 2, \ldots, n-1\}$; in fact H is isomorphic to S_{n-1}, the correspondence $\alpha \to \alpha'$, where $(i)\alpha' = (i)\alpha$ for $1 \le i < n$, being an isomorphism (α' is the restriction of α to $\{1, 2, \ldots, n-1\}$). As α in H does not move the last integer n, none of the inversions (9.22) in α involve n; indeed α and α' contain the same inversions and so have the same parity.

From (9.29), all the terms in $|A|$ corresponding to permutations $\alpha \notin H$ are zero: for $\alpha \notin H$ implies $(i)\alpha = n$ for some i with $1 \le i < n$, and so the term in $|A|$ corresponding to α has factor $a_{i\,(i)\alpha} = a_{in} = 0$. As $a_{nn} = 1$, we obtain

$$|A| = \sum_{\alpha \in H} (-1)^{N(\alpha)} a_{1\,(1)\alpha} \ldots a_{n-1\,(n-1)\alpha}$$

$$= \sum_{\alpha' \in S_{n-1}} (-1)^{N(\alpha')} a_{1\,(1)\alpha'} \ldots a_{n-1\,(n-1)\alpha'}$$

$$= |B|. \qquad \square$$

Proposition 9.33

Let $A = [a_{ij}]$ be an $n \times n$ matrix over a field ($n > 1$). Then

(a) $|A| = a_{k1}A_{k1} + a_{k2}A_{k2} + \ldots + a_{kn}A_{kn}$

(expansion of $|A|$ along row k),

(b) $|A| = A_{1l}a_{1l} + A_{2l}a_{2l} + \ldots + A_{nl}a_{nl}$

(expansion of $|A|$ along column l).

Proof

(b) For typographical reasons we treat the column expansion case (it is easier to print a row of columns than a column of rows). Let c_j denote column j of A; then $c_l = \sum_{i=1}^n e_i^T a_{il}$ where e_i^T is column i of the $n \times n$ identity matrix I. Using the column version of property I, holding all columns (except column l) constant, we obtain

$$|A| = \left| c_1, \ldots, \sum_{i=1}^n e_i^T a_{il}, \ldots, c_n \right|$$

$$= \sum_{i=1}^n \left| c_1, \ldots, e_i^T, \ldots, c_n \right| a_{il}.$$

Now $\left| c_1, \ldots, e_i^T, \ldots, c_n \right|$ is the determinant of the matrix obtained from A by substituting e_i^T in column l; with help from (9.30)(a) and (9.32), we shall recognize this determinant as being the cofactor A_{il}: interchange e_i^T successively with the $n - l$ columns following it to obtain

$$(-1)^{n-l} \left| c_1, \ldots, c_l, \ldots, c_n, e_i^T \right|.$$

Next interchange row i successively with the $n - i$ rows following it, to get

$$(-1)^{n-i}(-1)^{n-l} \begin{vmatrix} B & \vdots \\ \cdots\cdots\cdots & \vdots \\ *\ldots* & 1 \end{vmatrix}$$

where B is the $(n - 1) \times (n - 1)$ matrix obtained from A by deleting row i and column l; in this case the entries $*$ are $(a_{i1}, \ldots, a_{il}, \ldots, a_{in})$. As $n - i + n - l \equiv i + l \pmod{2}$, the above scalar becomes

$$(-1)^{i+l} |B| = A_{il}$$

using (9.31) and (9.32). Therefore $\left| c_1, \ldots, e_i^T, \ldots, c_n \right| = A_{il}$ where e_i^T replaces c_l; substituting back we obtain $|A| = \sum_{i=1}^n A_{il}a_{il}$ which is the expansion of $|A|$ along column l.

(a) The row expansion formula can be deduced from the column case, using matrix transposition and property IV. $\square$

Property I and (9.30) tell us how the determinant changes when an elementary row operation (8.16) is applied to the underlying matrix; the row-reduction algorithm (8.22) together with the expansion formulae (9.33) provide a systematic method of evaluating determinants. For example, using row-reduction and expansion along column 1:

$$\begin{vmatrix} 1 & 2 & 3 & 5 \\ 2 & 7 & 3 & 4 \\ 3 & 9 & 7 & 6 \\ 1 & 4 & 6 & 7 \end{vmatrix} = \begin{vmatrix} 1 & 2 & 3 & 5 \\ 0 & 3 & -3 & -6 \\ 0 & 3 & -2 & -9 \\ 0 & 2 & 3 & 2 \end{vmatrix} = \begin{vmatrix} 3 & -3 & -6 \\ 3 & -2 & -9 \\ 2 & 3 & 2 \end{vmatrix}$$

$$= 3 \begin{vmatrix} 1 & -1 & -2 \\ 3 & -2 & -9 \\ 2 & 3 & 2 \end{vmatrix} = 3 \begin{vmatrix} 1 & -1 & -2 \\ 0 & 1 & -3 \\ 0 & 5 & 6 \end{vmatrix}$$

$$= 3 \begin{vmatrix} 1 & -3 \\ 5 & 6 \end{vmatrix} = 63.$$

Exercises 9.3

1. Evaluate the following determinants:

$$\begin{vmatrix} 3 & 1 & 4 \\ 2 & 2 & 5 \\ 6 & 3 & 3 \end{vmatrix}, \quad \begin{vmatrix} 7 & 4 & 5 \\ 6 & 3 & 8 \\ 9 & 3 & 25 \end{vmatrix}, \quad \begin{vmatrix} 5 & 6 & 3 & 4 \\ 5 & 7 & 4 & 5 \\ 6 & 7 & 4 & 7 \\ 2 & 2 & 3 & 2 \end{vmatrix}.$$

2. Find the rational numbers x such that

$$\begin{vmatrix} 6-x & 5 & 2 \\ 8 & 7+x & 3 \\ 1 & 1 & 1 \end{vmatrix} = 0.$$

For each such x, verify that the rows of the underlying matrix are linearly dependent.

3. Factorize the following determinants into linear polynomials:

$$\begin{vmatrix} 4-x & -1 & 3 \\ 2 & 2-x & 2 \\ -1 & 1 & -x \end{vmatrix}, \quad \begin{vmatrix} 1-x & -2 & 2 \\ 1 & 2-x & 1 \\ -1 & 1 & -2-x \end{vmatrix}.$$

4. (a) Factorize the **Vandermonde determinant**

$$\begin{vmatrix} 1 & x & x^2 \\ 1 & y & y^2 \\ 1 & z & z^2 \end{vmatrix}$$

into linear factors, where x, y, z belong to a field F. In what circumstances is this determinant zero?

(b) Let $x_1, x_2, \ldots, x_n$ be elements of a field. Factorize into linear factors the determinant of the $n \times n$ matrix with (i, j)-entry x_i^{j-1}.

5. (a) Let J denote the $n \times n$ matrix having all its entries equal to 1. Factorize the determinant $|xI + J|$ into linear polynomials.

(b) Let Δ_n denote the determinant of the $n \times n$ matrix having $|i - j|$ as (i, j)-entry. Evaluate Δ_2 and Δ_3. For $n \geqslant 3$, express Δ_n in terms of Δ_{n-1}, and hence find a formula for Δ_n.

6. Let $A = [a_{ij}]$ be a 9×9 matrix over a field. Find the sign prefixing the following terms in $|A|$:

$$a_{13}a_{24}a_{35}a_{46}a_{57}a_{68}a_{79}a_{82}a_{91}, \qquad a_{12}a_{21}a_{34}a_{43}a_{55}a_{67}a_{76}a_{89}a_{98}.$$

7. (a) The $n \times n$ matrix $A = [a_{ij}]$ over a field is lower triangular ($a_{ij} = 0$ for $i < j$). Prove directly from (9.29) (or by using (9.33)(a)) that $|A| = a_{11}a_{22} \ldots a_{nn}$.

(b) Let $A = [a_{ij}]$ be an $n \times n$ matrix over a field and let r be a given integer with $1 \leqslant r < n$. Suppose $a_{ij} = 0$ whenever $i > r$ and $j \leqslant r$, that is,

$$A = \begin{pmatrix} B & \vdots & * \\ \cdots & \cdots & \cdots \\ & \vdots & C \end{pmatrix}$$

where B and C are $r \times r$ and $(n - r) \times (n - r)$ matrices, and the lower-left block is the $(n - r) \times r$ zero matrix. By considering $H = \{\alpha \in S_n : (i)\alpha \leqslant r$ for all $i \leqslant r\}$, show, directly from (9.29), that $|A| = |B| |C|$.

8. Let $\delta : M_2(F) \to F$ be a scalar-valued function of the set of 2×2 matrices $A = (c_1, c_2)$ over the field F. Suppose

(i) $\delta(c_1, xa + yb) = \delta(c_1, x)a + \delta(c_1, y)b$ where $x, y \in {}^2F$, $(a, b \in F)$,
(ii) $\delta(c_1, c_1) = 0$ for all $c_1 \in {}^2F$,
(iii) $\delta(I) = 1$.

Show that $\delta(A) = |A|$ for all A in $M_2(F)$.

Multiplicative properties of determinants

Having learnt how to evaluate determinants, we now meet their multiplicative properties. We begin by introducing a matrix having cofactors (9.31) as entries, which is closely related to the inverse matrix.

Definition 9.34   Let $A = [a_{ij}]$ be an $n \times n$ matrix over a field F ($n > 1$). The $n \times n$ matrix over F with (i, j)-entry A_{ji} is called the **adjugate** (or **adjoint**) of A and denoted by adj A. (By convention, $1 = $ adj A if A is a 1×1 matrix.)

In the case of a 3×3 matrix A, we have

$$\text{adj } A = \begin{pmatrix} A_{11} & A_{21} & A_{31} \\ A_{12} & A_{22} & A_{32} \\ A_{13} & A_{23} & A_{33} \end{pmatrix}.$$

Notice that the suffix scheme is the transpose of the usual one; thus A_{12} appears in row 2 and column 1. The reader may verify that if

$$A = \begin{pmatrix} 2 & 3 & 4 \\ 4 & 5 & 1 \\ 2 & 4 & 3 \end{pmatrix} \quad \text{then} \quad \text{adj } A = \begin{pmatrix} 11 & 7 & -17 \\ -10 & -2 & 14 \\ 6 & -2 & -2 \end{pmatrix},$$

the $(1, 3)$-entry in $\text{adj } A$ being $A_{31} = (-1)^{3+1} \begin{vmatrix} 3 & 4 \\ 5 & 1 \end{vmatrix} = -17$, etc. It is routine to verify that $|A| = 16$ and (more surprisingly) that

$$A(\text{adj } A) = \begin{pmatrix} 16 & 0 & 0 \\ 0 & 16 & 0 \\ 0 & 0 & 16 \end{pmatrix}.$$

This is too nice to be merely a coincidence!

Proposition 9.35 Let A be an $n \times n$ matrix over a field. Then

$$A(\text{adj } A) = |A| I = (\text{adj } A)A.$$

Proof The (k, l)-entry in $(\text{adj } A)A$ is the result of multiplying row k of $\text{adj } A$ into column l of $A = [a_{ij}]$ and so is

$$A_{1k}a_{1l} + A_{2k}a_{2l} + \ldots + A_{nk}a_{nl} \quad (= b_{kl} \text{ say}).$$

Now b_{ll} is the expansion of $|A|$ along column l by (9.33)(b), and so $b_{ll} = |A|$. To recognize b_{kl} for $k \neq l$, consider the $n \times n$ matrix A' having the same columns as A except that (column k of A') = (column l of A), that is,

columns k and l

$$A' = (c_1, \ldots, c_l, \ldots, c_l, \ldots, c_n).$$

As A and A' agree except possibly in column k, and column k is deleted when forming the cofactors A_{ik} by (9.31), we see that b_{kl} is the expansion of $|A'|$ along column k. Therefore $b_{kl} = |A'| = 0$, as A' has two equal columns. So $|A| I = (\text{adj } A)A$ by (7.2), as the (k, l)-entries in these matrices agree for $1 \leq k, l \leq n$.

Replacing A by A^T gives $|A^T|I = (\text{adj } A^T)A^T$, which on transposing by (7.11)(b) becomes $|A|I = A(\text{adj } A)$, since $|A| = |A^T|$ and adj $A^T = (\text{adj } A)^T$. □

The product (in either order) of A and adj A is the scalar matrix $|A|I$ by (9.35); we use this result (which, incidentally, holds for any $n \times n$ matrix A over a commutative ring—the above proof goes through unchanged) in (10.25) to show that A satisfies a polynomial equation; the polynomial involved has degree n and is called the **characteristic polynomial** of A.

Because of its simple statement, the reader might think our next theorem is 'obvious': the determinant of a product is the product of the individual determinants. In fact this fundamental property of determinants is not particularly easy to prove!

Theorem 9.36

Let A and B be $n \times n$ matrices over a field. Then

$$|AB| = |A|\,|B|.$$

Proof

Write $c_j = (\text{column } j \text{ of } A)$ and $B = [b_{jk}]$. Then $\sum_{j=1}^n c_j b_{jk} =$ (column k of AB). It is necessary to use a different summation suffix for each of the n columns of AB, and so we replace j by j_k, obtaining

$$(\text{column } k \text{ of } AB) = \sum_{j_k} c_{j_k} b_{j_k k}$$

the summation range being $1 \le j_k \le n$. We apply the column version of property I successively to columns $1, 2, \ldots, n$ as follows:

$$|AB| = \left| \sum_{j_1} c_{j_1} b_{j_1 1}, \sum_{j_2} c_{j_2} b_{j_2 2}, \ldots, \sum_{j_n} c_{j_n} b_{j_n n} \right|$$

$$= \sum_{j_1} \left| c_{j_1}, \sum_{j_2} c_{j_2} b_{j_2 2}, \ldots, \sum_{j_n} c_{j_n} b_{j_n n} \right| b_{j_1 1} = \ldots$$

$$= \sum_{j_1, \ldots, j_n} |c_{j_1}, c_{j_2}, \ldots, c_{j_n}| \, b_{j_1 1} b_{j_2 2} \ldots b_{j_n n}.$$

There are n^n terms in the last summation above, as each of the n summation indices j_k has range $1 \le j_k \le n$. Do not lose heart, because if two of $j_1, j_2, \ldots, j_n$ are equal, then $|c_{j_1}, c_{j_2}, \ldots, c_{j_n}| = 0$ being the determinant of a matrix with two equal columns. So the summation can be restricted to terms with $j_1, j_2, \ldots, j_n$ distinct, that is, with $j_1, j_2, \ldots, j_n$ being a rearrangement of $1, 2, \ldots, n$; for such terms

$$\begin{pmatrix} j_1 & j_2 & \cdots & j_n \\ 1 & 2 & \cdots & n \end{pmatrix}$$

is a permutation α in S_n, and so $(k)\alpha^{-1} = j_k$ $(1 \leqslant k \leqslant n)$. Using this notation,

$$|c_{j_1}, c_{j_2}, \ldots, c_{j_n}| = |c_{(1)\alpha^{-1}}, c_{(2)\alpha^{-1}}, \ldots, c_{(n)\alpha^{-1}}|$$
$$= (-1)^{N(\alpha)} |c_1, c_2, \ldots, c_n| = (-1)^{N(\alpha)} |A|$$

for the columns $c_{(1)\alpha^{-1}}, c_{(2)\alpha^{-1}}, \ldots, c_{(n)\alpha^{-1}}$ can be arranged in natural order using an even/odd number of transpositions according as α is even/odd by (9.26), each transposition producing a sign change by (9.30)(a). Therefore

$$|AB| = \sum_{\alpha \in S_n} (-1)^{N(\alpha)} |A| \, b_{(1)\alpha^{-1}1} b_{(2)\alpha^{-1}2} \cdots b_{(n)\alpha^{-1}n}$$

$$= |A| \left(\sum_{\alpha \in S_n} (-1)^{N(\alpha)} b_{1\,(1)\alpha} b_{2\,(2)\alpha} \cdots b_{n\,(n)\alpha} \right)$$

$$= |A| \, |B|$$

on applying (9.29) to B. □

Do not *misuse* (9.36) to find the determinant of a scalar matrix (generally $|aI| \neq a |I| = a$); in fact, $|aI| = a^n$ where I is the $n \times n$ identity matrix, by properties I and III.

Beginning with matrix inverses, we now review some of the concepts of Chapter 8.

Corollary 9.37

Let P be an $n \times n$ matrix over a field F. Then P is invertible over F if and only if $|P| \neq 0$, in which case $P^{-1} = (1/|P|) \, \mathrm{adj} \, P$.

Proof

Suppose that P is invertible over F. There is a matrix P^{-1} over F such that $P^{-1}P = I$; taking determinants of this matrix equation by (9.36), we obtain $|P^{-1}| |P| = 1$, showing that the scalar $|P|$ has inverse $|P^{-1}|$ in F. Therefore $|P| \neq 0$ (and incidentally $|P^{-1}| = 1/|P|$).

Suppose $|P| \neq 0$. Multiplying the matrix equations (9.35), with P in place of A, through by $1/|P|$, we see that P is invertible over F with inverse $P^{-1} = (1/|P|) \, \mathrm{adj} \, P$ by (8.11). □

The above formula for P^{-1} does *not* provide a practical method of inverting $n \times n$ matrices over a field, except for $n \leq 3$; determining adj P is itself a daunting task involving n^2 evaluations of $(n-1)$-order determinants.

The units (2.11) of the field F are the non-zero elements of F. In fact:

> The $n \times n$ matrix P over the commutative ring R
> is invertible over R if and only if $|P|$ is a unit of R.

This is because the proof of (9.37) goes through unchanged with F replaced by R. In particular a square matrix P over $\mathbb{Z}$ is invertible over $\mathbb{Z}$ (that is, the entries in P^{-1} are integers) if and only if $|P| = \pm 1$; for instance $P = \begin{pmatrix} 3 & 5 \\ 8 & 13 \end{pmatrix}$ is invertible over $\mathbb{Z}$ as $|P| = -1$, whereas $\begin{pmatrix} 3 & 4 \\ 8 & 13 \end{pmatrix}$ is invertible over $\mathbb{Q}$ but not over $\mathbb{Z}$.

Definition 9.38
Let P be a square matrix over a field. Then P is called **singular** if $|P| = 0$ and called **non-singular** if $|P| \neq 0$.

From (9.37), a square matrix over a field is invertible if and only if it is non-singular (the terminology 'non-singular' is commonly used instead of 'invertible' in this context). From (8.12) we obtain:

> The rows of the $n \times n$ matrix P over the field
> F form a basis of F^n if and only if $|P| \neq 0$.

For example, do the vectors $r_1 = (9, 5, 22)$, $r_2 = (7, 11, 26)$, $r_3 = (11, 3, 21)$ form a basis of $\mathbb{R}^3$? Form the matrix P having these vectors as its rows, and expand $|P|$ along a column (column 3 in this case):

$$|P| = \begin{vmatrix} 9 & 5 & 22 \\ 7 & 11 & 26 \\ 11 & 3 & 21 \end{vmatrix} = \begin{vmatrix} 7 & 11 \\ 11 & 3 \end{vmatrix} 22 - \begin{vmatrix} 9 & 5 \\ 11 & 3 \end{vmatrix} 26 + \begin{vmatrix} 9 & 5 \\ 7 & 11 \end{vmatrix} 21$$

$$= (-100)22 + (28)26 + (64)21 = 0.$$

So r_1, r_2, r_3 do *not* form a basis of $\mathbb{R}^3$; in fact $(-100, 28, 64)$ is row 1 of adj P, and comparing first rows in the matrix equation (adj $P)P = 0$ (using (9.35) with $A = P$, $|P| = 0$) gives the linear dependence relation $-100r_1 + 28r_2 + 64r_3 = 0$, that is, $25r_1 = 7r_2 + 16r_3$. On the other hand, as $\begin{vmatrix} 5 & 2 & 3 \\ 4 & 7 & 4 \\ 9 & 2 & 7 \end{vmatrix} = 56$ is non-zero, the rows

$r_1 = (5, 2, 3)$, $r_2 = (4, 7, 4)$, $r_3 = (9, 2, 7)$ of the underlying matrix form a basis of $\mathbb{R}^3$; in Chapter 11 we shall see that 56 measures the volume of the parallelepiped (box with parallelogram faces) having edges r_1, r_2, r_3.

Combining (8.12) and (9.37), the $n \times n$ matrix P over a field has rank n if and only if $|P| \neq 0$. In fact the rank of an arbitrary matrix over a field can be expressed in terms of determinants.

Definition 9.39

Let A be an $m \times n$ matrix over a field and let s be an integer with $1 \leqslant s \leqslant \min\{m, n\}$. Suppose $m - s$ rows and $n - s$ columns of A are deleted. The determinant of the remaining $s \times s$ matrix is called an s-**minor** of A.

For instance, deleting column 3 of $A = \begin{pmatrix} 1 & 2 & 3 & 4 \\ 8 & 7 & 6 & 5 \\ 9 & 10 & 11 & 12 \end{pmatrix}$ we

see that $\begin{vmatrix} 1 & 2 & 4 \\ 8 & 7 & 5 \\ 9 & 10 & 12 \end{vmatrix} = 0$ is a 3-minor of A; in fact all the

3-minors of A are zero. Deleting row 1 and columns 2 and 4, we

obtain the 2-minor $\begin{vmatrix} 8 & 6 \\ 9 & 11 \end{vmatrix} = 34$. In general, the 1-minors of A are

the entries in A; in the case of an $n \times n$ matrix A, the cofactors A_{ij} are, apart from sign, the $(n - 1)$-minors of A, and $|A|$ is the unique n-minor of A.

Proposition 9.40

Let A be a non-zero matrix of rank r over a field. Then A has a non-zero r-minor and all s-minors of A are zero for $s > r$.

Proof

Let us suppose that A is an $m \times n$ matrix. Select r linearly independent rows of A and delete the remaining $m - r$ rows. The resulting $r \times n$ matrix is of rank r and so has r linearly independent columns. On deleting the remaining $n - r$ columns, we are left with an $r \times r$ matrix of rank r; by (8.12) and (9.37), the determinant of this $r \times r$ matrix is non-zero, showing that A has a non-zero r-minor by (9.39). For $s > r$, every s rows of A are linearly dependent, and hence every s-minor of A is zero by (8.12) and (9.37).   □

Notice that rank $A = 0 \Leftrightarrow A = 0 \Leftrightarrow$ all s-minors of A are zero. Therefore the rank of an arbitrary matrix A over a field is the *smallest* non-negative integer r such that all s-minors of A are zero for $s > r$ (if the rows or columns of A are linearly independent, there will not be any s-minors of A with $s > r$).

Our next (and last) corollary is useful in connection with diagonalization.

Corollary 9.41   Let A be an $n \times n$ matrix over the field F. The homogeneous system $xA = 0$ has a non-zero solution x in F^n if and only if $|A| = 0$.

Proof   By (8.10) the solutions x of $xA = 0$ form an $(n - r)$-dimensional subspace of F^n where $r = \text{rank } A$. So there is a non-zero solution x (at least one entry in x is non-zero) if and only if $n - r > 0$, that is, if and only if $|A| = 0$ by (8.12) and (9.37). $\qquad\square$

To find the general solution of $xA = 0$ one must transpose and row-reduce as in (8.29) (or column-reduce A). Although (9.41) does not help the search for non-trivial solutions, it does tell us explicitly whether or not there are any to be found.

Exercises 9.4

1. For each of the following matrices A over $\mathbb{Z}$, determine $\text{adj } A$ and verify that $A(\text{adj } A) = |A| I$. Which of these matrices are invertible over $\mathbb{Z}$ and which are invertible over $\mathbb{Q}$? Find A^{-1} where appropriate.

$$\begin{pmatrix} 3 & 4 \\ 8 & 11 \end{pmatrix}, \quad \begin{pmatrix} 3 & 6 & 2 \\ 3 & 2 & 3 \\ 4 & 3 & 4 \end{pmatrix}, \quad \begin{pmatrix} 3 & 2 & 7 \\ 4 & 1 & 6 \\ 4 & 2 & 8 \end{pmatrix}, \quad \begin{pmatrix} 4 & 4 & 3 \\ 1 & 3 & 1 \\ 7 & 7 & 6 \end{pmatrix}.$$

2. Let $A = \begin{pmatrix} a & b \\ c & d \end{pmatrix}$ and $A' = \begin{pmatrix} a' & b' \\ c' & d' \end{pmatrix}$ have entries belonging to a commutative ring. Verify by direct calculation that $|AA'| = |A| \, |A'|$.

3. (a) Express $\begin{pmatrix} a^2 + b^2 & bc & ac \\ bc & a^2 + c^2 & ab \\ ac & ab & b^2 + c^2 \end{pmatrix}$ as a product of two matrices and hence factorize its determinant into linear factors.

   (b) Factorize $\begin{vmatrix} s_1 & s_2 & s_3 \\ s_2 & s_3 & s_4 \\ s_3 & s_4 & s_5 \end{vmatrix}$ into linear factors, where $s_i = a^i + b^i + c^i$.

4. The entries in $A = \begin{pmatrix} \bar{4} & \bar{1} & \bar{1} \\ \bar{5} & \bar{6} & \bar{2} \\ \bar{9} & \bar{7} & \bar{1} \end{pmatrix}$ belong to the ring $\mathbb{Z}_n$. Form $\text{adj } A$

and find $|A|$. For which positive integers n is A invertible over $\mathbb{Z}_n$? If $n = 115$, determine the entries in A^{-1} in the form $\bar{r}$ $(0 \leqslant r < 115)$. If $n = p$ (a prime), for which p is A singular? For these primes, find a linear dependence relation between the rows of A.

5. (a) Evaluate the 3-minors of $\begin{pmatrix} 1 & -2 & 1 & 2 \\ -2 & 4 & -2 & 1 \\ 3 & -6 & 4 & 3 \end{pmatrix}$ and determine its rank over $\mathbb{Q}$.

(b) All the s-minors of the matrix A are zero. Show that all the $(s + 1)$-minors of A are also zero.

(c) The $n \times n$ matrix A over the field F has a non-zero $(s - 1)$-minor for some integer s with $1 < s < n$, and all except (possibly) one of the s-minors of A are zero. Must all s-minors of A be zero? Answer the same question if all $(s - 1)$-minors of A are non-zero.

6. Let A be an $n \times n$ matrix over a field $(n \geqslant 2)$.

(a) Show $|\text{adj } A| = |A|^{n-1}$.

(b) Show that

$$\text{rank(adj } A) = \begin{cases} 0 \\ 1 \\ n \end{cases} \text{ according as rank } A = \begin{cases} r < n - 1 \\ n - 1 \\ n \end{cases}.$$

(c) Use (9.35) with A replaced by $\text{adj } A$ to show that $\text{adj}(\text{adj } A) = |A|^{n-2}A$.

(d) Show that $\text{adj}(AB) = (\text{adj } B)(\text{adj } A)$, where A and B are non-singular $n \times n$ matrices over a field.

7. (a) Let A be a non-singular $n \times n$ matrix over the field F, let $b \in {}^nF$, and let B_j be the matrix obtained from A on replacing column j by b. Prove **Cramer's rule:** the system $Ax = b$ has unique solution $x \in {}^nF$, where $|B_j|/|A|$ is the jth entry in x. (Hint: compare entries in $x = A^{-1}b$ using $A^{-1} = (1/|A|) \text{adj } A$.)

(b) By expanding the determinant of the coefficient matrix along row 3 find the unique solution of the system over $\mathbb{Q}$:

$$x_1 + 3x_2 + x_3 + 4x_4 = 0$$
$$x_1 + 5x_2 + 2x_3 + 5x_4 = 0$$
$$2x_1 + x_2 + 2x_3 + x_4 = 1$$
$$2x_1 + 4x_2 + x_3 + 6x_4 = 0.$$

8. Let G denote the group of non-singular $n \times n$ matrices over the field F.

(a) Show that $\alpha : G \to F^*$, defined by $(A)\alpha = |A|$ for all A in G, is a surjective group homomorphism (9.6). Show that α is injective if and only if $n = 1$.

(b) Show that the $n \times n$ matrices of determinant 1 form a subgroup K of G. (K is called the **special linear group** $SL_n(F)$). Show that the matrices A and B in G belong to the same coset (left or right) of K in G if and only if $|A| = |B|$.

(c) Show that $SL_2(\mathbb{Z}_3)$ has order 24. Is this group isomorphic to the symmetric group S_4?

(d) If F is a finite field, show that $SL_n(F)$ has index $q - 1$ in G where $q = |F|$, and express $|SL_n(F)|$ in terms of n and q.

10 Diagonalization and duality

Diagonalization

Let $\alpha : V \to V$ be a linear mapping of the n-dimensional vector space V. It is appropriate to specify α in terms of a basis b of V; in other words, α is defined by its matrix A relative to b (7.29). The matrix of α relative to another basis b' of V is PAP^{-1} where P is the transition matrix (8.13) from b' to b. The matrices A and PAP^{-1}, which represent α relative to different bases of V, are said to be **similar**; in fact, similarity of matrices is a relationship of fundamental importance in linear algebra which we cannot fully investigate here. Rather we deal with a particular aspect: given the matrix A, can an invertible matrix P be found such that PAP^{-1} is **diagonal** (that is, the (i, j)-entries in PAP^{-1} are zero for $i \neq j$)? We shall see that such a matrix P may or may not exist, but if P can be found, then a very satisfactory and simple description of α results: the individual vectors in the basis b' are each mapped into scalar multiples of themselves by α (such vectors are called **characteristic** (or **eigen-**) **vectors** of α), which means that α decomposes completely into linear mappings of 1-dimensional subspaces of V. Although linear mappings of this type are to some extent contrived, nevertheless we shall become familiar with the important idea of simultaneously decomposing α and V into components (the direct sum (6.28) again, but with a special purpose in mind) which occurs in other contexts.

Definition 10.1   The $n \times n$ matrices A and B over the field F are called **similar** if there is an invertible $n \times n$ matrix P over F such that $PAP^{-1} = B$.

Following (7.30) we discussed an example of similar matrices. As a further illustration, write $V = \{f \in \mathbb{Q}[x] : \deg f \leqslant 3\}$ and let $\delta : V \to V$ denote formal differentiation, that is,

$$(a_0 + a_1 x + a_2 x^2 + a_3 x^3)\delta = a_1 + 2a_2 x + 3a_3 x^2 \qquad (a_i \in \mathbb{Q}).$$

V has basis b consisting of the powers x^3, x^2, x, x^0 and basis b' consisting of the factorials $(x)_3 = x(x-1)(x-2)$, $(x)_2 = x(x-1)$, $(x)_1 = x$, $(x)_0 = x^0$. Let A be the matrix of δ relative to b (7.29), let P be the transition matrix from b' to b, that is, P is the matrix of the identity mapping ι relative to $\mathscr{b}'$ and $\mathscr{b}$ (7.22), and let B be the matrix of δ relative to $\mathscr{b}'$. Then

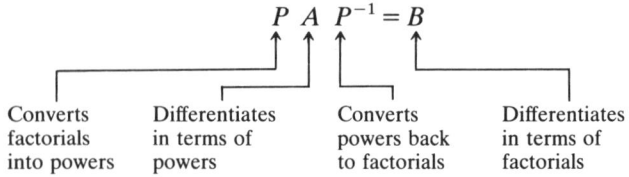

$$P \ A \ P^{-1} = B$$

Converts	Differentiates	Converts	Differentiates
factorials	in terms of	powers back	in terms of
into powers	powers	to factorials	factorials

that is, in explicit form:

$$\begin{pmatrix} 1 & -3 & 2 & 0 \\ 0 & 1 & -1 & 0 \\ 0 & 0 & 1 & 0 \\ 0 & 0 & 0 & 1 \end{pmatrix} \begin{pmatrix} 0 & 3 & 0 & 0 \\ 0 & 0 & 2 & 0 \\ 0 & 0 & 0 & 1 \\ 0 & 0 & 0 & 0 \end{pmatrix} \begin{pmatrix} 1 & 3 & 1 & 0 \\ 0 & 1 & 1 & 0 \\ 0 & 0 & 1 & 0 \\ 0 & 0 & 0 & 1 \end{pmatrix} = \begin{pmatrix} 0 & 3 & -3 & 2 \\ 0 & 0 & 2 & -1 \\ 0 & 0 & 0 & 1 \\ 0 & 0 & 0 & 0 \end{pmatrix}.$$

It is straightforward to verify, directly from (10.1), that similarity is an equivalence relation on the set $M_n(F)$ of $n \times n$ matrices over F. We show next how the concept of similarity arises.

Proposition 10.2 Let the linear mapping α have matrix A relative to the basis $\mathscr{b}$ of the vector space V. Then B is the matrix of α relative to some basis of V if and only if A and B are similar.

Proof Suppose first that α has matrix B relative to the basis $\mathscr{b}'$ of V. Let P be the matrix of the identity mapping ι of V relative to $\mathscr{b}'$ and $\mathscr{b}$; then P is invertible by (8.13) and P^{-1} is the matrix of ι^{-1} $(=\iota)$ relative to $\mathscr{b}$ and $\mathscr{b}'$. By (7.27), the composite linear mapping $\iota\alpha\iota^{-1}$ (which is α thinly disguised!) has matrix PAP^{-1} relative to $\mathscr{b}'$, for ι and ι^{-1} allow us to move from $\mathscr{b}'$ to $\mathscr{b}$, and then from $\mathscr{b}$ back to $\mathscr{b}'$ again; therefore $PAP^{-1} = B$, showing that A and B are similar by (10.1).

Conversely, let A and B be similar; then $PAP^{-1} = B$ where P is invertible. Supposing, as usual, that V is n-dimensional over the field F, let us consider the basis $\mathscr{b}'$ of V which is the image of the standard basis of F^n by the isomorphism $\mu_P \kappa_{\mathscr{b}}^{-1} : F^n \cong V$ (see (7.13) and (7.24)); so $\mathscr{b}'$ consists of $v_1, v_2, \ldots, v_n$ where $(e_i)\mu_P\kappa_{\mathscr{b}}^{-1} = v_i$ $(1 \leq i \leq n)$. As v_i is the i th vector in $\mathscr{b}'$, we see $(v_i)\kappa_{\mathscr{b}'} = e_i$ and hence $(v_i)\kappa_{\mathscr{b}'}\mu_P\kappa_{\mathscr{b}}^{-1} = v_i$ $(1 \leq i \leq n)$. Therefore $\kappa_{\mathscr{b}'}\mu_P\kappa_{\mathscr{b}}^{-1} = \iota$ (for ι is the only linear mapping of V which fixes each of the vectors in the basis $\mathscr{b}'$ by (7.16)) and so ι has matrix P relative to $\mathscr{b}'$ and $\mathscr{b}$ by (7.25). As before, $\alpha = \iota\alpha\iota^{-1}$ has matrix $B = PAP^{-1}$ relative to $\mathscr{b}'$ by (7.27). □

Therefore the matrices which arise on referring $\alpha : V \to V$ to different bases of V are precisely the matrices of a similarity class. The following concrete version of (10.2) will be useful.

**Corollary
10.3**

Let A and P be $n \times n$ matrices over the field F where P is invertible. Then the linear mapping $\mu_A : F^n \to F^n$ has matrix PAP^{-1} relative to the basis of F^n consisting of the rows of P.

Proof

We apply the second part of the proof of (10.2) to $\mu_A : F^n \to F^n$, taking ℓ to be the standard basis of F^n. Writing $A = [a_{ij}]$, the equations

$$(e_i)\mu_A = e_i A = \sum_{j=1}^{n} a_{ij} e_j \qquad (1 \leq i \leq n)$$

show that μ_A has matrix A relative to ℓ. As κ_ℓ is the identity mapping of F^n (each vector of F^n coincides with its co-ordinate vector relative to the standard basis), we see that ℓ' consists of $v_1, v_2, \ldots, v_n$ where $v_i = (e_i)\mu_P \kappa_\ell^{-1} = (e_i)P = $ (row i of P), and so μ_A has matrix PAP^{-1} relative to the basis ℓ' consisting of the rows of P. □

For instance, let $F = \mathbb{Q}$, $A = \begin{pmatrix} 1 & 2 \\ 3 & 4 \end{pmatrix}$, $P = \begin{pmatrix} 0 & 1 \\ 1 & 1 \end{pmatrix}$. Then, relative to ℓ' consisting of e_2 and $e_1 + e_2$ (the rows of P), the linear mapping μ_A has matrix $PAP^{-1} = \begin{pmatrix} 1 & 3 \\ 2 & 4 \end{pmatrix} = A^T$. (Actually, every square matrix over a field is similar to its transpose, though this simple fact is not easy to prove.)

**Definition
10.4**

The $n \times n$ matrix $D = [d_{ij}]$ is called **diagonal** if $d_{ij} = 0$ whenever $i \neq j$.

We now concentrate on linear mappings $\alpha : V \to V$ such that, relative to some basis of V, the matrix of α is diagonal; many, but by no means all, linear mappings of V have this property.

**Example
10.5**

Consider $\mu_A : \mathbb{Q}^2 \to \mathbb{Q}^2$ where $A = \begin{pmatrix} 7 & 4 \\ -2 & 1 \end{pmatrix}$. Let us find the matrix of μ_A relative to the basis $v_1 = (1, 2)$ and $v_2 = (1, 1)$ of $\mathbb{Q}^2$. from (10.3) this matrix is PAP^{-1} where $P = \begin{pmatrix} 1 & 2 \\ 1 & 1 \end{pmatrix}$; it is routine to find P^{-1} and verify that $PAP^{-1} = \begin{pmatrix} 3 & 0 \\ 0 & 5 \end{pmatrix}$, showing that PAP^{-1} is diagonal. So the basis v_1, v_2 of $\mathbb{Q}^2$ has the remarkable property that μ_A has a diagonal matrix relative to v_1, v_2. However, the question in the reader's mind should be: given A, how can a basis such as v_1, v_2 be found? It will take a little time to unravel the 'mystery', but as a first step notice:

$$(v_1)\mu_A = v_1 A = (1, 2)\begin{pmatrix} 7 & 4 \\ -2 & 1 \end{pmatrix} = (3, 6) = 3v_1$$

that is, μ_A maps v_1 into a scalar multiple of v_1; in fact μ_A simply multiplies v_1 by 3, which accounts for row 1 of PAP^{-1} being $3e_1$. Also

$$(v_2)\mu_A = v_2 A = (1, 1)\begin{pmatrix} 7 & 4 \\ -2 & 1 \end{pmatrix} = (5, 5) = 5v_2$$

shows that μ_A multiplies v_2 by 5 and accounts for the entries in row 2 of PAP^{-1}. Generally, vectors such as v_1 and v_2 which are mapped into scalar multiples of themselves, play an important role in the analysis of linear mappings.

Definition 10.6
Let $\alpha : V \to V$ be a linear mapping of the vector space V over the field F. Any *non-zero* vector v in V such that

$$(v)\alpha = \lambda v \quad \text{for some } \lambda \in F$$

is called a **characteristic (eigen-) vector** of α, and the scalar λ is called the corresponding **characteristic root (eigenvalue)** of α.

A characteristic vector of α is therefore any non-zero vector which is mapped into a scalar multiple of itself by α, the scalar involved being a characteristic root of α. As (10.5) shows, characteristic vectors are closely related to diagonalization.

Definition 10.7
Let A be an $n \times n$ matrix over the field F. Any *non-zero* vector v in F^n, such that $vA = \lambda v$ where $\lambda \in F$, is called a **characteristic (row) vector** of A and λ is called the corresponding **characteristic root** of A.

From (10.5) we see that $v_1 = (1, 2)$ and $v_2 = (1, 1)$ are characteristic vectors of $A = \begin{pmatrix} 7 & 4 \\ -2 & 1 \end{pmatrix}$, the corresponding characteristic roots being $\lambda_1 = 3$ and $\lambda_2 = 5$.

It follows directly from (10.6) and (10.7) that μ_A and A have the same characteristic vectors and characteristic roots. Our next lemma establishes the connection between characteristic vectors and characteristic roots of α and A, where A is any matrix representing α.

Lemma 10.8

Let the linear mapping $\alpha : V \to V$ have matrix A relative to the basis ℓ of the vector space V. Then α and A have the same characteristic roots, and v is a characteristic vector of α if and only if $(v)\kappa_\ell$ is a characteristic vector of A.

Proof

By (7.24), $(v)\kappa_\ell$ denotes the co-ordinate vector of v relative to ℓ; in fact $\kappa_\ell : V \cong F^n$ where V is n-dimensional over F, and the relationship between α and A is expressed by $\kappa_\ell\mu_A = \alpha\kappa_\ell$ (7.25).

Let v be a characteristic vector of α and λ the corresponding characteristic root; so $v \neq 0$ and $(v)\alpha = \lambda v$ by (10.6). Then $u = (v)\kappa_\ell \neq 0$ and

$$uA = (v)\kappa_\ell\mu_A = (v)\alpha\kappa_\ell = (\lambda v)\kappa_\ell = \lambda u$$

which shows that u is a characteristic vector of A by (10.7) with λ being the corresponding characteristic root of A. Conversely, if u in F^n satisfies $uA = \lambda u$ with $u \neq 0$, then, reversing the preceding argument, we see that v in V satisfies $(v)\alpha = \lambda v$, $v \neq 0$ where $(v)\kappa_\ell = u$. $\qquad \square$

So the characteristic roots of α can be found from any matrix A representing α. However, the characteristic roots of A are the zeros of the polynomial we now introduce using determinants.

Definition 10.9

Let A be a square matrix over a field. The polynomial

$$|xI - A|$$

is called the **characteristic polynomial** of A.

For instance, $A = \begin{pmatrix} 7 & 4 \\ -2 & 1 \end{pmatrix}$ as in (10.5) has characteristic polynomial

$$|xI - A| = \begin{vmatrix} x - 7 & -4 \\ 2 & x - 1 \end{vmatrix} = x^2 - 8x + 15 = (x - 3)(x - 5)$$

and the zeros 3 and 5 of this polynomial over $\mathbb{Q}$ are characteristic roots of A.

More generally, if A is an $n \times n$ matrix over the field F, then $|xI - A|$ is a monic polynomial of degree n over F.

Proposition Let A be an $n \times n$ matrix over the field F. The characteristic
10.10 roots of A are precisely the zeros (in F) of the characteristic
polynomial of A.

Proof Let λ be a characteristic root of A. By (10.7) there is a non-zero
vector v in F^n with $vA = \lambda v$; this equation can be rearranged to
give $v(\lambda I - A) = 0$. The homogeneous system $x(\lambda I - A) = 0$ has
the non-zero solution $x = v$, and so $|\lambda I - A| = 0$ by (9.41), show-
ing that λ is a zero (4.23) of the polynomial $|xI - A|$.
  Conversely let λ in F be a zero of $|xI - A|$. Then $|\lambda I - A| = 0$ and
so the homogeneous system $x(\lambda I - A) = 0$ has a solution $x = v \neq 0$
by (9.41). Therefore $v(\lambda I - A) = 0$ and so $vA = \lambda v$, showing that v
is a characteristic vector of A and λ is the corresponding characteristic root. □

  Therefore first find the characteristic roots of A by factorizing
$|xI - A|$, and then, for each characteristic root λ, find non-zero row
vectors v satisfying $v(\lambda I - A) = 0$, for such vectors v are characteristic vectors of A.
  Returning to (10.5), we know $A = \begin{pmatrix} 7 & 4 \\ -2 & 1 \end{pmatrix}$ has characteristic
roots 3 and 5. A characteristic vector corresponding to 3 is any
non-zero solution (x_1, x_2) of

$$(x_1, x_2)(3I - A) = (x_1, x_2)\begin{pmatrix} -4 & -4 \\ 2 & 2 \end{pmatrix} = 0$$

which boils down to the equation $2x_1 = x_2$; so $v_1 = (1, 2)$ (or any
non-zero multiple of v_1) is a characteristic vector of A with
corresponding characteristic root 3. Similarly, a characteristic vector
corresponding to 5 is any non-zero solution (x_1, x_2) of

$$(x_1, x_2)(5I - A) = (x_1, x_2)\begin{pmatrix} -2 & -4 \\ 2 & 4 \end{pmatrix} = 0$$

that is, $x_1 = x_2$; so $v_2 = (1, 1)$ (or any non-zero multiple of v_2) is a
characteristic vector of A with corresponding characteristic root 5.
  The procedure for finding characteristic vectors of a given $n \times n$
matrix A over F should now be clear (see (10.12) for more worked
examples). Our next proposition tells us that A can be diagonalized
(that is, A is similar to a diagonal matrix) if and only if F^n has a
basis consisting of characteristic vectors of A.

Proposition 10.11 Let A and P be $n \times n$ matrices, P being invertible over F. Then PAP^{-1} is diagonal if and only if the rows of P are characteristic vectors of A.

Proof Let $v_i = (\text{row } i \text{ of } P)$. By (8.12) $v_1, v_2, \ldots, v_n$ form a basis (which we call $\mathscr{b}'$ as in (10.3)) of F^n since P is invertible over F.

Suppose PAP^{-1} is diagonal having (i, i)-entry λ_i (we use this notation because, as we shall see in a moment, λ_i is a characteristic root of A). By (10.3) the linear mapping μ_A has matrix PAP^{-1} relative to $\mathscr{b}'$, which means $(v_i)\mu_A = \lambda_i v_i$ by (7.29), for (row i of $PAP^{-1}) = \lambda_i e_i$. Therefore $v_i A = \lambda_i v_i$ $(1 \le i \le n)$, showing that v_i is a characteristic vector of A (with λ_i being the corresponding characteristic root).

Conversely suppose $v_i A = \lambda_i v_i$ $(1 \le i \le n)$. The equations $(v_i)\mu_A = \lambda_i v_i$ tell us that the matrix of μ_A relative to $\mathscr{b}'$ is diagonal with (i, i)-entry λ_i by (7.29), that is, PAP^{-1} is diagonal by (10.3). $\square$

Having found the characteristic vectors of the $n \times n$ matrix A over F, select (if possible) a basis $v_1, v_2, \ldots, v_n$ of F^n from among them; the matrix P having the vectors v_i as its rows is then invertible and, by (10.11), such that PAP^{-1} is diagonal.

Example 10.12(a) Consider $A = \begin{pmatrix} 6 & -2 & 3 \\ 1 & 5 & 1 \\ 1 & 3 & 3 \end{pmatrix}$ over $\mathbb{Q}$.

We require the zeros of $|xI - A|$, but in practice it is easier to work with $|A - xI|$ $(= -|xI - A|$ as A is a 3×3 matrix). In this case

$$|A - xI| = \begin{vmatrix} 6-x & -2 & 3 \\ 1 & 5-x & 1 \\ 1 & 3 & 3-x \end{vmatrix} = (2 - x)(5 - x)(7 - x)$$

on subtracting row 3 from row 2, then adding columns 1 and 2 to column 3, etc.; therefore 2, 5, 7 are the characteristic roots of A. To find the characteristic vectors of A, we find non-zero solutions of $x(A - \lambda I) = 0$ taking $\lambda = 2, 5, 7$ in turn, where $x = (x_1, x_2, x_3) \in \mathbb{Q}^3$.

$$x(A - 2I) = (x_1, x_2, x_3) \begin{pmatrix} 4 & -2 & 3 \\ 1 & 3 & 1 \\ 1 & 3 & 1 \end{pmatrix} = 0$$

has non-zero solution $v_1 = (0, 1, -1)$.

$$x(A - 5I) = (x_1, x_2, x_3) \begin{pmatrix} 1 & -2 & 3 \\ 1 & 0 & 1 \\ 1 & 3 & -2 \end{pmatrix} = 0$$

has non-zero solution $v_2 = (3, -5, 2)$.

$$x(A - 7I) = (x_1, x_2, x_3) \begin{pmatrix} -1 & -2 & 3 \\ 1 & -2 & 1 \\ 1 & 3 & -4 \end{pmatrix} = 0,$$

that is, $-x_1 + x_2 + x_3 = 0$ and $-2x_1 - 2x_2 + 3x_3 = 0$ (the third equation can be ignored as it is a linear combination of the first two, the coefficient matrix having rank 2). Eliminating x_1 gives $4x_2 = x_3$ and so taking $x_2 = 1$ we see $x_3 = 4$ and hence $x_1 = 5$; so $v_3 = (5, 1, 4)$ is a non-zero solution.

We show in (10.18) that characteristic vectors of A corresponding to *distinct* characteristic roots are linearly independent. Therefore v_1, v_2, v_3 as above form a basis of $\mathbb{Q}^3$ and so are the rows of an invertible matrix P over $\mathbb{Q}$ such that PAP^{-1} is diagonal by (10.11); explicitly

$$PAP^{-1} = \begin{pmatrix} 2 & 0 & 0 \\ 0 & 5 & 0 \\ 0 & 0 & 7 \end{pmatrix} \quad \text{where} \quad P = \begin{pmatrix} 0 & 1 & -1 \\ 3 & -5 & 2 \\ 5 & 1 & 4 \end{pmatrix}.$$

Having completed a calculation of this kind, it is a good policy to check that no numerical mistakes have crept in by verifying that P is non-singular (in this case $|P| = -30$) and that $PA = DP$ where D is the diagonal matrix of characteristic roots of A; there is no need to find P^{-1}.

Example 10.12(b)   How can *multiple* zeros (zeros of multiplicity greater than 1) of the characteristic polynomial of A be dealt with? Either the technique of the previous example can be modified (as here) to diagonalize A as before, or (as in (c) below) A cannot be diagonalized.

Consider $A = \begin{pmatrix} 6 & 1 & 1 \\ -2 & 3 & -1 \\ 6 & 3 & 7 \end{pmatrix}$ over $\mathbb{Q}$. In this case $|A - xI| = (4 - x)^2(8 - x)$ and so the 3×3 matrix A has only two characteristic roots, namely 4 and 8. As before, the characteristic vectors of A are non-zero solutions of $x(A - \lambda I) = 0$ where $\lambda = 4, 8$.

$$x(A - 4I) = (x_1, x_2, x_3)\begin{pmatrix} 2 & 1 & 1 \\ -2 & -1 & -1 \\ 6 & 3 & 3 \end{pmatrix} = 0$$

amounts to the single equation $x_1 - x_2 + 3x_3 = 0$ for the coefficient matrix has rank 1. The solutions form the 2-dimensional subspace $\langle v_1, v_2 \rangle$ where $v_1 = (1, 1, 0)$ and $v_2 = (-3, 0, 1)$; then v_1 and v_2 are linearly independent characteristic vectors of A corresponding to the repeated zero $\lambda = 4$ (we are 'lucky' in the sense that v_1 and v_2 can be found). As in (10.12)(a), $v_3 = (2, 1, 1)$ is a non-zero solution of $x(A - 8I) = 0$. Then v_1, v_2, v_3 are linearly independent (this is a consequence of (10.17), but may be directly verified) and so form a basis of $\mathbb{Q}^3$ consisting of characteristic vectors of A; therefore the matrix P with $v_i = $ (row i of P) is such that PAP^{-1} is diagonal:

$$PAP^{-1} = \begin{pmatrix} 4 & 0 & 0 \\ 0 & 4 & 0 \\ 0 & 0 & 8 \end{pmatrix} \quad \text{where} \quad P = \begin{pmatrix} 1 & 1 & 0 \\ -3 & 0 & 1 \\ 2 & 1 & 1 \end{pmatrix}.$$

Notice that the *order* in which the characteristic vectors are used as rows of P determines the order in which the characteristic roots appear in PAP^{-1}.

Example 10.12(c) Consider $A = \begin{pmatrix} 6 & 1 & 1 \\ 6 & 7 & 3 \\ 1 & -2 & 3 \end{pmatrix}$ over $\mathbb{Q}$. As in (10.12)(b) we find $|A - xI| = (4 - x)^2(8 - x)$ and so A has characteristic roots 4 and 8.

$$x(A - 4I) = (x_1, x_2, x_3)\begin{pmatrix} 2 & 1 & 1 \\ 6 & 3 & 3 \\ 1 & -2 & -1 \end{pmatrix} = 0$$

has coefficient matrix of rank 2 and so, by (8.10), the solutions form a 1-dimensional subspace of $\mathbb{Q}^3$; in fact $\langle (3, -1, 0) \rangle$ is the solution space. In other words, there do *not* exist two linearly independent characteristic vectors of A corresponding to the repeated zero $\lambda = 4$ (this time we are 'unlucky'). As before the solutions x in $\mathbb{Q}^3$ of $x(A - 8I) = 0$ can be found: they form the 1-dimensional subspace $\langle (11, 3, 4) \rangle$ of $\mathbb{Q}^3$. Therefore each characteristic vector of A is either proportional to $(3, -1, 0)$ or proportional to $(11, 3, 4)$, and so A does not have three linearly independent characteristic vectors; by (10.11), A cannot be diagonalized: there is no invertible matrix P over $\mathbb{Q}$ (or any larger field) such that PAP^{-1} is diagonal.

**Example
10.12(d)**

The diagonalization process can be upset if the characteristic polynomial fails to factorize into linear factors. This problem can be
avoided, as illustrated here, by extending the field of scalars.
(There is a more serious limitation: the characteristic polynomial
may factorize, but it may be impossible, in practice, to find the
factors.)

For instance $A = \begin{pmatrix} 2 & 5 \\ -1 & -2 \end{pmatrix}$ has characteristic polynomial $x^2 + 1$.

Regarding A as a matrix over $\mathbb{R}$, we see that A has no characteristic
vectors in $\mathbb{R}^2$, for $x^2 + 1$ has no real zeros and hence A has no real
characteristic roots by (10.10). On the other hand, regarding A as
a matrix over $\mathbb{C}$, we may proceed as before because $x^2 + 1 =
(x - i)(x + i)$. The system $x(A - iI) = 0$ has non-zero solution
$v_1 = (1, 2 - i)$, and $x(A + iI) = 0$ has the (conjugate) non-zero
solution $v_2 = (1, 2 + i)$. As v_1 and v_2 are linearly independent
characteristic vectors with corresponding characteristic roots i and
$-i$, we obtain

$$PAP^{-1} = \begin{pmatrix} i & 0 \\ 0 & -i \end{pmatrix} \quad \text{where} \quad P = \begin{pmatrix} 1 & 2 - i \\ 1 & 2 + i \end{pmatrix}.$$

We now return to the theory.

**Lemma
10.13**

Let A and B be similar $n \times n$ matrices over the field F. Then A
and B have identical characteristic polynomials.

Proof

By (10.1) there is an invertible matrix P over F with $PAP^{-1} = B$.
Working in the ring of $n \times n$ matrices over $F[x]$ (matrices having
polynomial entries),

$$xI - B = P(xI)P^{-1} - PAP^{-1} = P(xI - A)P^{-1}$$

and taking determinants:

$$|xI - B| = |P|\,|xI - A|\,|P^{-1}| = |P|\,|P^{-1}|\,|xI - A| = |xI - A|$$

as these determinants belong to the commutative ring $F[x]$, and
$|P|\,|P^{-1}| = |PP^{-1}| = |I| = 1$. □

Notice that the converse of (10.13) is false: two matrices with
identical characteristic polynomials need not be similar. For instance, both

$$\begin{pmatrix} 0 & 0 \\ 0 & 0 \end{pmatrix} \quad \text{and} \quad \begin{pmatrix} 0 & 1 \\ 0 & 0 \end{pmatrix}$$

have characteristic polynomial x^2, but these matrices are not similar as their ranks are unequal.

Definition 10.14

Let the linear mapping $\alpha : V \to V$ of the vector space V over F have characteristic root $\lambda \in F$. Then $U_\lambda = \{v \in V : (v)\alpha = \lambda v\}$ is called the **characteristic subspace (eigenspace)** of α corresponding to λ.

It is straightforward using (6.6), to verify that U_λ is a subspace of V. The non-zero vectors in U_λ are characteristic vectors of α having λ as corresponding characteristic root. If λ is a characteristic root of the $n \times n$ matrix A over F, we refer to the characteristic subspace U_λ of A (meaning the characteristic subspace of $\mu_A : F^n \to F^n$).

The matrix A of (10.12)(a) has characteristic subspaces $U_2 = \langle (0, 1, -1) \rangle$, $U_5 = \langle (3, -5, 2) \rangle$, $U_7 = \langle (5, 1, 4) \rangle$. Using $\oplus$ in the sense of internal direct sum (6.28), it is legitimate to write $\mathbb{Q}^3 = U_2 \oplus U_5 \oplus U_7$; we shall say more about the iterated use of $\oplus$ shortly, but the above equation means that every vector v in $\mathbb{Q}^3$ is *uniquely* expressible $v = u_2 + u_5 + u_7$ where $u_i \in U_i$ $(i = 2, 5, 7)$.

The matrix A of (10.12)(b) has characteristic subspaces $U_4 = \langle (1, 1, 0), (-3, 0, 1) \rangle$ and $U_8 = \langle (2, 1, 1) \rangle$. $\mathbb{Q}^3 = U_4 \oplus U_8$, as U_4 and U_8 are complementary subspaces of $\mathbb{Q}^3$.

The matrix A of (10.12)(c) has characteristic subspaces $U_4 = \langle (3, -1, 0) \rangle$ and $U_8 = \langle (11, 3, 4) \rangle$. In this case it is legitimate to write $U_4 \oplus U_8$ in place of $U_4 + U_8$ as $U_4 \cap U_8 = 0$; however $\mathbb{Q}^3 \neq U_4 \oplus U_8$.

Definition 10.15

The subspaces $U_1, U_2, \ldots, U_t$ of the vector space V are called **independent** if $u_1 + u_2 + \ldots + u_t = 0$ where $u_i \in U_i$ $(1 \leq i \leq t)$ holds *only* in the case $u_1 = u_2 = \ldots = u_t = 0$.

Therefore the t subspaces U_i are independent if and only if whenever t vectors, one from each U_i, have sum zero, then each of these t vectors is zero. The characteristic subspaces of a given linear mapping of a vector space are independent (10.17)—hence our interest in subspaces of this type.

Let v belong to $U_1 + U_2 + \ldots + U_t$ where $U_1, U_2, \ldots, U_t$ are independent subspaces of V. Then $v = u_1 + u_2 + \ldots + u_t$ $(u_i \in U_i)$ by (6.11). Let us suppose $v = u_1' + u_2' + \ldots + u_t'$ $(u_i' \in U_i)$. Subtracting gives

$$(u_1 - u_1') + (u_2 - u_2') + \ldots + (u_t - u_t') = 0.$$

As $u_i - u_i' \in U_i$, by (10.15) each bracketed term above is zero, that is, $u_i = u_i'$ $(1 \leqslant i \leqslant t)$. Therefore v is *uniquely* expressible as $v = u_1 + u_2 + \ldots + u_t$ where $u_i \in U_i$; this uniqueness property is exactly what is meant by using the iterated direct sum notation $U_1 \oplus U_2 \oplus \ldots \oplus U_t$ in place of $U_1 + U_2 + \ldots + U_t$.

Conversely let $U_1, U_2, \ldots, U_t$ be subspaces of V such that each vector v in $U_1 + U_2 + \ldots + U_t$ can be expressed uniquely in the form $v = u_1 + u_2 + \ldots + u_t$ where $u_i \in U_i$ (that is, it is legitimate to form $U_1 \oplus U_2 \oplus \ldots \oplus U_t$). Taking $v = 0$ we see that $u_1 = u_2 = \ldots = u_t = 0$ (we are extracting the last drop of meaning from 'uniquely') and so $U_1, U_2, \ldots, U_t$ are independent by (10.15). Summing up:

It is legitimate to use $U_1 \oplus U_2 \oplus \ldots \oplus U_t$
in place of $U_1 + U_2 + \ldots + U_t$ if and only
if $U_1, U_2, \ldots, U_t$ are independent.

A word of caution! The internal direct sum $U_1 \oplus U_2 \oplus U_3$ makes sense only if the subspaces U_1, U_2, U_3 are independent, and their independence can be expressed (unsymmetrically) using (6.27) and (6.28) by the equations $U_1 \cap U_2 = (U_1 + U_2) \cap U_3 = 0$; but it is *not* enough for U_1, U_2, U_3 to be independent in pairs. For example, the subspaces $U_1 = \langle e_1 \rangle$, $U_2 = \langle e_2 \rangle$, $U_3 = \langle e_1 + e_2 \rangle$ of $\mathbb{Q}^2$ are not independent as $e_1 + e_2 + (-e_1 - e_2) = 0$, although $U_1 \cap U_2 = U_2 \cap U_3 = U_3 \cap U_1 = 0$; in this case $U_1 \oplus U_2 = U_2 \oplus U_3 = U_3 \oplus U_1 = \mathbb{Q}^2$, but $U_1 \oplus U_2 \oplus U_3$ does not make sense.

Our next corollary, which is an extension of (6.29), says that a basis of the direct sum of independent subspaces can be built up from bases of the individual components.

Corollary 10.16 Let $U_1, U_2, \ldots, U_t$ be independent subspaces of the vector space V and let U_i have basis b_i $(1 \leqslant i \leqslant t)$. Then $U_1 \oplus U_2 \oplus \ldots \oplus U_t$ has basis $b_1 \cup b_2 \cup \ldots \cup b_t$ and dimension dim $U_1 +$ dim $U_2 + \ldots +$ dim U_t.

Proof We use induction on t, there being nothing to do if $t = 1$. Taking $t > 1$, as $U_2, \ldots, U_t$ are independent, we may suppose that $W = U_2 \oplus \ldots \oplus U_t$ has basis $b_2 \cup \ldots \cup b_t$. As U_1 and W are independent, $U_1 \oplus W = U_1 \oplus U_2 \oplus \ldots \oplus U_t$ has basis $b_1 \cup (b_2 \cup \ldots \cup b_t) = b_1 \cup b_2 \cup \ldots \cup b_t$ by (6.27), which completes the induction. Each b_i consists of dim U_i vectors, that is, $|b_i| =$ dim U_i. Therefore

$$\dim(U_1 \oplus U_2 \oplus \ldots \oplus U_t) = |b_1 \cup b_2 \cup \ldots \cup b_t|$$
$$= |b_1| + |b_2| + \ldots + |b_t| = \dim U_1 + \dim U_2 + \ldots + \dim U_t. \qquad \square$$

We are on the point of justifying in general the procedure, used in (10.12), of building up a linearly independent set of characteristic vectors by considering each characteristic root in turn.

Proposition 10.17   Let $\alpha : V \to V$ be a linear mapping of the n-dimensional vector space V. Then the characteristic subspaces of α are independent.

Proof   From (10.8) and (10.10), α has $t\ (\leqslant n)$ *distinct* characteristic roots $\lambda_1, \lambda_2, \ldots, \lambda_t$. Let U_i denote the characteristic subspace of α corresponding to λ_i. The single subspace U_1 is independent, and so let us assume inductively that $U_1, U_2, \ldots, U_{s-1}$ are independent where $1 < s \leqslant t$. To show $U_1, U_2, \ldots, U_{s-1}, U_s$ are independent suppose

$$u_1 + u_2 + \ldots + u_{s-1} + u_s = 0 \quad \text{where} \quad u_i \in U_i \ (1 \leqslant i \leqslant s).$$

Applying α to this equation and using $(u_i)\alpha = \lambda_i u_i$ produces

$$\lambda_1 u_1 + \lambda_2 u_2 + \ldots + \lambda_{s-1} u_{s-1} + \lambda_s u_s = 0.$$

Using the above equations to eliminate u_s gives

$$(\lambda_1 - \lambda_s)u_1 + (\lambda_2 - \lambda_s)u_2 + \ldots + (\lambda_{s-1} - \lambda_s)u_{s-1} = 0.$$

As $(\lambda_i - \lambda_s)u_i \in U_i\ (1 \leqslant i < s)$, this equation and the independence of $U_1, U_2, \ldots, U_{s-1}$ tell us that $(\lambda_i - \lambda_s)u_i = 0$ for $1 \leqslant i < s$. As $\lambda_i - \lambda_s \neq 0$, we deduce $u_1 = u_2 = \ldots = u_{s-1} = 0$, and hence $u_s = 0$ also. Therefore $U_1, U_2, \ldots, U_{s-1}, U_s$ are independent. The induction is now complete, and so $U_1, U_2, \ldots, U_t$ are independent. $\square$

Therefore, as the examples (10.12) suggest, it is always legitimate to refer to the internal direct sum of the characteristic subspaces of the matrix A.

Corollary 10.18   Let A be an $n \times n$ matrix over the field F. Characteristic vectors of A corresponding to distinct characteristic roots are linearly independent. If A has n distinct characteristic roots, then A can be diagonalized.

Proof   Let $v_1, v_2, \ldots, v_t$ be characteristic vectors of A corresponding to the distinct characteristic roots $\lambda_1, \lambda_2, \ldots, \lambda_t$. Let U_i denote the characteristic subspace of A corresponding to λ_i. To show that $v_1, v_2, \ldots, v_t$ are linearly independent, suppose

$$a_1 v_1 + a_2 v_2 + \ldots + a_t v_t = 0 \quad \text{where} \quad a_i \in F.$$

As $a_i v_i \in U_i$ and $U_1, U_2, \ldots, U_t$ are independent by (10.17), we deduce $a_i v_i = 0$. But $v_i \neq 0$ and so $a_i = 0$ by (6.4)(f). Therefore $v_1, v_2, \ldots, v_t$ are linearly independent.

Suppose that A has n distinct characteristic roots $\lambda_1, \lambda_2, \ldots, \lambda_n$. There is a characteristic vector v_i of A corresponding to λ_i, and by the previous paragraph $v_1, v_2, \ldots, v_n$ are linearly independent vectors in F^n. Let P be the invertible matrix over F with $v_i = $ (row i of P). Then PAP^{-1} is diagonal by (10.11). □

Lastly, we cover the case of multiple zeros of the characteristic polynomial.

Theorem 10.19

Let A be an $n \times n$ matrix over the field F. Then A is similar to a diagonal matrix over F if and only if

(i) $|xI - A|$ factorizes into linear factors over F and
(ii) rank $(\lambda I - A) = n - m$ whenever λ is a zero of multiplicity m of $|xI - A|$.

Proof

Suppose first that there is an invertible matrix P over F such that $PAP^{-1} = D$, where D is diagonal with (i, i)-entry λ_i. By (10.13)

$$|xI - A| = |xI - D| = (x - \lambda_1)(x - \lambda_2) \ldots (x - \lambda_n)$$

as the determinant of the diagonal matrix $xI - D$ is the product of its diagonal entries (its (i, i)-entries) $x - \lambda_i$ for $1 \leq i \leq n$. Therefore $|xI - A|$ factorizes into linear factors over F. What is more, m_i of the diagonal entries in $xI - D$ are $x - \lambda_i$, where m_i is the multiplicity of λ_i as a zero of $|xI - A|$; in other words, λ_i occurs m_i times in the diagonal of D. As rank is unchanged on multiplying by invertible matrices and $P(\lambda_i I - A)P^{-1} = \lambda_i I - D$, we see

$$\text{rank}(\lambda_i I - A) = \text{rank}(\lambda_i I - D) = n - m_i$$

since $\lambda_i I - D$ is a diagonal matrix with $n - m_i$ non-zero entries (and the number of non-zero entries in a diagonal matrix is its rank). Therefore conditions (i) and (ii) above are satisfied.

Conversely, suppose that conditions (i) and (ii) are satisfied. Let $\lambda_1, \lambda_2, \ldots, \lambda_t$ denote the distinct zeros of $|xI - A|$. Collecting together equal factors of $|xI - A|$ gives

$$|xI - A| = (x - \lambda_1)^{m_1}(x - \lambda_2)^{m_2} \ldots (x - \lambda_t)^{m_t}$$

by condition (i), and so $n = m_1 + m_2 + \ldots + m_t$ on comparing degrees. Let U_i denote the characteristic subspace of A corresponding to λ_i; then

$$U_i = \{v \in F^n : vA = \lambda_i v\} = \{v \in F^n : v(\lambda_i I - A) = 0\},$$

that is, U_i is the solution space of the homogeneous system $v(\lambda_i I - A) = 0$, and so $\dim U_i = n - \text{rank}(\lambda_i I - A) = n - (n - m_i) = m_i$, using (8.10) and condition (ii). We are ready for the *coup de grâce*: let b_i denote a basis of U_i ($1 \leq i \leq t$); by (10.16), (10.17),

$\ell = \ell_1 \cup \ldots \cup \ell_t$ is a basis of $U_1 \oplus U_2 \oplus \ldots \oplus U_t$ and consists of

$$\dim U_1 + \dim U_2 + \ldots + \dim U_t = m_1 + m_2 + \ldots + m_t = n$$

vectors. So in fact $U_1 \oplus U_2 \oplus \ldots \oplus U_t = F^n$ and ℓ is a basis of F^n. Let P be the invertible matrix over F having the vectors in ℓ as its rows; as these vectors are characteristic vectors of A, the matrix PAP^{-1} is a diagonal matrix over F by (10.11). □

The matrix A of (10.12)(b) satisfies (10.19)(i) and (ii): this follows from $|xI - A| = (x - 4)^2(x - 8)$, $\operatorname{rank}(4I - A) = 3 - 2$, and $\operatorname{rank}(8I - A) = 3 - 1$ (condition (ii) tells us that the rank of $\lambda I - A$ drops (from n) by the multiplicity m of λ); and so, as we saw, A can be diagonalized over the rational field. On the other hand the matrix A of (10.12)(c) satisfies (10.19)(i), but not (10.19)(ii) as $\operatorname{rank}(4I - A) = 2 \neq 3 - 2$, and so A cannot be diagonalized.

Exercises 10.1

1. For each of the following matrices A, find, as in (10.12)(a), an invertible matrix P over $\mathbb{Q}$ with PAP^{-1} diagonal.

$$\begin{pmatrix} 2 & 2 & 1 \\ 2 & 2 & 3 \\ 1 & -1 & 3 \end{pmatrix}, \quad \begin{pmatrix} -1 & 7 & -4 \\ 3 & 3 & -4 \\ 3 & -11 & 10 \end{pmatrix}, \quad \begin{pmatrix} 3 & 1 & -1 \\ 2 & 3 & 2 \\ 5 & 3 & 5 \end{pmatrix}.$$

2. For each of the following matrices A, find, as in (10.12)(b), an invertible matrix P over $\mathbb{Q}$ with PAP^{-1} diagonal.

$$\begin{pmatrix} 7 & 2 & 1 \\ -1 & 4 & -1 \\ 2 & 4 & 8 \end{pmatrix}, \quad \begin{pmatrix} 1 & 2 & -4 \\ 6 & 2 & -8 \\ -6 & -4 & 6 \end{pmatrix}, \quad \begin{pmatrix} 1 & 2 & 3 \\ -8 & 9 & 6 \\ -4 & 2 & 8 \end{pmatrix}.$$

3. Diagonalize if possible (working over $\mathbb{R}$ or $\mathbb{C}$ as necessary) the following matrices:

$$\begin{pmatrix} 3 & -1 & 1 \\ 1 & 3 & -2 \\ -1 & 3 & -2 \end{pmatrix}, \quad \begin{pmatrix} 1 & 3 & 1 \\ 1 & 4 & 2 \\ -3 & -2 & -2 \end{pmatrix}, \quad \begin{pmatrix} -1 & 2 & -4 \\ 1 & -1 & 4 \\ 1 & -1 & 3 \end{pmatrix}.$$

4. (a) Partition the following matrices over $\mathbb{Q}$ into similarity classes. (Determine which pairs of matrices are similar and which not.)

$$\begin{pmatrix} 1 & 0 \\ 0 & 2 \end{pmatrix}, \quad \begin{pmatrix} 2 & 0 \\ 0 & 1 \end{pmatrix}, \quad \begin{pmatrix} 2 & 1 \\ 0 & 1 \end{pmatrix}, \quad \begin{pmatrix} -1 & -2 \\ 3 & 4 \end{pmatrix}, \quad \begin{pmatrix} -1 & 3 \\ -2 & 4 \end{pmatrix},$$

$$\begin{pmatrix} -9 & 1 \\ -110 & 12 \end{pmatrix}, \quad \begin{pmatrix} 1 & -3 \\ 2 & 4 \end{pmatrix}, \quad \begin{pmatrix} 2 & 1 \\ 0 & 2 \end{pmatrix}, \quad \begin{pmatrix} 2 & 0 \\ 0 & 2 \end{pmatrix}, \quad \begin{pmatrix} 2 & 2 \\ 0 & 2 \end{pmatrix}.$$

(b) Partition the sixteen 2×2 matrices over $\mathbb{Z}_2$ into similarity classes (there are six classes).

5. (a) Show that the matrices $\begin{pmatrix} a & 1 \\ 0 & b \end{pmatrix}$ and $\begin{pmatrix} a & 0 \\ 0 & b \end{pmatrix}$ over the field F are similar if and only if $a \neq b$.

(b) Show, directly from (10.1), that the matrices

$$\begin{pmatrix} a & b \\ c & d \end{pmatrix}, \quad \begin{pmatrix} a & -b \\ -c & d \end{pmatrix}, \quad \begin{pmatrix} d & c \\ b & a \end{pmatrix}$$

over the field F are similar.

(c) Let v_1 and v_2 be characteristic vectors of the linear mapping $\alpha : V \to V$ corresponding to distinct characteristic roots λ_1 and λ_2. Prove directly from (10.6) that v_1 and v_2 are linearly independent.

6. Show that the matrix $C = \begin{pmatrix} 0 & 1 & 0 \\ 0 & 0 & 1 \\ -a_0 & -a_1 & -a_2 \end{pmatrix}$ over the field F has

characteristic polynomial $f = x^3 + a_2 x^2 + a_1 x + a_0$. ($C$ is called the **companion matrix** of the monic polynomial f.) Use (10.18), (10.19) to show that C is similar to a diagonal matrix if and only if f factorizes into distinct linear factors.

If $f = x(x-1)(x-2)$ over $\mathbb{Q}$, find an invertible matrix P such that PCP^{-1} is diagonal.

7. Let $U_1, U_2, \ldots, U_t$ be subspaces of the vector space V.

(a) Prove that U_1, U_2, U_3, U_4 are independent if and only if

$$U_1 \cap U_2 = U_3 \cap U_4 = (U_1 + U_2) \cap (U_3 + U_4) = 0.$$

(b) Prove that $U_1, U_2, \ldots, U_t$ are independent if and only if $U_1, U_2, \ldots, U_{t-1}$ are independent and $(U_1 + U_2 + \ldots + U_{t-1}) \cap U_t = 0$.

8. (a) Let A and P be $n \times n$ matrices over a field with P invertible. Show that v is a characteristic vector of A if and only if vP^{-1} is a characteristic vector of PAP^{-1}.

(b) Let D be a diagonal $n \times n$ matrix over F with distinct diagonal entries. Show that the characteristic subspaces of D are $\langle e_i \rangle$ $(1 \leq i \leq n)$ where $e_1, e_2, \ldots, e_n$ form the standard basis of F^n.

(c) Describe (in terms of P) the characteristic vectors of $P^{-1}DP$, where P and D are as above. Find the 3×3 matrix over $\mathbb{Q}$ having characteristic vectors $(7, -3, 6)$, $(4, -3, 4)$, $(6, -2, 5)$ corresponding respectively to the characteristic roots 1, 2, 3.

9. (a) Let A and B be diagonal $n \times n$ matrices over a field. Show that A and B are similar if and only if their characteristic polynomials are identical.

(b) Let F be a finite field. Find a formula (in terms of $q = |F|$) for the number of similarity classes of 3×3 matrices over F containing diagonal matrices. Write down 10 diagonal 3×3 matrices over $\mathbb{Z}_3$ no two of which are similar.

10. Consider $A = \begin{pmatrix} \bar{4} & \bar{2} & \bar{3} \\ \bar{4} & -\bar{3} & \bar{4} \\ \bar{2} & -\bar{7} & \bar{3} \end{pmatrix}$ over the field $\mathbb{Z}_p$ (p prime). Find the

characteristic roots of A and show that A is similar to a diagonal matrix for just one prime p; for this prime, find an invertible matrix P over $\mathbb{Z}_p$ with PAP^{-1} diagonal.

11. Let λ be a characteristic root of the $n \times n$ matrix A over the field F, and write $l = \dim U_\lambda$, where $U_\lambda = \{v \in F^n : vA = \lambda v\}$. By extending a basis of U_λ to a basis of F^n, show that A is similar to a matrix of the form $\begin{pmatrix} \lambda I & \vdots \\ \cdots & \cdots \\ * & \vdots & * \end{pmatrix}$ where I is the $l \times l$ identity matrix (and the top-right is the $l \times (n - l)$ zero matrix). If λ is a zero of multiplicity m of $|xI - A|$, use (10.13) to deduce that $l \leqslant m$.

Diagonalization (continued) and the characteristic polynomial

We first apply the diagonalization process to solve a system of differential equations and then discuss further properties of the characteristic polynomial, including a most striking fact, namely, the Cayley–Hamilton theorem: every square matrix over a field satisfies its own characteristic equation; for example, $A = \begin{pmatrix} 3 & 6 \\ 2 & 7 \end{pmatrix}$ has characteristic polynomial $x^2 - 10x + 9$ and so (miraculously!) $A^2 - 10A + 9I = 0$.

Example 10.20

Consider the system of linear differential equations

$$\ddot{x}_1 = -4x_1 + 2x_2 + x_3$$
$$\ddot{x}_2 = 3x_1 - 3x_2 - x_3$$
$$\ddot{x}_3 = -6x_1 + 4x_2 + x_3$$

where x_1, x_2, x_3 are real functions of the real variable t and $\ddot{x}_i = \mathrm{d}^2 x_i / \mathrm{d}t^2$. These equations may be regarded as expressing the acceleration vector $\ddot{x} = (\ddot{x}_1, \ddot{x}_2, \ddot{x}_3)$ of a particle in terms of its co-ordinate vector $x = (x_1, x_2, x_3)$. The variables x_1, x_2, x_3 are present in each equation and appear to be inextricably mixed up. However, the coefficient matrix has been carefully chosen to be similar to a diagonal matrix, which means that the above system can be expressed

$$\ddot{y}_1 = \lambda_1 y_1$$
$$\ddot{y}_2 = \qquad \lambda_2 y_2$$
$$\ddot{y}_3 = \qquad\qquad \lambda_3 y_3$$

where y_1, y_2, y_3 are new variables related to x_1, x_2, x_3 by an invertible 3×3 matrix over $\mathbb{R}$. Each y_i occurs in just one equation, and so these equations can be solved separately and their solutions combined to produce the solution of the original system, as we now explain.

Writing $A = \begin{pmatrix} -4 & 3 & -6 \\ 2 & -3 & 4 \\ 1 & -1 & 1 \end{pmatrix}$, the given system is expressed by the

row equation $\ddot{x} = xA$. Let P be an invertible 3×3 matrix over $\mathbb{R}$ and let $y = (y_1, y_2, y_3)$ be a vector of new variables related to $x = (x_1, x_2, x_3)$ by $x = yP$. As the entries in P are constants (independent of t), differentiating twice with respect to t gives $\ddot{x} = \ddot{y}P$. On substituting $x = yP$ and $\ddot{x} = \ddot{y}P$, the equation $\ddot{x} = xA$ becomes $\ddot{y}P = yPA$, that is,

$$\ddot{y} = yPAP^{-1}$$

which is the original system of differential equations expressed in terms of the new variables. The matrix P is at our disposal; let us choose P so that PAP^{-1} is as simple as possible. In fact A can be diagonalized as in (10.12)(b); the reader may verify $|xI - A| = (x + 1)^2(x + 4)$ and

$$PAP^{-1} = \begin{pmatrix} -1 & 0 & 0 \\ 0 & -1 & 0 \\ 0 & 0 & -4 \end{pmatrix} \quad \text{where} \quad P = \begin{pmatrix} 2 & 3 & 0 \\ 1 & 0 & 3 \\ 1 & -1 & 2 \end{pmatrix}.$$

So $\ddot{y} = yPAP^{-1}$ consists of the differential equations

$$\ddot{y}_1 = -y_1$$
$$\ddot{y}_2 = \qquad -y_2$$
$$\ddot{y}_3 = \qquad\qquad -4y_3.$$

The general solution of $\ddot{y}_1 = -y_1$ is $y_1 = a_1 \cos t + b_1 \sin t$ and similarly we see $y_2 = a_2 \cos t + b_2 \sin t$ and $y_3 = a_3 \cos 2t + b_3 \sin 2t$ $(a_i, b_i \in \mathbb{R})$. As $x = yP$ we obtain

$$x_1 = 2y_1 + y_2 + y_3, \qquad x_2 = 3y_1 - y_3, \qquad x_3 = 3y_2 + 2y_3,$$

Fig. 10.1

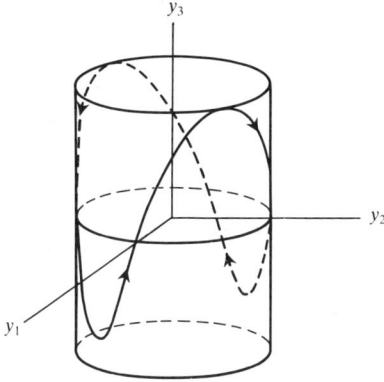

Path of particle in y-space

and substituting for the y_i produces the general solution of the given system in terms of cosine, sine, and the constants a_i and b_i.

To get an idea of the motion, suppose initially (at $t = 0$), that the particle sets off from $x = (2, 3, 0)$ with velocity $\ddot{x} = (3, -2, 7)$. A little calculation shows $a_1 = b_2 = b_3 = 1$ and $b_1 = a_2 = a_3 = 0$, and so

$$y = (y_1, y_2, y_3) = (\cos t, \sin t, \sin 2t).$$

Treating y_1, y_2, y_3 as spatial co-ordinates (and referring to y-space), from the above expression we see that the particle follows the sine wave $y_3 = \sin 2t$ wrapped round the cylinder $y_1^2 + y_2^2 = 1$ (Fig. 10.1). The actual path of the particle in x-space (that is, in terms of x_1, x_2, x_3) is a *distortion* of this picture, for we must apply $\mu_P : \mathbb{R}^3 \cong \mathbb{R}^3$ (which maps y-space isomorphically to x-space) in order to describe the motion using the original co-ordinates. Now μ_P maps $\langle e_1, e_2 \rangle$ (the plane $y_3 = 0$) to the characteristic subspace $U_{-1} = \langle v_1, v_2 \rangle$ of A, and μ_P maps $\langle e_3 \rangle$ (the y_3-axis) to the characteristic subspace $U_{-4} = \langle v_3 \rangle$ of A, where $v_i = $ row i of P. Without going into the details, we note some properties of the motion in x-space:

(i) the particle moves on a cylinder with axis $U_{-4} = \langle (1, -1, 2) \rangle$ having an elliptic intersection (the image by μ_P of the circle $y_1^2 + y_2^2 = 1$, $y_3 = 0$) with the plane U_{-1} having equation $3x_1 - 2x_2 - x_3 = 0$,

(ii) the particle moves between the planes $3x_1 - 2x_2 - x_3 = \pm 3$ (the images of the planes $y_3 = \pm 1$ by μ_P),

(iii) the path of the particle crosses U_{-1} at $\pm v_1$, $\pm v_2$ (as in y-space the path crosses $y_3 = 0$ at $\pm e_1$, $\pm e_2$).

We next investigate the coefficients in the characteristic polynomial.

Definition 10.21 Let A be an $n \times n$ matrix over a field. Suppose that rows $i_1, i_2, \ldots, i_{n-r}$ and columns $i_1, i_2, \ldots, i_{n-r}$ of A are deleted $(1 \leqslant i_1 < i_2 < \ldots < i_{n-r} \leqslant n)$. Then the determinant of the remaining $r \times r$ matrix is called a **principal r-minor** of A.

The reader should compare (10.21) with (9.39); to form a principal r-minor, delete any $n - r$ rows and the same $n - r$ columns of A, and take the determinant of what is left. The principal 1-minors of the 3×3 matrix $A = [a_{ij}]$ are the diagonal entries a_{11}, a_{22}, a_{33}; the principal 2-minors of A are

$$\begin{vmatrix} a_{11} & a_{12} \\ a_{21} & a_{22} \end{vmatrix}, \quad \begin{vmatrix} a_{11} & a_{13} \\ a_{31} & a_{33} \end{vmatrix}, \quad \begin{vmatrix} a_{22} & a_{23} \\ a_{32} & a_{33} \end{vmatrix}$$

and $|A|$ is the only principal 3-minor of A.

Proposition 10.22 Let A be an $n \times n$ matrix over a field and let s_r denote the sum of the principal r-minors of A. Then $(-1)^{n-r}s_{n-r}$ is the coefficient of x^r in the characteristic polynomial $|xI - A|$ of A $(0 \leqslant r < n)$.

Proof Let $A = [a_{ij}]$ and write $B = [b_{ij}] = xI - A$; so $b_{ii} = x - a_{ii}$, and $b_{ij} = -a_{ij}$ for $i \neq j$. Denoting by c_r the coefficient of x^r in $|xI - A|$, we have

$$\sum_{r=0}^{n} c_r x^r = |xI - A| = |B| = \sum_{\alpha \in S_n} (-1)^{N(\alpha)} b_{1\,(1)\alpha} b_{2\,(2)\alpha} \cdots b_{n\,(n)\alpha}.$$

For each subset X of r distinct integers in $\{1, 2, \ldots, n\}$, let M_X be the principal $(n - r)$-minor of A obtained by deleting row i and column i for all $i \in X$; then M_X is the sum of $(n - r)!$ terms of the form

$$(-1)^{N(\alpha')} \prod_{i \in X'} a_{i\,(i)\alpha'}$$

where α' ranges over all permutations of the complement X' of X in $\{1, 2, \ldots, n\}$.

How do contributions to $c_r x^r$ arise from terms $(-1)^{N(\alpha)} b_{1\,(1)\alpha} b_{2\,(2)\alpha} \cdots b_{n\,(n)\alpha}$ of $|B|$? We may refer to the (X, α)-**contribution**, meaning the product obtained on multiplying x (selected from each of the r factors $b_{i\,(i)\alpha}$ with $i \in X$) by $-a_{i\,(i)\alpha}$ (selected from $b_{i\,(i)\alpha}$ with $i \in X'$); such a selection is possible only if $(i)\alpha = i$ for all $i \in X$, in which case α is effectively a permutation α'

of X' and the (X, α)-contribution is

$$(-1)^{N(\alpha')} \prod_{i \in X'} (-a_{i\,(i)\alpha'}) x^r$$

for α and α' (the restriction of α to X') have the same parity by (9.26). Keeping X fixed, the sum of the (X, α)-contributions for α in S_n is therefore

$$\sum_{\alpha'} \left((-1)^{N(\alpha')} \prod_{i \in X'} -a_{i\,(i)\alpha'} \right) x^r = (-1)^{n-r} M_X x^r$$

where the summation is over all permutations α' of X'. As $c_r x^r$ is the sum of all the (X, α)-contributions, we see $c_r = (-1)^{n-r} \sum_X M_X = (-1)^{n-r} s_{n-r}$, the summation being over all subsets X of r distinct integers in $\{1, 2, \ldots, n\}$. □

As similar matrices have identical characteristic polynomials by (10.13), we see, on comparing coefficients by (10.22), that similar $n \times n$ matrices have equal sums s_r of principal r-minors $(1 \le r \le n)$.

Definition 10.23 Let $A = [a_{ij}]$ be an $n \times n$ matrix over a field. The sum $s_1 = a_{11} + a_{22} + \ldots + a_{nn}$ of the diagonal entries in A is called the **trace** of A.

As similar matrices have equal traces, one can often see at a glance that two matrices are *not* similar; for instance

$$A = \begin{pmatrix} 5 & 1 \\ 4 & 2 \end{pmatrix} \quad \text{and} \quad B = \begin{pmatrix} 3 & 0 \\ 7 & 2 \end{pmatrix}$$

over $\mathbb{Q}$ are not similar as trace $A = 7$ and trace $B = 5$.

Corollary 10.24 Let A be an $n \times n$ matrix over the field F and suppose $|xI - A| = (x - \lambda_1)(x - \lambda_2) \ldots (x - \lambda_n)$ where $\lambda_i \in F$. Then s_r, the sum of the principal r-minors of A, is the sum of the $\binom{n}{r}$ products of r characteristic roots from among $\lambda_1, \lambda_2, \ldots, \lambda_n$ $(1 \le r \le n)$. In particular

$$\text{trace } A = \lambda_1 + \lambda_2 + \ldots + \lambda_n \quad \text{and} \quad |A| = \lambda_1 \lambda_2 \ldots \lambda_n.$$

Proof The coefficient of x^{n-r} in $|xI - A|$ is $(-1)^r s_r$ by (10.22). On the other hand, multiplying out $(x - \lambda_1)(x - \lambda_2) \ldots (x - \lambda_n)$, we see that this coefficient is $(-1)^r \sum \lambda_{i_1} \lambda_{i_2} \ldots \lambda_{i_r}$, where the summation is over all integers $i_1, i_2, \ldots, i_r$ with $1 \le i_1 < i_2 < \ldots < i_r \le n$.

Therefore $s_r = \sum \lambda_{i_1}\lambda_{i_2}\ldots\lambda_{i_r}$; the extreme cases are: trace $A = s_1 = \lambda_1 + \lambda_2 + \ldots + \lambda_n$ and $|A| = s_n = \lambda_1\lambda_2\ldots\lambda_n$. $\square$

The trace being the sum of the characteristic roots (counted according to their multiplicities as zeros of the characteristic polynomial) provides an immediate check. For instance, having found that

$$A = \begin{pmatrix} 4 & 1 & 3 \\ -22 & -1 & 6 \\ 9 & 1 & -2 \end{pmatrix}$$

has characteristic roots $3, 3, -5$, it is reassuring (but nothing more) to note:

$$3 + 3 - 5 = 4 - 1 - 2.$$

One thing *is* certain: if the calculated roots don't add up to the trace, then there's a mistake somewhere! In this case, $|A| = s_3 = 3 \times 3 \times (-5) = -45$, and
$s_2 = 3 \times 3 + 3 \times (-5) + 3 \times (-5) = -21$, which agrees with the direct calculation

$$s_2 = \begin{vmatrix} 4 & 1 \\ -22 & -1 \end{vmatrix} + \begin{vmatrix} 4 & 3 \\ 9 & -2 \end{vmatrix} + \begin{vmatrix} -1 & 6 \\ 1 & -2 \end{vmatrix}.$$

Notation Let A be an $n \times n$ matrix over the field F and let $f = c_m x^m + \ldots + c_1 x + c_0$ be a polynomial over F. Write $f(A) = c_m A^m + \ldots + c_1 A + c_0 I$ and $F[A] = \{f(A) : f \in F[x]\}$.

So $f(A)$ is itself an $n \times n$ matrix over F, and the set $F[A]$ of all such matrices is a commutative subring of $M_n(F)$; in fact $F[A]$ is the smallest subring of $M_n(F)$ containing the scalar matrices aI (for all $a \in F$) and the given matrix A.

To get the idea of the next theorem (and its proof) consider

$$A = \begin{pmatrix} 2 & 1 & 3 \\ 3 & 1 & 4 \\ 1 & 2 & 5 \end{pmatrix}$$

over $\mathbb{Q}$. Denoting the characteristic polynomial of A by χ, it is routine to show $\chi = |xI - A| = x^3 - 8x^2 + 3x + 2$, and hence verify the remarkable fact

$$\chi(A) = A^3 - 8A^2 + 3A + 2I = 0$$

that is, $\chi(A)$ is the zero matrix! The proof, in general, makes use of $xI - A$ and $\mathrm{adj}(xI - A)$, which are matrices with polynomial entries. In this case

$$\mathrm{adj}(xI - A) = \begin{pmatrix} x^2 - 6x - 3 & x + 1 & 3x + 1 \\ 3x - 11 & x^2 - 7x + 7 & 4x + 1 \\ x + 5 & 2x - 3 & x^2 - 3x - 1 \end{pmatrix}$$

$$= x^2 \begin{pmatrix} 1 & 0 & 0 \\ 0 & 1 & 0 \\ 0 & 0 & 1 \end{pmatrix} + x \begin{pmatrix} -6 & 1 & 3 \\ 3 & -7 & 4 \\ 1 & 2 & -3 \end{pmatrix} + \begin{pmatrix} -3 & 1 & 1 \\ -11 & 7 & 1 \\ 5 & -3 & -1 \end{pmatrix}$$

and so $\mathrm{adj}(xI - A)$ is a polynomial in x with matrix coefficients; these coefficients, which are written on the right of the powers of x, belong to $\mathbb{Q}[A]$ and are closely related to χ, for

$$\mathrm{adj}(xI - A) = x^2 I + x(A - 8I) + (A^2 - 8A + 3I).$$

In fact $\mathrm{adj}(xI - A)$ is the quotient on dividing $\chi I = x^3 I + x^2(-8I) + x(3I) + 2I$ by $xI - A$, working with polynomials over $\mathbb{Q}[A]$, for

$$\chi I = (xI - A)(\mathrm{adj}(xI - A))$$

by (9.35); this division is exact (the remainder is zero), and so comparing constant terms (or using (4.25)) gives $2I = (A^2 - 8A + 3I)(-A)$, which rearranges to $\chi(A) = 0$. We have reached the same point, but by a route which suggests how the general case can be tackled.

Theorem 10.25

(The Cayley–Hamilton theorem). Let A be an $n \times n$ matrix over a field F. Then $\chi(A) = 0$, where χ is the characteristic polynomial of A.

Proof

We use polynomials with coefficients in the ring $M_n(F)$. The scalar matrices aI $(a \in F)$ form a subring (of $M_n(F)$) isomorphic to F, and so $\chi = x^n + c_{n-1}x^{n-1} + \ldots + c_1 x + c_0$ over F can be 'promoted' to the polynomial

$$\chi I = x^n I + x^{n-1}(c_{n-1}I) + \ldots + x(c_1 I) + c_0 I$$

over $F[A]$. Let us divide χI by the monic polynomial $xI - A$ over $F[A]$: by (4.8) there are unique polynomials q and r over $F[A]$ satisfying

$$\chi I = (xI - A)q + r, \qquad r \text{ constant},$$

and, what is more, q and r are the only polynomials over $M_n(F)$

satisfying the above conditions. Comparing the above equation with

$$\chi I = (xI - A)(\mathrm{adj}(xI - A))$$

from (9.35), as $\mathrm{adj}(xI - A)$ is a polynomial over $M_n(F)$ we conclude $q = \mathrm{adj}(xI - A)$ and $r = 0$. Therefore $\chi I = (xI - A)q$ is an equality of polynomials over the commutative ring $F[A]$, and so replacing x by A, as in (4.25), gives $\chi(A) = (AI - A)q(A) = 0$. $\square$

The Cayley–Hamilton theorem is valid for square matrices A having entries from a commutative ring, for the above proof goes through unchanged.

We show next how real 2×2 matrices may be partitioned into similarity classes; this classification is used in the theory of differential equations.

Proposition 10.26 Let A be a 2×2 matrix over the real field $\mathbb{R}$. Then there is an invertible 2×2 matrix P over $\mathbb{R}$ such that PAP^{-1} is one of:

$$\begin{pmatrix} \lambda_1 & 0 \\ 0 & \lambda_2 \end{pmatrix}, \quad \begin{pmatrix} \lambda_1 & 1 \\ 0 & \lambda_1 \end{pmatrix}, \quad \begin{pmatrix} a & b \\ -b & a \end{pmatrix}$$

where λ_1, λ_2, a, b are real numbers, with $b \neq 0$.

Proof Suppose that there is *no* invertible matrix P over $\mathbb{R}$ such that PAP^{-1} is diagonal. There remain two cases depending on whether or not the characteristic polynomial χ of A factorizes into linear factors over $\mathbb{R}$.

Suppose first that $\chi = (x - \lambda_1)^2$, for χ cannot have two distinct real zeros by (10.18). Let v_1 be any vector in $\mathbb{R}^2$ which is *not* a characteristic vector of A and write $v_2 = v_1 A - \lambda_1 v_1$; then $v_2 \neq 0$ and $v_2 A - \lambda_1 v_2 = v_2(A - \lambda_1 I) = v_1(A - \lambda_1 I)(A - \lambda_1 I) = v_1 \chi(A) = 0$ as $\chi(A) = 0$ by (10.25) (it is an easy exercise to directly verify (10.25) in the 2×2 case). Therefore v_2 is a characteristic vector of A and so v_1 and v_2, being non-proportional, form a basis of $\mathbb{R}^2$. The equations

$$\begin{aligned} v_1 A &= \lambda_1 v_1 + v_2 \\ v_2 A &= \quad\ \ \lambda_1 v_2 \end{aligned} \quad \text{imply} \quad PAP^{-1} = \begin{pmatrix} \lambda_1 & 1 \\ 0 & \lambda_1 \end{pmatrix} \quad \text{where} \quad P = \begin{pmatrix} v_1 \\ v_2 \end{pmatrix}.$$

Suppose that χ does not factorize over $\mathbb{R}$. On completing the square, we obtain real numbers a, $b \neq 0$ such that $\chi = (x - a)^2 + b^2$. Let v be any non-zero vector in $\mathbb{R}^2$; then v and vA form a basis of $\mathbb{R}^2$ since A has no real characteristic roots. Write

$v_1 = vA - (a - b)v$ and $v_2 = vA - (a + b)v$; as $v_1 - v_2 = 2bv$, we see $\langle v_1, v_2 \rangle = \langle v, vA \rangle$ and so v_1 and v_2 form a basis of $\mathbb{R}^2$. Substituting for v_1 and v_2 and using $\chi(A) = A^2 - 2aA + (a^2 + b^2)I = 0$ gives

$$v_1 A = \quad av_1 + bv_2$$
$$v_2 A = -bv_1 + av_2$$

and so $PAP^{-1} = \begin{pmatrix} a & b \\ -b & a \end{pmatrix}$ where

$$P = \begin{pmatrix} v_1 \\ v_2 \end{pmatrix}.$$

□

Instead of diagonalizing $A = \begin{pmatrix} 2 & 5 \\ -1 & -2 \end{pmatrix}$ as in (10.12)(d), it is more appropriate to use the real invertible matrix P above; taking $v = (1, 0)$ gives $v_1 = (3, 5)$, $v_2 = (1, 5)$, and $PAP^{-1} = \begin{pmatrix} 0 & 1 \\ -1 & 0 \end{pmatrix}$ where $P = \begin{pmatrix} 3 & 5 \\ 1 & 5 \end{pmatrix}$.

In conclusion, let us note that diagonalization, though attractive, does have certain inherent limitations. Even when theoretically possible, diagonalization may not be a practical proposition because of the difficulty in factorizing the characteristic polynomial. Further, diagonalization is something of a red herring as far as similarity is concerned! For just as the g.c.d. of two integers can be found without knowing their prime factorization, so the basic questions about similarity can be answered by an algorithmic technique (involving row and column operations, but too complicated for discussion here) in which diagonalization plays no part and factorization problems do not arise.

Exercises 10.2

1. For each matrix A below, find an invertible matrix P over $\mathbb{R}$ such that PAP^{-1} is as in (10.26).

$$\begin{pmatrix} 4 & 2 \\ 1 & 3 \end{pmatrix}, \quad \begin{pmatrix} 7 & 1 \\ -4 & 3 \end{pmatrix}, \quad \begin{pmatrix} 1 & 8 \\ -1 & -3 \end{pmatrix}.$$

2. Verify the Cayley–Hamilton theorem (10.25) for the matrices A:

$$\begin{pmatrix} 2 & 3 \\ 4 & 5 \end{pmatrix}, \quad \begin{pmatrix} 6 & 9 & 1 \\ -4 & -6 & -1 \\ 3 & 4 & -1 \end{pmatrix}.$$

In each case express $\text{adj}(xI - A)$ as a polynomial in x with matrix coefficients, and express each coefficient as a polynomial in A.

3. (a) Find the general solution of the system of differential equations:

$$\ddot{x}_1 = -2x_1 + 4x_2 + 2x_3, \qquad \ddot{x}_2 = x_1 - 2x_2 + x_3, \qquad \ddot{x}_3 = -x_1 - 2x_2 - 5x_3.$$

Find the particular solution with initial position vector $(4, 0, -1)$ and initial velocity vector $(0, -2, 4)$. Determine where this solution crosses the plane $x_1 + 2x_2 + x_3 = 0$ and where it is at maximum distance from this plane.

(b) Find the general solution of the system:

$$\dot{x}_1 = 4x_1 + x_2 + 8x_3, \quad \dot{x}_2 = 7x_1 - x_2 + 9x_3, \quad \dot{x}_3 = -5x_1 - x_2 - 9x_3.$$

If the initial position vector is $(2, 3, -2)$, show that $x = (x_1, x_2, x_3)$ remains in a 2-dimensional subspace of $\mathbb{R}^3$. What happens to x if $x = (9, 11, -7)$ initially?

4. Let A be a diagonal 3×3 matrix over F with distinct diagonal entries. Show that every diagonal matrix in $M_3(F)$ can be expressed in the form $a_0 I + a_1 A + a_2 A^2$ $(a_i \in F)$.

Let A be a 3×3 matrix over F having three distinct characteristic roots.

Show that $F[A] = \{X \in M_3(F) : XA = AX\}$.

5. (a) Let A be a 2×2 matrix over $\mathbb{R}$ having no real characteristic roots. Show, as in (10.26), that A is similar over $\mathbb{R}$ to a rotation matrix if and only if $|A| = 1$.

(b) Show that the 2×2 rotation matrices A, B are similar if and only if either $A = B$ or $AB = I$.

(c) Show that the reflection matrix $\begin{pmatrix} \cos \theta & \sin \theta \\ \sin \theta & -\cos \theta \end{pmatrix}$ is similar to $\begin{pmatrix} 1 & 0 \\ 0 & -1 \end{pmatrix}$.

6. Let A and B be $n \times n$ matrices over F. Working in the ring of polynomials over $M_n(F)$, use (9.35) to show

$$\mathrm{adj}((xI + A)(xI + B)) = \mathrm{adj}(xI + B)\mathrm{adj}(xI + A)$$

and deduce $\mathrm{adj}(AB) = (\mathrm{adj}\, B)(\mathrm{adj}\, A)$.

7. Let A be an $n \times n$ matrix over the field F with characteristic polynomial χ.

(a) Let $f = a_m x^m + \ldots + a_1 x + a_0$ be a polynomial over F with $f(A) = 0$. Show that $xI - A$ is a factor of $fI = x^m(a_m I) + \ldots + x(a_1 I) + a_0 I$ as polynomials over $F[A]$. Deduce $\chi \mid f^n$ by taking determinants.

(b) Show that $\alpha : F[x] \to F[A]$, defined by $(f)\alpha = f(A)$ for all f in $F[x]$, is a surjective ring homomorphism with $\chi \in \ker \alpha$. Let μ denote the monic generator of $\ker \alpha$ (μ, being the monic polynomial of least degree over F with $\mu(A) = 0$, is called the **minimum polynomial** of A). Show that $\mu \mid \chi$ and $\chi \mid \mu^n$, and deduce that μ and χ have the same irreducible factors over F.

(c) By testing I, A, A^2 for linear dependence, find the minimum polynomial of each matrix A over $\mathbb{Q}$ below:

$$\begin{pmatrix} 2 & 3 & 1 \\ -2 & -5 & -2 \\ 4 & 12 & 5 \end{pmatrix}, \quad \begin{pmatrix} 4 & 3 & 3 \\ 2 & 3 & 2 \\ -7 & -7 & -6 \end{pmatrix}, \quad \begin{pmatrix} 4 & 2 & 1 \\ 2 & 7 & 2 \\ 1 & 2 & 4 \end{pmatrix}.$$

By considering trace A, factorize (without further calculation) the characteristic polynomial of A.

Duality

A finite-dimensional vector space V is a very symmetrical object: not only does it possess 'side-to-side' symmetry (V, viewed from any of its bases, looks the same), but V has an 'upside-down' form of symmetry called **duality**. Here we 'invert' V obtaining its dual space denoted by $\hat{V}$ (rather than Λ) and study the connection between these spaces.

Definition 10.27

Let V be a vector space over the field F. A linear mapping $\alpha : V \to F$ is called a **linear form** on V. The vector space $\text{Hom}(V, F)$, of all linear forms on F, is called the **dual space** of V and denoted by $\hat{V}$.

A linear form on V is therefore a scalar-valued linear mapping. By (7.21), we know that $\hat{V} = \text{Hom}(V, F)$ is a vector space over F; if V is a left vector space, we regard $\hat{V}$ as a right vector space (the mapping $\alpha a : V \to F$, defined by $(v)(\alpha a) = ((v)\alpha)a$ for $v \in V$, is linear for $\alpha \in \hat{V}$ and $a \in F$).

Our discussion of duality will be confined to finite-dimensional vector spaces. Let ℓ denote the basis $v_1, v_2, \ldots, v_n$ of V. In the vector space F, every scalar is a vector (and vice-versa); in particular the vector 1 is the standard basis ℓ_s of F. We make use of the isomorphism

$$\hat{V} = \text{Hom}(V, F) \cong {}^n F$$

$$\alpha \to c$$

of (7.26), where $c = ((v_1)\alpha, (v_2)\alpha, \ldots, (v_n)\alpha)^{\text{T}}$ is the matrix of α relative to ℓ and ℓ_s.

Definition 10.28

The basis $\hat{v}_1, \hat{v}_2, \ldots, \hat{v}_n$ of $\hat{V}$, which maps by the above isomorphism to the standard basis $e_1^{\text{T}}, e_2^{\text{T}}, \ldots, e_n^{\text{T}}$ of ${}^n F$, is called the basis $\hat{\ell}$ **dual** to $v_1, v_2, \ldots, v_n$.

So dim $V = \dim \hat{V}$ and each basis $\mathscr{b}$ of V has a corresponding dual basis $\hat{\mathscr{b}}$ of $\hat{V}$. As $e_1^T, e_2^T, \ldots, e_n^T$ are the columns of the $n \times n$ identity matrix, the linear forms $\hat{v}_1, \hat{v}_2, \ldots, \hat{v}_n$ are related to $v_1, v_2, \ldots, v_n$ by

$$(v_i)\hat{v}_j = \delta_{ij} \qquad (1 \leqslant i, j \leqslant n),$$

that is, $\hat{v}_j$ takes the value 1 on the vector v_j and the value 0 on all other vectors in $\mathscr{b}$; therefore $\hat{v}_j$ depends on the whole basis $\mathscr{b}$ and not (as the notation might suggest) solely on v_j.

By (7.17) the linear forms on F^n are the mappings μ_c for $c \in {}^nF$ (c is an $n \times 1$ column matrix over F). The above equations connecting v_i and $\hat{v}_j$ tell us that if the vectors $v_1, v_2, \ldots, v_n$ in the basis $\mathscr{b}$ of F^n are the rows of the invertible matrix P, then $\hat{v}_j = \mu_{c_j}$ $(1 \leqslant j \leqslant n)$ where c_j is column j of P^{-1}. So finding a dual basis is essentially the same as inverting a matrix.

For instance, the vectors $v_1 = (1, \ 2, \ 2)$, $v_2 = (3, \ 1, \ 2)$, $v_3 = (2, -2, -1)$ in the basis $\mathscr{b}$ of $\mathbb{R}^3$ are the rows of

$$P = \begin{pmatrix} 1 & 2 & 2 \\ 3 & 1 & 2 \\ 2 & -2 & -1 \end{pmatrix} \quad \text{and} \quad P^{-1} = \begin{pmatrix} 3 & -2 & 2 \\ 7 & -5 & 4 \\ -8 & 6 & -5 \end{pmatrix} = (c_1, c_2, c_3).$$

So $(x_1, x_2, x_3)\hat{v}_1 = (x_1, x_2, x_3)\mu_{c_1} = 3x_1 + 7x_2 - 8x_3$, which satisfies $(v_1)\hat{v}_1 = 1$, $(v_2)\hat{v}_1 = 0$, and $(v_3)\hat{v}_1 = 0$, and similarly for $\hat{v}_2 = \mu_{c_2}$ and $\hat{v}_3 = \mu_{c_3}$.

Returning to (10.28), each v in V is uniquely expressible as $v = a_1 v_1 + \ldots + a_n v_n$; evaluating $\hat{v}_i$ at v produces

$$(v)\hat{v}_i = (a_1 v_1 + \ldots + a_n v_n)\hat{v}_i = a_i((v_i)\hat{v}_i) = a_i$$

giving the formula

$$v = \sum_{i=1}^{n} ((v)\hat{v}_i)v_i.$$

Similarly, each α in $\hat{V}$ is uniquely expressible as $\alpha = \hat{v}_1 a_1 + \ldots + \hat{v}_n a_n$; evaluation at v_i gives

$$(v_i)\alpha = (v_i)(\hat{v}_1 a_1 + \ldots + \hat{v}_n a_n) = ((v_i)\hat{v}_i)a_i = a_i$$

and so we obtain

$$\alpha = \sum_{i=1}^{n} \hat{v}_i((v_i)\alpha).$$

We now investigate $\hat{\hat{V}}$ (the dual space of $\hat{V}$) which turns out to be a left vector space closely related to V. The elements (vectors) of $\hat{\hat{V}}$

are linear mappings $f : \hat{V} \to F$ (here $\hat{V}$ and F are regarded as right vector spaces and the functional notation is used; so f satisfies: $f(\alpha + \beta) = f(\alpha) + f(\beta)$ and $f(\alpha a) = f(\alpha)a$ for α, $\beta \in \hat{V}$ and $a \in F$). The sum $f + g$ and scalar multiple af of f and g in $\hat{V}$ are defined by:

$$(f + g)(\alpha) = f(\alpha) + g(\alpha), \qquad (af)(\alpha) = a(f(\alpha)).$$

It is routine to verify that $\hat{V}$ is a vector space over F, although it's difficult to visualize a typical vector f in $\hat{V}$; knowing that f is a linear form on the space $\hat{V}$ of linear forms on V doesn't really help! Luckily there is a more helpful approach to $\hat{V}$: let us write

$$(v)\alpha = [v, \alpha] \quad (v \in V, \alpha \in \hat{V})$$

so that the vector v and the linear form α acquire equal status. The equations

$$[u + v, \alpha] = [u, \alpha] + [v, \alpha], \qquad [av, \alpha] = a[v, \alpha],$$
$$[v, \alpha + \beta] = [v, \alpha] + [v, \beta], \qquad [v, \alpha a] = [v, \alpha]a,$$
$$(u, v \in V, \ \alpha, \beta \in \hat{V}, \ a \in F)$$

show the similarities between the first and second entries in the square bracket, and are summed up by saying that the mapping

$$V \times \hat{V} \to F$$

$$(v, \alpha) \to [v, \alpha]$$

is **bilinear**. If the second entry α is kept fixed, then the resulting mapping of V to F is simply α. However, if the first entry v is held constant, then a mapping $\hat{v} : \hat{V} \to F$ is obtained, namely $\hat{v}(\alpha) = [v, \alpha]$ for all α in $\hat{V}$; in other words $\hat{v}$ evaluates each linear form α at the given vector v. For instance, let $v = (2, 3, 4)$ in $\mathbb{R}^3$ and $\alpha = \mu_c$ in $\hat{\mathbb{R}}^3$ where $c = (3, 1, 5)^T$; then $\hat{v}(\alpha) = [v, \alpha] = (v)\mu_c = vc = 29$.

Proposition 10.29 Let V be a finite-dimensional vector space. Then $\eta : V \cong \hat{V}$, defined by $(v)\eta = \hat{v}$ for $v \in V$, is an isomorphism called the natural isomorphism of V to $\hat{V}$.

Proof As the above mapping $(v, \alpha) \to [v, \alpha]$ is linear in the second entry α, the mapping $\hat{v} : \hat{V} \to F$ is itself linear; therefore, as the notation suggests, $\hat{v}$ belongs to $\hat{V}$. So η maps V to $\hat{V}$, and as $(v, \alpha) \to [v, \alpha]$ is linear in the first entry v, we see that η is linear.

Let v_1 be any non-zero vector of V. Then v_1 belongs to some basis b of V and hence $\hat{v}_1$ (in $\hat{b}$) is a linear form on V with

$(v_1)\hat{v}_1 = 1$. On the other hand,

$$v \in \ker \eta \iff \hat{v} = 0 \iff [v, \alpha] = (v)\alpha = 0$$

for all $\alpha \in \hat{V}$. Taking $\alpha = \hat{v}_1$ gives $v_1 \notin \ker \eta$; we have shown that ker η doesn't contain any non-zero vectors, that is, ker $\eta = 0$. So rank $\eta = \dim V$ by (8.5). As dim $\hat{\hat{V}} = \dim \hat{V} = \dim V$, we see im $\eta = \hat{\hat{V}}$, showing that η is an isomorphism. $\qquad \square$

So each linear form on $\hat{V}$ is of the type $\hat{v}$ for a unique v in V. Consider a basis $\hat{\ell}$ of $\hat{V}$ consisting of $\alpha_1, \alpha_2, \ldots, \alpha_n$. As in (10.28), we may construct the basis $\hat{\alpha}_1, \hat{\alpha}_2, \ldots, \hat{\alpha}_n$ of $\hat{\hat{V}}$ which is dual to $\hat{\ell}$, that is, $\hat{\alpha}_i$ is the linear form on $\hat{V}$ such that $\hat{\alpha}_i(\alpha_j) = \delta_{ij}$ $(1 \le i, j \le n)$. By (10.29) the vectors $v_1, v_2, \ldots, v_n$ of V with $\hat{v}_i = \hat{\alpha}_i$ $(1 \le i \le n)$ constitute the basis ℓ of V, for the equations

$$(v_i)\alpha_j = [v_i, \alpha_j] = \hat{v}_i(\alpha_j) = \hat{\alpha}_i(\alpha_j) = \delta_{ij} \quad (1 \le i, j \le n)$$

show that $\hat{\ell}$ is the basis dual to ℓ. Therefore $\alpha_i = \hat{v}_i$ $(1 \le i \le n)$ and the basis $\hat{v}_1, \hat{v}_2, \ldots, \hat{v}_n$ of $\hat{\hat{V}}$ is dual to the basis $\hat{v}_1, \hat{v}_2, \ldots, \hat{v}_n$ of $\hat{V}$. (It makes practical sense to identify $\hat{\hat{V}}$ with V by means of η and write $\hat{\hat{v}} = v$; in other words, inverting V twice gets us precisely nowhere!)

Definition 10.30　Let U be a subspace of the vector space V. Then $\mathscr{A}(U) = \{\alpha \in \hat{V} : (u)\alpha = 0$ for all $u \in U\}$ is called the **annihilator** of U.

So $\mathscr{A}(U)$ consists of the linear forms on V which are zero on the vectors of U; $\mathscr{A}(U)$ is a subspace of $\hat{V}$ by (6.6). In the case $V = F^n$, annihilators and solutions of linear equations are closely connected: for let $v_1, v_2, \ldots, v_m$ span the subspace U of F^n and let A be the $m \times n$ matrix over F with $v_i = (\text{row } i \text{ of } A)$; then $\alpha = \mu_x$ $(x \in {}^nF)$ belongs to $\mathscr{A}(U)$ if and only if $Ax = 0$. Therefore finding an annihilator is the same as solving a homogeneous system of equations.

For instance, the subspace $U = \langle (1, 2, 2), (3, 1, 2) \rangle$ of $\mathbb{R}^3$ has annihilator $\mathscr{A}(U) = \langle \hat{v}_3 \rangle$ where $(x_1, x_2, x_3)\hat{v}_3 = 2x_1 + 4x_2 - 5x_3$; in fact $2x_1 + 4x_2 - 5x_3 = 0$ is the equation of the plane U. More generally, as we show next, $(n-1)$-dimensional subspaces of F^n are conveniently specified by their 1-dimensional annihilators.

Notation　Let $\mathscr{L}(V)$ denote the set of subspaces of the vector space V.

The subspaces of V form a system called a **lattice** (hence the notation), the essential property being that each pair U_1 and U_2 of

Fig. 10.2

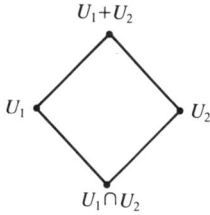

subspaces gives rise to two further subspaces (Fig. 10.2), namely $U_1 \cap U_2$ (the largest subspace contained in U_1 and U_2) and $U_1 + U_2$ (the smallest subspace containing U_1 and U_2).

The properties of the annihilator mapping, proved below, should remind the reader of complements (1.3) of subsets and, in particular, of the mapping $P(Y) \to P(Y)$ defined by $X \mapsto X'$ for all subsets X of a set Y of n elements (2.3).

Proposition 10.31 Let U be a subspace of the n-dimensional vector space V. Then $\dim \mathcal{A}(U) = n - \dim U$. The annihilator mapping $\mathcal{A} : \mathcal{L}(V) \to \mathcal{L}(\hat{V})$ is bijective and inclusion-reversing, that is, $U_1 \subseteq U_2 \Leftrightarrow \mathcal{A}(U_2) \subseteq \mathcal{A}(U_1)$.

Proof Extend the basis $v_1, \ldots, v_m$ of U to a basis $v_1, \ldots, v_m, v_{m+1}, \ldots, v_n$ (let us call it $\textit{b}$) of V. Consider the forms $\hat{v}_{m+1}, \ldots, \hat{v}_n$ in the basis $\hat{b}$ dual to $\textit{b}$. As each α in $\hat{V}$ is expressible $\alpha = \sum_{i=1}^{n} \hat{v}_i((v_i)\alpha)$, we see that

$$\alpha \in \mathcal{A}(U) \quad \Leftrightarrow \quad (v_i)\alpha = 0 \ (1 \leq i \leq m) \quad \Leftrightarrow \quad \alpha \in \langle \hat{v}_{m+1}, \ldots, \hat{v}_n \rangle$$

and so $\mathcal{A}(U) = \langle \hat{v}_{m+1}, \ldots, \hat{v}_n \rangle$ has dimension $n - m = n - \dim U$.

Let W be a subspace of $\hat{V}$ and write

$$\mathcal{B}(W) = \{v \in V : (v)\alpha = 0 \text{ for all } \alpha \in W\}.$$

Then $\mathcal{B}(W)$ is a subspace of V. We may find a basis of $\mathcal{B}(W)$ as in the preceding paragraph: extend the basis $\alpha_{m+1}, \ldots, \alpha_n$ of W to the basis $\alpha_1, \ldots, \alpha_m, \alpha_{m+1}, \ldots, \alpha_n$ (which we call $\hat{b}$) of $\hat{V}$. Using the basis $\textit{b}$ of V (that is, $v_1, \ldots, v_m, v_{m+1}, \ldots, v_n$ where $\hat{v}_i = \alpha_i$ $(1 \leq i \leq m)$) and the formula $v = \sum_{i=1}^{n} ((v)\alpha_i)v_i$, we see that $\mathcal{B}(W) = \langle v_1, \ldots, v_m \rangle$.

Taking $W = \mathcal{A}(U)$, we obtain $U = \langle v_1, \ldots, v_m \rangle = \mathcal{B}(\mathcal{A}(U))$ for all subspaces U of V. Taking $U = \mathcal{B}(W)$, we obtain $W = \langle \alpha_{m+1}, \ldots, \alpha_n \rangle = \mathcal{A}(\mathcal{B}(W))$ for all subspaces W of $\hat{V}$. In other words $\mathcal{A} : \mathcal{L}(V) \to \mathcal{L}(\hat{V})$ is bijective because its inverse is $\mathcal{B} : \mathcal{L}(\hat{V}) \to \mathcal{L}(V)$.

Directly from (10.30) we obtain: $U_1 \subseteq U_2 \Rightarrow \mathcal{A}(U_2) \subseteq \mathcal{A}(U_1)$, and similarly $W_2 \subseteq W_1 \Rightarrow \mathcal{B}(W_1) \subseteq \mathcal{B}(W_2)$ where W_1 and W_2 are subspaces of $\hat{V}$. Taking $W_i = \mathcal{A}(U_i)$ now gives: $\mathcal{A}(U_2) \subseteq \mathcal{A}(U_1) \Rightarrow U_1 \subseteq U_2$ as $\mathcal{B}(\mathcal{A}(U_i)) = U_i$ for $i = 1, 2$. □

Let U be an $(n-1)$-dimensional subspace of F^n. Then $\mathcal{A}(U)$ is 1-dimensional by (10.31), and so $\mathcal{A}(U) = \langle \mu_c \rangle$ for some non-zero $c \in {}^nF$; in other words

$$U = \{x \in {}^nF : xc = 0\}$$

which provides a concise description of U. The column vector c (or any non-zero vector proportional to c) is called the **dual co-ordinate vector** of U.

Definition 10.32 Let A be an $n \times n$ matrix over the field F. Any non-zero vector c in ${}^nF$, with $Ac = c\lambda$ where $\lambda \in F$, is called a **characteristic (column) vector** of A.

So A has characteristic vectors of two types, namely rows (10.7) and columns (10.32). What is the significance of these vectors?

Definition 10.33 Let $\alpha : V \to V$ be a linear mapping. The subspace U of V is called α-**invariant** if $(u)\alpha \in U$ for all $u \in U$.

Therefore the subspace U being α-invariant means that U is mapped into itself by α. Such subspaces are important in the further theory of linear mappings. For example, consider $\alpha = \mu_A : \mathbb{Q}^3 \to \mathbb{Q}^3$ where

$$A = \begin{pmatrix} 6 & 1 & 1 \\ 6 & 7 & 3 \\ 1 & -2 & 3 \end{pmatrix}.$$

From (10.12)(c) we know that $(3, -1, 0)$ is a characteristic row vector of A; in fact $(3, -1, 0)\alpha = (3, -1, 0)A = 4(3, -1, 0)$. Therefore α multiplies every vector in $W_1 = \langle (3, -1, 0) \rangle$ by 4, showing that W_1 is α-invariant. As $(11, 3, 4)$ is a characteristic row vector of A, $W_2 = \langle (11, 3, 4) \rangle$ is α-invariant. We have used all the characteristic row vectors of A, and so W_1 and W_2 are the only 1-dimensional α-invariant subspaces.

Now $\chi = (x - 4)^2(x - 8)$ and so the characteristic column vectors c of A satisfy $(A - \lambda I)c = 0$ where $\lambda = 4, 8$. Taking $\lambda = 4$ we obtain $c = (1, 3, -5)^T$, which means (see (10.34)) that the 2-dimensional subspace W_3 with dual co-ordinates c (W_3 has equation $x_1 + 3x_2 -$

Fig. 10.3

$$Q^3$$

W_3 W_4

W_2 W_1

0

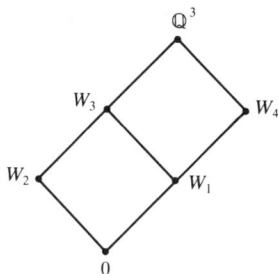

$5x_3 = 0$) is α-invariant. Taking $\lambda = 8$ leads to $c = (1, 3, -1)^T$ and the α-invariant subspace W_4 with equation $x_1 + 3x_2 - x_3 = 0$. We have found all the α-invariant subspaces (these include Q^3 and 0); notice that they can be arranged in the form of a lattice (Fig. 10.3) with $W_3 = W_1 + W_2$ and $W_1 = W_3 \cap W_4$.

Generally it is difficult to determine the r-dimensional μ_A-invariant subspaces F^n for $1 < r < n - 1$, where A is a given $n \times n$ matrix over F (however, see question 6 below). By contrast, it is easy to show that the 1-dimensional subspace $\langle v \rangle$ of F^n is μ_A-invariant if and only if v is a characteristic row vector of A; we now treat the dual case $r = n - 1$.

Proposition 10.34

Let A be an $n \times n$ matrix over the field F and let $c \in {}^nF$ be non-zero. Then the $(n - 1)$-dimensional subspace $W = \{x \in F^n : xc = 0\}$ is μ_A-invariant if and only if c is a characteristic column vector of A.

Proof

Suppose W to be μ_A-invariant. So $(x)\mu_A \in W$ for all $x \in W$ by (10.33). Therefore $xAc = 0$ for all $x \in W$, that is, $\mu_{Ac} \in \mathscr{A}(W)$ by (10.30). As $\mathscr{A}(W) = \langle \mu_c \rangle$, there is λ in F with $\mu_{Ac} = \mu_c\lambda$; so $\mu_{Ac} = \mu_{c\lambda}$ and hence $Ac = c\lambda$ by (7.13). Therefore c is a characteristic column vector of A by (7.32).

Conversely suppose $Ac = c\lambda$ and let $x \in W$. Then $xc = 0$ and hence $xAc = xc\lambda = 0\lambda = 0$, showing $(x)\mu_A = xA \in W$. By (10.33), W is μ_A-invariant. □

Exercises 10.3

1. In each case find the linear forms μ_c belonging to the dual of the given basis b of Q^3, and hence express $(17, 39, -24)$ as a linear combination of the vectors in b.

   (i) $v_1 = (1, 1, 1)$, $v_2 = (1, 0, 1)$, $v_3 = (0, 0, 1)$.
   (ii) $v_1 = (3, 3, 2)$, $v_2 = (5, 6, 5)$, $v_3 = (3, 4, 4)$.
   (iii) $v_1 = (4, -5, 2)$, $v_2 = (-4, 6, -3)$, $v_3 = (3, -5, 3)$.

2. Let v_1 be a non-zero vector of V and K a subspace of V such that $V = \langle v_1 \rangle \oplus K$. Show that there is a unique α in $\hat{V}$ with $(v_1)\alpha = 1$ and $K = \ker \alpha$. If α' in $\hat{V}$ is such that $K = \ker \alpha'$, show that $\alpha' = \alpha a$ for some non-zero scalar a.

3. For each of the following matrices A, find the α-invariant subspaces, where $\alpha = \mu_A : \mathbb{Q}^3 \to \mathbb{Q}^3$, and arrange them in the form of a lattice.

$$\begin{pmatrix} 4 & 1 & 3 \\ 1 & 1 & 1 \\ -2 & -1 & -1 \end{pmatrix}, \quad \begin{pmatrix} 6 & 1 & 4 \\ 3 & 2 & 3 \\ -4 & -1 & -2 \end{pmatrix}, \quad \begin{pmatrix} 5 & 1 & 4 \\ 1 & 2 & 1 \\ -2 & -1 & -1 \end{pmatrix}.$$

4. Let v and c be characteristic row and column vectors of the square matrix A. If the characteristic roots corresponding to v and c are distinct, show that $vc = 0$.

5. Let U and W be subspaces of the finite-dimensional vector space V. Use (10.31) to show that $\mathscr{A}(U + W) = \mathscr{A}(U) \cap \mathscr{A}(W)$ and $\mathscr{A}(U \cap W) = \mathscr{A}(U) + \mathscr{A}(W)$.

6. Let $\alpha : V \to V$ be a linear mapping with distinct characteristic roots $\lambda_1, \lambda_2, \ldots, \lambda_t$ and suppose $V = U_1 \oplus U_2 \oplus \ldots \oplus U_t$ where U_i is the characteristic subspace of α corresponding to λ_i. Show that the subspace U of V is α-invariant if and only if $U = U_1' \oplus U_2' \oplus \ldots \oplus U_t'$ where $U_i' = U \cap U_i$ $(1 \leqslant i \leqslant t)$.
Find the μ_A-invariant subspaces of F^4, where A is the 4×4 diagonal matrix over F with (i, i)-entry i, and F is

(i) $\mathbb{Q}$,   (ii) $\mathbb{Z}_3$,   (iii) $\mathbb{Z}_2$.

7. Let A and B be $n \times n$ matrices over the field F with $AB = BA$. Show that the characteristic subspaces U_i $(1 \leqslant i \leqslant t)$ of A are μ_B-invariant and that the characteristic subspaces W_j $(1 \leqslant j \leqslant s)$ of B are μ_A-invariant. If $F^n = U_1 \oplus \ldots \oplus U_t$ and $F^n = W_1 \oplus \ldots \oplus W_s$, deduce (using question 6) that F^n is the direct sum of the subspaces $U_i \cap W_j$. Hence prove that if A and B are each diagonalizable, then there is an invertible $n \times n$ matrix P over F such that both PAP^{-1} and PBP^{-1} are diagonal. Find such a matrix P in the case $F = \mathbb{Q}$,

$$A = \begin{pmatrix} 3 & 2 & -4 \\ 4 & 1 & -4 \\ 8 & 4 & -9 \end{pmatrix}, \quad B = \begin{pmatrix} 4 & 9 & -6 \\ 2 & 7 & -4 \\ 4 & 12 & -7 \end{pmatrix}.$$

Bilinear forms

The annihilator $\mathscr{A}$ maps subspaces of V to subspaces of $\hat{V}$. We now ask: are there sensible ways of mapping vectors in V to vectors in $\hat{V}$? As multiplication of scalars is commutative, we may consider linear mappings $\alpha : V \to \hat{V}$, which are essentially the same as **bilinear forms** (10.35) on V. To get a glimpse of where we are going, take $V = F^n$ and so $\hat{V} \cong {}^nF$ (using $\mu_c \to c$); in terms of co-ordinate vectors, α maps row vectors to column vectors, that is, $\alpha : F^n \to {}^nF$. The most obvious (and when $F = \mathbb{R}$, the most important) mapping of this type is matrix transposition: $(x)^\alpha = x^T$ for all $x \in {}^nF$. In the case $F = \mathbb{C}$, it is appropriate to waive the linearity of α and combine complex conjugation with matrix transposition (α is matching $\mathbb{C}^n$ with its upside-down version $\widehat{\mathbb{C}^n}$, and so it is reasonable to allow conjugation, which inverts the Argand diagram) to give $(x)^\alpha = x^*$ for all $x = (x_1, x_2, \ldots, x_n)$ in $\mathbb{C}^n$, where $x^* = (x_1^*, x_2^*, \ldots, x_n^*)^T$.

Returning to the abstract setting, as V and $\hat{V}$ are left and right vector spaces over F respectively, we use $(v)^\alpha$ for the image of v by $\alpha : V \to \hat{V}$ ($(v)^\alpha$ is something of a compromise between $(v)\alpha$ and $\alpha(v)$). Let us assume that α is linear, which in this context means

$$(u + v)^\alpha = (u)^\alpha + (v)^\alpha, \quad (av)^\alpha = (v)^\alpha a \quad (u, v \in V, a \in F).$$

Associated with α there is a mapping $\beta : V \times V \to F$ defined by

$$(u, v)^\beta = [u, (v)^\alpha] \quad \text{for all } u, v \in V,$$

that is, $(u, v)^\beta$ is the result of evaluating the linear form $(v)^\alpha$ at u. It is conceptually easier (and just as good) to deal with β rather than α. As α is linear, β satisfies

$$(u + u', v)^\beta = (u, v)^\beta + (u', v)^\beta, \quad (au, v)^\beta = a(u, v)^\beta,$$
$$(u, v + v')^\beta = (u, v)^\beta + (u, v')^\beta, \quad (u, av)^\beta = (u, v)^\beta a,$$
$$(u, u', v, v' \in V, a \in F).$$

Definition 10.35   Any mapping $\beta : V \times V \to F$ satisfying the above equations, is called a **bilinear form** on V.

In fact the above correspondence between linear mappings $\alpha : V \to \hat{V}$ and bilinear forms $\beta : V \times V \to F$ is bijective: for if β is given, then α can be 'recaptured', $(v)^\alpha$ being, for each v in V, the linear form defined by $[u, (v)^\alpha] = (u, v)^\beta$ for all $u \in V$. We now concentrate on β and allow α to recede into the background.

Let $A = [a_{ij}]$ be an $n \times n$ matrix over F. Then β defined by $(x, y)^\beta = xAy^T$ for $x, y \in F^n$ is a bilinear form on F^n (in this case $(y)^\alpha = \mu_c$ where $c = Ay^T$). Multiplying out the matrix product xAy^T

gives

$$xAy^T = \sum_{i,j} x_i a_{ij} y_j$$

where

$$x = (x_1, \ldots, x_n), \qquad y = (y_1, \ldots, y_n),$$

and so a_{ij} is the coefficient of $x_i y_j$. For instance

$$(x_1, x_2)\begin{pmatrix} 5 & 7 \\ 6 & 8 \end{pmatrix}\begin{pmatrix} y_1 \\ y_2 \end{pmatrix} = 5x_1y_1 + 7x_1y_2 + 6x_2y_1 + 8x_2y_2$$

is a bilinear form on $\mathbb{Q}^2$.

We show next that every bilinear form β on an n-dimensional vector space V is of the above type, by referring β to a basis of V.

Definition 10.36

Let β be a bilinear form on the vector space V with basis ℓ consisting of $v_1, v_2, \ldots, v_n$. The $n \times n$ matrix A with (i, j)-entry $(v_i, v_j)^\beta$ is called the **matrix of β relative to** ℓ and is said to **represent** β.

Let u and v in V have co-ordinate vectors $x = (x_1, \ldots, x_n)$ and $y = (y_1, \ldots, y_n)$ respectively relative to ℓ. Writing $a_{ij} = (v_i, v_j)^\beta$ we obtain

$$(u, v)^\beta = \left(\sum_i x_i v_i, \sum_j y_j v_j \right)^\beta = \sum_{i,j} x_i (v_i, v_j)^\beta y_j$$

$$= \sum_{i,j} x_i a_{ij} y_j = xAy^T$$

for bilinear forms may be 'multiplied out' in the usual way.

How are the matrices of β relative to different bases ℓ and ℓ' of V related to each other? This question should ring familiarly in the reader's ear! Let $P = [p_{ij}]$ be the transition matrix from ℓ' to ℓ, and so $v_i' = \sum_j p_{ij} v_j$, where $v_1', v_2', \ldots, v_n'$ constitute ℓ'. Then

$$(v_i', v_l')^\beta = \left(\sum_j p_{ij} v_j, \sum_k p_{lk} v_k \right)^\beta = \sum_{j,k} p_{ij} (v_j, v_k)^\beta p_{lk}$$

$$= \sum_{j,k} p_{ij} a_{jk} p_{lk} = (i, l)\text{-entry in } PAP^T.$$

Therefore PAP^T is the matrix of β relative to ℓ'.

Definition 10.37

The $n \times n$ matrices A, B over the field F are called **congruent** if there is an invertible $n \times n$ matrix P over F with $PAP^T = B$.

There is *no* connection between congruent integers (3.16) and congruent matrices (or congruent triangles), except that in each case congruence is an equivalence relation. Thus congruence (10.37) is an equivalence relation on $M_n(F)$, two matrices being congruent if and only if they represent the same bilinear form on V. Congruent matrices have the same rank; the **rank of a bilinear form** is (by definition) the rank of any matrix representing it.

If P is elementary (8.17), then PAP^T is the result of applying the corresponding elementary row and column operations to A; in matrix congruence, any elementary row operation may be applied to A, provided that the same column operation is immediately applied also. For instance, applying $r_2 - 3r_1$ followed by $c_2 - 3c_1$ to

$$A = \begin{pmatrix} 2 & 6 \\ 6 & 5 \end{pmatrix} \text{ produces } PAP^T = \begin{pmatrix} 2 & 0 \\ 0 & -13 \end{pmatrix} \text{ where } P = \begin{pmatrix} 1 & 0 \\ -3 & 1 \end{pmatrix}.$$

By (8.27), any matrix may be changed into any other congruent matrix by a sequence of coupled row and column operations of this kind.

The problem of determining the congruence classes of matrices over an arbitrary field is to a large extent unsolved; we concentrate on three important cases.

Alternating bilinear forms

Definition 10.38
The bilinear form β on the vector space V is called **alternating** if $(v, v)^\beta = 0$ for all $v \in V$.

Alternating forms are important in geometry—they give rise to the family of **symplectic groups**—and they are closely connected with determinants (9.30). The classification (10.39) of such forms is exceptionally easy, if V finite-dimensional; once again, *rank* is the all-important number.

Let β be an alternating form on V. Then

$$0 = (u + v, u + v)^\beta = (u, u)^\beta + (u, v)^\beta + (v, u)^\beta + (v, v)^\beta$$
$$= (u, v)^\beta + (v, u)^\beta \quad \text{where } u, v \in V.$$

Therefore β is **antisymmetric**, that is

$$(u, v)^\beta = -(v, u)^\beta \quad \text{for} \quad u, v \in V.$$

Let $A = [a_{ij}]$ be the matrix of β relative to the basis $v_1, v_2, \ldots, v_n$ of V. Then $a_{ii} = (v_i, v_i)^\beta = 0$ $(1 \le i \le n)$ and $a_{ij} = (v_i, v_j)^\beta = -(v_j, v_i)^\beta = -a_{ji}$ $(1 \le i < j \le n)$. Conversely, let $A = [a_{ij}]$ be an

alternating $n \times n$ matrix over F (that is, each $a_{ii} = 0$ and $a_{ij} = -a_{ji}$ for $i \neq j$). Then β, defined by $(x, y)^\beta = xAy^T$, is an alternating form on F^n: for $xAx^T = \sum_{i,j} x_i a_{ij} x_j = 0$, as $x_i a_{ii} x_i = 0$ and the 'off-diagonal' terms occur in pairs $x_i a_{ij} x_j + x_j a_{ji} x_i = 0$. The following matrices are alternating:

$$J = \begin{pmatrix} 0 & 1 \\ -1 & 0 \end{pmatrix}, \quad \begin{pmatrix} 0 & 2 & -3 \\ -2 & 0 & 4 \\ 3 & -4 & 0 \end{pmatrix}, \quad \begin{pmatrix} 0 & 0 & 2 & 3 \\ 0 & 0 & 4 & 0 \\ -2 & -4 & 0 & 5 \\ -3 & 0 & -5 & 0 \end{pmatrix}.$$

The fundamental alternating form is the 2×2 determinant mapping δ given by

$$(x, y)^\delta = \begin{vmatrix} x_1 & x_2 \\ y_1 & y_2 \end{vmatrix} = x_1 y_2 - x_2 y_1 = (x_1, x_2) \begin{pmatrix} 0 & 1 \\ -1 & 0 \end{pmatrix} \begin{pmatrix} y_1 \\ y_2 \end{pmatrix} = xJy^T.$$

Notation

For $0 \leq r \leq n$ (r even), let J_r denote the $n \times n$ matrix with $(i, i+1)$-entries 1 (i odd), $(i, i-1)$-entries -1 (i even) where $1 \leq i \leq r$, all other entries being zero.

J_r is the simplest alternating $n \times n$ matrix of rank r. There are three 4×4 matrices of this type, namely J_0 (the zero matrix),

$$J_2 = \begin{pmatrix} J & \vdots \\ \cdots & \cdots \\ \vdots & \end{pmatrix} = \begin{pmatrix} 0 & 1 & 0 & 0 \\ -1 & 0 & 0 & 0 \\ 0 & 0 & 0 & 0 \\ 0 & 0 & 0 & 0 \end{pmatrix}, \quad J_4 = \begin{pmatrix} J & \vdots \\ \cdots & \cdots \\ \vdots & J \end{pmatrix} = \begin{pmatrix} 0 & 1 & 0 & 0 \\ -1 & 0 & 0 & 0 \\ 0 & 0 & 0 & 1 \\ 0 & 0 & -1 & 0 \end{pmatrix}.$$

Theorem 10.39

Let β be an alternating form on the n-dimensional vector space V. Then β has even rank r, and is represented by the $n \times n$ matrix J_r.

Proof

If rank $\beta = 0$, then $(u, v)^\beta = 0$ for all u, $v \in V$, and β is represented by J_0. Suppose rank $\beta \neq 0$. Then there are u' and v' in V with $(u', v')^\beta \neq 0$; hence $v_1 = au'$ and $v_2 = v'$ satisfy $(v_1, v_2)^\beta = 1$ where $a^{-1} = (u', v')^\beta$; as $(v_1, v_1)^\beta = 0$, the vectors v_1 and v_2 are linearly independent. Write $U = \langle v_1, v_2 \rangle$ and consider the **orthogonal complement** W of U, that is

$$W = \{w \in V : (w, v_1)^\beta = (w, v_2)^\beta = 0\}.$$

Then W is a subspace of V. Further $V = U \oplus W$, since each v in V can be expressed uniquely as $v = u + w$ ($u \in U$, $w \in W$), namely with $u = (v, v_2)^\beta v_1 - (v, v_1)^\beta v_2$ and $w = v - (v, v_2)^\beta v_1 + (v, v_1)^\beta v_2$.

Let W have basis $w_3, \ldots, w_n$. Then β has matrix

$$\begin{pmatrix} J & \vdots \\ \cdots\cdots & \cdots\cdots \\ \vdots & A' \end{pmatrix}$$

relative to the basis $v_1, v_2, w_3, \ldots, w_n$ of V, where A' is the matrix of β' (the restriction of β to $W \times W$) relative to $w_3, \ldots, w_n$. Now β' is an alternating form on W and rank $\beta' = $ rank $A' = ($rank $\beta) - 2 = r - 2$ since rank $J = 2$. We may inductively assume (10.39) to be true for β': so $r - 2$ is even and there is a basis $v_3, \ldots, v_n$ of W relative to which β' has matrix J_{r-2}. Therefore r is even and β has matrix J_r relative to the basis $v_1, v_2, v_3, \ldots, v_n$ of V, completing the induction. $\square$

Therefore rank β tells us all there is to know, in abstract terms, about the alternating bilinear form β.

Corollary 10.40

The alternating $n \times n$ matrices A and B over the field F are congruent if and only if rank $A = $ rank B.

Proof

If A and B are congruent (10.37), then rank $A = $ rank B as multiplication by invertible matrices leaves rank unchanged. On the other hand, we may apply (10.39) to the alternating form β on F^n defined by $(x, y)^\beta = xAy^T$: as β has matrix A relative to the standard basis of F^n, $r = $ rank A is even and there is a basis $v_1, v_2, \ldots, v_n$ of F^n relative to which β has matrix $PAP^T = J_r$, where P is the invertible matrix with $v_i = $ row i of P ($1 \le i < n$). Assuming rank $A = $ rank B, arguing with B in place of A produces an invertible matrix P_1 with $P_1 B P_1^T = J_r$, and so $(P_1^{-1}P)A(P_1^{-1}P)^T = B$, showing that A and B are congruent. $\square$

Given an alternating matrix A, then P as above can be found by applying coupled row and column operations to $(A : I)$ producing $(PAP^T : P) = (J_r : P)$, as in the following illustration over $\mathbb{Q}$:

$$\begin{pmatrix} 0 & 2 & -3 & 1 & 0 & 0 \\ -2 & 0 & 4 & 0 & 1 & 0 \\ 3 & -4 & 0 & 0 & 0 & 1 \end{pmatrix} \underset{\frac{1}{2}c_2}{\overset{\frac{1}{2}r_2}{\equiv}} \begin{pmatrix} 0 & 1 & -3 & 1 & 0 & 0 \\ -1 & 0 & 2 & 0 & \frac{1}{2} & 0 \\ 3 & -2 & 0 & 0 & 0 & 1 \end{pmatrix}$$

$$\underset{\substack{r_3 + 3r_2 \\ c_3 + 3c_2}}{\equiv} \begin{pmatrix} 0 & 1 & 0 & 1 & 0 & 0 \\ -1 & 0 & 2 & 0 & \frac{1}{2} & 0 \\ 0 & -2 & 0 & 0 & \frac{3}{2} & 1 \end{pmatrix} \underset{\substack{r_3 + 2r_1 \\ c_3 + 2c_1}}{\equiv} \begin{pmatrix} 0 & 1 & 0 & 1 & 0 & 0 \\ -1 & 0 & 0 & 0 & \frac{1}{2} & 0 \\ 0 & 0 & 0 & 2 & \frac{3}{2} & 1 \end{pmatrix}$$

from which we read off

$$PAP^{\mathrm{T}} = J_2 = \begin{pmatrix} 0 & 1 & 0 \\ -1 & 0 & 0 \\ 0 & 0 & 0 \end{pmatrix},$$

where $P = \begin{pmatrix} 1 & 0 & 0 \\ 0 & \frac{1}{2} & 0 \\ 2 & \frac{3}{2} & 1 \end{pmatrix}$, $A = \begin{pmatrix} 0 & 2 & -3 \\ -2 & 0 & 4 \\ 3 & -4 & 0 \end{pmatrix}$.

Although the above technique is satisfactory, it is often easier to find $Q = P^{-1}$ satisfying $A = QJ_rQ^{\mathrm{T}}$ by the method of **completing the determinant** (analogous to 'completing the square') as follows. Suppose $A = [a_{ij}]$ is an alternating $n \times n$ matrix with $a_{12} \neq 0$; writing $x = (x_1, \ldots, x_n)$ and $y = (y_1, \ldots, y_n)$, we match the terms involving x_1, y_1, x_2, y_2 in xAy^{T} with those in a suitable second order determinant: for instance let X_1 be the coefficient of y_2 in xAy^{T} and let Y_2 be the coefficient of $a_{12}x_1$ in xAy^{T}, that is,

$$X_1 = a_{12}x_1 + a_{32}x_3 + \ldots + a_{n2}x_n, \quad Y_2 = y_2 + \frac{a_{31}}{a_{21}} y_3 + \ldots + \frac{a_{n1}}{a_{21}} y_n.$$

Similarly, let X_2 and Y_1 be the coefficients in xAy^{T} of $a_{21}y_1$ and $-x_2$ respectively; so

$$X_2 = x_2 + \frac{a_{31}}{a_{21}} x_3 + \ldots + \frac{a_{n1}}{a_{21}} x_n, \quad Y_1 = a_{12}y_1 + a_{32}y_3 + \ldots + a_{n2}y_n.$$

Then

$$xAy^{\mathrm{T}} = X_1Y_2 - X_2Y_1 + xA'y^{\mathrm{T}}$$

where $xA'y^{\mathrm{T}}$ does not involve x_1, y_1, x_2, y_2 as these variables are accounted for in the determinant $X_1Y_2 - X_2Y_1$. Writing $X = xQ$, we have found the first two entries in $X = (X_1, X_2, \ldots, X_n)$; in other words, columns 1 and 2 of Q are

$$(a_{12}, 0, a_{32}, \ldots, a_{n2})^{\mathrm{T}} = (\text{column 2 of } A),$$
$$(0, 1, a_{31}/a_{21}, \ldots, a_{n1}/a_{21})^{\mathrm{T}} = (1/a_{21})(\text{column 1 of } A).$$

The first step is now complete (if $a_{12} = 0$ but some $a_{ij} \neq 0$, then $xA'y^{\mathrm{T}}$ is arranged to have no terms involving x_i, y_i, x_j, y_j). The procedure is now repeated on $xA'y^{\mathrm{T}}$ and continued until nothing remains. The r columns obtained may be completed to a basis of ^nF and Q taken as the matrix having these basis vectors as its columns.

Only one step is required in the reduction over $\mathbb{Q}$ of

$$2(x_1y_2 - x_2y_1) - 3(x_1y_3 - x_3y_1) + 4(x_2y_3 - x_3y_2)$$
$$= (2x_1 - 4x_3)(y_2 - \tfrac{3}{2}y_3) - (x_2 - \tfrac{3}{2}x_3)(2y_1 - 4y_3).$$

Here $X_1 = 2x_1 - 4x_3$, $X_2 = x_2 - \tfrac{3}{2}x_3$ and

$$\begin{pmatrix} 0 & 2 & -3 \\ -2 & 0 & 4 \\ 3 & -4 & 0 \end{pmatrix} = \begin{pmatrix} 2 & 0 & 0 \\ 0 & 1 & 0 \\ -4 & -\tfrac{3}{2} & 1 \end{pmatrix} \begin{pmatrix} 0 & 1 & 0 \\ -1 & 0 & 0 \\ 0 & 0 & 0 \end{pmatrix} \begin{pmatrix} 2 & 0 & -4 \\ 0 & 1 & -\tfrac{3}{2} \\ 0 & 0 & 1 \end{pmatrix}.$$

If, as below, the coefficient of x_1y_2 is zero, the reduction begins with x_1y_3 instead:

$$2(x_1y_3 - x_3y_1) + 3(x_1y_4 - x_4y_1) + 4(x_2y_3 - x_3y_2)$$
$$= X_1Y_2 - X_2Y_1 - 6(x_2y_4 - x_4y_2)$$
$$= (X_1Y_2 - X_2Y_1) + (X_3Y_4 - X_4Y_3)$$

where

$$X_1 = 2x_1 + 4x_2, \quad X_2 = x_3 + \tfrac{3}{2}x_4, \quad X_3 = -6x_2, \quad X_4 = x_4,$$
$$Y_1 = 2y_1 + 4y_2, \quad Y_2 = y_3 + \tfrac{3}{2}y_4, \quad Y_3 = -6y_2, \quad Y_4 = y_4.$$

In terms of matrices, $A = QJ_4Q^{\mathrm{T}}$ where

$$A = \begin{pmatrix} 0 & 0 & 2 & 3 \\ 0 & 0 & 4 & 0 \\ -2 & -4 & 0 & 0 \\ -3 & 0 & 0 & 0 \end{pmatrix} \quad \text{and} \quad Q = \begin{pmatrix} 2 & 0 & 0 & 0 \\ 4 & 0 & -6 & 0 \\ 0 & 1 & 0 & 0 \\ 0 & \tfrac{3}{2} & 0 & 1 \end{pmatrix}.$$

Symmetric bilinear forms

Definition 10.41   The bilinear form β, on the vector space V over the field F, is called **symmetric** if $(u, v)^\beta = (v, u)^\beta$ for all $u, v \in V$.

For example, β defined by

$$(x, y)^\beta = xAy^{\mathrm{T}} = (x_1, x_2) \begin{pmatrix} 2 & 3 \\ 3 & 4 \end{pmatrix} \begin{pmatrix} y_1 \\ y_2 \end{pmatrix}$$
$$= 2x_1y_1 + 3x_1y_2 + 3x_2y_1 + 4x_2y_2 \quad \text{for } x, y \in \mathbb{R}^2$$

is a symmetric bilinear form on $\mathbb{R}^2$.

Let β have matrix A relative to the basis $\mathscr{E}$ of V. Using (10.36) it is straightforward to show that β is symmetric if and only if A is symmetric $(A^{\mathrm{T}} = A)$. The theory of symmetric bilinear forms depends heavily on the nature of the field F, and we focus on the case $F = \mathbb{R}$. However, our next proposition is valid over any field F with $\chi(F) \neq 2$ (5.23), that is, $1 + 1 \neq 0$ in F.

Proposition 10.42

Let β be a symmetric bilinear form on the n-dimensional vector space V over the field F where $\chi(F) \neq 2$. Then β is represented by a diagonal matrix.

Proof

We use induction on $r = \operatorname{rank} \beta$ to show that V has a basis $v_1, v_2, \ldots, v_n$ with $(v_i, v_j)^\beta = 0$ for all $i \neq j$. If $r = 0$, then every basis of V has this property. So suppose $r \neq 0$. Then β has a **non-isotropic vector** v_1, that is, $(v_1, v_1)^\beta \neq 0$: for $\operatorname{rank} \beta \neq 0$ means that there are u and v in V with $(u, v)^\beta \neq 0$; if both u and v are isotropic $((u, u)^\beta = (v, v)^\beta = 0)$, then $v_1 = u + v$ is non-isotropic as

$$(v_1, v_1)^\beta = (u + v, u + v)^\beta = (u, v)^\beta + (v, u)^\beta = 2(u, v)^\beta \neq 0$$

since $\chi(F) \neq 2$. Mimicking the proof of (10.39), write $U = \langle v_1 \rangle$ and $W = \{w \in V : (w, v_1)^\beta = 0\}$. For each v in V there is a unique $a \in F$ and a unique $w \in W$ with $v = av_1 + w$, namely $a = (v, v_1)^\beta/(v_1, v_1)^\beta$, $w = v - ((v, v_1)^\beta/(v_1, v_1)^\beta)v_1$; therefore $V = U \oplus W$.

Let $\beta' : W \times W \to F$ denote the restriction of β to $W \times W$. Then β' is a symmetric bilinear form on W. As $\operatorname{rank} \beta' = r - 1$, we may apply the inductive hypothesis: there is a basis $v_2, \ldots, v_n$ of W such that $(v_i, v_j)^\beta = 0$ whenever $2 \leq i < j \leq n$. Then β has a diagonal matrix relative to the basis $v_1, v_2, \ldots, v_n$ of V, completing the induction. □

In matrix terms, (10.42) tells us that each symmetric matrix A over F (where $\chi(F) \neq 2$) is congruent to a diagonal matrix D; as before, an invertible matrix P with $PAP^T = D$ can be found by applying coupled row and column operations to $[A : I]$ obtaining $[D : P]$.

Let β be a given form (as in (10.42)) and suppose $F = \mathbb{R}$. We show next that the diagonal matrices representing β have a simple property in common: the number (p say) of *positive* entries is the same in each case! This remarkable fact is crucial in the classification of real symmetric bilinear forms.

Theorem 10.43

(Sylvester's law of inertia.) Let β be a symmetric bilinear form of rank r on the n-dimensional vector space V over the real field $\mathbb{R}$. Then there is a unique non-negative integer p such that

β is represented by the diagonal matrix $D = [d_{ij}]$ with $d_{ii} = 1$ $(1 \leqslant i \leqslant p)$, $d_{ii} = -1$ $(p < i \leqslant r)$.

Proof

By (10.42), V has basis $v_1, v_2, \ldots, v_n$ with $(v_i, v_j)^\beta = 0$ for all $i \neq j$. These vectors may be ordered so that the p 'positive' ones come first, the $r - p$ 'negative' ones come next, followed by the $n - r$ isotropic vectors in the basis, that is,

$$(v_i, v_i)^\beta = \begin{cases} a_i^2 & (1 \leqslant i \leqslant p) \\ -a_i^2 & (p < i \leqslant r) \\ 0 & (r < i \leqslant n) \end{cases}$$

where $a_1, \ldots, a_r \in \mathbb{R}^*$, for each positive real number has a real square-root. Let $v_i' = (1/a_i)v_i$ $(1 \leqslant i \leqslant r)$ and $v_i' = v_i$ $(r < i \leqslant n)$; then

$$(v_i', v_i')^\beta = \begin{cases} 1 & (1 \leqslant i \leqslant p) \\ -1 & (p < i \leqslant r) \\ 0 & (r < i \leqslant n) \end{cases}$$

showing that β has matrix D (as above) relative to the **normalized basis** $v_1', v_2', \ldots, v_n'$ of V.

To show p unique, suppose to the contrary that β is represented by the diagonal matrix $D' = [d_{ij}']$ where $d_{ii}' = 1$ $(1 \leqslant i \leqslant q)$, $d_{ii}' = -1$ $(q < i \leqslant r)$ where $p \neq q$. As p and q have equal status, we may take $p < q$. As D and D' represent β, there is an invertible matrix P over $\mathbb{R}$ with $PDP^T = D'$. Writing $x = (x_1, \ldots, x_n)$, $X = (X_1, \ldots, X_n)$ for vectors of $\mathbb{R}^n$ related by $xP = X$, consider

$$U = \{x \in \mathbb{R}^n : x_{q+1} = \ldots = x_n = 0\},$$
$$W = \{x \in \mathbb{R}^n : X_1 = \ldots = X_p = 0\}.$$

Then U and W are subspaces of $\mathbb{R}^n$. What can be said of vectors belonging to both U and W? For x in $U \cap W$, the scalar equation $XDX^T = xPDP^Tx^T = xD'x^T$ gives

$$-X_{p+1}^2 - \ldots - X_r^2 = x_1^2 + \ldots + x_q^2.$$

The left-hand side cannot be positive and the right-hand side cannot be negative; there is only one way out: both sides are zero! Therefore $x_1^2 + \ldots + x_q^2 = 0$ and hence $x_1 = \ldots = x_q = 0$, showing that $x = 0$ (as all the entries in x are zero). We have shown $U \cap W = 0$. However, dim $U = q$ and, as P is invertible, dim $W = n - p$. Using (6.31) and $p < q$, we see dim$(U + W) = q + n - p > n$, which is contrary to (6.26) as $U + W$ is a subspace of $\mathbb{R}^n$. Therefore p is unique. $\qquad \square$

Definition 10.44 The integer $2p - r$ is called the **signature** of the real symmetric bilinear form β and of every matrix representing β.

For instance, the real symmetric matrix

$$\begin{pmatrix} 1 & 0 & 0 \\ 0 & -1 & 0 \\ 0 & 0 & -1 \end{pmatrix}$$

has signature -1 as $r = 3$ and $p = 1$. More generally the signature of a real diagonal matrix D of rank r is the number p of positive entries minus the number $r - p$ of negative entries.

Corollary 10.45 The real symmetric $n \times n$ matrices A and B are congruent if and only if they have the same rank and signature.

Proof Let A have rank r and signature s; then $|s| \leq r$ and r and s have the same parity. By (10.42) and (10.43), A is congruent to the diagonal matrix $D = [d_{ij}]$ with $d_{ii} = 1$ $(1 \leq i \leq (r+s)/2)$, $d_{ii} = -1$ $((r+s)/2 < i \leq r)$, and $d_{ij} = 0$ otherwise. If A is congruent to B, then B, being congruent to D, has rank r and signature s also. Conversely if B has rank r and signature s, then B is congruent to D by (10.42) and (10.43); hence A and B, being congruent to D, are congruent to each other. □

It is time to introduce an important type of mapping closely connected with bilinear forms.

Definition 10.46 Let β' be a (not necessarily symmetric) bilinear form on the vector space V over the field F. Then $\gamma : V \rightarrow F$, defined by $(v)\gamma = (v, v)^{\beta'}$ for all $v \in V$, is called the **quadratic form** on V associated with β'.

Quadratic forms arise in many branches of mathematics; in real analysis, quadratic form approximations derived from Taylor's theorem are used to investigate stationary points of functions of several variables. Here we assume $\chi(F) \neq 2$ throughout, and show that there is a correspondence between quadratic forms and *symmetric* bilinear forms. In fact γ, as in (10.46), is associated with the unique symmetric bilinear form β defined by

$$(u, v)^{\beta} = \tfrac{1}{2}((u + v)\gamma - (u)\gamma - (v)\gamma) \quad \text{for all} \quad u, v \in V.$$

Notice that β is the symmetrized version of β':

$$(u, v)^{\beta} = \tfrac{1}{2}((u, v)^{\beta'} + (v, u)^{\beta'}) \quad \text{for all} \quad u, v \in V.$$

If β' has matrix B relative to the basis b of V, then β has the symmetric matrix $A = \frac{1}{2}(B + B^T)$ relative to b. We refer to the **rank** (**signature** if $F = \mathbb{R}$) of γ, meaning the rank (signature) of β. For example, β' on $\mathbb{R}^2$ where

$$(x, y)^{\beta'} = (x_1, x_2)\begin{pmatrix} 2 & 6 \\ 0 & 4 \end{pmatrix}\begin{pmatrix} y_1 \\ y_2 \end{pmatrix} = 2x_1y_1 + 6x_1y_2 + 4x_2y_2$$

has associated quadratic form γ where

$$(x)\gamma = 2x_1^2 + 6x_1x_2 + 4x_2^2 = (x_1, x_2)\begin{pmatrix} 2 & 3 \\ 3 & 4 \end{pmatrix}\begin{pmatrix} x_1 \\ x_2 \end{pmatrix}$$

and γ is associated with the symmetric β given by

$$(x, y)^{\beta} = (x_1, x_2)\begin{pmatrix} 2 & 3 \\ 3 & 4 \end{pmatrix}\begin{pmatrix} y_1 \\ y_2 \end{pmatrix} = 2x_1y_1 + 3x_1y_2 + 3x_2y_1 + 4x_2y_2.$$

Every quadratic form γ on F^n can be expressed

$$(x)\gamma = xAx^T = \sum_{i,j} a_{ij}x_ix_j$$

where $x = (x_1, \ldots, x_n)$ and $A = [a_{ij}]$ is a symmetric $n \times n$ matrix. To find an invertible matrix Q over F with $A = QDQ^T$, D being diagonal (10.42), we **complete the square** as follows: suppose $a_{11} \neq 0$ and write $X_1 = a_{11}x_1 + a_{12}x_2 + \ldots + a_{1n}x_n$. Then the terms involving x_1 in $(x)\gamma$ agree with the terms involving x_1 in the 'complete square' $(1/a_{11})X_1^2$ and so

$$(x)\gamma = (1/a_{11})X_1^2 + (x)\gamma'$$

where $(x)\gamma'$ does not involve x_1; if $a_{11} = 0$ but some $a_{ii} \neq 0$, then we arrange instead that $(x)\gamma'$ does not involve x_i.

How do we proceed if $a_{ii} = 0$ for all i? Assuming $a_{12} \neq 0$, all the terms involving x_1 and x_2 in $(x)\gamma$ may be accounted for in a difference of two complete squares: write

$$X_1 = a_{12}(x_1 + x_2) + \Sigma_{i>2}(a_{1i} + a_{2i})x_i,$$
$$X_2 = a_{12}(x_1 - x_2) - \Sigma_{i>2}(a_{1i} - a_{2i})x_i.$$

Then $(x)\gamma = (2a_{12})^{-1}(X_1^2 - X_2^2) + (x)\gamma'$ where $(x)\gamma'$ does not involve either x_1 or x_2; similarly, if $a_{ii} = 0$ for all i, but some $a_{ij} \neq 0$, then $(x)\gamma'$ is arranged to involve neither x_i nor x_j. Writing $x = XQ$, we have found X_1 (X_1 and X_2) in terms of x and hence found the first column (first two columns) of Q.

This process is repeated on γ' until the zero quadratic form is obtained; the r ($=\text{rank } A$) columns of Q formed may be completed to a basis of ^nF, giving the complete matrix Q.

In the case of the real quadratic form γ below:

$$x_1^2 + 4x_1x_2 + 6x_1x_3 + 4x_2^2 + 10x_2x_3 + 6x_3^2$$
$$= (x_1 + 2x_2 + 3x_3)^2 - 2x_2x_3 - 3x_3^2$$
$$= (x_1 + 2x_2 + 3x_3)^2 - \tfrac{1}{3}(3x_3 + x_2)^2 + \tfrac{1}{3}x_2^2.$$

In terms of matrices:

$$A = \begin{pmatrix} 1 & 2 & 3 \\ 2 & 4 & 5 \\ 3 & 5 & 6 \end{pmatrix} = \begin{pmatrix} 1 & 0 & 0 \\ 2 & 1 & 1 \\ 3 & 3 & 0 \end{pmatrix} \begin{pmatrix} 1 & 0 & 0 \\ 0 & -\tfrac{1}{3} & 0 \\ 0 & 0 & \tfrac{1}{3} \end{pmatrix} \begin{pmatrix} 1 & 2 & 3 \\ 0 & 1 & 3 \\ 0 & 1 & 0 \end{pmatrix}$$

showing that A and γ have rank 3 and signature 1. The following real quadratic form contains no squares:

$$2x_1x_2 + 3x_1x_3 + 4x_2x_3$$
$$= \tfrac{1}{2}(x_1 + x_2 + \tfrac{7}{2}x_3)^2 - \tfrac{1}{2}(x_1 - x_2 + \tfrac{1}{2}x_3)^2 - 6x_3^2$$

and so has rank 3 and signature -1.

Definition 10.47

The quadratic form γ, on the real vector space V, is called **positive definite** if $(v)\gamma > 0$ for all non-zero $v \in V$.

As $(0)\gamma = 0$, we see $\gamma : V \to \mathbb{R}$ is positive definite if and only if γ has a strict minimum at 0 (the value of γ on the zero vector is less than its value on each non-zero vector).

Corollary 10.48

The quadratic form γ, on the n-dimensional vector space V over $\mathbb{R}$, is positive definite if and only if γ has signature n.

Proof

Let γ be associated with the symmetric bilinear form β. By (10.42) V has basis $v_1, v_2, \ldots, v_n$ with $(v_i, v_j)^\beta = 0$ for all $i \neq j$. If γ is positive definite, then $(v_i, v_i)^\beta = (v_i, v_i)\gamma > 0$ for all i and so γ has signature $2p - r = 2n - n = n$.

Conversely suppose γ has signature $2p - r = n$. As $p \leqslant r \leqslant n$, we see $n = 2p - r \leqslant r$ and so $p = r = n$. Therefore V has basis $v_1', v_2', \ldots, v_n'$ with $(v_i', v_j')^\beta = \delta_{ij}$ by (10.43). Let $0 \neq v = x_1v_1' + x_2v_2' + \ldots + x_nv_n'$; then $(v, v)\gamma = x_1^2 + x_2^2 + \ldots + x_n^2 > 0$, showing γ to be positive definite. $\qquad \square$

We take up the study of **orthonormal** bases, such as $v_1', v_2', \ldots, v_n'$ above, in Chapter 11.

Hermitian forms

Definition 10.49

Let V be a vector space over the complex field $\mathbb{C}$. A mapping $\beta : V \times V \to \mathbb{C}$ is called an **hermitian form** on V if

(i) $\left. \begin{array}{l} (u + u', v)^\beta = (u, v)^\beta + (u', v)^\beta \\ (au, v)^\beta = a(u, v)^\beta \end{array} \right\}$ for all $u, u', v \in V, a \in \mathbb{C}$,

(ii) $(v, u)^\beta = ((u, v)^\beta)^*$ for all $u, v \in V$, where $*$ denotes complex conjugation.

The theory of hermitian forms (named after the mathematician Hermite) is almost identical to the theory of real symmetric forms. Notice that non-zero hermitian forms are not bilinear; however from (10.49) we see

$$\left. \begin{array}{l} (u, v + v')^\beta = (u, v)^\beta + (u, v')^\beta \\ (u, av)^\beta = (u, v)^\beta a^* \end{array} \right\} \text{ for all } u, v, v' \in V, a \in \mathbb{C}.$$

Definition 10.50

Let A be the $m \times n$ matrix over $\mathbb{C}$ with (i, j)-entry a_{ij}. The $n \times m$ matrix A^* with (j, i)-entry a_{ij}^* is called the **hermitian conjugate** of A. The complex matrix A is called **hermitian** if $A^* = A$.

The hermitian conjugate combines matrix transposition with complex conjugation, and so $(AB)^* = B^*A^*$ using (7.11)(b). If $B = \begin{pmatrix} 2 + i & 3 + i \\ 5 + i & 1 + i \end{pmatrix}$, then $B^* = \begin{pmatrix} 2 - i & 5 - i \\ 3 - i & 1 - i \end{pmatrix}$.

The matrix $A = \begin{pmatrix} 2 & 1 + i \\ 1 - i & 3 \end{pmatrix}$ is hermitian and β, defined by

$$(x, y)^\beta = xAy^* = 2x_1y_1^* + (1 + i)x_1y_2^* + (1 - i)x_2y_1^* + 3x_2y_2^*$$

where $x = (x_1, x_2)$ and $y = (y_1, y_2)$, is an hermitian form on $\mathbb{C}^2$.

Let $\mathscr{b}$ consisting of $v_1, v_2, \ldots, v_n$ be a basis of the vector space V over $\mathbb{C}$ and let β be an hermitian form on V. As in (10.36), the $n \times n$ matrix A with (i, j)-entry $(v_i, v_j)^\beta$ is called the matrix of β relative to $\mathscr{b}$. Notice that A is hermitian, and if $u = \sum_i x_i v_i$ and $v = \sum_j y_j v_j$, then

$$(u, v)^\beta = \sum_{i,j} x_i (v_i, v_j)^\beta y_j^* = xAy^*$$

where $x = (x_1, \ldots, x_n)$ and $y = (y_1, \ldots, y_n)$. Let u and v have co-ordinate vectors X and Y relative to the basis $\mathscr{b}'$ of V. Then $x = XP$ and $y = YP$ where P is invertible and so $xAy^* = XPA(YP)^* = XPAP^*Y^*$, showing that β has matrix PAP^* relative to ℓ'. In short, the hermitian conjugate appears throughout in place of the matrix transpose.

Definition 10.51

The $n \times n$ matrices A and B over $\mathbb{C}$ are called **hermitian congruent** if there is an invertible $n \times n$ matrix P over $\mathbb{C}$ with $PAP^* = B$.

The reader may check that (10.42) is valid for β an hermitian form, as the proof goes through unchanged; therefore, given the hermitian matrix A, there is an invertible matrix P with $PAP^* = D$, where D is diagonal and hence real (the diagonal entries in every hermitian matrix are real). Such a matrix P may be found by applying 'conjugate pairs' of row and column operations to $[A : I]$ obtaining $[PAP^* : P] = [D : P]$; for instance

$$\begin{pmatrix} 1 & 1-i & 0 & 1 & 0 & 0 \\ 1+i & 4 & 2i & 0 & 1 & 0 \\ 0 & -2i & 7 & 0 & 0 & 1 \end{pmatrix}$$

$$\underset{\substack{c_2 - (1-i)c_1}}{\overset{\substack{\equiv \\ r_2 - (1+i)r_1}}{\sim}} \begin{pmatrix} 1 & 0 & 0 & 1 & 0 & 0 \\ 0 & 2 & 2i & -1-i & 1 & 0 \\ 0 & -2i & 7 & 0 & 0 & 1 \end{pmatrix}$$

$$\underset{\substack{c_3 - ic_2}}{\overset{\substack{\equiv \\ r_3 + ir_2}}{\sim}} \begin{pmatrix} 1 & 0 & 0 & 1 & 0 & 0 \\ 0 & 2 & 0 & -1-i & 1 & 0 \\ 0 & 0 & 5 & 1-i & i & 1 \end{pmatrix} \text{ showing that } PAP^* = D \text{ whe}$$

$$A = \begin{pmatrix} 1 & 1-i & 0 \\ 1+i & 4 & 2i \\ 0 & -2i & 7 \end{pmatrix}, \quad P = \begin{pmatrix} 1 & 0 & 0 \\ -1-i & 1 & 0 \\ 1-i & i & 1 \end{pmatrix}, \quad D = \begin{pmatrix} 1 & 0 & 0 \\ 0 & 2 & 0 \\ 0 & 0 & 5 \end{pmatrix}$$

Sylvester's law (10.43) is valid for hermitian forms, although a minor change in the proof is necessary (in place of the equation $-X_{p+1}^2 - \ldots - X_r^2 = x_1^2 + \ldots + x_q^2$ between real numbers, one obtains $-|X_{p+1}|^2 - \ldots - |X_r|^2 = |x_1|^2 + \ldots + |x_q|^2$ involving complex numbers, which, as before, implies $x_1 = \ldots = x_q = 0$). So it is legitimate to refer to the signature of an hermitian matrix (or hermitian form). What is more, modification of (10.45) shows that:

Hermitian $n \times n$ matrices are hermitian congruent if and only if they have the same rank and signature.

The mapping $\gamma : V \to \mathbb{R}$, defined by $(v)\gamma = (v, v)^\beta$ for all $v \in V$, is called the **hermitian quadratic form** associated with the hermitian form β (notice that $(v)\gamma$ is real by (10.50)(ii)). Completing the

square of the modulus of terms in $(x)\gamma = xAx^*$, provides a direct method of calculating Q and D with $A = QDQ^*$, and of determining the rank and signature of the hermitian matrix A. For instance

$$2x_1x_1^* + (1+i)x_1x_2^* + (1-i)x_2x_1^* + 3x_2x_2^*$$
$$= \tfrac{1}{2}(2x_1 + (1-i)x_2)(2x_1 + (1-i)x_2)^* + 2x_2x_2^*$$
$$= \tfrac{1}{2}|X_1|^2 + 2|X_2|^2 \quad \text{where} \quad X_1 = 2x_1 + (1-i)x_2, \; X_2 = x_2.$$

Therefore $A = \begin{pmatrix} 2 & 1+i \\ 1-i & 3 \end{pmatrix} = QDQ^*$ where $Q = \begin{pmatrix} 2 & 0 \\ 1-i & 1 \end{pmatrix}$ and $D = \begin{pmatrix} \tfrac{1}{2} & 0 \\ 0 & 2 \end{pmatrix}$, showing that A has rank 2 and signature 2.

The hermitian quadratic form $\gamma : V \to \mathbb{R}$ is called **positive definite** if $(v)\gamma > 0$ for all non-zero $v \in V$. If $\dim V = n$, then adapting (10.48) shows that γ is positive definite if and only if $n = \text{signature } \gamma$ (the signature of the unique hermitian form β associated with γ).

On completing the square, the real quadratic form

$$ax^2 + 2bxy + cy^2 = (x, y)\begin{pmatrix} a & b \\ b & c \end{pmatrix}\begin{pmatrix} x \\ y \end{pmatrix}$$

is seen to be positive definite if and only if $a > 0$ and $\begin{vmatrix} a & b \\ b & c \end{vmatrix} > 0$. This familiar condition (used in connection with maxima and minima) extends to real and hermitian quadratic forms in n variables.

Let A be an $n \times n$ matrix over a field. The determinant of the $r \times r$ matrix, formed by deleting the last $n - r$ rows and last $n - r$ columns of A, is called the **leading principal r-minor** of A.

Proposition 10.52 The hermitian quadratic form $\gamma : \mathbb{C}^n \to \mathbb{R}$, defined by $(x)\gamma = xAx^*$, is positive definite if and only if all the leading principal minors of the $n \times n$ hermitian matrix $A = [a_{ij}]$ are positive.

Proof Let $x = (x_1, \ldots, x_n)$. The statement is clearly true for $n = 1$. Taking $n > 1$ and assuming $a_{11} \neq 0$, we may complete the square of the modulus of terms involving x_1 to obtain

$$xAx^* = a_{11}^{-1}|a_{11}x_1 + a_{21}x_2 + \ldots + a_{n1}x_n|^2 + \sum_{i,j \geqslant 2} b_{ij}x_ix_j^*$$

where $b_{ij} = a_{ij} - a_{i1}a_{1j}/a_{11}$. The $(n-1) \times (n-1)$ matrix B with $(i-1, j-1)$-entry b_{ij} is hermitian and its leading principal minors are related to those of A by the equation

$$\begin{vmatrix} b_{22} \ldots b_{2r} \\ \vdots \quad \vdots \\ b_{r2} \ldots b_{rr} \end{vmatrix} = a_{11}^{-1} \begin{vmatrix} a_{11} \ldots a_{1r} \\ \vdots \quad \vdots \\ a_{r1} \ldots a_{rr} \end{vmatrix}$$

on starting the row-reduction of A. We are now ready to take the inductive step, making use of $\gamma' : \mathbb{C}^{n-1} \to \mathbb{R}$, defined by $(x_2, \ldots, x_n)\gamma' = \sum_{i,j \geqslant 2} b_{ij} x_i x_j^*$.

Suppose γ is positive definite. Then $(e_1)\gamma = a_{11} > 0$ and γ' is positive definite: for if $(x_2, \ldots, x_n)\gamma' \leqslant 0$ for some $x_2, \ldots, x_n$ in $\mathbb{C}$, we see $(x)\gamma = xAx^* = (x_2, \ldots, x_n)\gamma' \leqslant 0$ on taking $x_1 = -a_{11}^{-1}(a_{21}x_2 + \ldots + a_{n1}x_n)$; this implies $x = 0$, and so $x_2 = x_3 = \ldots = x_n = 0$. Therefore all the leading principal minors of B, and hence those of A, are positive.

Conversely, suppose all the leading principal minors of A are positive. Using the above determinant equality, the same is true of B and so, by inductive hypothesis, γ' is positive definite. Suppose $xAx^* \leqslant 0$ for some $x \in \mathbb{C}^n$. As $a_{11} > 0$, we deduce $a_{11}x_1 + a_{21}x_2 + \ldots + a_{n1}x_n = 0$ and $(x_2, \ldots, x_n)\gamma' = 0$; therefore $x_2 = \ldots = x_n = 0$ and hence $x_1 = 0$ also, showing $x = 0$, that is, γ is positive definite. $\square$

Exercises 10.4

1. (a) By applying coupled row and column operations to $[A \vdots I]$, find, for each alternating matrix A over $\mathbb{Q}$ below, an invertible matrix P with $PAP^T = J_r$.

$$\begin{pmatrix} 0 & -1 & 2 \\ 1 & 0 & -3 \\ -2 & 3 & 0 \end{pmatrix}, \quad \begin{pmatrix} 0 & 0 & 1 \\ 0 & 0 & 2 \\ -1 & -2 & 0 \end{pmatrix}, \quad \begin{pmatrix} 0 & 2 & 1 & 2 \\ -2 & 0 & 3 & 0 \\ -1 & -3 & 0 & 0 \\ -2 & 0 & 0 & 0 \end{pmatrix}.$$

(b) By 'completing the determinant' in xAy^T, find an invertible matrix Q with $A = QJ_rQ^T$ for each of the above alternating matrices A.

2. (a) Use the method of question 1(a) to find, for each real symmetric matrix A below, an invertible matrix P with PAP^T diagonal.

$$\begin{pmatrix} 1 & 2 & 3 \\ 2 & 5 & 4 \\ 3 & 4 & 9 \end{pmatrix}, \quad \begin{pmatrix} 1 & 2 & 3 \\ 2 & 3 & 5 \\ 3 & 5 & 8 \end{pmatrix}, \quad \begin{pmatrix} 0 & 1 & 2 \\ 1 & 0 & 3 \\ 2 & 3 & 0 \end{pmatrix}, \quad \begin{pmatrix} 1 & 2 & 3 \\ 2 & 4 & 6 \\ 3 & 6 & 9 \end{pmatrix}.$$

(b) By completing the square in xAx^T, express each of the above matrices A in the form QDQ^T, where Q is invertible and D is diagonal. Show that no two of these matrices are congruent over $\mathbb{R}$.

3. Determine the rank and signature of the hermitian matrices below:

$$\begin{pmatrix} 1 & 1+i & i \\ 1-i & 3 & 2+i \\ -i & 2-i & 2 \end{pmatrix}, \quad \begin{pmatrix} 1 & 1+i & i \\ 1-i & 3 & 2+i \\ -i & 2-i & 1 \end{pmatrix}, \quad \begin{pmatrix} 1 & 1+i & i \\ 1-i & 4 & i \\ -i & -i & 2 \end{pmatrix}.$$

Express the last matrix above in the form QQ^*, where Q is invertible over $\mathbb{C}$. Can the first two matrices above be expressed in this way? Why not?

4. (a) Let the matrix A be congruent to a diagonal matrix. Show that A is symmetric.

   (b) Let Q be an invertible matrix over $\mathbb{C}$. Show that QQ^* is hermitian and positive definite.

   (c) Let A be a square matrix over a field. Show that $A - A^{\mathrm{T}}$ is alternating.

   (d) Let A be matrix over a field F with $\chi(F) \neq 2$. Show that A is alternating if and only if $A^{\mathrm{T}} = -A$.

5. (a) Show that there are $(n+1)(n+2)/2$ congruence classes of real symmetric $n \times n$ matrices. Write down ten real symmetric 3×3 matrices, no two of which are hermitian congruent.

   (b) How many congruence classes of alternating $n \times n$ matrices over the field F are there?

6. Show that the symmetric $n \times n$ matrices over F form a subspace of $M_n(F)$ of dimension $n(n+1)/2$. What is the dimension of the subspace of alternating $n \times n$ matrices over F? Do the hermitian $n \times n$ matrices form a subspace of $M_n(\mathbb{C})$? Show that the hermitian $n \times n$ matrices form a vector space of dimension n^2 over $\mathbb{R}$.

7. Let A be a hermitian $n \times n$ matrix with each $a_{ii} = 0$, but $a_{12} \neq 0$. Writing $(x)\gamma = \sum_{i,j} a_{ij} x_i x_j^*$, find independent linear combinations X_1 and X_2 of the complex variables $x_1, x_2, \ldots, x_n$ such that

$$(x)\gamma = X_1 X_1^* - X_2 X_2^* + (x)\gamma'$$

where $(x)\gamma'$ does not involve either x_1 or x_2.

Find the rank and signature of $A = \begin{pmatrix} 0 & 1+i & i \\ 1-i & 0 & 1 \\ -i & 1 & 0 \end{pmatrix}$.

8. (a) Let $A = [a_{ij}]$ be an alternating 4×4 matrix over a field. Show that $|A| = (a_{12}a_{34} + a_{13}a_{42} + a_{14}a_{23})^2$.

   (b) Let A be an alternating $n \times n$ matrix over the field F. Show that F^n has a subspace U with $\dim U \geqslant n/2$ and $xAy^{\mathrm{T}} = 0$ for all $x, y \in U$. Find bases of the three such subspaces U through $(1, 0, 0, 0)$ in the case $A = J_4$, $n = 4$, $F = \mathbb{Z}_2$.

9. Let A be a hermitian $n \times n$ matrix.
   (a) Show that the principal minors of A are real.
   (b) Show that A is positive definite ($xAx^* > 0$ for all non-zero $x \in \mathbb{C}^n$) if and only if all the principal minors of A are positive.
   (c) Show that A is **negative definite** (meaning $-A$ is positive definite) if and only if the leading principal r-minors of A are positive (r even) and negative (r odd).
   (d) Show that one of the functions f, $g : \mathbb{R}^2 \to \mathbb{R}$ has a maximum and the other a minimum at $(0, 0)$, where

$$f(x_1, x_2) = 2x_1^2 + 8x_1x_2 + 9x_2^2,$$
$$g(x_1, x_2) = -2x_1^2 + 8x_1x_2 - 9x_2^2.$$

10. Find the rank and signature of the real quadratic forms:
   (a) $x_1^2 + x_2^2 + \ldots + x_n^2 - (x_1 + x_2 + \ldots + x_n)^2$,
   (b) $x_1x_2 + x_2x_3 + \ldots + x_{n-1}x_n$.

11 Euclidean and unitary spaces

Geometric considerations have, so far, played only an incidental part in our development of linear algebra. Here we restrict our attention to an n-dimensional vector space V over the real field $\mathbb{R}$ and show that the concepts of 'angle', 'length', 'area', and 'volume' can be expressed in terms of a positive definite quadratic form γ (10.47) on V and its associated symmetric bilinear form β (10.41); in fact γ is interpreted as the **square of the norm**, and β as the **scalar** (or **dot**) **product**, that is, in co-ordinates:

$$\left.\begin{array}{l}(x)\gamma = \|x\|^2 = x_1^2 + x_2^2 + \ldots + x_n^2 \\ (x, y)^\beta = x \cdot y = x_1 y_1 + x_2 y_2 + \ldots + x_n y_n\end{array}\right\} \text{ for } x, y \in \mathbb{R}^n,$$

formulae which will be familiar to the reader. So our starting-point is a **Euclidean space**, meaning a pair (V, β) as above. However, it is only a small step, but nevertheless a helpful one, to replace $\mathbb{R}$ by the complex field $\mathbb{C}$ and γ by a positive definite hermitian quadratic form, obtaining the **unitary space** (V, β) and the analogous formulae:

$$\left.\begin{array}{l}(x)\gamma = \|x\|^2 = |x_1|^2 + |x_2|^2 + \ldots + |x_n|^2 \\ (x, y)^\beta = x \cdot y = x_1 y_1^* + x_2 y_2^* + \ldots + x_n y_n^*\end{array}\right\} \text{ for } x, y \in \mathbb{C}^n.$$

Euclidean spaces

Definition 11.1 Let V be an n-dimensional vector space over $\mathbb{R}$ and let β be a symmetric bilinear form on V such that $(v, v)^\beta > 0$ for all v in V with $v \neq 0$. Then the pair (V, β) is called a **Euclidean space**.

We refer simply to the Euclidean space V, with β being understood, and use the scalar product notation $u \cdot v$ in place of $(u, v)^\beta$; in particular $\mathbb{R}^n$, with β defined by $(x, y)^\beta = x \cdot y = xy^T = x_1 y_1 + \ldots + x_n y_n$ where $x = (x_1, \ldots, x_n)$, $y = (y_1, \ldots, y_n)$, satisfies (11.1), and is called the **standard n-dimensional Euclidean space** $\mathbb{R}^n$. We proceed to borrow terminology from $\mathbb{R}^n$ and use it in the abstract n-dimensional Euclidean space V; this is not as outrageous as may appear, for it turns out (11.11) that $\mathbb{R}^n$ and V are abstractly identical Euclidean spaces.

Definition 11.2 Let u and v belong to the Euclidean space V. Then $\|v\| = \sqrt{(v \cdot v)}$ is called the **norm (length)** of v, and v is called a **unit vector** if $\|v\| = 1$. The non-negative real number $\|u - v\|$ is called the **distance** between u and v.

We establish the basic property of distance, namely the triangle inequality, in (11.25); in the case of $\mathbb{R}^n$

$$\|x - y\| = \sqrt{((x_1 - y_1)^2 + \ldots + (x_n - y_n)^2)}$$

which is the Euclidean distance formula (based on Pythagoras' theorem). For $v \neq 0$, the unit vector $(1/\|v\|)v$ is called the result of **normalizing** v; in fact $(1/\|v\|)v$ is the unique unit vector in the same direction as v.

Fig. 11.1

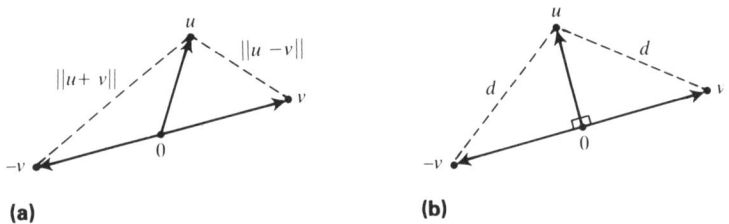

(a)   (b)

Let u and v belong to the Euclidean space V and consider the triangle formed by u, v, $-v$ (Fig. 11.1(a)). As $\|u - v\|$ is the distance between u and v, and $\|u + v\|$ is the distance between u and $-v$, we see that u and v are **orthogonal** (at right-angles) if and only if this triangle is isosceles (Fig. 11.1(b)), that is, $\|u + v\| = d = \|u - v\|$. Therefore $\|u + v\|^2 = \|u - v\|^2$, which by (11.2) gives $(u + v) \cdot (u + v) = (u - v) \cdot (u - v)$; multiplying the scalar product out in the usual way:

$$u \cdot u + u \cdot v + v \cdot u + v \cdot v = u \cdot u - u \cdot v - v \cdot u + v \cdot v$$

which, as $u \cdot v = v \cdot u$, simplifies to $4(u \cdot v) = 0$, that is, $u \cdot v = 0$. The steps in the argument can be reversed, and so:

u and v are orthogonal $\Leftrightarrow$ $u \cdot v = 0$.

One could scarcely hope for a simpler condition! Indeed, from our point of view, it is the fulcrum balancing geometry and algebra.

Definition 11.3 Let u and v belong to the Euclidean space V where $v \neq 0$. The vector $((u \cdot v)/(v \cdot v))v$ is called the **orthogonal projection (resolution)** of u along v.

Fig. 11.2

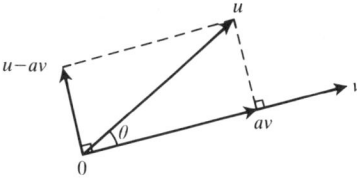

The justification of (11.3) is given in Fig. 11.2. The vector av is the orthogonal projection of u along v, where the scalar a is chosen so that $u - av$ is orthogonal to v; so $(u - av) \cdot v = 0$ giving $a = (u \cdot v)/(v \cdot v)$. If $V = \mathbb{R}^n$, let θ be the angle between the non-zero vectors u and v; resolving u along v gives $\|u\| \cos \theta = a \|v\|$ and hence

$$u \cdot v = \|u\| \, \|v\| \cos \theta \quad \text{in } \mathbb{R}^n.$$

Lemma 11.4

Let $v_1, v_2, \ldots, v_m$ be mutually orthogonal non-zero vectors of the Euclidean space V. Then $v_1, v_2, \ldots, v_m$ are linearly independent.

Proof

The vectors $v_1, v_2, \ldots, v_m$ being mutually orthogonal means $v_i \cdot v_j = 0$ for all $i \neq j$. Suppose $a_1 v_1 + a_2 v_2 + \ldots + a_m v_m = 0$ and form the scalar product of this vector with v_j:

$$0 = 0 \cdot v_j = (a_1 v_1 + a_2 v_2 + \ldots + a_m v_m) \cdot v_j$$
$$= a_1(v_1 \cdot v_j) + a_2(v_2 \cdot v_j) + \ldots + a_m(v_m \cdot v_j)$$
$$= a_j(v_j \cdot v_j) = a_j \|v_j\|^2.$$

Therefore $a_j = 0$ as $\|v_j\| \neq 0$, and so $a_1 = a_2 = \ldots = a_m = 0$. □

We now introduce the special types of basis appropriate to Euclidean spaces.

Definition 11.5

Let b denote the basis $v_1, v_2, \ldots, v_n$ of the Euclidean space V. Then b is called **orthogonal** if $v_i \cdot v_j = 0$ for all $i \neq j$. If $v_i \cdot v_j = \delta_{ij}$ ($= 1$ for $i = j$, and zero otherwise), then b is called an **orthonormal basis**.

So an orthogonal basis is a basis of mutually orthogonal vectors, while an orthonormal basis is a basis of mutually orthogonal unit vectors. For instance $v_1 = (1, 1, 1)$, $v_2 = (1, -2, 1)$, $v_3 = (1, 0, -1)$ form an orthogonal basis of $\mathbb{R}^3$; normalizing each of these vectors, that is, dividing each vector by the square root of the sum of the squares of its entries, produces the orthonormal basis $(1/\sqrt{3})v_1$, $(1/\sqrt{6})v_2$, $(1/\sqrt{2})v_3$ of $\mathbb{R}^3$. The standard basis $e_1, e_2, \ldots, e_n$ of $\mathbb{R}^n$ is orthonormal because $e_i \cdot e_j = e_i e_j^{\mathrm{T}} = \delta_{ij}$.

It follows from (10.42) that each Euclidean space has an orthogonal basis and that such a basis may be built up vector by vector; however (11.7) provides an independent proof of this fact. Normalizing, we see that each Euclidean space has an orthonormal basis.

We show next that each vector is the sum of its orthogonal projections along the vectors of an orthogonal basis.

Lemma 11.6 Let $v_1, v_2, \ldots, v_n$ form an orthogonal basis of the Euclidean space V. Then $v = \sum_i ((v \cdot v_i)/(v_i \cdot v_i))v_i$ for each $v \in V$.

Proof Each v in V can be expressed $v = \sum_i a_i v_i$ where $a_i \in \mathbb{R}$. We may isolate a_j by taking the scalar product of v and v_j:

$$v \cdot v_j = \left(\sum_i a_i v_i \right) \cdot v_j = \sum_i a_i (v_i \cdot v_j)$$

$$= a_j (v_j \cdot v_j)$$

as $v_1, v_2, \ldots, v_n$ are mutually orthogonal. Therefore

$$a_j = (v \cdot v_j)/(v_j \cdot v_j) \quad \text{for} \quad 1 \leqslant j \leqslant n. \qquad \square$$

So it is a simple matter to express a given vector in terms of a given orthogonal basis—this is the advantage of such bases over arbitrary bases. For example, taking $v = (7, 2, -4)$ we obtain

$$v = \tfrac{5}{3}v_1 - \tfrac{1}{6}v_2 + \tfrac{11}{2}v_3$$

where $v_1 = (1, 1, 1)$, $v_2 = (1, -2, 1)$, $v_3 = (1, 0, -1)$.

Our next theorem describes how an arbitrary basis may be modified to give an orthogonal basis.

Theorem 11.7 (The Gram–Schmidt orthogonalization process.) Let the Euclidean space V have basis $v_1, v_2, \ldots, v_n$. Then $v'_1, v'_2, \ldots, v'_n$ form an orthogonal basis of V where

$$v'_i = v_i - \sum_{j=1}^{i-1} ((v_i \cdot v'_j)/\|v'_j\|^2)v'_j \quad (1 \leqslant i \leqslant n).$$

Proof The above formula expresses v'_i as a linear combination of v_i and $v'_1, v'_2, \ldots, v'_{i-1}$, and suggests an inductive approach. So suppose that $v'_1, v'_2, \ldots, v'_{i-1}$ form an orthogonal basis of

Fig. 11.3

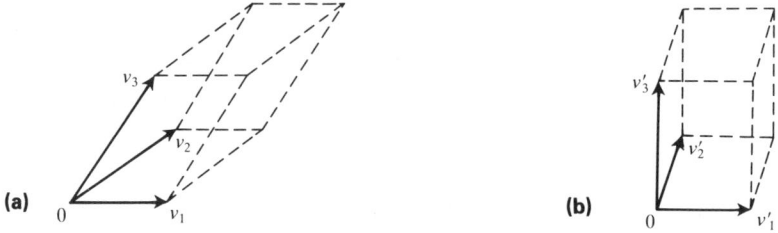

(a)   (b)

Before and after orthogonalization

$\langle v_1, v_2, \ldots, v_{i-1} \rangle$ where $1 < i \leq n$ (v_1' is an orthogonal basis of $\langle v_1 \rangle$ as $v_1' = v_1$). It is now routine to show that v_i' is orthogonal to $v_1', v_2', \ldots, v_{i-1}'$; in fact

$$v_i' \cdot v_j' = v_i \cdot v_j' - ((v_i \cdot v_j')/\|v_j'\|^2)(v_j' \cdot v_j') = 0 \quad \text{for } j < i.$$

As $v_i \notin \langle v_1, v_2, \ldots, v_{i-1} \rangle = \langle v_1', v_2', \ldots, v_{i-1}' \rangle$, we see $v_i' \neq 0$. Therefore $v_1', v_2', \ldots, v_i'$ are mutually orthogonal non-zero vectors belonging to the i-dimensional subspace $\langle v_1, v_2, \ldots, v_i \rangle$, and so they form a basis of this subspace by (11.4). The induction is now complete and terminates in the orthogonal basis $v_1', v_2', \ldots, v_n'$ of V. $\qquad\qquad\square$

The geometric significance of the Gram–Schmidt process will be discussed later (11.44) in connection with volume, but to get the idea consider the basis $v_1 = (0, 1, 0)$, $v_2 = (-3, 2, 0)$, $v_3 = (-1, 3, 4)$ of $\mathbb{R}^3$. Applying (11.7), we obtain the orthogonal basis $v_1' = v_1$, $v_2' = v_2 - 2v_1' = (-3, 0, 0)$, $v_3' = v_3 - 3v_1' - \frac{1}{3}v_2' = (0, 0, 4)$. The vectors v_1, v_2, v_3 are the edges of a parallelepiped (Fig. 11.3(a)) while v_1', v_2', v_3' are the edges of a rectangular parallelepiped (Fig. 11.3(b)); these parallelepipeds have the same 'frontage' ($v_1 = v_1'$) and, as we shall see, the same 'ground' area (the parallelogram with edges v_1 and v_2 has the same area as the rectangle with edges v_1' and v_2'), and the same volume. In other words, the Gram–Schmidt process makes drunken boxes stand to attention and become rectangular, the volume of 'leading sub-boxes' remaining unchanged!

Corollary 11.8

Let $v_1', v_2', \ldots, v_m'$ be mutually orthogonal non-zero vectors of the Euclidean space V. Then V has an orthogonal basis of the form $v_1', v_2', \ldots, v_m', v_{m+1}', \ldots, v_n'$.

Proof

As $v_1', v_2', \ldots, v_m'$ are linearly independent by (11.4), V has a basis of the form $v_1', v_2', \ldots, v_m', v_{m+1}, \ldots, v_n$. Applying the orthogonalization process (11.7) to this basis leaves the mutually orthogonal vectors $v'_1, v'_2, \ldots, v'_m$ unchanged, and so an orthogonal basis of the form $v'_1, v'_2, \ldots, v'_m, v'_{m+1}, \ldots, v'_n$ results.

□

For example, $v'_1 = (1, 1, 1, 1)$ and $v'_2 = (1, -1, 1, -1)$ are orthogonal vectors of $\mathbb{R}^4$. The vector $x = (x_1, x_2, x_3, x_4)$ is orthogonal to both v'_1 and v'_2 if and only if

$$x \cdot v'_1 = x_1 + x_2 + x_3 + x_4 = 0, \qquad x \cdot v'_2 = x_1 - x_2 + x_3 - x_4 = 0,$$

which are satisfied by $v'_3 = (1, 1, -1, -1)$. Solving these equations simultaneously with $x \cdot v'_3 = x_1 + x_2 - x_3 - x_4 = 0$ gives $v'_4 = (1, -1, -1, 1)$. Therefore v'_1, v'_2 extends to the orthogonal basis v'_1, v'_2, v'_3, v'_4 of $\mathbb{R}^4$.

Normalizing the vectors of an orthogonal basis produces an orthonormal basis (11.5). So every Euclidean space has an orthonormal basis by (11.7) and from (11.8) we see that:

Each set of mutually orthogonal unit vectors
can be extended to an orthonormal basis.

Let the Euclidean space V have orthonormal basis $v_1, v_2, \ldots, v_n$. Then (11.6) becomes

$$v = \sum_{i=1}^{n} (v \cdot v_i) v_i \quad \text{for each } v \in V$$

which should remind the reader of the formulae involving linear forms following (10.28) (here $v \cdot v_i$ replaces $(v)\hat{v}_i$).

Definition 11.9

The Euclidean spaces (V, β) and (V', β') are called **isometric** if there is a linear bijection $\alpha : V \cong V'$ with

$$(u, v)^\beta = ((u)\alpha, (v)\alpha)^{\beta'} \quad \text{for all } u, v \in V.$$

Such a mapping α is called an **isometry**.

The above condition may be expressed: $u \cdot v = (u)\alpha \cdot (v)\alpha$ for all $u, v \in V$; therefore an isometry is a vector space isomorphism which respects scalar products. Also

$$\|u - v\|^2 = (u - v) \cdot (u - v) = (u - v)\alpha \cdot (u - v)\alpha$$
$$= ((u)\alpha - (v)\alpha) \cdot ((u)\alpha - (v)\alpha)$$
$$= \|(u)\alpha - (v)\alpha\|^2$$

showing that the isometry α does indeed preserve distances, that is,

$$\|u - v\| = \|(u)\alpha - (v)\alpha\| \quad \text{for all } u, v \in V.$$

We show next that isometries can be recognized by their effect on orthonormal bases.

Proposition 11.10 Let $\alpha : V \to V'$ be a linear mapping where V and V' are Euclidean spaces, and let V have orthonormal basis $v_1, v_2, \ldots, v_n$. Then α is an isometry if and only if $(v_1)\alpha, (v_2)\alpha, \ldots, (v_n)\alpha$ form an orthonormal basis of V'.

Proof Let α be an isometry. As α is bijective, $(v_1)\alpha, (v_2)\alpha, \ldots, (v_n)\alpha$ is a basis of V' by (7.16). As α respects scalar products, we have $(v_i)\alpha \cdot (v_j)\alpha = v_i \cdot v_j = \delta_{ij}$, showing that $(v_1)\alpha, (v_2)\alpha, \ldots, (v_n)\alpha$ form an orthonormal basis of V'.

Conversely, suppose $(v_1)\alpha, (v_2)\alpha, \ldots, (v_n)\alpha$ is an orthonormal basis of V'. Then α is bijective by (7.16). The vectors u and v in V can be expressed

$$u = \sum_i x_i v_i, \qquad v = \sum_j y_j v_j \qquad (x_i, y_j \in \mathbb{R});$$

as α is linear,

$$(u)\alpha = \sum_i x_i(v_i)\alpha \quad \text{and} \quad (v)\alpha = \sum_j y_j(v_j)\alpha.$$

Therefore

$$u \cdot v = \left(\sum_i x_i v_i\right) \cdot \left(\sum_j y_j v_j\right) = \sum_{i,j} x_i \delta_{ij} y_j$$

$$= \sum_i x_i y_i$$

using the bilinearity of the scalar product on V and the orthonormality of $v_1, v_2, \ldots, v_n$. In the same way we obtain $(u)\alpha \cdot (v)\alpha = \sum_i x_i y_i$, using the bilinearity of the scalar product on V' and the orthonormality of $(v_1)\alpha, (v_2)\alpha, \ldots, (v_n)\alpha$. So $u \cdot v = (u)\alpha \cdot (v)\alpha$ for all u and v in V, showing α to be an isometry. $\quad\square$

Let b denote the orthonormal basis $v_1, v_2, \ldots, v_n$ of the Euclidean space V. Then $u = \sum_i x_i v_i$ and $v = \sum_j y_j v_j$, as above, have scalar product $u \cdot v = \sum_i x_i y_i = x \cdot y$, the scalar product of $x = (x_1, \ldots, x_n)$ and $y = (y_1, \ldots, y_n)$ in the standard Euclidean space $\mathbb{R}^n$. So working relative to an orthonormal basis of V is the same as working in $\mathbb{R}^n$; more precisely, the co-ordinatizing isomorphism $\kappa_{\measuredangle} : V \cong \mathbb{R}^n$ (7.24) is an isometry, for $(u)\kappa_{\measuredangle} = x$, $(v)\kappa_{\measuredangle} = y$ and so $u \cdot v = x \cdot y = (u)\kappa_{\measuredangle} \cdot (v)\kappa_{\measuredangle}$, showing that:

Every n-dimensional Euclidean space V is
isometric to the standard Euclidean space $\mathbb{R}^n$.

Because of the existence of orthonormal bases, it is easy to find out whether two Euclidean spaces are isometric or not: the dimension, and only the dimension, matters.

Corollary 11.11
The Euclidean spaces V and V' are isometric if and only if $\dim V = \dim V'$.

Proof
If V and V' are isometric, then V and V' are isomorphic vector spaces and so $\dim V = \dim V'$. Suppose conversely that $\dim V = \dim V'$ $(= n$ say). Then V has an orthonormal basis $v_1, v_2, \ldots, v_n$ and V' has an orthonormal basis $v'_1, v'_2, \ldots, v'_n$. By (7.16) there is a unique linear bijection $\alpha : V \cong V'$ with $(v_i)\alpha = v'_i (1 \leqslant i \leqslant n)$. By (11.10) α is an isometry, and so V and V' are isometric.   □

We now introduce the special matrices associated with isometries of Euclidean spaces; the reader should take heart, for it turns out that such isometries are merely rotations and reflections.

Definition 11.12
The $n \times n$ matrix P over the real field $\mathbb{R}$ is called **orthogonal** if $PP^{\mathrm{T}} = I$.

Every rotation matrix $\begin{pmatrix} \cos\phi & \sin\phi \\ -\sin\phi & \cos\phi \end{pmatrix}$ is orthogonal (7.14), and so also is every reflection matrix $\begin{pmatrix} \cos\phi & \sin\phi \\ \sin\phi & -\cos\phi \end{pmatrix}$ (the 'mirror' is the characteristic subspace corresponding to the characteristic root 1; geometrically, it is the line of gradient $\tan\phi/2$ through the origin of $\mathbb{R}^2$). Taking determinants of $PP^{\mathrm{T}} = I$, we see $|P| = \pm 1$ for each orthogonal matrix P; the orthogonal 2×2 matrix P represents a rotation/reflection according as $|P| = \pm 1$, this simple test being valid in general (11.40).

Let r_i denote row i of the $n \times n$ matrix P over $\mathbb{R}$. Regarding r_i as belonging to the standard Euclidean space $\mathbb{R}^n$, the matrix equation $PP^{\mathrm{T}} = I$ becomes $r_i \cdot r_j = \delta_{ij}$ $(1 \leqslant i, j \leqslant n)$. Therefore:

A real $n \times n$ matrix is orthogonal if and only if
its rows form an orthonormal basis of $\mathbb{R}^n$.

Let P be an orthogonal $n \times n$ matrix. The equation $PP^T = I$ tells us that $P^{-1} = P^T$ and so $P^T P = P^{-1}P = I$ showing that P^T is orthogonal; therefore the rows of P^T form an orthonormal basis of $\mathbb{R}^n$. For instance, the reader may verify that P and P^T are orthogonal where

$$P = \begin{pmatrix} 1/3 & 2/3 & 2/3 \\ 0 & 1/\sqrt{2} & -1/\sqrt{2} \\ 4/\sqrt{18} & -1/\sqrt{18} & -1/\sqrt{18} \end{pmatrix}.$$

Let P and Q be orthogonal $n \times n$ matrices. Then PQ is orthogonal as

$$(PQ)(PQ)^T = PQQ^T P^T = PIP^T = PP^T = I.$$

As the identity matrix is orthogonal, by (9.10) the orthogonal $n \times n$ matrices form a subgroup (called the **orthogonal group** $O_n(\mathbb{R})$) of the group of invertible $n \times n$ matrices over $\mathbb{R}$.

Corollary 11.13

Let ℓ and ℓ' be orthonormal bases of the Euclidean spaces V and V' and let $\alpha : V \to V'$ be linear. Then α is an isometry if and only if the matrix P of α relative to ℓ and ℓ' is orthogonal.

Proof

We use the equation $\kappa_\ell \mu_P = \alpha \kappa_{\ell'}$ (see (7.25)) and the fact that κ_ℓ and $\kappa_{\ell'}$ are isometries. Let α be an isometry. Then $\dim V = \dim V'$ $(= n$ say) by (11.11), and $\mu_P = \kappa_\ell^{-1} \alpha \kappa_{\ell'}$ is an isometry of $\mathbb{R}^n$, for the composition and inverses of isometries are again isometries. As the standard basis of $\mathbb{R}^n$ is orthonormal and $(e_i)\mu_P = e_i P = r_i$ $(1 \leq i \leq n)$, by (11.10) the rows $r_1, r_2, \ldots, r_n$ of P form an orthonormal basis of $\mathbb{R}^n$; therefore P is orthogonal. Conversely, if P is orthogonal, then μ_P is an isometry of $\mathbb{R}^n$ by (11.10) and hence $\alpha = \kappa_\ell \mu_P \kappa_{\ell'}^{-1}$ is also an isometry.   □

The isometries of $\mathbb{R}^n$ are therefore the mappings μ_P where P is an orthogonal $n \times n$ matrix, and so they form a group isomorphic to $O_n(\mathbb{R})$. Taking $V = V'$ and $\alpha = \iota$ (the identity mapping, which is certainly an isometry of V) we see that:

The orthonormal bases ℓ and ℓ' of the Euclidean space V are related by an orthogonal matrix P.

Definition 11.14

The real $n \times n$ matrices A and B are called **orthogonally similar** if there is an orthogonal $n \times n$ matrix P such that $PAP^T = B$.

As $P^T = P^{-1}$, orthogonally similar matrices are similar by (10.1), and congruent by (10.37). Let $\alpha : V \to V'$ be linear; mimicking

Fig. 11.4

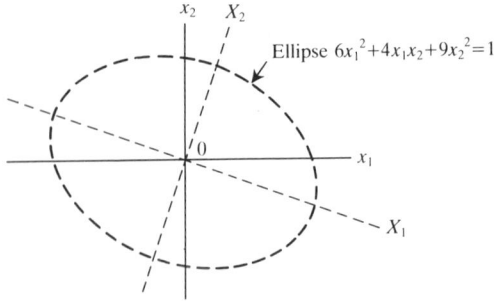

Ellipse $6x_1{}^2 + 4x_1x_2 + 9x_2{}^2 = 1$

(10.2), we see that orthogonally similar matrices arise on referring α to different orthonormal bases of the Euclidean space V.

For example $PAP^T = D$ where

$$P = (1/\sqrt{5})\begin{pmatrix} 2 & -1 \\ 1 & 2 \end{pmatrix}, \qquad A = \begin{pmatrix} 6 & 2 \\ 2 & 9 \end{pmatrix}, \qquad D = \begin{pmatrix} 5 & 0 \\ 0 & 10 \end{pmatrix}.$$

Geometrically, the change of variables $x = XP$, where $x = (x_1, x_2)$ and $X = (X_1, X_2)$, amounts to a rotation of axes; the X_1- and X_2-axes are the major and minor axes of the ellipse with equation $xAx^T = 1$, that is, $6x_1^2 + 4x_1x_2 + 9x_2^2 = 1$ (Fig. 11.4). In terms of X_1 and X_2, this ellipse has equation $XPAP^TX^T = XDX^T = 5X_1^2 + 10X_2^2 = 1$.

Let the real $n \times n$ matrix A be orthogonally similar to the diagonal matrix D. Then $PAP^T = D$ where P is orthogonal and so $A = P^TDP$ which, on transposing, gives $A^T = P^TD^TP = P^TDP = A$ as D is symmetric. Therefore A is symmetric. It is an important fact (11.20) of geometric significance, generalizing the above illustration, that the converse is also true, namely:

Every real symmetric matrix is orthogonally similar to a diagonal matrix.

The proof involves decomposing the Euclidean space V into orthogonal components, which we now discuss. Notice that if U is a subspace of V, then, by restricting the scalar product, U itself becomes a Euclidean space; so every subspace of V has an orthonormal basis.

Definition 11.15

Let U be a subspace of the Euclidean space V. Then

$$U^\perp = \{v \in V : v \cdot u = 0 \text{ for all } u \in U\}$$

is called the **orthogonal complement** of U.

Fig. 11.5

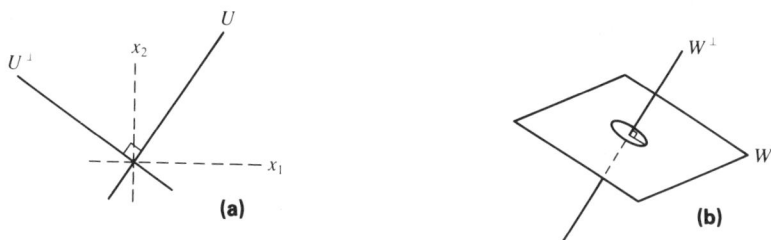

(a)   (b)

So $U^\perp$ consists of those vectors which are orthogonal to all the vectors in U. Using (6.6) and the linearity of $v \cdot u$ in v (for given u, the mapping $v \mapsto v \cdot u$ of V to $\mathbb{R}$ is linear), we see that $U^\perp$ is a subspace of V.

Let U be the 1-dimensional subspace of $\mathbb{R}^2$ spanned by $(2, 3)$. Then $U = \langle (2, 3) \rangle$ and $U^\perp = \{(x_1, x_2) \in \mathbb{R}^2 : (x_1, x_2) \cdot (2, 3) = 0\} = \langle (3, -2) \rangle$; geometrically, U and $U^\perp$ are perpendicular lines through the origin of $\mathbb{R}^2$ (Fig. 11.5(a)). Let W be the 2-dimensional subspace of $\mathbb{R}^3$ with equation $2x_1 + 3x_2 + 4x_3 = 0$; this equation tells us directly that $W^\perp = \langle (2, 3, 4) \rangle$, that is, $(2, 3, 4)$ is *normal* to the plane W (Fig. 11.5(b)).

We show next that the orthogonal complement deserves its name.

Proposition 11.16

Let U be a subspace of the Euclidean space V. Then U and $U^\perp$ are complementary subspaces of V and $(U^\perp)^\perp = U$.

Proof

The subspace U has an orthonormal basis $v_1, \ldots, v_m$ which extends to an orthonormal basis $v_1, \ldots, v_m, v_{m+1}, \ldots, v_n$ of V. It is routine to verify that $v = \sum_{i=1}^n (v \cdot v_i) v_i$ belongs to $U^\perp$ if and only if $v \cdot v_1 = v \cdot v_2 = \ldots = v \cdot v_m = 0$, that is, if and only if $v \in \langle v_{m+1}, \ldots, v_n \rangle$; therefore $U^\perp = \langle v_{m+1}, \ldots, v_n \rangle$ and so $U^\perp$ has basis $v_{m+1}, \ldots, v_n$. By (6.29), $V = U \oplus U^\perp$ which means U and $U^\perp$ are complementary subspaces of V. By symmetry, $(U^\perp)^\perp$ has basis $v_1, \ldots, v_m$ and so $(U^\perp)^\perp = U$. ☐

Decompositions of the type $V = U \oplus U^\perp$ are used to analyse linear mappings of V.

Proposition 11.17

Let A be a real symmetric $n \times n$ matrix.

(a) The characteristic polynomial of A factorizes into linear factors over $\mathbb{R}$.

(b) Characteristic vectors of A corresponding to distinct characteristic roots are orthogonal vectors of $\mathbb{R}^n$.

Proof

(a) By (4.26) the characteristic polynomial $\chi = |xI - A|$ factorizes into linear factors over the complex field $\mathbb{C}$. Let $x - \lambda$ be a factor of χ; we show that λ is, in fact, real. As λ is a zero of χ, the (possibly complex) matrix $\lambda I - A$ is singular. By (9.41) there is a non-zero vector $x = (x_1, \ldots, x_n)$ in $\mathbb{C}^n$ with $x(\lambda I - A) = 0$, that is, $xA = \lambda x$. Postmultiplying by $x^* = (x_1^*, \ldots, x_n^*)^T$ gives the scalar equation

$$xAx^* = \lambda xx^*.$$

Conjugating $w = xAx^*$ produces $w^* = xA^*x^* = xAx^* = w$, as $A^* = A$ (the crucial property of A is that it is equal to its hermitian conjugate A^* (10.50)), and so w is real. But $\|x\|^2 = xx^* = |x_1|^2 + \ldots + |x_n|^2$ is real and positive as $x \neq 0$ (unitary spaces are founded on this fact). So $\lambda = w/\|x\|^2$ is real. Therefore χ factorizes into linear factors over $\mathbb{R}$.

(b) The complex field $\mathbb{C}$ has, for the moment, done its job; we now revert to working within $\mathbb{R}$. Let x and y in $\mathbb{R}^n$ satisfy $xA = \lambda x$ and $yA = \mu y$ where λ and μ are distinct real numbers. Then

$$\lambda xy^T = xAy^T = xA^Ty^T = x(yA)^T = \mu xy^T$$

and so $(\lambda - \mu)xy^T = 0$. As $\lambda - \mu \neq 0$ we deduce $x \cdot y = xy^T = 0$, showing that x and y are orthogonal vectors of $\mathbb{R}^n$. $\square$

Let $PAP^T = D$ where the real $n \times n$ matrices A, D, P are respectively symmetric, diagonal, orthogonal. As $P^T = P^{-1}$, the rows of P are characteristic vectors of A by (10.11); therefore $\mathbb{R}^n$ has an orthonormal basis (the rows of P) consisting of characteristic vectors of A.

We are now in a position to reverse the argument as in (10.18). Let A be a real symmetric $n \times n$ matrix having n distinct characteristic roots $\lambda_1, \lambda_2, \ldots, \lambda_n$. By (11.17)(a) each λ_i is real; for each λ_i, pick a unit characteristic vector r_i in $\mathbb{R}^n$, and so $r_iA = \lambda_ir_i$ $(1 \leqslant i \leqslant n)$. By (11.17)(b) $r_1, r_2, \ldots, r_n$ are mutually orthogonal and so they form an orthonormal basis of $\mathbb{R}^n$. Finally, $PAP^T = D$ where P is the orthogonal matrix having r_i as its ith row and D is the diagonal matrix with (i, i)-entry λ_i.

Example 11.18(a)

Let $A = \begin{pmatrix} 7 & 1 & -3 \\ 1 & 3 & 1 \\ -3 & 1 & 7 \end{pmatrix}$. Then A has characteristic polynomial $\chi = (x - 2)(x - 5)(x - 10)$. We refine the procedure of (10.12)(a): for each zero λ of χ, find a non-zero solution $x = (x_1, x_2, x_3)$ in

$\mathbb{R}^3$ of the system $x(A - \lambda I) = 0$ and then normalize this solution. The reader may verify the entries in the table below:

characteristic root λ	characteristic vector x	$(1/\|x\|)x$
2	$(1, -2, 1)$	$(1/\sqrt{6})(1, -2, 1)$
5	$(1, 1, 1)$	$(1/\sqrt{3})(1, 1, 1)$
10	$(-1, 0, 1)$	$(1/\sqrt{2})(-1, 0, 1)$

So in this case $PAP^{\mathrm{T}} = D$ where

$$P = \begin{pmatrix} 1/\sqrt{6} & -2/\sqrt{6} & 1/\sqrt{6} \\ 1/\sqrt{3} & 1/\sqrt{3} & 1/\sqrt{3} \\ -1/\sqrt{2} & 0 & 1/\sqrt{2} \end{pmatrix} \quad \text{and} \quad D = \begin{pmatrix} 2 & 0 & 0 \\ 0 & 5 & 0 \\ 0 & 0 & 10 \end{pmatrix}.$$

Geometrically, the rows of P are unit vectors along the principal axes of the ellipsoid $xAx^{\mathrm{T}} = c$ (c positive). As $|P| = 1$, the change of variables $x = XP$ amounts to a rotation (11.40); in terms of the variables $X = (X_1, X_2, X_3)$ the ellipsoid has equation $2X_1^2 + 5X_2^2 + 10X_3^2 = c$.

Example 11.18(b)

Consider $A = \begin{pmatrix} 4 & 2 & -3 \\ 2 & 1 & 6 \\ -3 & 6 & -4 \end{pmatrix}$. We show in (11.19) that multiple zeros of the characteristic polynomial of a real symmetric matrix present no difficulties. Here $\chi = (x - 5)^2(x + 9)$. Let U denote the characteristic subspace (10.14) corresponding to 5; so $U = \{x \in \mathbb{R}^3 : x(A - 5I) = 0\}$, that is, U is the plane with equation $x_1 - 2x_2 + 3x_3 = 0$. Choose any non-zero solution of this equation, e.g. $(1, -1, -1)$; next choose a second non-zero solution orthogonal to $(1, -1, -1)$ and so satisfying $x_1 - x_2 - x_3 = 0$, e.g. $(5, 4, 1)$. Then $(1, -1, -1)$, $(5, 4, 1)$ form an orthogonal basis of U. In this case there is no need for further calculation as the characteristic subspace corresponding to -9 is $U^{\perp} = \langle (1, -2, 3) \rangle$ by (11.17)(b). So $(1, -1, -1)$, $(5, 4, 1)$, $(1, -2, 3)$ are characteristic vectors of A forming an orthogonal basis of $\mathbb{R}^3$. Normalizing produces an orthogonal matrix P with $PAP^{\mathrm{T}} = D$ (diagonal); specifically,

$$P = \begin{pmatrix} 1/\sqrt{3} & -1/\sqrt{3} & -1/\sqrt{3} \\ 5/\sqrt{42} & 4/\sqrt{42} & 1/\sqrt{42} \\ 1/\sqrt{14} & -2/\sqrt{14} & 3/\sqrt{14} \end{pmatrix} \quad \text{and} \quad D = \begin{pmatrix} 5 & 0 & 0 \\ 0 & 5 & 0 \\ 0 & 0 & -9 \end{pmatrix}.$$

The points (x_1, x_2, x_3) of $\mathbb{R}^3$ satisfying $xAx^{\mathrm{T}} = 1$ form an hyperboloid of one sheet, two of the characteristic roots of A being positive and one negative; as the two positive roots are equal, it is an hyperboloid of revolution, obtained by rotating the hyperbola $5X_2^2 - 9X_3^2 = 1$ in the plane $X_1 = 0$ about the X_3-axis where $x = XP$.

In contrast to arbitrary square matrices, (10.12)(c) for example, we show next that real symmetric matrices A are invariably well-behaved—they are model citizens of the real matrix world! Not only does the characteristic polynomial $|xI - A|$ factorize completely over $\mathbb{R}$, but each of its zeros λ of multiplicity m corresponds to a characteristic subspace of dimension m.

Proposition 11.19 Let the characteristic polynomial $|xI - A|$ of the real symmetric $n \times n$ matrix A have zero λ of multiplicity m. Then $\mathrm{rank}(\lambda I - A) = n - m$.

Proof Let the characteristic subspace $U = \{x \in \mathbb{R}^n : xA = \lambda x\}$ have dimension m'. As $U = \{x \in \mathbb{R}^n : x(\lambda I - A) = 0\}$, we see $\mathrm{rank}(\lambda I - A) = n - m'$ by (8.10). Now U is μ_A-invariant (10.33) as $(x)\mu_A = xA = \lambda x \in U$ for all $x \in U$. Further, $U^\perp$ is μ_A-invariant also: for let $y \in U^\perp$; then $yx^\mathrm{T} = 0$ for all $x \in U$ and so $yAx^\mathrm{T} = yA^\mathrm{T}x^\mathrm{T} = y(xA)^\mathrm{T} = \lambda yx^\mathrm{T} = 0$ for all $x \in U$, showing that $(y)\mu_A \in U^\perp$. Let ℓ and ℓ' be bases of U and $U^\perp$ respectively. Then μ_A has matrix

$$B = \begin{pmatrix} \lambda I & \vdots & \\ \cdots & \vdots & \cdots \\ & \vdots & A' \end{pmatrix}$$

relative to the basis $\ell \cup \ell'$ of $U \oplus U^\perp = \mathbb{R}^n$, where I denotes the $m' \times m'$ identity matrix and A' is the matrix of μ_A' (the restriction of μ_A to $U^\perp$) relative to ℓ'. By (10.2), A and B are similar and so

$$|xI - A| = |xI - B| = (x - \lambda)^{m'} |xI - A'|$$

using (10.13) and determinant expansion (9.33). However, λ is *not* a characteristic root of $\mu_A' : U^\perp \to U^\perp$ as $U \cap U^\perp = 0$ (in fact $U^\perp$ is the direct sum of the characteristic subspaces $\neq U$ of A). By (10.8), (10.10), λ is not a zero of $|xI - A'|$, showing that λ is a zero of $|xI - A|$ of multiplicity m'. Therefore $m' = m$ and hence $\mathrm{rank}(\lambda I - A) = n - m$. $\qquad\square$

As $\dim U = m$, we see that the technique (11.18)(b) for dealing with repeated zeros of $|xI - A|$ always works: an orthonormal basis of U provides us with m rows of P, and the complete orthogonal matrix P, with PAP^T diagonal, is built up using, in turn, the zeros of $|xI - A|$.

Theorem 11.20

Every real symmetric matrix is orthogonally similar to a diagonal matrix. Two real symmetric matrices are orthogonally similar if and only if their characteristic polynomials are identical.

Proof

By (11.17)(a) the characteristic polynomial χ of the real symmetric $n \times n$ matrix A factorizes: $\chi = |xI - A| = \prod_{i=1}^{t}(x - \lambda_i)^{m_i}$ where $\lambda_1, \ldots, \lambda_t$ are distinct real numbers. Let b_i denote an orthonormal basis of the characteristic subspace U_i of A corresponding to λ_i. By (11.17)(b) the vectors in $b = b_1 \cup \ldots \cup b_t$ are mutually orthogonal and so linearly independent (11.4). As $\dim U_i = m_i$ by (11.19), and $m_1 + m_2 + \ldots + m_t = n$, we see b contains n vectors and so b is an orthonormal basis of $\mathbb{R}^n$. Let P be the orthogonal $n \times n$ matrix having the vectors of b as its rows. Then PAP^T is diagonal by (10.11), for the rows of P are characteristic vectors of A.

Let A and B be real symmetric matrices with $\chi = |xI - A| = |xI - B|$. Then A and B are $n \times n$ matrices where $n = \deg \chi$. Using the above factorization of χ and the above paragraph, both A and B are orthogonally similar to the diagonal matrix having (i, i)-entry λ_i for $m_1 + \ldots + m_{i-1} < i \leqslant m_1 + \ldots + m_{i-1} + m_i$ ($1 \leqslant i \leqslant t$). Hence A and B are orthogonally similar to each other. Conversely, if A and B are orthogonally similar, then their characteristic polynomials are identical by (10.13). □

From the theoretical point of view, (11.20) provides a complete solution to the problem of orthogonal similarity of real symmetric matrices. However, except for contrived examples, the factorization of the characteristic polynomial presents practical difficulties.

As for the geometric interpretation of (11.20): let A be an invertible real symmetric 3×3 matrix and c a positive real number. The points x of $\mathbb{R}^3$ satisfying $xAx^T = c$ form a **quadric surface** S, the nature of which depends on the sign of the zeros $\lambda_1, \lambda_2, \lambda_3$ of $|xI - A|$. By (11.20) there is an orthogonal matrix P with $PAP^T = D$, where D is diagonal with (i, i)-entry λ_i. In terms of $(X_1, X_2, X_3) = xP^T$, S has equation $\lambda_1 X_1^2 + \lambda_2 X_2^2 + \lambda_3 X_3^2 = c$. Taking $\lambda_1 \geqslant \lambda_2 \geqslant \lambda_3$, we see that there are four cases, as follows.

$\lambda_1, \lambda_2, \lambda_3$ positive:	S an ellipsoid.
λ_1, λ_2 positive; λ_3 negative:	S an hyperboloid of one sheet.
λ_1 positive; λ_2, λ_3 negative:	S an hyperboloid of two sheets.
$\lambda_1, \lambda_2, \lambda_3$ negative:	$S = \varnothing$ (a virtual quadric).

In fact P can be chosen with $|P| = 1$ showing that the x_i-axes are

Fig. 11.6

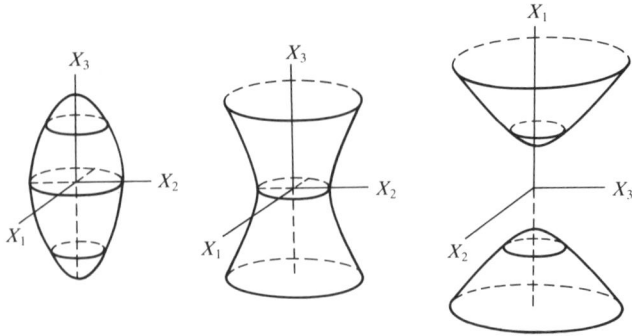

Principal axes of the ellipsoid and hyperboloids of one and two sheets

related to the principal X_i-axes of S (Fig. 11.6) by the rotation $XP = x$.

We show next that (11.20) leads to the simultaneous reduction of two real quadratic forms, one of which is positive definite; problems of this kind arise in mechanics, kinetic and potential energy giving rise to quadratic forms, the first being positive definite. For convenience, we present a pure matrix formulation using (10.48).

Corollary 11.21   Let A and B be real symmetric $n \times n$ matrices and suppose A has signature n. Then there is a real invertible $n \times n$ matrix P such that $PAP^{\mathrm{T}} = I$ and PBP^{T} is diagonal.

Proof   By (10.43) there is a real invertible $n \times n$ matrix P_1 with $P_1AP_1^{\mathrm{T}} = I$. The matrix $C = P_1BP_1^{\mathrm{T}}$ is real and symmetric, and so there is an orthogonal $n \times n$ matrix P_2 with $P_2CP_2^{\mathrm{T}} = D$ (diagonal) by (11.20). Then $P = P_2P_1$ is invertible over $\mathbb{R}$ and

$$PAP^{\mathrm{T}} = P_2P_1AP_1^{\mathrm{T}}P_2^{\mathrm{T}} = P_2P_2^{\mathrm{T}} = I,$$
$$PBP^{\mathrm{T}} = P_2P_1BP_1^{\mathrm{T}}P_2^{\mathrm{T}} = P_2CP_2^{\mathrm{T}} = D. \qquad \square$$

The matrix P of (11.21) can be found directly by working in the Euclidean space $V = (\mathbb{R}^n, \beta)$ where $(x, y)^\beta = xAy^{\mathrm{T}}$, for the equation $PAP^{\mathrm{T}} = I$ tells us that the rows of P form an orthonormal basis of V. To ensure that $PBP^{\mathrm{T}} = D$ is diagonal, the technique (11.18) is modified as follows:

1. Factorize the polynomial $|xA - B|$; because

$$|xI - D| = |xPAP^{\mathrm{T}} - PBP^{\mathrm{T}}| = |PP^{\mathrm{T}}| \, |xA - B|$$

shows that the diagonal entries in D are the zeros of $|xA - B|$, each zero λ_i of $|xA - B|$ with multiplicity m_i occurring m_i times in D.

2. For each such λ_i, find an orthonormal basis b_i (in the sense of V above) of the m_i-dimensional subspace

$$U_i = \{x \in \mathbb{R}^n : x(\lambda_i A - B) = 0\}$$

as the matrix $P(\lambda_i A - B)P^T = \lambda_i I - D$ has m_i rows of zeros and so the corresponding rows of P form an orthonormal basis of U_i.

3. Form P having the vectors of $b_1 \cup b_2 \cup \ldots \cup b_t$ as its rows, where $\lambda_1, \lambda_2, \ldots, \lambda_t$ are the distinct zeros of $|xA - B|$. The U_i are mutually orthogonal subspaces of V and hence P satisfies (11.21).

For example, let $A = \begin{pmatrix} 3 & 4 & 4 \\ 4 & 6 & 5 \\ 4 & 5 & 6 \end{pmatrix}$, $B = \begin{pmatrix} 2 & 6 & 2 \\ 6 & 11 & 7 \\ 2 & 7 & 3 \end{pmatrix}$. The three

leading principal minors of A are positive and so the quadratic form xAx^T is positive definite by (10.52). The polynomial

$$|xA - B| = \begin{vmatrix} 3x - 2 & 4x - 6 & 4x - 2 \\ 4x - 6 & 6x - 11 & 5x - 7 \\ 4x - 2 & 5x - 7 & 6x - 3 \end{vmatrix} = (x - 2)^2(x + 4)$$

has zero $\lambda_1 = 2$ of multiplicity $m_1 = 2$ and zero $\lambda_2 = -4$ of multiplicity $m_2 = 1$. So $U_1 = \{x \in \mathbb{R}^3 : x(2A - B) = 0\}$, that is, U_1 has equation $2x_1 + x_2 + 3x_3 = 0$. Then $v_1 = (1, 1, -1)$ belongs to U_1, and using the scalar product $x \cdot y = xAy^T$, we see that $x \cdot v_1 = 3x_1 + 5x_2 + 3x_3 = 0$ is the condition for x to be orthogonal to v_1. Therefore U_1 has an orthogonal basis v_1, v_2 where $v_2 = (12, -3, -7)$. As $v_1 A v_1^T = 5$ and $v_2 A v_2^T = 30$, normalizing produces the orthonormal basis $(1/\sqrt{5})v_1$, $(1/\sqrt{30})v_2$ of U_1. Similarly $U_2 = \{x \in \mathbb{R}^3 : x(-4A - B) = 0\} = \langle v_3 \rangle$ where $v_3 = (6, -3, -1)$; as $v_3 A v_3^T = 6$, we see that $(1/\sqrt{6})v_3$ is an orthonormal basis of U_2. Finally $PAP^T = I$ and $PBP^T = D$ where

$$P = \begin{pmatrix} 1/\sqrt{5} & 1/\sqrt{5} & -1/\sqrt{5} \\ 12/\sqrt{30} & -3/\sqrt{30} & -7/\sqrt{30} \\ 6/\sqrt{6} & -3/\sqrt{6} & -1/\sqrt{6} \end{pmatrix} \text{ and } D = \begin{pmatrix} 2 & 0 & 0 \\ 0 & 2 & 0 \\ 0 & 0 & -4 \end{pmatrix}.$$

Exercises 11.1

1. Normalize $(8, 4, 1)$ and $(-1, 4, -8)$, obtaining v_1 and v_2. Find the vectors v_3 such that v_1, v_2, v_3 form an orthonormal basis of $\mathbb{R}^3$.

2. Determine the angle θ $(0 \le \theta \le \pi)$ between the following pairs of vectors in $\mathbb{R}^3$.

$(1, 2, -2), (1, -1, 4)$; $(2, 1, -1), (1, 2, 1)$; $(3, 5, 8), (1, 4, -9)$.

3. Test the triangles in $\mathbb{R}^3$ with vertices as below for being equilateral, isosceles, right-angled.

(i) $(7, 2, 6)$, $(5, 1, 2)$, $(4, 1, 2)$.
(ii) $(5, 2, 6)$, $(4, 4, 3)$, $(2, 1, 4)$.
(iii) $(7, 1, -1)$, $(1, 2, 1)$, $(3, 3, 7)$.

4. Find the co-ordinates (x_1, x_2, x_3) of the points in $\mathbb{R}^3$ such that $(1, 0, 0)$, $(0, 1, 0)$, $(0, 0, 1)$, (x_1, x_2, x_3) are the vertices of a regular tetrahedron (a pyramid with equilateral triangular faces).

5. (a) Obtain orthogonal bases of $\mathbb{R}^3$ by applying the Gram–Schmidt process (11.7) to the following bases:

(i) $(1, 0, 0)$, $(2, 1, 0)$, $(3, 4, 1)$.
(ii) $(1, 1, 1)$, $(2, 1, 2)$, $(1, 3, 5)$.

(b) Find orthonormal bases of the following subspaces of $\mathbb{R}^4$.

(i) $\langle (0, 1, 1, 1), (1, 0, 1, 1), (1, 1, 0, 1) \rangle$.
(ii) $\{(x_1, x_2, x_3, x_4) \in \mathbb{R}^4 : x_1 - x_2 + x_3 - x_4 = 0\}$.

6. (a) Find the orthogonal matrix P such that the isometry μ_P of $\mathbb{R}^3$ maps the basis $(1/\sqrt{2})(1, 1, 0)$, $(1/\sqrt{3})(1, -1, 1)$, $(1/\sqrt{6})(1, -1, -2)$ to the basis $\frac{1}{3}(-1, 2, 2)$, $\frac{1}{3}(2, -1, 2)$, $\frac{1}{3}(2, 2, -1)$.
(b) Show that the composition of compatible isometries is itself an isometry.
(c) Show that the inverse of each isometry is an isometry.
(d) Let $\alpha : V \cong V'$ be a vector space isomorphism where V and V' are Euclidean spaces. If $\|v\| = \|(v)\alpha\|$ for all v in V, show that α is an isometry.
(e) Let V and V' be Euclidean spaces and $\alpha : V \to V'$ a bijection satisfying $u \cdot v = (u)\alpha \cdot (v)\alpha$ for all $u, v \in V$. By considering an orthonormal basis of V, show that α is linear and hence an isometry.

7. For each real symmetric matrix A below, find an orthogonal matrix P with PAP^T diagonal and state the type of the quadric $xAx^T = 1$.

$$\begin{pmatrix} 2 & 3 & 1 \\ 3 & 2 & 1 \\ 1 & 1 & 6 \end{pmatrix}, \begin{pmatrix} -1 & -1 & -3 \\ -1 & -1 & 3 \\ -3 & 3 & 3 \end{pmatrix}, \begin{pmatrix} 7 & -2 & 1 \\ -2 & 10 & -2 \\ 1 & -2 & 7 \end{pmatrix}, \begin{pmatrix} -11 & 3 & -1 \\ 3 & -3 & -3 \\ -1 & -3 & -11 \end{pmatrix}.$$

8. Let λ_l (λ_m) denote the least (greatest) of the real numbers $\lambda_1, \ldots, \lambda_n$. If the real variables X_i satisfy $X_1^2 + \ldots + X_n^2 = 1$ show that the least and greatest values of $\lambda_1 X_1^2 + \ldots + \lambda_n X_n^2$ are λ_l and λ_m respectively.

Using $x = XP$ (P as above), find the minimum and maximum values of xAx^T subject to $\|x\| = 1$ for each of the matrices A of question 7. Determine the co-ordinates x of points on the unit sphere $\|x\| = 1$ at which these values are achieved.

9. (a) Verify that the characteristic polynomial of the real symmetric matrix $\begin{pmatrix} a & b \\ b & c \end{pmatrix}$ factorizes over $\mathbb{R}$.

(b) The real $n \times n$ matrix A has n mutually orthogonal characteristic vectors in $\mathbb{R}^n$. Prove that A is symmetric.

(c) Let P be an orthogonal $n \times n$ matrix and D a diagonal $n \times n$ matrix over $\mathbb{R}$. Describe, in terms of D and P, the characteristic roots and vectors of the real symmetric matrix $P^T D P$.

Find the real 3×3 matrix A having characteristic vectors $(1, -1, 0)$, $(1, 1, 1)$, $(1, 1, -2)$ with corresponding characteristic roots $4, -3, 6$.

10. For each pair A and B of real matrices below, verify that A has signature 3 and find an invertible matrix P over $\mathbb{R}$ with $PAP^T = I$ and PBP^T diagonal.

(i) $\begin{pmatrix} 3 & 2 & -1 \\ 2 & 2 & 1 \\ -1 & 1 & 5 \end{pmatrix}$, $\begin{pmatrix} 6 & 3 & -5 \\ 3 & 3 & 1 \\ -5 & 1 & 13 \end{pmatrix}$.

(ii) $\begin{pmatrix} 3 & 2 & 2 \\ 2 & 2 & 1 \\ 2 & 1 & 2 \end{pmatrix}$, $\begin{pmatrix} 2 & 1 & 4 \\ 1 & 1 & 3 \\ 4 & 3 & -2 \end{pmatrix}$.

11. The real symmetric $n \times n$ matrix A is called **positive** (short for positive semi-definite) if $xAx^T \geq 0$ for all $x \in \mathbb{R}^n$.

(a) Show that the real symmetric matrix A is positive if and only if its characteristic roots are non-negative.

(b) Show that BB^T is positive for all real matrices B.

(c) Let $B^2 = A$ where B is real, symmetric and positive. Show that A and B have the same characteristic subspaces and $|xI - B| = \prod_{i=1}^{t} (x - \sqrt{\lambda_i})^{m_i}$ where $|xI - A| = \prod_{i=1}^{t} (x - \lambda_i)^{m_i}$. Deduce that each real, symmetric, and positive matrix A has a unique real symmetric, and positive square root $\sqrt{A}$ satisfying $(\sqrt{A})^2 = A$.

(d) Find an orthogonal matrix P with $PAP^T = \begin{pmatrix} 4 & 0 \\ 0 & 9 \end{pmatrix}$ where $A = \begin{pmatrix} 5 & 2 \\ 2 & 8 \end{pmatrix}$. Hence calculate $\sqrt{A} = P^T \begin{pmatrix} 2 & 0 \\ 0 & 3 \end{pmatrix} P$.

(e) Find $\sqrt{A}$ where $A = \begin{pmatrix} 6 & 2 & 2 \\ 2 & 6 & 2 \\ 2 & 2 & 1 \end{pmatrix}$.

Unitary spaces

The foregoing theory of Euclidean spaces generalizes—painlessly—to unitary spaces: for instance every unitary space has an orthonormal basis, and this property leads directly to their classification. As well as discussing the analogy between Euclidean and unitary spaces, we introduce some new aspects, such as inequalities arising from the positive definite form β (11.1) and its complex analogue (11.22), and study the type of diagonalization appropriate to unitary spaces.

Definition 11.22

Let V be an n-dimensional vector space over $\mathbb{C}$ and let β be an hermitian form on V such that $(v, v)^\beta > 0$ for all v in V with $v \neq 0$. Then the pair (V, β) is called a **unitary space**.

Let (V, β) be a unitary space. As before we write $u \cdot v = (u, v)^\beta$ and $\|v\| = \sqrt{(v \cdot v)}$, and call $\|u - v\|$ the distance (11.2) between u and v in V, while not forgetting the **hermitian commutation law** (10.49):

$$v \cdot u = (u \cdot v)^* \quad \text{for all } u, v \in V.$$

Of course there is no need for the star if $u \cdot v$ is real, but in general commutation involves complex conjugation. $\mathbb{C}^n$, with hermitian form β defined by $(x, y)^\beta = x \cdot y = xy^* = x_1 y_1^* + \ldots + x_n y_n^*$ where $x = (x_1, \ldots, x_n)$ and $y = (y_1, \ldots, y_n)$ is called the **standard n-dimensional unitary space** $\mathbb{C}^n$.

Proposition 11.23

(Schwarz's inequality.) Let V be a unitary space. Then

$$|u \cdot v| \leq \|u\| \, \|v\| \quad \text{for all } u, v \in V.$$

Proof

The above inequality is valid for $u = 0$, as both sides are zero in this case. So suppose $u \neq 0$ and let $w = au - v$ where $a = (v \cdot u)/(u \cdot u)$ as in (11.3); then $w \cdot u = 0$, and therefore

$$0 \leq w \cdot w = w \cdot (au - v) = (w \cdot u)a^* - w \cdot v = -w \cdot v$$

on conjugating $(au) \cdot w = a(u \cdot w)$ by (2.23) and (10.49). Substituting for w in the above inequality and multiplying out produces $a(u \cdot v) \leq v \cdot v$, that is, $|u \cdot v|^2 \leq \|u\|^2 \, \|v\|^2$, using $(v \cdot u)(u \cdot v) = (u \cdot v)^*(u \cdot v) = |u \cdot v|^2$. Taking non-negative square roots gives $|u \cdot v| \leq \|u\| \, \|v\|$. $\qquad \square$

We mention two consequences of (11.23).

Corollary 11.24

(Cauchy's inequality.) Let z_i and w_i be complex numbers ($1 \leq i \leq n$). Then

$$|z_1 w_1^* + \ldots + z_n w_n^*| \leq \sqrt{(|z_1|^2 + \ldots + |z_n|^2)} \sqrt{(|w_1|^2 + \ldots + |w_n|^2)}.$$

Proof   Applying (11.23) to the vectors $z = (z_1, \ldots, z_n)$, $w = (w_1, \ldots, w_n)$ of the standard unitary space $\mathbb{C}^n$ gives $|z \cdot w| \leq \|z\| \, \|w\|$, which is the above inequality.   □

Our next corollary generalizes (2.27).

Corollary 11.25   (The triangle inequality.) Let V be a unitary space. Then

$$\|v_1 - v_3\| \leq \|v_1 - v_2\| + \|v_2 - v_3\| \quad \text{where } v_1, v_2, v_3 \in V.$$

Proof   Write $u = v_1 - v_2$ and $v = v_2 - v_3$; then $u + v = v_1 - v_3$, and so we must prove $\|u + v\| \leq \|u\| + \|v\|$. Now $\|u + v\|^2 = (u + v) \cdot (u + v) = u \cdot u + u \cdot v + v \cdot u + v \cdot v$ and so

$$\|u + v\|^2 - (\|u\| + \|v\|)^2 = u \cdot v + v \cdot u - 2\|u\| \, \|v\|$$

$$\leq u \cdot v + (u \cdot v)^* - 2|u \cdot v|$$

by (11.23). However $z + z^* \leq 2|z|$ for all $z \in \mathbb{C}$, and so $u \cdot v + (u \cdot v)^* - 2|u \cdot v| \leq 0$. Hence $\|u + v\|^2 - (\|u\| + \|v\|)^2 \leq 0$ which gives $\|u + v\| \leq \|u\| + \|v\|$.   □

Schwarz's inequality and the triangle inequality hold in Euclidean spaces, as the above proofs apply without change. In fact (11.25) is the basic property of distance and is required of all **metrics** (distance functions).

The vectors u, v of the unitary space V are said to be **orthogonal** if $u \cdot v = 0$. With this definition (which we shall soon explain in real terms), the theory of orthogonal and orthonormal bases of Euclidean spaces extends unchanged to unitary spaces.

For example, $v_1 = (i, 1, 1 + i)$ and $v_2 = (i, -2 - i, i)$ in $\mathbb{C}^3$ are orthogonal as $v_1 \cdot v_2 = i(-i) + 1(-2 + i) + (1 + i)(-i) = 0$; do not forget to conjugate the *second* vector when forming a scalar product in $\mathbb{C}^n$. To find a non-zero vector $v_3 = (z_1, z_2, z_3)$ orthogonal to v_1 and v_2, we must solve $v_3 \cdot v_1 = v_3 \cdot v_2 = 0$, that is,

$$-z_1 i + z_2 + z_3(1 - i) = -z_1 i + z_2(-2 + i) - z_3 i = 0$$

which are satisfied by $v_3 = (4 + i, 1, -3 + i)$. Therefore v_1, v_2, v_3 form an orthogonal basis of $\mathbb{C}^3$. Now $\|v_1\|^2 = v_1 \cdot v_1 = 4$, $\|v_2\|^2 = 7$, $\|v_3\|^2 = 28$ and so normalizing produces the orthonormal basis $\frac{1}{2}v_1$, $(1/\sqrt{7})v_2$, $(1/\sqrt{28})v_3$ of $\mathbb{C}^3$.

Let (V, β) be a unitary space. Simply by restricting the scalars to be real numbers, V becomes a vector space V_r over $\mathbb{R}$; what is more, (V_r, β_r) is a Euclidean space, called the **realization** of (V, β), where $(u, v)^{\beta_r}$ is defined to be the *real part* of $(u, v)^\beta$. Sometimes it is possible to visualize "what's going on" in a unitary space by considering its realization. Notice that if V has orthonormal basis $v_1, v_2, \ldots, v_n$, then V_r has orthonormal basis $v_1, iv_1, v_2, iv_2, \ldots, v_n, iv_n$, and so the realization of an n-dimensional unitary space V is the $2n$-dimensional Euclidean space V_r; the price of restricting the scalars is the doubling of the dimension!

Let u and w be non-zero vectors in the unitary space V. Then $U = \langle u \rangle$ and $W = \langle w \rangle$ are 1-dimensional subspaces of V, and their realizations $U_r = \langle u, iu \rangle$ and $W_r = \langle w, iw \rangle$ are 2-dimensional subspaces of the Euclidean space V_r. It is routine to show

$$u \cdot w = 0 \quad \Leftrightarrow \quad U_r \text{ and } W_r \text{ are mutually orthogonal}$$

that is, u and w are orthogonal (in the unitary sense) if and only if each vector u' in U_r is orthogonal (in the Euclidean sense) to each vector w' in W_r (Fig. 11.7), in which case u, iu, w, iw are mutually orthogonal vectors of V_r.

Fig. 11.7

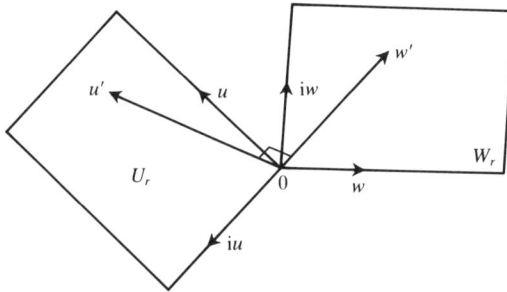

4–dimensional realization of $u \cdot w = 0$

The concept of an isometry (11.9) extends unchanged to unitary spaces and (11.10) and (11.11) are valid for unitary spaces; in particular:

Every n-dimensional unitary space V is isometric to the standard unitary space $\mathbb{C}^n$.

We now introduce the complex analogue of an orthogonal matrix (11.12).

Definition 11.26 The $n \times n$ matrix P over the complex field $\mathbb{C}$ is called **unitary** if $PP^* = I$.

As $P^* = P^T$ for real $n \times n$ matrices, a real matrix is unitary if and only if it is orthogonal. More generally:

A complex $n \times n$ matrix is unitary if and only if
its rows form an orthonormal basis of $\mathbb{C}^n$.

Remembering (10.50) that P^* is the result of conjugating every entry in P followed by matrix transposition, we see that

$$\frac{1}{2}\begin{pmatrix} 2i & 0 \\ 0 & 1 + \sqrt{3}\,i \end{pmatrix} \quad \text{and} \quad (1/\sqrt{3})\begin{pmatrix} i & 1 + i \\ 1 - i & i \end{pmatrix}$$

are unitary. It is straightforward to verify that the unitary $n \times n$ matrices form a subgroup, called the **unitary group** $U_n(\mathbb{C})$, of the group of all invertible $n \times n$ matrices over $\mathbb{C}$. The complex version of (11.13) says that a linear mapping of unitary spaces is an isometry if and only if its matrix relative to orthonormal bases is unitary; in particular:

Orthonormal bases of the unitary space V
are related by unitary matrices.

Definition 11.27

The complex $n \times n$ matrices A and B are called **unitarily similar** if there is a unitary $n \times n$ matrix P such that $PAP^* = B$.

As $P^* = P^{-1}$ for unitary matrices P, unitarily similar matrices are similar and hermitian congruent (10.51). Unitarily similar matrices arise on referring a given linear mapping $\alpha : V \to V$ to different orthonormal bases of the unitary space V.

Let U be a subspace of the unitary space V; then $U^{\perp}$ is defined as in (11.15) and satisfies (11.16). Replacing matrix transposition by hermitian conjugation throughout (so real symmetric matrices are replaced by hermitian matrices, and orthogonal matrices by unitary matrices), we arrive at the complex analogues of (11.17), (11.19), (11.20).

Theorem 11.28

Let A be an hermitian $n \times n$ matrix.

(a) The characteristic polynomial of A factorizes into linear factors over $\mathbb{R}$.

(b) Characteristic vectors of A corresponding to distinct characteristic roots are orthogonal vectors of $\mathbb{C}^n$.

(c) A is unitarily similar to a diagonal matrix.

(d) Two hermitian matrices are unitarily similar if and only if their characteristic polynomials are identical.

Rather than spoil the fun for the reader by proving (11.28)—a modification of the theory of real symmetric matrices is all that is required—we work an analogous example to (11.18)(b). Consider the hermitian matrix

$$A = \begin{pmatrix} 3 & 1+i & 1+2i \\ 1-i & 4 & 3+i \\ 1-2i & 3-i & 7 \end{pmatrix}$$

which has characteristic polynomial $|xI - A| = (x-2)^2(x-10)$. The characteristic subspace corresponding to 2 is $U = \{z \in \mathbb{C}^3 : z(A - 2I) = 0\}$, that is, U is the 2-dimensional subspace of $\mathbb{C}^3$ with equation $z_1 + (1-i)z_2 + (1-2i)z_3 = 0$. So $(1-i, -1, 0)$ and $(1-2i, 3-i, -3)$ form an orthogonal basis of U; as $U^\perp$ is the characteristic subspace corresponding to 10, we see that $(1, 1+i, 1+2i)$ is a basis of $U^\perp$. On normalizing these vectors, we obtain the rows of a unitary matrix P with PAP^* diagonal, that is,

$$PAP^* = \begin{pmatrix} 2 & 0 & 0 \\ 0 & 2 & 0 \\ 0 & 0 & 10 \end{pmatrix}$$

where

$$P = \begin{pmatrix} (1-i)/\sqrt{3} & -1/\sqrt{3} & 0 \\ (1-2i)/\sqrt{24} & (3-i)/\sqrt{24} & -3/\sqrt{24} \\ 1/\sqrt{8} & (1+i)/\sqrt{8} & (1+2i)/\sqrt{8} \end{pmatrix}.$$

The complex analogue of (11.21), stated below without proof, tells us a fact about simultaneous hermitian congruence (10.51).

Corollary 11.29   Let A and B be hermitian $n \times n$ matrices and suppose A has signature n. Then there is an invertible $n \times n$ matrix P over $\mathbb{C}$ such that $PAP^* = I$ and PBP^* is diagonal.

Having acquired some experience of unitary spaces, let us 'take the bull by the horns' and ask: *which complex $n \times n$ matrices are unitarily similar to diagonal matrices?* Hermitian matrices have this property (11.28)(c), and so too do unitary matrices as we shall see. To pose the above question in its proper setting, we return briefly to dual spaces.

Let $\alpha : V \to V'$ be linear, where V and V' are vector spaces over F, and let $f : V' \to F$ be a linear form (10.27) on V'. The composition $\alpha f : V \to F$ is a linear form on V, and further, $\hat{\alpha} : \hat{V}' \to \hat{V}$ defined by $\hat{\alpha}(f) = \alpha f$ for all f in $\hat{V}'$, is a linear mapping called the **dual** of α. In the notation of (10.29)

$$[(v)\alpha, f] = [v, \hat{\alpha}(f)] \qquad \text{for all } v \in V, f \in \hat{V}'.$$

Suppose now that V and V' are unitary spaces. Then V may be identified with its dual $\hat{V}$ by means of the scalar product: we write $V \approx \hat{V}$ to denote the (non-linear) bijection $w \mapsto \hat{w}$ where $(v)\hat{w} = v \cdot w$ for all $v, w \in V$. Let $\alpha^* : V' \to V$ be the mapping which makes the diagram

$$\hat{V} \xleftarrow{\hat{\alpha}} \hat{V}'$$
$$\wr\wr \qquad \wr\wr$$
$$V \xleftarrow{\alpha^*} V'$$

commutative, that is, $\hat{\alpha}$ becomes α^* on identifying $\hat{V}$ with V and $\hat{V}'$ with V'.

Definition 11.30   Let $\alpha : V \to V'$ be linear where V and V' are unitary spaces. Then $\alpha^* : V' \to V$, as above, is called the **adjoint** of α.

It is routine to verify that α^* is linear; placing two commutative diagrams of the above type side by side, we obtain $(\alpha\beta)^* = \beta^*\alpha^*$. Notice that $(v')\alpha^* = w \Leftrightarrow \hat{\alpha}(\hat{v}') = \hat{w} \Leftrightarrow (v)\alpha \cdot v' = v \cdot w$ for all v in V, that is, substituting for w:

$$(v)\alpha \cdot v' = v \cdot (v')\alpha^* \qquad \text{for all } v \in V, v' \in V' \qquad (\star)$$

which expresses the connection between α and its adjoint α^*. The reader may suspect that there is something special about α^{**}, the adjoint of the adjoint of α; in fact $\alpha^{**} : V \to V'$ and

$$(v)\alpha^{**} = v' \Leftrightarrow \hat{\alpha}^*(\hat{v}) = \hat{v}'$$
$$\Leftrightarrow (u')\alpha^* \cdot v = u' \cdot v' \qquad \text{for all } u' \in V',$$
$$\Leftrightarrow v \cdot (u')\alpha^* = v' \cdot u' \qquad \text{on conjugating,}$$
$$\Leftrightarrow (v)\alpha \cdot u' = v' \cdot u' \qquad \text{by the equation } (\star).$$

Taking $u' = (v)\alpha - v'$ gives $\|(v)\alpha - v'\|^2 = 0$, and so $(v)\alpha = v' = (v)\alpha^{**}$ for all v in V, showing that $\alpha^{**} = \alpha$.

Comparing the equation $(\star)$ above with the unitary version of (11.9) we deduce that:

$\alpha : V \to V'$ is an isometry if and only if $\alpha^* = \alpha^{-1}$.

As the notation suggests, there is a close connection between α^* and the hermitian conjugate A^*.

Lemma
11.31

Let the linear mapping $\alpha : V \rightarrow V'$ have matrix A relative to the orthonormal bases ℓ and ℓ' of the unitary spaces V and V'. Then $\alpha^* : V' \rightarrow V$ has matrix A^* relative to ℓ' and ℓ.

Proof

Let $v_1, \ldots, v_m$ form the basis ℓ of V, and let $v'_1, \ldots, v'_n$ form the basis ℓ' of V'. Write $A = [a_{ij}]$ and let α^* have $n \times m$ matrix $B = [b_{ji}]$ relative to ℓ' and ℓ. By (7.22) and the orthonormality of ℓ and ℓ',

$$a_{ij} = \left(\sum_{k=1}^{n} a_{ik} v'_k \right) \cdot v'_j = (v_i)\alpha \cdot v'_j = v_i \cdot (v'_j)\alpha^*$$

$$= v_i \cdot \left(\sum_{l=1}^{m} b_{jl} v_l \right) = b_{ji}^* \quad (1 \leqslant i \leqslant m, \ 1 \leqslant j \leqslant n).$$

Therefore $b_{ji} = a_{ij}^*$ and so $B = A^*$ by (10.50). □

So forming the adjoint α^* of the linear mapping α (a construction which is co-ordinate-free, as it is done without reference to a basis) corresponds, by (11.31), to the familiar process of forming the hermitian conjugate A^* of the complex matrix A. We now take $V = V'$ and show that there is a simple relationship between the α-invariant subspaces and the α^*-invariant subspaces.

Lemma
11.32

Let $\alpha : V \rightarrow V$ be linear where V is a unitary space. Then the subspace U of V is α-invariant if and only if $U^{\perp}$ is α^*-invariant.

Proof

Suppose that U is α-invariant, which by (10.33) means $(u)\alpha \in U$ for all $u \in U$. Using the adjoint equation $(\star)$ for $u \in U$ and $v \in U^{\perp}$ we have

$$0 = (u)\alpha \cdot v = u \cdot (v)\alpha^*$$

showing that $(v)\alpha^*$ is orthogonal to every u in U, that is, $(v)\alpha^* \in U^{\perp}$. Therefore $U^{\perp}$ is α^*-invariant by (10.33).

Conversely, suppose $U^{\perp}$ is α^*-invariant. By the preceding paragraph, $(U^{\perp})^{\perp}$ is α^{**}-invariant; in other words, U is α-invariant. □

Definition
11.33

Let $\alpha : V \rightarrow V$ be linear where V is a unitary space. If $\alpha = \alpha^*$, then α is called **self-adjoint**. If $\alpha\alpha^* = \alpha^*\alpha$, then α is called **normal**.

By (11.31), α is self-adjoint if and only if the matrix A of α

relative to the orthonormal basis b of V is hermitian; in symbols: $\alpha = \alpha^* \Leftrightarrow A = A^*$. So 'self-adjoint linear mapping' is the co-ordinate-free abstraction of 'hermitian matrix'.

The significance of normal mappings is made clear in (11.35); in fact, they are *exactly what we are looking for,* namely those mappings of a unitary space having an orthonormal basis of characteristic vectors. Notice that self-adjoint mappings and iso-metries of V are normal.

We now generalize (11.17), (11.19), (11.20), replacing 'real symmetric matrix' by 'normal mapping'.

Proposition 11.34 Let α be a normal mapping of the unitary space V. Then $(v)\alpha = \lambda v$ if and only if $(v)\alpha^* = \lambda^* v$ where $v \in V$ and $\lambda \in \mathbb{C}$. Characteristic vectors corresponding to distinct characteristic roots of α are orthogonal.

Proof Consider $\beta : V \to V$ defined by $\beta = \alpha - \lambda \iota$ where ι is the identity mapping of V. So $(v)\beta = (v)\alpha - \lambda v$ for all $v \in V$. It is routine to show $\beta^* = \alpha^* - \lambda^* \iota$ as $\iota^* = \iota$. Since α is normal, by multiplying out we see $\beta\beta^* = \beta^*\beta$, showing that β is normal. Hence

$$\|(v)\alpha - \lambda v\|^2 = \|(v)\beta\|^2 = (v)\beta \cdot (v)\beta = v \cdot (v)\beta\beta^*$$
$$= v \cdot (v)\beta^*\beta = (v)\beta^*\beta \cdot v = (v)\beta^* \cdot (v)\beta^*$$
$$= \|(v)\beta^*\|^2 = \|(v)\alpha^* - \lambda^* v\|^2$$

using the adjoint equation $(\star)$ connecting β and β^* and the fact that $v \cdot (v)\beta^*\beta$ is real. Therefore $(v)\alpha = \lambda v \Leftrightarrow (v)\alpha^* = \lambda^* v$.

Suppose $(v)\alpha = \lambda v$, $(w)\alpha = \mu w$ where $\lambda \neq \mu$. Then $(w)\alpha^* = \mu^* w$ by the paragraph above. Hence

$$\lambda(v \cdot w) = (\lambda v) \cdot w = (v)\alpha \cdot w = v \cdot (w)\alpha^*$$
$$= v \cdot (\mu^* w) = \mu(v \cdot w)$$

and so $v \cdot w = 0$. □

By (11.34), the normal mapping α has the same characteristic vectors as its adjoint α^*. All is now prepared for the final assault on normality!

Theorem 11.35 Let $\alpha : V \to V$ be a linear mapping of the unitary space V. Then V has an orthonormal basis b consisting of characteristic vectors of α if and only if α is normal.

Proof   Let us do the easy part first: suppose $v_1, v_2, \ldots, v_n$ form an orthonormal basis b of V with $(v_i)\alpha = \lambda_i v_i$ $(1 \le i \le n)$. Then α has matrix D relative to b where D is diagonal with (i, i)-entry λ_i. By (11.31), α^* has matrix D^* relative to b, that is, $(v_i)\alpha^* = \lambda_i^* v_i$ $(1 \le i \le n)$. Therefore $(v_i)\alpha\alpha^* = \lambda_i \lambda_i^* v_i = \lambda_i^* \lambda_i v_i = (v_i)\alpha^*\alpha$, and so $\alpha\alpha^* = \alpha^*\alpha$ as these composite linear mappings agree on b. By (11.33), α is normal.

Suppose now that α is normal. Let $\lambda_1, \ldots, \lambda_t$ denote the distinct zeros of the characteristic polynomial χ of α, that is, the characteristic polynomial of any matrix representing α relative to a basis of V. Let $U_i = \{v \in V : (v)\alpha = \lambda_i v\}$. Then the characteristic subspaces $U_1, \ldots, U_t$ of α are mutually orthogonal and α^*-invariant by (11.34). Hence $U = U_1 \oplus \ldots \oplus U_t$ is also α^*-invariant and so $U^\perp$ is α-invariant by (11.32).

Let us suppose $\dim U^\perp$ $(= m$ say$)$ is positive. Then $\deg \chi' = m$ where χ' is the characteristic polynomial of α restricted to $U^\perp$. By (4.26) χ' has a zero in $\mathbb{C}$, and hence $U^\perp$ contains a characteristic vector v of α. But U contains all the characteristic vectors of α, and so $v \in U \cap U^\perp = 0$ which is a contradiction as $v \ne 0$. Therefore in fact $m = 0$, showing $U^\perp = 0$ and $U = V$.

It is now plain sailing to the finish: let b_i be an orthonormal basis of U_i. As $V = U_1 \oplus \ldots \oplus U_t$, we see that $b = b_1 \cup \ldots \cup b_t$ is an orthonormal basis of V made up of characteristic vectors of α.   □

The matrix version of (11.35) tells us which matrices can be unitarily diagonalized and so provides a complete answer to the question posed earlier (following (11.29)).

Definition   The complex $n \times n$ matrix A is called **normal** if $AA^* = A^*A$.
11.36

Corollary   The complex $n \times n$ matrix A is unitarily similar to a diagonal
11.37   matrix if and only if A is normal. Two normal matrices are unitarily similar if and only if their characteristic polynomials are identical.

Proof   We leave the reader to fill in the following outline: A is unitarily similar to a diagonal matrix if and only if $\mathbb{C}^n$ has an orthonormal basis of characteristic vectors of μ_A, which by (11.35) is the case if and only if μ_A is normal. But μ_A is normal if and only if A is normal by (11.31). The proof may be completed as in (11.20).
     □

It is straightforward to verify that

$$A = \begin{pmatrix} 2+5i & -2-i & -1-i \\ -2-i & 2+5i & 1+i \\ 1-i & -1+i & 4+6i \end{pmatrix}$$

is normal, that is, A commutes with its hermitian conjugate A^*. By (11.37) there is a unitary matrix P with PAP^* diagonal. In this case such a matrix P can be explicitly found since A has characteristic roots $4i$, $4+4i$, $4+8i$, with corresponding characteristic vectors $(1,1,0)$, $(1,-1,1-i)$, $(1,-1,-1+i)$. These vectors are 'automatically' orthogonal by (11.34), and so, on normalizing them, we obtain

$$P = \tfrac{1}{2} \begin{pmatrix} \sqrt{2} & \sqrt{2} & 0 \\ 1 & -1 & 1-i \\ 1 & -1 & -1+i \end{pmatrix}.$$

As unitary matrices P are normal ($P^* = P^{-1}$ commutes with P), (11.37) tells us that two unitary matrices are unitarily similar if and only if their characteristic polynomials are identical. One could not hope for more! However, we may deduce only that orthogonal matrices with the same characteristic polynomial are unitarily similar, whereas in fact they are *orthogonally* similar (11.39).

Exercises 11.2

1. Verify that $v_1 = (1-3i, 2i, 1+i)$ and $v_2 = (1-2i, 1-3i, -1)$ are orthogonal vectors of $\mathbb{C}^3$. Find v_3 such that v_1, v_2, v_3 form an orthogonal basis of $\mathbb{C}^3$. Find the unitary 3×3 matrix of determinant 1 having $(1/\|v_1\|)v_1$ and $(1/\|v_2\|)v_2$ as its first two rows.

2. Let u and v belong to the unitary space V where $u \neq 0$. Adapt (11.23) to show that $|u \cdot v| = \|u\| \|v\|$ if and only if v is a multiple of u.

   Verify that $u = \tfrac{1}{2}(1+i, i, -i)$ is a unit vector of $\mathbb{C}^3$. Find the vectors v in $\mathbb{C}^3$ such that $u \cdot v$ is pure imaginary and $|u \cdot v| = \|v\| = 2$.

3. (a) Let u and v belong to the Euclidean space V. Show that $\|u+v\|^2 = \|u\|^2 + \|v\|^2$ if and only if $u \cdot v = 0$ and interpret this result geometrically.

   (b) Verify that $u = (1+i, 2i, 1+3i)$ and $v = (3i, 1-i, -1)$ in the unitary space $\mathbb{C}^3$ satisfy $\|u+v\|^2 = \|u\|^2 + \|v\|^2$ and calculate $u \cdot v$.

   State and prove the complex analogue of part (a) above.

4. Show that products and inverses of unitary $n \times n$ matrices are themselves unitary matrices; deduce, using (9.9), that $U_n(\mathbb{C})$ is a group.

Let P be unitary. Show that P^{T} is unitary and find the connection between the (i, j)-entry in P and the (i, j)-entry in $(P^{\mathrm{T}})^{-1}$. Show that the determinant $|P|$ is a complex number of modulus 1.

Let $v_1, \ldots, v_{n-1}$ be mutually orthogonal unit vectors of $\mathbb{C}^n$ and let z in $\mathbb{C}$ satisfy $|z| = 1$. Show that there is a unique v_n in $\mathbb{C}^n$ such that the matrix P with $e_i P = v_i$ $(1 \leqslant i \leqslant n)$ is unitary with $|P| = z$.

5. For each hermitian matrix A below, find a unitary matrix P such that PAP^* is diagonal. Hence find vectors x in $\mathbb{C}^3$ with $\|x\| = 1$ such that the value of xAx^* is extreme (either maximum or minimum).

$$\begin{pmatrix} 2 & 4i & 4i \\ -4i & 1 & -3 \\ -4i & -3 & 1 \end{pmatrix}, \quad \begin{pmatrix} 2 & i & -i \\ -i & 2 & -1 \\ i & -1 & 2 \end{pmatrix}.$$

6. (a) Verify that the following matrices A are normal, and in each case find a unitary matrix P with PAP^* diagonal.

$$\begin{pmatrix} 2i & 2 & 1-2i \\ 2 & 2i & -1+2i \\ -1+2i & 1-2i & -4+i \end{pmatrix}, \quad \begin{pmatrix} 1+i & -i & 1-i \\ i & 1+i & 1+i \\ 1+i & 1-i & 2+i \end{pmatrix}.$$

(b) By 'undiagonalizing', determine the normal 3×3 matrix A with characteristic vectors $(i, 1, 1)$, $(1, 0, i)$, $(i, -2, 1)$ corresponding to the characteristic roots $3i$, $2 + 2i$, 6.

(c) Let A be a normal $n \times n$ matrix. If rank $A = 1$, show that $A = \lambda v^* v$ for some $v \in \mathbb{C}^n$ and $\lambda \in \mathbb{C}$.

(d) Show that every real and normal 2×2 matrix is either symmetric or proportional to a rotation matrix.

7. (a) Find an invertible 3×3 matrix P over $\mathbb{C}$ such that $PAP^* = I$ and PBP^* is diagonal where

$$A = \begin{pmatrix} 4 & 2+i & i \\ 2-i & 2 & i \\ -i & -i & 1 \end{pmatrix}, \quad B = \begin{pmatrix} 5 & 4+3i & i \\ 4-3i & 4 & i \\ -i & -i & 1 \end{pmatrix}.$$

(b) Using (11.21) as a guide, write down a proof of (11.29) starting with (11.28)(c).

8. (a) Let A and B be hermitian $n \times n$ matrices such that A is invertible and the characteristic polynomial of BA^{-1} has n distinct real zeros $\lambda_1, \lambda_2, \ldots, \lambda_n$. By (10.18) there is an invertible $n \times n$ matrix P over $\mathbb{C}$ with $PBA^{-1}P^{-1} = D$, the diagonal matrix with (i, i)-entry λ_i. Show that PAP^* and PBP^* are both diagonal. (Hint: every matrix which commutes with D is diagonal.)

(b) Verify that $|xA - B| = |xI - BA^{-1}|\,|A|$ has zeros $0, \pm 1$ where

$$A = \begin{pmatrix} 0 & 1-2i & -1-i \\ 1+2i & 0 & -1-i \\ -1+i & -1+i & 1 \end{pmatrix}, \qquad B = \begin{pmatrix} 3 & -1+i & -2+i \\ -1-i & 1 & 1 \\ -2-i & 1 & 2 \end{pmatrix}$$

and hence find an invertible matrix P over $\mathbb{C}$ with PAP^* and PBP^* diagonal.

9. (a) Show that a normal matrix is hermitian (unitary) if and only if all its characteristic roots are real (of modulus 1).

(b) Let $A = P^*DP$ where P is unitary and D is diagonal. Show that A is normal with unit characteristic vector v_i (row i of P) corresponding to the characteristic root λ_i (the (i, i)-entry in D).

(c) Let $v_1 = (i, 1, 0)$, $v_2 = (1, i, 1+i)$, $v_3 = (1, i, -1-i)$. Calculate the hermitian matrix A having characteristic vectors v_1, v_2, v_3 with corresponding characteristic roots 2, 4, 16. Find $\sqrt{A}$, the hermitian matrix with positive characteristic roots satisfying $(\sqrt{A})^2 = A$.

Find the unitary matrix P with characteristic roots 1, i, $-$i corresponding to v_1, v_2, v_3.

Isometries and volume

We return to Euclidean spaces and analyse their isometries; it turns out that every isometry μ_P of $\mathbb{R}^n$ with $|P| = 1$ can be orthogonally decomposed into planar rotations, and so it is reasonable to call μ_P itself a rotation. Also, determinants of real $n \times n$ matrices are interpreted as volumes of parallelepipeds (called **boxes** for short).

Let $\alpha : V \to V$ be linear where V and V' are Euclidean spaces. The adjoint α^* of α may be introduced as before (see (11.30)); thus $\alpha^* : V' \to V$ is the linear mapping satisfying $(v)\alpha \cdot v' = v \cdot (v')\alpha^*$ for $v \in V$ and $v' \in V'$. The real version of (11.31) states that α^* has matrix A^T relative to the orthonormal bases b' and b of V' and V where A is the matrix of α relative to b and b'. Although (11.32) is valid for Euclidean spaces, the analogue of (11.35) is little more than an abstraction of (11.20), namely:

The Euclidean space V has an orthonormal basis
of characteristic vectors of the linear mapping
$\alpha : V \to V$ if and only if $\alpha = \alpha^*$.

The concept (11.33) of a normal mapping is therefore significant only in the context of unitary spaces.

Suppose now that $\alpha : V \to V$ is an isometry of the Euclidean space V and let $(v)\alpha = \lambda v$ for some $v \in V$ and $\lambda \in \mathbb{R}$. By (11.9)

$$\|v\|^2 = v \cdot v = (v)\alpha \cdot (v)\alpha = \lambda v \cdot \lambda v = |\lambda|^2 \|v\|^2$$

and so $v \neq 0$ implies $|\lambda| = 1$. Therefore:

The characteristic roots of every isometry
have modulus 1.

This is valid for unitary spaces also; in the Euclidean case, λ is real
and so ± 1 are the only possibilities. Write $U_1 = \{v \in V : (v)\alpha = v\}$
and $U_{-1} = \{v \in V : (v)\alpha = -v\}$; then U_1 and U_{-1} are mutually
orthogonal by (11.34). Our next theorem tells us the effect of α on
$W = (U_1 \oplus U_{-1})^{\perp}$.

Theorem
11.38

Let α be an isometry of the Euclidean space V. Then W, as
above, can be expressed $W = W_1 \oplus \ldots \oplus W_t$ where the W_i are
mutually orthogonal α-invariant 2-dimensional subspaces, and α
acts as a rotation on each W_i.

Proof

The α-invariant subspace $U_1 \oplus U_{-1}$ contains all the characteristic
vectors of α. As $\alpha^* = \alpha^{-1}$, it follows that α^*-invariance coincides
with α-invariance; so $W = (U_1 \oplus U_{-1})^{\perp}$ is α-invariant by (11.32)
and contains no characteristic vectors of α. We use induction on
$\dim W$, the case $W = 0$ being covered by $t = 0$. Suppose $\dim W > 0$
and consider the self-adjoint mapping $\beta = \alpha + \alpha^*$ of V. As W is
α-invariant, we see that W is β-invariant also, and so there is a
unit characteristic vector w of β in W; we denote the corresponding real characteristic root by $2c$, and so $(w)(\alpha + \alpha^*) =
2cw$ which gives $(w)\alpha^2 + w = 2c(w)\alpha$ as $\alpha^* = \alpha^{-1}$. Hence using
(11.9),

$$2w \cdot (w)\alpha = (w)\alpha \cdot (w)\alpha^2 + w \cdot (w)\alpha$$
$$= ((w)\alpha^2 + w) \cdot (w)\alpha = 2c(w)\alpha \cdot (w)\alpha = 2c$$

showing that c is the cosine of the angle between the unit vectors w
and $(w)\alpha$. As $(w)\alpha$ is not proportional to w, we see that $c^2 \neq 1$. Let
$W_1 = \langle w, (w)\alpha \rangle$. Then $\dim W_1 = 2$ and W_1 is α-invariant: writing
$s = \sqrt{(1 - c^2)}$, it is routine to verify that $(\sqrt{2s})^{-1}w_1$ and $(\sqrt{2s})^{-1}w_2$
form an orthonormal basis b_1 of W_1 where

$$w_1 = (w)\alpha - (c - s)w, \qquad w_2 = (w)\alpha - (c + s)w.$$

What is more, α restricted to W_1 has the rotation matrix

$$A_\theta = \begin{pmatrix} \cos\theta & \sin\theta \\ -\sin\theta & \cos\theta \end{pmatrix}$$

relative to b_1 where θ is the unique real number satisfying $\cos\theta = c$
and $\sin\theta = s$ $(0 < \theta < \pi)$.

Finally, let $W' = (U_1 \oplus U_{-1} \oplus W_1)^\perp$. Then $W = W_1 \oplus W'$ and W_1, W' are mutually orthogonal. As W' is α-invariant and dim $W' =$ dim $W - 2$, we may assume inductively that α restricted to W' decomposes as stated, that is, $W' = W_2 \oplus \ldots \oplus W_t$. Therefore $W = W_1 \oplus W_2 \oplus \ldots \oplus W_t$ and the induction is complete.   □

Notice that α restricted to W_1 (as above) has matrix $A_{-\theta}$ relative to the basis $(\sqrt{2}\,s)^{-1}w_2$, $(\sqrt{2}\,s)^{-1}w_1$ of W_1. Therefore A_θ and $A_{-\theta}$ are orthogonally similar and so have the same characteristic polynomial

$$\chi_\theta = x^2 - (2\cos\theta)x + 1.$$

We show next that each orthogonal matrix P with characteristic polynomial χ is orthogonally similar to the *simplest* such matrix.

Corollary 11.39   Let P be an orthogonal $n \times n$ matrix with characteristic polynomial χ. Then P is orthogonally similar to the block-diagonal matrix

$$D = \begin{pmatrix} I & \vdots & \vdots \\ \cdots & -I & \cdots \\ \vdots & \vdots & A \end{pmatrix} \quad \text{where} \quad A = \begin{pmatrix} A_{\theta_1} & \vdots & \vdots \\ \cdots & \ddots & \cdots \\ \vdots & \vdots & A_{\theta_t} \end{pmatrix},$$

$\pm I$ denotes $m_{\pm 1} \times m_{\pm 1}$ matrices, $0 < \theta_i < \pi$ $(1 \le i \le t)$, and $\chi = (x-1)^{m_1}(x+1)^{m_{-1}}\chi_{\theta_1} \ldots \chi_{\theta_t}$ is the factorization of χ over $\mathbb{R}$.

Proof   We apply (11.38) to the isometry μ_P of $\mathbb{R}^n$, utilizing the decomposition $\mathbb{R}^n = U_1 \oplus U_{-1} \oplus W_1 \oplus \ldots \oplus W_t$. Let b be an orthonormal basis of $\mathbb{R}^n$ made up of orthonormal bases of these component subspaces; by interchanging the vectors in b_i (an orthonormal basis of W_i) if necessary, we may assume that μ_P restricted to W_i has matrix A_{θ_i} relative to b_i where $0 < \theta_i < \pi$. Therefore μ_P has matrix D relative to b, where m_1, m_{-1}, t are non-negative integers with $n = m_1 + m_{-1} + 2t$. By (10.3), $QPQ^T = D$ where the rows of the orthogonal matrix Q form the basis b. As P and D have identical characteristic polynomials, $\chi = |xI - D|$ factorizes as stated.   □

From (11.39) we deduce directly:

Orthogonal matrices are orthogonally similar if and only if their characteristic polynomials are identical.

Whether or not an orthogonal matrix is built up entirely from 2×2 rotation matrices is decided by the value ± 1 of its determinant.

Corollary 11.40

Let P be an orthogonal $n \times n$ matrix with $|P| = 1$. Then the isometry μ_P of $\mathbb{R}^n$ is expressible as the composition of $(n - m_1)/2$ rotations in mutually perpendicular planes.

Proof

Using the notation of (11.39), as each A_θ has determinant 1 we obtain $1 = |P| = |D| = (-1)^{m-1}$ and so m_{-1} is even in this case. But

$$A_\pi = \begin{pmatrix} -1 & 0 \\ 0 & -1 \end{pmatrix}$$

and hence the $m_{-1} \times m_{-1}$ matrix $-I$ represents $(m_{-1})/2$ rotations through π in perpendicular planes. Taking into account the t rotation matrices A_{θ_i}, we see that μ_P is made up of $(m_{-1})/2 + t = (n - m_1)/2$ plane rotations, each pair of planes being orthogonal. $\square$

Because of (11.40), the term **rotation matrix** is applied to any orthogonal matrix P with $|P| = 1$. The case $n = 3$ is worth noting: let P be a 3×3 rotation matrix; then $m_1 = 1$ or 3 as $3 - m_1$ is even. As $m_1 = 3 \Leftrightarrow P = I$, we see that every non-identity 3×3 rotation matrix P represents a rotation of the plane $U_1^\perp$ about its perpendicular axis U_1. For instance

$$P = \tfrac{1}{3}\begin{pmatrix} 2 & 1 & 2 \\ 1 & 2 & -2 \\ -2 & 2 & 1 \end{pmatrix}$$

has determinant 1 and trace $\tfrac{5}{3}$, which are sufficient to determine

$$D = \begin{pmatrix} 1 & 0 & 0 \\ 0 & 1/3 & \sqrt{8}/3 \\ 0 & -\sqrt{8}/3 & 1/3 \end{pmatrix}.$$

In this case U_1 (the characteristic subspace of P corresponding to 1) is the line $\langle (1, 1, 0) \rangle$ and $W_1 = U_1^\perp$ is the plane $x_1 + x_2 = 0$. Taking $w = (0, 0, 1)$ and $\alpha = \mu_P$ in the proof of (11.38), produces the orthonormal basis $(-\tfrac{1}{2}, \tfrac{1}{2}, 1/\sqrt{2})$, $(-\tfrac{1}{2}, \tfrac{1}{2}, -1/\sqrt{2})$ of W_1. Hence

$$Q = \begin{pmatrix} 1/\sqrt{2} & 1/\sqrt{2} & 0 \\ -1/2 & 1/2 & 1/\sqrt{2} \\ -1/2 & 1/2 & -1/\sqrt{2} \end{pmatrix}$$

is orthogonal and satisfies $QPQ^T = D$.

Definition 11.41 The orthonormal bases b and b' of the Euclidean space V are said to have the same or opposite **orientation** according as $|P| = +1$ or $|P| = -1$, where P is the orthogonal matrix relating b to b'.

By (11.40), b as above can be rotated into coincidence with b' if and only if these bases have the same orientation. On the other hand, interchanging two vectors of b, or changing the sign of one of the vectors in b, produces a basis b' of opposite orientation.

Definition 11.42 Let the Euclidean space V have basis $v_1, v_2, \ldots, v_n$. We call the set of all vectors of the form $t_1 v_1 + t_2 v_2 + \ldots + t_n v_n$, where $0 \le t_i < 1$ $(1 \le i \le n)$, the **box** $[v_1, v_2, \ldots, v_n]$, and say that it is **rectangular** if $v_1, v_2, \ldots, v_n$ are mutually orthogonal.

If each real number t_i above satisfies $0 \le t_i \le 1$, the resulting region is the parallelepiped with edges $v_1, v_2, \ldots, v_n$. For technical reasons we exclude the faces of the parallelepiped corresponding to $t_i = 1$; this does not affect what concerns us, namely its volume (area if $n = 2$).

To get the idea of the next proof, consider the box $B = [v_1, v_2]$ in $\mathbb{R}^2$, where $v_1 = (3, 0)$ and $v_2 = (8, 6)$. Applying the Gram–Schmidt process (11.7) to v_1 and v_2 produces the rectangular box $B' = [v_1', v_2']$ where

$$v_1' = v_1 = (3, 0)$$
$$v_2' = -\tfrac{8}{3}v_1 + v_2 = (0, 6)$$

(Fig. 11.8). In fact B and B' have equal areas, since B can be cut up as indicated and the pieces reassembled to form B'; so area $B =$ area $B' = \|v_1'\| \times \|v_2'\| = 18$. Notice that B is partitioned into regions by lines parallel to v_2' spaced at intervals of length $\|v_1'\|$; further, region j' of B' is the translation of region j of B by the vector $-jv_1$. Conversely, if a B-shaped area of wall (such as a stair-well) is to be papered, then Fig. 11.8 tells us how to cut up the roll of paper B'!

Fig. 11.8

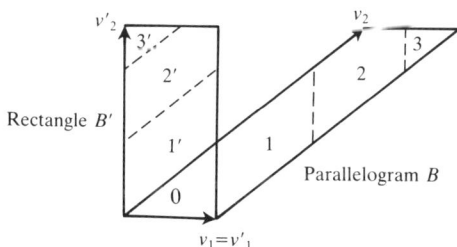

Rectangle B'

Parallelogram B

$v_1 = v_1'$

We assume the following 'self-evident' properties of volume:

1. If R_1 and R_2 are disjoint regions, then

$$\text{vol}(R_1 \cup R_2) = \text{vol } R_1 + \text{vol } R_2.$$

2. If R_1 is a translate of R_2, then $\text{vol } R_1 = \text{vol } R_2$.
3. The volume of a rectangular box is the product of its edge lengths.

Lemma 11.43

Let $v_1, \ldots, v_n$ and $v'_1, \ldots, v'_n$ form bases of the Euclidean space V related by a lower triangular matrix $L = [l_{ij}]$ where $l_{ii} = 1$ $(1 \leqslant i \leqslant n)$. Then the boxes $B = [v_1, \ldots, v_n]$ and $B' = [v'_1, \ldots, v'_n]$ have the same volume.

Proof

In fact we partition B into a finite number of regions which are then reassembled using translations to form B'. By (11.42), a typical vector in B has the form $v = t_1 v_1 + \ldots + t_n v_n$ $(0 \leqslant t_i < 1)$. Substituting $v_i = \sum_{j=1}^{n} l_{ij} v'_j$, we obtain $v = s_1 v'_1 + \ldots + s_n v'_n$ where $(s_1, \ldots, s_n) = (t_1, \ldots, t_n)L$. Each real number s_i is the sum of its integer part $\lfloor s_i \rfloor$ and its fractional part $r_i = s_i - \lfloor s_i \rfloor$; notice that $0 \leqslant r_i < 1$, and $\lfloor s_n \rfloor = 0$ as $s_n = t_n$, the last column of L being e_n^T. We say that

$v = s_1 v'_1 + \ldots + s_n v'_n$ belongs to the
region $(\lfloor s_1 \rfloor, \ldots, \lfloor s_{n-1} \rfloor)$ of B.

In effect, the box B is 'sliced up' into the above regions by certain translates of $(n-1)$-dimensional faces of B', namely those of the form

$$\langle v'_1, \ldots, \not\!v'_j, \ldots, v'_n \rangle + m v'_j \quad \text{where } m \in \mathbb{Z}, \ (1 \leqslant j < n).$$

Let us translate the region $(\lfloor s_1 \rfloor, \ldots, \lfloor s_{n-1} \rfloor)$ of B by the vector $w = -\lfloor s_1 \rfloor v'_1 - \ldots - \lfloor s_n \rfloor v'_n$; as $v + w = r_1 v'_1 + \ldots + r_n v'_n$ belongs to B', we obtain a region of B'. The proof is completed by showing that the mapping $\alpha : B \to B'$, defined by $(v)\alpha = v + w$ for all $v \in B$, is bijective; as w depends on the region of B containing v, we see that α is built up from translations.

To show that α is bijective, let $v' = r_1 v'_1 + \ldots + r_n v'_n$ in B' be given. Notice that $v = t_1 v_1 + \ldots + t_n v_n$ in B satisfies $(v)\alpha = v'$ if and only if $(t_1, \ldots, t_n)L - (r_1, \ldots, r_n)$ is a vector with *integer* entries. Because of the form of L, the entries in $(t_1, \ldots, t_n)$ can be determined in reverse order: as $0 \leqslant r_n$, $t_n < 1$ and $t_n - r_n \in \mathbb{Z}$, we see $t_n = r_n$. Suppose $t_{k+1}, t_{k+2}, \ldots, t_n$ have been found $(1 \leqslant k < n)$; then the stipulation that $t_k + \sum_{i>k} t_i l_{ik} - r_k$ be an integer together with $0 \le t_k < 1$ shows that t_k is the fractional part of $r_k - \sum_{i>k} t_i l_{ik}$. By induction, the vector $(t_1, \ldots, t_n)$ is uniquely determined, that is, there is a unique v in B with $(v)\alpha = v'$. So α is bijective and hence vol B = vol B' using properties 1 and 2 of volume. □

Algebra and geometry combine to produce our last theorem.

Theorem 11.44 Let A be a real $m \times n$ matrix of rank m, and let v_i denote row i of A. Then $\text{vol}[v_1, \ldots, v_m] = \sqrt{|AA^T|}$.

Proof Applying (11.7) to $v_1, \ldots, v_m$ leads to the orthogonal basis $v'_1, \ldots, v'_m$ of the Euclidean space $V = \langle v_1, \ldots, v_m \rangle$. Write $l_{ij} = (v_i \cdot v'_j)/\|v'_j\|^2$ for $i \ge j$, and let $l_{ij} = 0$ for $i < j$; then $L = [l_{ij}]$ is a lower triangular $m \times m$ matrix with $l_{ii} = 1$. The equations of the Gram–Schmidt process (11.7) rearrange to give $v_i = \sum_{j=1}^{m} l_{ij} v'_j$ ($1 \le i \le m$), that is, $A = LA'$ where A' is the $m \times n$ matrix with row $j = v'_j$. Since $v'_1, \ldots, v'_m$ are mutually orthogonal, $A'A'^T$ is diagonal with (j, j)-entry $\|v'_j\|^2$. Therefore taking determinants of $AA^T = LA'A'^T L^T$ and using $|L| = 1$,

$$|AA^T| = |A'A'^T| = \|v'_1\|^2 \|v'_2\|^2 \ldots \|v'_m\|^2$$

which is the square of the volume of the rectangular box $[v'_1, \ldots, v'_m]$; but this box has the same volume as $[v_1, \ldots, v_m]$ by (11.43), and hence $\text{vol}[v_1, \ldots, v_m] = \sqrt{|AA^T|}$. □

Consider $v_1 = (1, 2, 4)$ and $v_2 = (3, 1, 5)$ in $\mathbb{R}^3$. To find the area of the parallelogram $[v_1, v_2]$, we form $A = \begin{pmatrix} 1 & 2 & 4 \\ 3 & 1 & 5 \end{pmatrix}$ and use (11.44);

$AA^T = \begin{pmatrix} 21 & 25 \\ 25 & 35 \end{pmatrix}$ and $|AA^T| = 110$ showing that $[v_1, v_2]$ has area $\sqrt{110}$.

Finally, suppose the matrix A of (11.44) to be $n \times n$ and invertible. Then the box $[v_1, \ldots, v_n]$ has volume equal to the absolute value of the determinant $|A|$; further, the sign of $|A|$ is positive (negative) according as the orthonormal basis of $\mathbb{R}^n$, obtained by normalizing $v'_1, \ldots, v'_n$ has the same (opposite) orientation as the standard basis of $\mathbb{R}^n$.

Exercises 11.3 1. For each orthogonal matrix P below, find an orthogonal matrix Q such that $QPQ^T = D$ as in (11.39).

$$\frac{1}{3}\begin{pmatrix} 2 & -1 & 2 \\ -1 & 2 & 2 \\ 2 & 2 & -1 \end{pmatrix}, \quad \begin{pmatrix} 0 & 1 & 0 \\ 0 & 0 & 1 \\ 1 & 0 & 0 \end{pmatrix}, \quad \frac{1}{3}\begin{pmatrix} 2 & 2 & 1 \\ -1 & 2 & -2 \\ -2 & 1 & 2 \end{pmatrix}.$$

2. Decide which pairs of rotation matrices below are orthogonally similar:

$$\begin{pmatrix} 1/\sqrt{2} & 1/\sqrt{2} & 0 \\ 1/\sqrt{2} & -1/\sqrt{2} & 0 \\ 0 & 0 & -1 \end{pmatrix}, \quad \frac{1}{3}\begin{pmatrix} -1 & 2 & 2 \\ 2 & -1 & 2 \\ 2 & 2 & -1 \end{pmatrix},$$

$$\begin{pmatrix} -1 & 0 & 0 \\ 0 & 1/2 & \sqrt{3}/2 \\ 0 & \sqrt{3}/2 & -1/2 \end{pmatrix}, \quad \begin{pmatrix} -1 & 0 & 0 \\ 0 & 1 & 0 \\ 0 & 0 & -1 \end{pmatrix},$$

and find, where appropriate, suitable orthogonal matrices P as in (11.14).

3. Show that there are $n+1$ orthogonal similarity classes of $n \times n$ matrices which are orthogonal and symmetric. How many of these classes contain rotation matrices?

4. Let P, P', Q be orthogonal $n \times n$ matrices with $QP = P'Q$. If $|P| = -1$, show that there is a rotation matrix Q' such that $Q'P = P'Q'$. Can the same conclusion be drawn if

(i) $|P| = 1$ and n is odd,   (ii) $|P| = 1$ and $n = 2$?

5. Let P be an orthogonal $n \times n$ matrix with characteristic polynomial χ. Find a necessary and sufficient condition, in terms of the real factorization of χ, for P and $-P$ to be orthogonally similar.
   Write down a non-symmetric 4×4 rotation matrix which is orthogonally similar to its negative.

6. Find the volume of the box $[v_1, v_2, v_3]$ in $\mathbb{R}^3$ and the areas of its faces $[v_1, v_2]$, $[v_2, v_3]$, $[v_3, v_1]$, where $v_1 = (2, 2, 1)$, $v_2 = (5, 8, 1)$, $v_3 = (2, 3, 2)$.

7. Let B denote the box $[v_1, v_2, v_3]$ in $\mathbb{R}^3$ where $v_1 = (3, 0, 0)$, $v_2 = (4, 4, 0)$, $v_3 = (2, 2, 4)$. Show that the planes $x_1 = 3$, $x_1 = 6$, $x_2 = 4$ partition B into five regions as in (11.43). Sketch these regions, and verify that they reassemble to form $B' = [3e_1, 4e_2, 4e_3]$.

8. Let $\alpha : V \to V'$ be an invertible linear mapping where V and V' are Euclidean spaces, and let B be a box in V. Show that $(B)\alpha = \{(v)\alpha : v \in B\}$ is a box in V'. Show that $\mathrm{vol}(B)\alpha = \sqrt{|AA^T|}\,(\mathrm{vol}\,B)$ where A is the matrix of α relative to orthonormal bases of V and V'.

9. Let A be a real invertible $n \times n$ matrix. Use (11.7) to show that there are unique matrices $L = [l_{ij}]$ and P such that $A = LP$, with L lower triangular, $l_{ii} > 0$, and P orthogonal. State and prove the complex analogue.

Further reading

General

G. Birkhoff and S. Maclane (1977). *A survey of modern algebra.* 4th edn. Macmillan, New York.
P. M. Cohn (1974, 1977). *Algebra,* Vols. I and II. Wiley, Chichester.
I. N. Herstein (1975). *Topics in algebra.* Wiley, New York.

Set theory

P. R. Halmos (1960). *Naive set theory.* Van Nostrand Reinhold, New York.
R. R. Stoll (1961). *Sets, logic and axiomatic theories.* Freeman, San Francisco.

Field theory

I. N. Stewart (1973). *Galois theory.* Chapman and Hall, London.

Group theory

F. J. Budden (1972). *The fascination of groups.* Cambridge University Press, London.

Number theory

D. M. Burton (1980). *Elementary number theory.* Allyn and Bacon, Boston, Massachusetts.
G. H. Hardy and E. M. Wright (1980). *An introduction to the theory of numbers,* 5th edn. Oxford University Press, Oxford.

Geometry

H. S. M. Coxeter (1969). *Introduction to geometry,* 2nd edn. Wiley, New York.
K. W. Gruenberg and A. J. Weir (1977). *Linear geometry,* 2nd edn. Springer, New York.
I. Kaplansky (1969). *Linear algebra and geometry (a second course).* Allyn and Bacon, Boston, Massachusetts.

Linear algebra

P. Lancaster (1969). *Theory of matrices.* Academic Press, New York.
E. D. Nering (1970). *Linear algebra and matrix theory,* 2nd edn. Wiley, New York.

Applications

G. Birkhoff and T. C. Bartee (1970). *Modern applied algebra.* McGraw-Hill, New York.

Index